T0220455

Bose, Spin and Fermi Systems

PROBLEMS AND SOLUTIONS

Bose, Spin and
Fermi Systems

PROBLEMS AND SOLUTIONS

Willi-Hans Steeb
Yorick Hardy

University of Johannesburg and University of South Africa, South Africa

World Scientific

NEW JERSEY · LONDON · SINGAPORE · BEIJING · SHANGHAI · HONG KONG · TAIPEI · CHENNAI

Published by

World Scientific Publishing Co. Pte. Ltd.

5 Toh Tuck Link, Singapore 596224

USA office: 27 Warren Street, Suite 401-402, Hackensack, NJ 07601

UK office: 57 Shelton Street, Covent Garden, London WC2H 9HE

British Library Cataloguing-in-Publication Data
A catalogue record for this book is available from the British Library.

BOSE, SPIN AND FERMI SYSTEMS
Problems and Solutions

Copyright © 2015 by World Scientific Publishing Co. Pte. Ltd.

ISBN 978-981-4630-10-8
ISBN 978-981-4667-34-0 (pbk)

Printed in Singapore

Preface

The purpose of this book is to supply a collection of problems and solutions for Bose, spin and Fermi systems as well as coupled systems. So it covers essential parts of quantum theory and quantum field theory. For most of the problems the detailed solutions are provided which will prove to be valuable to graduate students as well as to research workers in these fields. Each chapter contains supplementary problems often with the solution provided. All the important concepts are provided either in the introduction or the problem and all relevant definitions are given. The topics range in difficulty from elementary to advanced. Almost all problems are solved in detail and most of the problems are self-contained. Students can learn important principles and strategies required for problem solving. Teachers will also find this text useful as a supplement, since important concepts and techniques are developed in the problems. The book can also be used as a text or a supplement for quantum theory, Hilbert space theory and linear and multilinear algebra or matrix theory. Computer algebra programs in SymbolicC++ and Maxima are also included. For Bose system number states, coherent states and squeezed states are covered. Applications to nonlinear dynamical systems and linear optics are given. The spin chapter concentrates mostly on spin-$\frac{1}{2}$ and spin-1 systems, but also higher order spins are included. The eigenvalue problem plays a central role. Exercises utilizing the spectral theorem and Cayley-Hamilton theorem are provided. For Fermi systems a special section on the Hubbard Hamilton operator is added. Chapter 4 is devoted to Lie algebras and their representation by Bose, Spin and Fermi operators. Superalgebras are also considered. Chapters 5 and 6 cover coupled Bose-Spin and coupled Bose-Fermi systems, respectively.

The material was tested in our lectures given around the world.

Any useful suggestions and comments are welcome.

The International School for Scientific Computing (ISSC) provides certificate courses for this subject. Please contact the first author if you want to do this course. More exercises can be found on the web page given below.

e-mail addresses of the authors:

steebwilli@gmail.com
yorickhardy@gmail.com

Home page of the first author:

http://issc.uj.ac.za

v

Contents

Notation

$\emptyset$	empty set
$A \subset B$	subset A of set B
$A \cap B$	the intersection of the sets A and B
$A \cup B$	the union of the sets A and B
f, g	maps
$f \circ g$	composition of two mappings $(f \circ g)(x) = f(g(x))$
$\mathbb{N}$	natural numbers
$\mathbb{N}_0$	natural numbers including 0
$\mathbb{Z}$	integers
$\mathbb{Q}$	rational numbers
$\mathbb{R}$	real numbers
$\mathbb{R}^+$	nonnegative real numbers
$\mathbb{C}$	complex numbers
$\mathbb{R}^n$	n-dimensional Euclidian space
$\mathbb{C}^n$	n-dimensional complex linear space
i	$:= \sqrt{-1}$
z	complex number
$\bar{z}, z^*$	complex conjugate of z
$\Re z$	real part of the complex number z
$\Im z$	imaginary part of the complex number z
$\mathbf{v} \in \mathbb{C}^n$	element $\mathbf{v}$ of $\mathbb{C}^n$ (column vector)
$\mathbf{v}^*$	transpose and complex conjugate of $\mathbf{v}$
$\mathbf{v}^* \mathbf{v}$	scalar product
$\mathbf{0}$	zero vector (column vector)
t	time variable
ω	frequency
$\mathbf{x}$	space variable
$\mathbf{v}^T = (v_1, v_2, \ldots, v_n)$	vector of independent variables, ^{T} means transpose
$\| \cdot \|$	norm
$\mathbf{x} \cdot \mathbf{y} \equiv \mathbf{x}^T \mathbf{y}$	scalar product (inner product) in vector space $\mathbb{R}^n$
$\mathbf{x} \times \mathbf{y}$	vector product in vector space $\mathbb{R}^3$
det	determinant of a square matrix
tr	trace of a square matrix
0_n	$n \times n$ zero matrix
I_n	$n \times n$ unit matrix (identity matrix)
A^T	transpose of matrix A
A^*	transpose and complex conjugate of matrix A
I	identity operator
$[,]$	commutator

$[\,,\,]_+$	anticommutator
$\oplus$	direct sum
$\oplus$	XOR operation
δ_{jk}	Kronecker delta with $\delta_{jk} = 1$ for $j = k$ and $\delta_{jk} = 0$ for $j \neq k$
$\epsilon_{jk\ell}$	total antisymmetric tensor $\epsilon_{123} = 1$
$\mathrm{sgn}(x)$	the sign of x, 1 if $x > 0$, -1 if $x < 0$, 0 if $x = 0$
λ	eigenvalue
ϵ	real parameter
$\otimes$	Kronecker product, Tensor product
$\wedge$	Grassmann product (exterior product, wedge product)
S_1, S_2, S_3	spin matrices for spin 1/2, 1, 3/2, 2, ...
$\sigma_1, \sigma_2, \sigma_3$	Pauli spin matrices
U	unitary operator, unitary matrix
Π	projection operator, projection matrix
ρ	density operator, density matrix
P	permutation matrix
$\mathbf{P}$	momentum operator
$\mathbf{Q}$	position operator
$\mathbf{k}$	wave vector $\mathbf{k} \cdot \mathbf{x} = k_1 x_1 + k_2 x_2 + k_3 x_3$
$b^\dagger, b$	Bose creation and annihilation operators
$|n\rangle$	number states $(n = 0, 1, \ldots)$
$|\beta\rangle$	coherent states $(\beta \in \mathbb{C})$
$|\zeta\rangle$	squeezed states $(\zeta \in \mathbb{C})$
$c^\dagger, c$	Fermi creation and annihilation operators
ρ	density operator, density matrix $\rho \geq 0$, $\mathrm{tr}(\rho) = 1$
$\hat{N}$	number operator
L	Lagrange function
H	Hamilton function
$\mathcal{L}$	Lagrange density
$\hat{H}$	Hamilton operator
$\mathcal{H}$	Hilbert space
$L_2(\Omega)$	Hilbert space of square integrable functions
$\ell_2(S)$	Hilbert space of square-summable (infinite) sequences
$\langle\,,\,\rangle$	scalar product in Hilbert space
δ	delta function
$\hbar$	Planck constant divided by 2π

The Pauli spin matrices σ_1, σ_2, σ_3 are used extensively in the book. They are given by

$$\sigma_1 := \begin{pmatrix} 0 & 1 \\ 1 & 0 \end{pmatrix}, \quad \sigma_2 := \begin{pmatrix} 0 & -i \\ i & 0 \end{pmatrix}, \quad \sigma_3 := \begin{pmatrix} 1 & 0 \\ 0 & -1 \end{pmatrix}.$$

They are hermitian and unitary matrices with eigenvalues $+1$ and -1. In some cases we also use σ_x, σ_y and σ_z to denote σ_1, σ_2 and σ_3. The matrices σ_+ and σ_- are defined by

$$\sigma_+ := \sigma_1 + i\sigma_2 = \begin{pmatrix} 0 & 2 \\ 0 & 0 \end{pmatrix}, \quad \sigma_- := \sigma_1 - i\sigma_2 = \begin{pmatrix} 0 & 0 \\ 2 & 0 \end{pmatrix}.$$

The spin matrices for spin-$\frac{1}{2}$ are defined as

$$S_1 = \frac{1}{2}\sigma_1, \quad S_2 = \frac{1}{2}\sigma_2, \quad S_3 = \frac{1}{2}\sigma_3$$

and
$$S_+ = S_1 + iS_2 = \begin{pmatrix} 0 & 1 \\ 0 & 0 \end{pmatrix}, \quad S_- = S_1 - iS_2 = \begin{pmatrix} 0 & 0 \\ 1 & 0 \end{pmatrix}.$$

The spin-1 matrices are defined as

$$S_1 = \frac{1}{\sqrt{2}} \begin{pmatrix} 0 & 1 & 0 \\ 1 & 0 & 1 \\ 0 & 1 & 0 \end{pmatrix}, \quad S_2 = \frac{1}{\sqrt{2}} \begin{pmatrix} 0 & -i & 0 \\ i & 0 & -i \\ 0 & i & 0 \end{pmatrix}, \quad S_3 = \begin{pmatrix} 1 & 0 & 0 \\ 0 & 0 & 0 \\ 0 & 0 & -1 \end{pmatrix}$$

with the eigenvalues $+1$, 0, -1. The matrices are hermitian. Then

$$S_+ = S_1 + iS_2 = \sqrt{2} \begin{pmatrix} 0 & 1 & 0 \\ 0 & 0 & 1 \\ 0 & 0 & 0 \end{pmatrix}, \quad S_- = S_1 - iS_2 = \sqrt{2} \begin{pmatrix} 0 & 0 & 0 \\ 1 & 0 & 0 \\ 0 & 1 & 0 \end{pmatrix}.$$

The Kronecker product is extensively used in the book. Let $A := (a_{ij})_{ij}$ be an $m \times n$ matrix and B be an $r \times s$ matrix over $\mathbb{C}$. The *Kronecker product* of A and B is defined as the $(m \cdot r) \times (n \cdot s)$ matrix

$$A \otimes B := \begin{pmatrix} a_{11}B & a_{12}B & \cdots & a_{1n}B \\ a_{21}B & a_{22}B & \cdots & a_{2n}B \\ \vdots & \vdots & \ddots & \vdots \\ a_{m1}B & a_{m2}B & \cdots & a_{mn}B \end{pmatrix}.$$

The Kronecker product is associative.

The spectral theorem for normal matrices will be utilized in the book:

Let $M_n(\mathbb{C})$ be the vector space of $n \times n$ matrices and $A \in M_n(\mathbb{C})$. Let $\lambda_1, \ldots, \lambda_n$ be the eigenvalues of A counted according to multiplicity. The following statements are equivalent
(i) A is normal
(ii) A is unitarily diagonalizable
(iii) $\sum_{j=1}^{n} \sum_{k=1}^{n} |a_{jk}|^2 = \sum_{k=1}^{n} |\lambda_k|^2$.
(iv) There exists an orthonormal set of n eigenvectors of A.

The Hermite and Laguerre polynomials are used in the book:

The functions
$$H_n(x) = (-1)^n e^{x^2} \frac{d^n}{dx^n} e^{-x^2}, \quad n = 0, 1, 2, \ldots$$

are called Hermite polynomial. The first four Hermite polynomial are $H_0(x) = 1$, $H_0(x) = 2x$, $H_2(x) = 4x^2 - 2$, $H_3(x) = 8x^3 - 12x$.

The functions
$$L_n(x) = e^x \frac{d^n}{dx^n} (x^n e^{-x})$$

are called Laguerre polynomials. The first four Laguerre polynomials are $L_0(x) = 1$, $L_1(x) = 1 - x$, $L_2(x) = x^2 - 4x + 2$, $L_3(x) = -x^3 + 9x^2 - 18x + 6$.

Chapter 1

Bose Systems

1.1 Commutators and Number States

Consider the Hamilton operator $\hat{H}$ for the quantized one-dimensional harmonic oscillator

$$\hat{H} = \frac{1}{2}\frac{\hat{p}^2}{m} + \frac{1}{2}m\omega^2\hat{q}^2, \qquad \hat{p} = -i\hbar\frac{d}{dq}$$

with

$$\hat{q}f(q) = qf(q)$$

where $f(q), qf(q) \in L_2(\mathbb{R})$. The differential operator

$$\hat{p} = -i\hbar\frac{d}{dq}$$

also acts in an appropriate subspace of the Hilbert space $L_2(\mathbb{R})$.

Define the Bose annihilation operator b and Bose creation operator $b^\dagger$ as

$$b := \frac{1}{\sqrt{2m\hbar\omega}}(m\omega\hat{q} + i\hat{p}), \qquad b^\dagger := \frac{1}{\sqrt{2m\hbar\omega}}(m\omega\hat{q} - i\hat{p}).$$

The commutator of b and $b^\dagger$ is given by $[b, b^\dagger] = I$, where I is the identity operator. Using the distributive law we have

$$
\begin{aligned}
[b, b^\dagger] &= \frac{1}{2m\hbar\omega}[m\omega\hat{q} + i\hat{p}, m\omega\hat{q} - i\hat{p}] \\
&= \frac{i}{2\hbar}([\hat{q}, -\hat{p}] + [\hat{p}, \hat{q}]) \\
&= -\frac{i}{\hbar}[\hat{q}, \hat{p}] \\
&= I
\end{aligned}
$$

where we used that $[\hat{q}, \hat{p}] = i\hbar I$. The operator $\hat{N} := b^\dagger b$ is called the *number operator* and has the eigenvalues 0, 1, 2, ..., i.e. the operator is unbounded.

Problem 1. (i) Let b, $b^\dagger$ be Bose annihilation and creation operators with $[b, b^\dagger] = I$. Find the commutator of the operators

$$\widetilde{b} = \epsilon_1 I + b, \qquad \widetilde{b^\dagger} = \epsilon_2 I + b^\dagger$$

where I is the identity operator and ϵ_1, ϵ_2 are constants.

(ii) Let $b^\dagger$, b be Bose creation and annihilation operators. Consider the operators

$$F_+ := b^\dagger e^{-i\phi} + b e^{i\phi}, \qquad F_- := b^\dagger e^{-i\phi} - b e^{i\phi}$$

and $\hat{N} := b^\dagger b$. Find the commutators

$$[F_+, F_-], \quad [F_\pm, \hat{N}], \quad [F_\pm, \hat{N} F_\pm], \quad [F_\pm, \hat{N} F_\mp].$$

(iii) Calculate the commutators

$$[b^2 (b^\dagger)^2, (b^\dagger)^2 b^2], \quad [b^3 (b^\dagger)^3, (b^\dagger)^3 b^3], \quad [b^4 (b^\dagger)^4, (b^\dagger)^4 b^4].$$

(iv) Calculate the commutator $[(b^\dagger)^2 + b^2, b^\dagger b]$.

(v) Calculate the commutators $[b^2, (b^\dagger)^2]$, $[b^3, (b^\dagger)^3]$. Extend the result to $[b^n, (b^\dagger)^n]$, where $n \in \mathbb{N}$. We can apply the formula

$$[f(b), g(b^\dagger)] = \sum_{j=1}^{\infty} \frac{1}{j!} \frac{\partial^j}{\partial b^{\dagger j}} g(b^\dagger) \frac{\partial^j}{\partial b^j} f(b)$$

where $f : \mathbb{R} \to \mathbb{R}$ and $g : \mathbb{R} \to \mathbb{R}$ are analytic functions and $f(b)$, $g(b^\dagger)$ denote the series expansion of $f(x)$ and $g(x)$, respectively with x replaced by the operators b and $b^\dagger$.

(vi) Let $n \in \mathbb{N}$ and $\beta \in \mathbb{C}$. Calculate the commutator $[b^\dagger, (\beta b^\dagger - \beta^* b)^n]$.

Solution 1. (i) We have

$$\begin{aligned}
[\widetilde{b}, \widetilde{b^\dagger}] &= [\epsilon_1 I + b, \epsilon_2 I + b^\dagger] \\
&= [\epsilon_1 I, \epsilon_2 I] + [\epsilon_1 I, b^\dagger] + [b, \epsilon_2 I] + [b, b^\dagger] \\
&= [b, b^\dagger] \\
&= I.
\end{aligned}$$

(ii) We obtain

$$[F_+, F_-] = 2I, \quad [F_\pm, \hat{N}] = -F_\mp, \quad [F_\pm, \hat{N} F_\pm] = -F_\mp F_\pm, \quad [F_\pm, \hat{N} F_\mp] = -F_\mp^2 \pm 2\hat{N}.$$

(iii) Using $bb^\dagger = I + b^\dagger b$ we find

$$\begin{aligned}
[b^2 (b^\dagger)^2, (b^\dagger)^2 b^2] &= 4b^\dagger b + 2I \\
[b^3 (b^\dagger)^3, (b^\dagger)^3 b^3] &= 9(b^\dagger)^2 b^2 + 18 b^\dagger b + 6I \\
[b^4 (b^\dagger)^4, (b^\dagger)^4 b^4] &= 16(b^\dagger)^3 b^3 + 72(b^\dagger)^2 b^2 + 96 b^\dagger b + 24I.
\end{aligned}$$

(iv) Using $bb^\dagger = I + b^\dagger b$ we find

$$[(b^\dagger)^2 + b^2, b^\dagger b] = [(b^\dagger)^2, b^\dagger b] + [b^2, b^\dagger b] = 2(-b^\dagger b^\dagger + bb).$$

(v) Using $bb^\dagger = I + b^\dagger b$ we obtain

$$[b^2, (b^\dagger)^2] = 2I + 4b^\dagger b, \qquad [b^3, (b^\dagger)^3] = 6I + 18 b^\dagger b + 9 b^\dagger b^\dagger bb.$$

To calculate $[b^n, (b^\dagger)^n]$ we apply the formula given above and obtain

$$[b^n, (b^\dagger)^n] = \sum_{j=1}^{n} \frac{(n!)^2}{j!((n-j)!)^2} (b^\dagger)^{n-j} b^{n-j}.$$

(vi) If f is an analytic function we have

$$[b^\dagger, f(b)] = -\frac{df(b)}{db}.$$

Thus

$$[b^\dagger, (\beta b^\dagger - \beta^* b)^n] = n\beta^* (\beta b^\dagger - \beta^* b)^{n-1}.$$

For $n = 1$ we have $[b^\dagger, (\beta b^\dagger - \beta^* b)] = \beta^* I$.

Problem 2. (i) Let $\gamma, \phi \in \mathbb{R}$. Consider the transformation

$$\begin{pmatrix} \tilde{b} \\ \tilde{b}^\dagger \end{pmatrix} = \begin{pmatrix} \cosh(\gamma) & e^{i\phi}\sinh(\gamma) \\ e^{-i\phi}\sinh(\gamma) & \cosh(\gamma) \end{pmatrix} \begin{pmatrix} b \\ b^\dagger \end{pmatrix}.$$

Find the commutator $[\tilde{b}, \tilde{b}^\dagger]$.
(ii) Let $t \in [0, 1)$. Define the new operators

$$\tilde{b}^\dagger = \frac{(b^\dagger - tb)}{(1 - t^2)^{1/2}}, \qquad \tilde{b} = \frac{(b - tb^\dagger)}{(1 - t^2)^{1/2}}.$$

Calculate the commutator $[\tilde{b}, \tilde{b}^\dagger]$.

Solution 2. (i) Utilizing that $\cosh^2(\gamma) - \sinh^2(\gamma) = 1$ we obtain $[\tilde{b}, \tilde{b}^\dagger] = I$.
(ii) We obtain

$$\begin{aligned}
[\tilde{b}, \tilde{b}^\dagger] &= \frac{1}{(1 - t^2)} [b - tb^\dagger, b^\dagger - tb] \\
&= \frac{1}{(1 - t^2)} [b, b^\dagger] + \frac{t^2}{(1 - t^2)} [b^\dagger, b] \\
&= \frac{1}{(1 - t^2)} I - \frac{t^2}{(1 - t^2)} I \\
&= I.
\end{aligned}$$

Problem 3. Let b, $b^\dagger$ be Bose annihilation and creation operators. Let $m, n \geq 1$. We have the *normal ordering formula*

$$b^m (b^\dagger)^n = \sum_{j=0}^{\min\{m,n\}} \binom{m}{j} \frac{n!}{(n-j)!} (b^\dagger)^{n-j} b^{m-j}.$$

(i) Apply it to the operator $b^2 b^\dagger$ and $b(b^\dagger)^2$.
(ii) Find the normal ordering of $((b^\dagger)^2 b^2)^2$.
(iii) Find the normal ordering of $(b^2 (b^\dagger)^2)^2$.
(iv) Calculate the state $b^m (b^\dagger)^n |0\rangle$, where $b|0\rangle = 0|0\rangle$.

Solution 3.   (i) Since $m = 2$, $n = 1$ for $b^2 b^\dagger$ we have $\min\{m, n\} = 1$. Thus we find

$$b^2 b^\dagger = \sum_{j=0}^{1} \binom{2}{j} \frac{1}{(1-j)!} (b^\dagger)^{1-j} b^{2-j} = b^\dagger b^2 + 2b.$$

Since $m = 1$, $n = 2$ for $b(b^\dagger)^2$ we have $\min\{m, n\} = 1$. Thus we find

$$b(b^\dagger)^2 = \sum_{j=0}^{1} \binom{1}{j} \frac{2}{(2-j)!} (b^\dagger)^{2-j} b^{1-j} = (b^\dagger)^2 b + 2b^\dagger.$$

(ii) Using the commutation relation $[b, b^\dagger] = I$ we find

$$((b^\dagger)^2 b^2)^2 = (b^\dagger)^4 b^4 + 4(b^\dagger)^3 b^3 + 2(b^\dagger)^2 b^2.$$

(iii) Using the commutation relation $[b, b^\dagger] = I$ we find

$$(b^2(b^\dagger)^2)^2 = (b^\dagger)^4 b^4 + 12(b^\dagger)^3 b^3 + 38(b^\dagger)^2 b^2 + 32 b^\dagger b + 4I.$$

(iv) If $m > n$ we find $0|0\rangle$ utilizing $b|0\rangle = 0|0\rangle$. If $m = n$ we find the state

$$b^m (b^\dagger)^m |0\rangle = m! |0\rangle$$

which is not normalized. If $n > m$ we obtain the state

$$\frac{n!}{(n-m)!} (b^\dagger)^{n-m} |0\rangle$$

which is not normalized.

Problem 4.   Let $b^\dagger$, b be Bose creation and annihilation operators, respectively. Let μ, ν, γ be complex numbers and

$$\widetilde{b} = \mu b + \nu b^\dagger + \gamma F(b, b^\dagger)$$

where F is an arbitrary analytic hermitian function. Find the conditions on μ, ν, γ and F such that

$$[\widetilde{b}, \widetilde{b}^\dagger] = I$$

where I is the identity operator. Find solution for F and parametrizations for μ, ν and γ.

Solution 4.   Using $[b, b^\dagger] = I$ and

$$[b, G(b, b^\dagger)] = \frac{\partial G(x, y)}{\partial y}\bigg|_{x=b, y=b^\dagger}, \qquad [b^\dagger, G(b, b^\dagger)] = -\frac{\partial G(x, y)}{\partial x}\bigg|_{x=b, y=b^\dagger}$$

where $G : \mathbb{R}^2 \to \mathbb{R}$ is an analytic function we obtain the condition

$$(|\mu|^2 - |\nu|^2)I + |\gamma|^2 [F, F^\dagger] + \mu\gamma^* \frac{\partial F^\dagger}{\partial b^\dagger} - \nu\gamma^* \frac{\partial F^\dagger}{\partial b} + \mu^*\gamma \frac{\partial F}{\partial b} - \nu^*\gamma \frac{\partial F}{\partial b^\dagger} = I.$$

If the analytic function F is of the form $F(\beta^* b + \beta b^\dagger)$, where $\beta = |\beta| e^{i\theta}$ it follows that

$$|\mu|^2 - |\nu|^2 = 1, \qquad \Re(e^{i\theta}(\mu\gamma^* - \nu^*\gamma)) = 0.$$

With the parametrization ($r \in \mathbb{R}$)

$$\mu = \cosh(r), \quad \nu = e^{i\phi} \sinh(r), \quad \gamma = |\gamma| e^{i\delta}$$

the equation $|\mu|^2 - |\nu|^2 = 1$ is satisfied identically. The equation

$$\Re(e^{i\theta}(\mu\gamma^* - \nu^*\gamma)) = 0$$

provides

$$\cosh(r)\cos(\theta - \delta) - \sinh(r)\cos(\delta + \theta - \phi) = 0.$$

This equation has to be solved numerically. If $\phi = 0$ we obtain the simplified form

$$\tan(\theta)\tan(\delta) = -e^{-2r}.$$

Problem 5. Let $b^\dagger$, b be Bose creation and annihilation operators, respectively. Let $f : \mathbb{C} \to \mathbb{C}$ be an analytic function and $z \in \mathbb{C}$. Let $\hat{N} := b^\dagger b$ be the number operator.
(i) Find the operators

$$e^{z\hat{N}} f(b), \quad e^{z\hat{N}} f(b^\dagger), \quad e^{zb^\dagger} f(b), \quad e^{zb} f(b^\dagger).$$

(ii) Let $z \in \mathbb{C}$. Calculate the commutators $[e^{zb^\dagger}, b]$, $[e^{zb}, b^\dagger]$.
(iii) Calculate $e^{zb^\dagger} b e^{-zb^\dagger}$.

Solution 5. (i) Since $\hat{N}b = b(\hat{N} - I)$ and $\hat{N}b^\dagger = b^\dagger(I + \hat{N})$ we find

$$e^{z\hat{N}} f(b) = f(e^{-z}b)e^{z\hat{N}}, \quad e^{z\hat{N}} f(b^\dagger) = f(e^z b^\dagger)e^{z\hat{N}}$$
$$e^{zb^\dagger} f(b) = f(b - zI)e^{zb^\dagger}, \quad e^{zb} f(b^\dagger) = f(b^\dagger + zI)e^{zb}.$$

(ii) We obtain

$$[e^{zb^\dagger}, b] = -ze^{zb^\dagger}, \qquad [e^{zb}, b^\dagger] = ze^{zb}.$$

(iii) We set

$$f(z) = e^{zb^\dagger} b e^{-z\lambda b^\dagger}, \qquad f(0) = b.$$

Then $df(z)/dz = -I$. Integration with the initial condition $f(0) = b$ gives $f(z) = -zI + b$ or

$$e^{zb^\dagger} b e^{-zb^\dagger} = -zI + b.$$

Problem 6. (i) Let $\hat{N} = b^\dagger b$. Find the sequence of commutators

$$[\hat{N}, b], \quad [\hat{N}, [\hat{N}, b]], \quad [\hat{N}, [\hat{N}, [\hat{N}, b]]].$$

Consider then the general case.
(ii) Let $\zeta \in \mathbb{C}$ and

$$G = \frac{1}{4}(\zeta b^\dagger b^\dagger - \zeta^* bb).$$

Find the sequence of commutators $[G, b]$, $[G, [G, b]]$, $[G, [G, [G, b]]]$.

Solution 6. (i) We obtain

$$[\hat{N}, b] = -b, \quad [\hat{N}, [\hat{N}, b]] = (-1)^2 b, \quad [\hat{N}, [\hat{N}, [\hat{N}, b]]] = (-1)^3 b.$$

For the general case with m operators $\hat{N}$ we have

$$[\hat{N}, [\hat{N}, \ldots [\hat{N}, [\hat{N}, b] \ldots] = (-1)^m b.$$

(ii) Since $[b^\dagger b^\dagger, b] = -2b^\dagger$ and $[bb, b^\dagger] = 2b$ we obtain

$$[G, b] = -\frac{\zeta}{2} b^\dagger, \quad [G, [G, b]] = \frac{\zeta}{2} \frac{\zeta^*}{2} b, \quad [G, [G, [G, b]]] = \frac{\zeta^2 \zeta^*}{8} b^\dagger.$$

Problem 7.   Let $b^\dagger$, b be Bose creation and annihilation operator with the commutator relation $[b, b^\dagger] = I$, where I is the identity operator and $z \in \mathbb{C}$.
(i) Show that one has a representation (*Bargmann representation*)

$$b^\dagger \mapsto z, \qquad b \mapsto \frac{d}{dz}.$$

(ii) Consider the map ($x \in \mathbb{R}$)

$$b^\dagger \mapsto x, \qquad b \mapsto \frac{1}{4} x^3 + \frac{d}{dx}.$$

Find the commutator of these operators on the right-hand side.

Solution 7.   (i) We calculate the commutator $[d/dz, z] f(z)$, where f is an analytic function $f : \mathbb{C} \to \mathbb{C}$. Using the product rule we obtain

$$[d/dz, z] f(z) = \frac{d}{dz}(zf(z)) - z \frac{df(z)}{dz} = f(z) + z \frac{df(z)}{dz} - z \frac{df(z)}{dz} = f(z).$$

(ii) Let $f : \mathbb{R} \to \mathbb{R}$ be an analytic function. We have

$$\begin{aligned}
\left[\frac{1}{4} x^3 + \frac{d}{dx}, x \right] f &= \left(\frac{1}{4} x^3 + \frac{d}{dx} \right) xf - x \left(\frac{1}{4} x^3 + \frac{d}{dx} \right) f \\
&= \frac{1}{4} x^4 f + f + x \frac{df}{dx} - \frac{1}{4} x^4 f - x \frac{df}{dx} \\
&= f.
\end{aligned}$$

Problem 8.   Let $S(\mathbb{R})$ be the vector space of all infinitely differentiable functions which decrease as $|x| \to \infty$, together with all their derivatives, faster than any power of $|x|^{-1}$. Note that $S(\mathbb{R})$ is a dense subspace of the Hilbert space $L_2(\mathbb{R})$. Consider the differential operators

$$b = \frac{1}{\sqrt{2}} \left(x + \frac{d}{dx} \right), \qquad b^\dagger = \frac{1}{\sqrt{2}} \left(x - \frac{d}{dx} \right)$$

acting in the vector space $S(\mathbb{R})$.
(i) Find the commutator $[b, b^\dagger]$.
(ii) Find the number operator $\hat{N} = b^\dagger b$ expressed with differential operators.

Solution 8.   (i) Let $f \in S(\mathbb{R})$. We have

$$\begin{aligned}
[b, b^\dagger] f &= bb^\dagger f - b^\dagger b f \\
&= \frac{1}{2} \left(x^2 f - \frac{d}{dx}(xf) - x \frac{df}{dx} - \frac{d^2 f}{dx^2} \right) - \frac{1}{2} \left(x^2 f - \frac{d}{dx}(xf) + x \frac{df}{dx} - \frac{d^2 f}{dx^2} \right) \\
&= \frac{d}{dx}(xf) - x \frac{df}{dx} \\
&= f.
\end{aligned}$$

(ii) We obtain the operator

$$\hat{N} = \frac{1}{2}\left(x^2 - 1 - \frac{d^2}{dx^2}\right)$$

which contains a second order derivative.

Problem 9. (i) Calculate the anticommutator $\hat{H} = [b, b^\dagger]_+$ utilizing $bb^\dagger = I + b^\dagger b$.
(ii) Let $\hat{N} = b^\dagger b$ be the number operator. Calculate the commutators

$$[b, \hat{N}^2], \qquad [b^\dagger, \hat{N}^2].$$

Express the result using $\hat{H}$.

Solution 9. (i) We find

$$\hat{H} = [b, b^\dagger]_+ = bb^\dagger + b^\dagger b = (I + b^\dagger b) + b^\dagger b = I + 2b^\dagger b = I + 2\hat{N}.$$

(ii) We have

$$\begin{aligned}
[b, \hat{N}^2] &= bb^\dagger bb^\dagger b - b^\dagger bb^\dagger bb = bb^\dagger + b^\dagger bbb^\dagger b - b^\dagger bb^\dagger b \\
&= bb^\dagger b + b^\dagger b(I + b^\dagger b)b - b^\dagger bb^\dagger bb = bb^\dagger b + b^\dagger bb \\
&= (I + b^\dagger b)b + b^\dagger bb = (I + 2b^\dagger b)b \\
&= \hat{H}b.
\end{aligned}$$

Analogously we find $[b^\dagger, b^\dagger bb^\dagger b] = -b^\dagger(I + 2b^\dagger b) = -b^\dagger \hat{H}$.

Problem 10. (i) Calculate the commutators

$$[b^\dagger, b^\dagger b], \quad [b, b^\dagger b], \quad [b^\dagger, (b^\dagger b)^2], \quad [b, (b^\dagger b)^2]$$

utilizing $[b, b^\dagger] = I$.
(ii) The commutators can also be calculated as follows. Let $f : \mathbb{R}^2 \to \mathbb{R}$ be an analytic function where in any term of $f(b, b^\dagger)$ the annihilation operators are on the left hand side and thus all creation operators are on the right hand side. Then the commutators are given by

$$[b^\dagger, f(b, b^\dagger)] = -\frac{\partial f}{\partial b}, \qquad [b, f(b, b^\dagger)] = \frac{\partial f}{\partial b^\dagger}.$$

Apply it to the cases $[b^\dagger, (b^\dagger b)^2]$ and $[b^\dagger, (b^\dagger b)^2]$.

Solution 10. (i) We have $[b^\dagger, b^\dagger b] = -b^\dagger$, $[b, b^\dagger b] = b$ and thus arrive at

$$\begin{aligned}
[b^\dagger, (b^\dagger b)^2] &= -2b^\dagger b^\dagger b - b^\dagger = -2bb^\dagger b^\dagger + 3b^\dagger \\
[b, (b^\dagger b)^2] &= 2b^\dagger bb + b = 2bbb^\dagger - 3b.
\end{aligned}$$

(ii) Since $b^\dagger bb^\dagger b = bbb^\dagger b^\dagger - 3bb^\dagger + I$ it follows that

$$f(b, b^\dagger) = bbb^\dagger b^\dagger - 3bb^\dagger + I$$

with

$$\frac{\partial f}{\partial b^\dagger} = 2bbb^\dagger - 3b, \qquad \frac{\partial f}{\partial b} = 2bb^\dagger b^\dagger - 3b^\dagger.$$

Problem 11. Let $|n\rangle$ $n = 0, 1, 2, \ldots$ be the *number states*. The number states $|n\rangle$ are defined as

$$|n\rangle := \frac{(b^\dagger)^n}{\sqrt{n!}}|0\rangle$$

where $|0\rangle$ is the vacuum state, i.e. $b|0\rangle = 0|0\rangle$ and $\langle 0|0\rangle = 1$. The number states are also called *Fock states*. They are normalized, i.e. $\langle n|n\rangle = 1$. The *number operator* $\hat{N}$ is defined as

$$\hat{N} := b^\dagger b.$$

(i) Calculate the state $[b^n, (b^\dagger)^n]|0\rangle$. Is the state normalized?
(ii) Calculate the states

$$b|n\rangle, \qquad b^\dagger|n\rangle, \qquad b^\dagger b|n\rangle \equiv \hat{N}|n\rangle.$$

(iii) Calculate the expectation value $\langle n|b^\dagger b|n\rangle \equiv \langle n|\hat{N}|n\rangle$.
(iv) Calculate the expectation value $\langle n|(b^\dagger + b)|n\rangle$.
(v) Find the matrix elements of b and $b^\dagger$

$$\langle n|b|n+1\rangle, \qquad \langle n+1|b^\dagger|n\rangle.$$

(vi) Find the operator

$$\sum_{n=0}^{\infty} |n\rangle\langle n|.$$

Solution 11. (i) Since $b|0\rangle = 0|0\rangle$ we obtain the state

$$[b^n, (b^\dagger)^n]|0\rangle = (n!)|0\rangle.$$

The state is not normalized.
(ii) Since $bb^\dagger = I + b^\dagger b$ we have

$$b|n\rangle = \sqrt{n}|n-1\rangle, \qquad b^\dagger|n\rangle = \sqrt{n+1}|n+1\rangle.$$

(iii) Since $b|n\rangle = \sqrt{n}|n-1\rangle$ we obtain $\langle n|b^\dagger b|n\rangle = n$.
(iv) Since $\langle n|n+1\rangle = 0$, $\langle n|n-1\rangle = 0$ we obtain

$$\langle n|(b^\dagger + b)|n\rangle = 0.$$

(v) Since $b|n+1\rangle = \sqrt{n+1}|n\rangle$ and $b^\dagger|n\rangle = \sqrt{n+1}|n+1\rangle$ we have

$$\langle n|b|n+1\rangle = \langle n|\sqrt{n+1}|n\rangle = \sqrt{n+1}$$
$$\langle n+1|b^\dagger|n\rangle = \langle n+1|\sqrt{n+1}|n+1\rangle = \sqrt{n+1}.$$

(vi) We obtain

$$\sum_{n=0}^{\infty} |n\rangle\langle n| = I$$

where I is the identity operator. If we consider $|n\rangle$ $(n = 0, 1, \ldots)$ as the standard basis in the Hilbert space $\ell_2(\mathbb{N}_0)$ then I is the infinite dimensional identity matrix.

Problem 12. Let $b^\dagger$, b be Bose creation and annihilation operators and $|0\rangle$ be the vacuum state.
(i) Calculate the states $bb^\dagger|0\rangle$, $b(b^\dagger)^2|0\rangle$, $b(b^\dagger)^3|0\rangle$.

(ii) Generalize to $b(b^\dagger)^m|0\rangle$, where $m \geq 1$.

Solution 12. (i) Since $b|0\rangle = 0$ and $bb^\dagger = I + b^\dagger b$ we obtain

$$bb^\dagger|0\rangle = |0\rangle, \quad b(b^\dagger)^2|0\rangle = 2b^\dagger|0\rangle, \quad b(b^\dagger)^3|0\rangle = 3(b^\dagger)^2|0\rangle.$$

(ii) In general we find
$$b(b^\dagger)^m|0\rangle = m(b^\dagger)^{m-1}|0\rangle, \qquad m \geq 1.$$

Problem 13. Find $n \times n$ matrices A and B such that $[A, B] = I_n$, where I_n is the $n \times n$ identity matrix. Discuss.

Solution 13. Since $\text{tr}([A, B]) = 0$ and $\text{tr}(I_n) = n$ such matrices cannot be found. The infinite dimensional matrices

$$b = \begin{pmatrix} 0 & \sqrt{1} & 0 & 0 & \cdots \\ 0 & 0 & \sqrt{2} & 0 & \cdots \\ 0 & 0 & 0 & \sqrt{3} & \cdots \\ \vdots & \vdots & \vdots & & \end{pmatrix}$$

and

$$b^\dagger = \begin{pmatrix} 0 & 0 & 0 & \cdots \\ \sqrt{1} & 0 & 0 & \cdots \\ 0 & \sqrt{2} & 0 & \cdots \\ 0 & 0 & \sqrt{3} & \cdots \\ \vdots & \vdots & \vdots & \end{pmatrix}$$

satisfy $[b, b^\dagger] = I$, where I is the identity operator, i.e. the infinite dimensional identity matrix.

Problem 14. (i) Using the number states $|n\rangle$ find the *matrix representation* of the number operator $\hat{N} = b^\dagger b$. The operator $\hat{N}$ is unbounded.
(ii) Using the number states $|n\rangle$ find the matrix representation of the unbounded operators $b^\dagger + b$.

Solution 14. (i) Since $b^\dagger b|n\rangle = n|n\rangle$ we obtain the infinite dimensional unbounded diagonal matrix

$$\text{diag}(0, 1, 2, \ldots).$$

(ii) Since $b^\dagger|n\rangle = \sqrt{n+1}|n+1\rangle$, $b|n\rangle = \sqrt{n}|n-1\rangle$ we obtain the infinite dimensional unbounded matrix

$$\begin{pmatrix} 0 & 1 & 0 & 0 & \cdots \\ 1 & 0 & \sqrt{2} & 0 & \cdots \\ 0 & \sqrt{2} & 0 & \sqrt{3} & \cdots \\ 0 & 0 & \sqrt{3} & 0 & \cdots \\ \vdots & \vdots & \vdots & & \ddots \end{pmatrix}.$$

Problem 15. Truncating the matrix representation of the unbounded operator $b^\dagger + b$ up the 4×4 matrix we obtain the symmetric matrices over $\mathbb{R}$

$$
B_2 = \begin{pmatrix} 0 & 1 \\ 1 & 0 \end{pmatrix}, \quad
B_3 = \begin{pmatrix} 0 & 1 & 0 \\ 1 & 0 & \sqrt{2} \\ 0 & \sqrt{2} & 0 \end{pmatrix}, \quad
B_4 = \begin{pmatrix} 0 & 1 & 0 & 0 \\ 1 & 0 & \sqrt{2} & 0 \\ 0 & \sqrt{2} & 0 & \sqrt{3} \\ 0 & 0 & \sqrt{3} & 0 \end{pmatrix}.
$$

Find the eigenvalues and eigenvectors of these matrices.

Solution 15.   Since the matrices are symmetric over the real number the eigenvalues must be real. Furthermore the sum of the eigenvalues must be 0 since the trace of the matrices is 0 and the eigenvalues are symmetric around 0. For B_n with n odd one of the eigenvalues is always 0. We order the eigenvalues from largest to smallest. For B_2 we obtain the eigenvalues 1, −1 with the eigenvectors

$$
\frac{1}{\sqrt{2}} \begin{pmatrix} 1 \\ 1 \end{pmatrix}, \quad \frac{1}{\sqrt{2}} \begin{pmatrix} 1 \\ -1 \end{pmatrix}.
$$

The matrix B_2 is the Pauli spin matrix σ_1. The eigenvalues of the matrix B_3 are $\sqrt{3}, 0, -\sqrt{3}$ with the corresponding unnormalized eigenvectors

$$
\begin{pmatrix} 1 \\ \sqrt{3} \\ \sqrt{2} \end{pmatrix}, \quad \begin{pmatrix} 1 \\ 0 \\ -1/\sqrt{2} \end{pmatrix}, \quad \begin{pmatrix} 1 \\ -\sqrt{3} \\ \sqrt{2} \end{pmatrix}.
$$

The eigenvalues of the matrix B_4 are

$$
\sqrt{3 + \sqrt{6}}, \quad \sqrt{3 - \sqrt{6}}, \quad -\sqrt{3 - \sqrt{6}}, \quad -\sqrt{3 + \sqrt{6}}
$$

with the corresponding normalized eigenvectors

$$
\begin{pmatrix} 1 \\ \sqrt{3 + \sqrt{2}\sqrt{3} + 3} \\ \sqrt{2} + \sqrt{3} \\ \sqrt{3 + \sqrt{2}\sqrt{3}} \end{pmatrix}, \quad
\begin{pmatrix} 1 \\ \sqrt{3 - \sqrt{2}\sqrt{3}} \\ \sqrt{2} - \sqrt{3} \\ -\sqrt{3 - \sqrt{2}\sqrt{3}} \end{pmatrix},
$$

$$
\begin{pmatrix} 1 \\ -\sqrt{3 - \sqrt{2}\sqrt{3}} \\ \sqrt{2} - \sqrt{3} \\ \sqrt{3 - \sqrt{2}\sqrt{3}} \end{pmatrix}, \quad
\begin{pmatrix} 1 \\ -\sqrt{3 + \sqrt{2}\sqrt{3}} \\ \sqrt{2} + \sqrt{3} \\ -\sqrt{3 + \sqrt{2}\sqrt{3}} \end{pmatrix}.
$$

Extend to the n-dimensional case. Then consider $n \to \infty$ and thus find the spectrum of $b + b^\dagger$.

Problem 16.   Let $b^\dagger$, b be Bose creation and annihilation operators, respectively.
(i) Let $z \in \mathbb{C}$. Calculate the states

$$
b \exp(zb^\dagger)|0\rangle, \qquad b^2 \exp(zb^\dagger)|0\rangle.
$$

Are the states normalized?
(ii) Let $g : \mathbb{C} \to \mathbb{C}$ be an analytic function of z. Calculate the state

$$
g(b) \exp(zb^\dagger)|0\rangle.
$$

(iii) Show that

$$
e^{zb^\dagger} b = (b - zI)e^{zb^\dagger}.
$$

(iv) Extend the identity to $e^{zb^\dagger} b^k$, where $k \in \mathbb{N}$.

Solution 16. (i) Since $bb^\dagger = I + b^\dagger b$ and $b|0\rangle = 0|0\rangle$ we find

$$b(b^\dagger)^n |0\rangle = n(b^\dagger)^{n-1}|0\rangle, \qquad n = 1, 2, \ldots .$$

Thus

$$
\begin{aligned}
b \exp(zb^\dagger)|0\rangle &= b(I + zb^\dagger + \frac{z^2}{2!}(b^\dagger)^2 + \frac{z^3}{3!}(b^\dagger)^3 + \cdots)|0\rangle \\
&= z(bb^\dagger + \frac{z}{2!}b(b^\dagger)^2 + \frac{z^2}{3!}b(b^\dagger)^3 + \cdots)|0\rangle \\
&= z(I + zb^\dagger + \frac{z^2}{2!}(b^\dagger)^2 + \cdots)|0\rangle \\
&= z \exp(zb^\dagger)|0\rangle.
\end{aligned}
$$

Analogously we obtain $b^2 \exp(zb^\dagger)|0\rangle = z^2 \exp(zb^\dagger)|0\rangle$.
(ii) Since g is an analytic function we can expand it as Taylor series. Using the result from (i) that

$$b^n \exp(zb^\dagger)|0\rangle = z^n \exp(zb^\dagger)|0\rangle$$

we have

$$g(b) \exp(zb^\dagger)|0\rangle = g(z) \exp(zb^\dagger)|0\rangle.$$

(iii) Using the *Baker Campbell Hausdorff formula* we have $e^{zb^\dagger} b e^{-zb^\dagger} = b - zI$. To find this result we can also use *parameter differentiation*. We set

$$f(z) := e^{zb^\dagger} b e^{-zb^\dagger}$$

with the "initial condition" $f(0) = b$. Then with $b^\dagger b - bb^\dagger = -I$ we obtain

$$\frac{df}{dz} = e^{zb^\dagger} b^\dagger b e^{-zb^\dagger} - e^{zb^\dagger} bb^\dagger e^{-zb^\dagger} = -I.$$

Now $f(z) = -zI + C$, where the operator constant of integration is determined by $f(0) = b$. Thus $f(z) = b - zI$. It follows that

$$e^{\epsilon b^\dagger} b = e^{\epsilon b^\dagger} b e^{-\epsilon b^\dagger} e^{\epsilon b^\dagger} = (b - \epsilon I)e^{\epsilon b^\dagger}.$$

(iv) Using the result from (iii) we obtain the operator

$$e^{\epsilon b^\dagger} b^k = (b - \epsilon I)^k e^{\epsilon b^\dagger}.$$

Problem 17. (i) Let $|n\rangle$ be a number state ($n = 0, 1, \ldots$). Calculate the state $b^n |n\rangle$.
(ii) Consider the operator

$$K = \left(I - \frac{b^\dagger b}{n}\right) \frac{b^n}{\sqrt{n!}} + h.c. .$$

Use the result from (i) to calculate the states $K|n\rangle$, $K|0\rangle$.

Solution 17. (i) We have $b^n |n\rangle = \sqrt{n}\sqrt{n-1}\cdots\sqrt{2}\sqrt{1}|0\rangle$.

(ii) We have

$$\left(I - \frac{b^\dagger b}{n}\right) \frac{b^n}{\sqrt{n!}} |n\rangle = \left(I - \frac{b^\dagger b}{n}\right) |0\rangle = |0\rangle.$$

Analogously we find

$$\frac{(b^\dagger)^n}{\sqrt{n!}} \left(I - \frac{b^\dagger b}{n}\right) |0\rangle = \frac{(b^\dagger)^n}{\sqrt{n!}} |0\rangle = |n\rangle.$$

Problem 18.   Let $|n\rangle$ be a number state $(n = 0, 1, \ldots)$ and $\hat{N} = b^\dagger b$. Consider the linear bounded operator

$$\hat{T} := \sum_{n=0}^{\infty} |n\rangle\langle n+1|$$

in the Hilbert space $\ell_2(\mathbb{N}_0)$.
(i) Show that the operator $\hat{T}$ is not unitary.
(ii) Find $[\hat{T}, \hat{N}]$, $[\hat{N}, \hat{T}^\dagger]$, $\hat{T}\hat{T}^\dagger$, $\hat{T}^\dagger\hat{T}$, $[\hat{T}, \hat{T}^\dagger]$, $\hat{T}^\dagger|n\rangle$, $\hat{T}|n\rangle$.
(iii) Calculate $\hat{T}\hat{N}^{1/2}$, $\hat{N}^{1/2}\hat{T}^\dagger$.

Solution 18.   (i) We find $\hat{T}\hat{T}^\dagger = I$, $\hat{T}^\dagger\hat{T} = I - |0\rangle\langle 0|$.
(ii) We have

$$[\hat{T}, \hat{N}] = \hat{T}, \quad [\hat{N}, \hat{T}^\dagger] = \hat{T}^\dagger, \quad \hat{T}\hat{T}^\dagger = I, \quad \hat{T}^\dagger\hat{T} = I - |0\rangle\langle 0|,$$

$$[\hat{T}, \hat{T}^\dagger] = |0\rangle\langle 0|, \quad \hat{T}^\dagger|n\rangle = |n+1\rangle$$

and $\hat{T}|n\rangle = |n-1\rangle$ if $n \geq 1$ and $\hat{T}|0\rangle = 0|0\rangle$.
(iii) Furthermore

$$\hat{T}\hat{N}^{1/2} = b, \quad \hat{N}^{1/2}\hat{T}^\dagger = b^\dagger.$$

Problem 19.   Let $\hat{N} = b^\dagger b$, $|n\rangle$ be the number states $(n = 0, 1, \ldots)$ and $\alpha > 0$.
(i) Consider the operator

$$\hat{T} = \sum_{n=0}^{\infty} |n\rangle e^{2\alpha n} \langle 2n|.$$

Is the operator $\hat{T}e^{-\alpha\hat{N}}$ bounded? Is the operator $e^{-\alpha\hat{N}/2}\hat{T}e^{-\alpha\hat{N}/2}$ bounded?
(ii) Consider the operator

$$\hat{T} = \sum_{n=0}^{\infty} |2n\rangle e^{3\alpha n/2} \langle n|.$$

Is $\hat{T}e^{-\alpha\hat{N}}$ bounded? Is the operator $e^{-\alpha\hat{N}/2}\hat{T}e^{-\alpha\hat{N}/2}$ bounded?

Solution 19.   (i) We have

$$\hat{T}e^{-\alpha\hat{N}} = \sum_{n=0}^{\infty} |n\rangle\langle 2n|$$

which is bounded. We have

$$e^{-\alpha\hat{N}/2}\hat{T}e^{-\alpha\hat{N}/2} = \sum_{n=0}^{\infty} |n\rangle e^{\alpha n/2} \langle 2n|$$

which is unbounded.

(ii) We have

$$\hat{T}e^{-\alpha\hat{N}} = \sum_{n=0}^{\infty} |2n\rangle e^{\alpha n/2}\langle n|$$

which is unbounded and

$$e^{-\alpha\hat{N}/2}\hat{T}e^{-\alpha\hat{N}/2} = \sum_{n=0}^{\infty} |2n\rangle\langle n|$$

which is bounded.

Problem 20. Let $\hat{N} := b^\dagger b$ and $\phi \in \mathbb{R}$. Consider the operator

$$\hat{M}(\phi) := \exp(-i\phi\hat{N}).$$

(i) Find the operator $\hat{M}(\phi)b\hat{M}(-\phi)$.
(ii) Find the operator $\hat{M}(\phi)b^\dagger\hat{M}(-\phi)$.
(iii) Let $\theta \in \mathbb{R}$. Consider the operator

$$G := -\frac{1}{2}i(\hat{p}\hat{q} + \hat{q}\hat{p})$$

and the unitary operator $U(\theta) = \exp(\theta G)$. Let $f : \mathbb{C}^2 \to \mathbb{C}$ be an analytic function. Show that

$$U^{-1}(\theta)f(\hat{p},\hat{q})U(\theta) = f(e^{-\theta}\hat{p}, e^{\theta}\hat{q})$$

i.e. G generates a scale transformation.

Solution 20. (i) We obtain

$$\hat{M}(\phi)b\hat{M}(-\phi) = e^{i\phi}b.$$

(ii) We obtain

$$\hat{M}(\phi)b^\dagger\hat{M}(-\phi) = e^{-i\phi}b^\dagger.$$

(iii) Let $\alpha, \beta \in \mathbb{R}$. Differentiating

$$f_{\alpha,\beta}(\theta) := U^{-1}(\theta)\hat{p}^\alpha\hat{q}^\beta U(\theta)$$

with respect to θ yields

$$\frac{df_{\alpha,\beta}(\theta)}{d\theta} = -U^{-1}(\theta)[G, \hat{p}^\alpha\hat{q}^\beta]U(\theta).$$

Now

$$[G, \hat{p}^\alpha\hat{q}^\beta] = (\alpha - \beta)\hat{p}^\alpha\hat{q}^\beta.$$

Thus we obtain the linear differential equation

$$\frac{df_{\alpha,\beta}(\theta)}{d\theta} = (\beta - \alpha)f_{\alpha,\beta}(\theta)$$

together with the initial condition $f_{\alpha,\beta}(\theta = 0) = \hat{p}^\alpha\hat{q}^\beta$. The solution is

$$f_{\alpha,\beta}(\theta) = (e^{-\theta}\hat{p})^\alpha(e^{\theta}\hat{q})^\beta.$$

Setting $\alpha = \beta = 1$ we obtain the desired result.

Problem 21.   Let $\hat{N} = b^\dagger b$ be the number operator.
(i) Consider the *parity operator*

$$P = \exp(i\pi b^\dagger b) \equiv \exp(i\pi \hat{N}).$$

What are the eigenvalues of P?
(ii) Show that

$$b\cos(\pi b^\dagger b) = -\cos(\pi b^\dagger b)b.$$

Solution 21.   (i) Since $\hat{N} = b^\dagger b$ is an infinite diagonal matrix with $0, 1, 2, \dots$ on the diagonal the eigenvalues can only be $+1$ and -1.
(ii) With $\hat{N} = b^\dagger b = \text{diag}(0, 1, 2, \dots)$ we find that the matrix representation of $\cos(\pi b^\dagger b)$ is the diagonal matrix $\text{diag}(1, -1, 1, -1, \dots)$. Thus the identity follows.

Problem 22.   Let $j, k = 0, 1, 2, \dots$. The matrix representations for the Bose creation and annihilation operators $b^\dagger$, b, respectively, are given by

$$(b^\dagger)_{jk} = \sqrt{k}\delta_{j+1,k}, \qquad (b)_{jk} = \sqrt{j}\delta_{j,k+1}.$$

Let $n = 2$ be a positive integer number. Find the matrix representation of the operators

$$((b^\dagger)^n(b)^n)_{jk}, \quad (b^\dagger)^{n+1}(b)^n)_{jk}, \quad ((b^\dagger)^n(b)^{n+1})_{jk}.$$

Solution 22.   We have

$$((b^\dagger)^n(b)^n)_{jk} = \frac{\Gamma(j)}{\Gamma(j-n)}\delta_{jk},$$

$$((b^\dagger)^{n+1}(b)^n)_{jk} = \frac{\Gamma(k)}{\Gamma(k-n)}\sqrt{k}\delta_{k,j+1}, \quad ((b^\dagger)^n(b)^{n+1})_{jk} = \frac{\Gamma(j)}{\Gamma(j-n)}\sqrt{j}\delta_{j,k+1}$$

where Γ is the *gamma function* with $\Gamma(1) = 1$, $\Gamma(2) = 1$ and

$$\Gamma(n) = \int_0^\infty t^{n-1}e^{-t}dt, \quad n > 0.$$

Problem 23.   The operator

$$\hat{x}(\phi) := \frac{1}{\sqrt{2}}(be^{-i\phi} + b^\dagger e^{i\phi})$$

is called *quadrature-component operator*. Solve the eigenvalue problem

$$\hat{x}(\phi)|x, \phi\rangle = x|x, \phi\rangle$$

where $x \in \mathbb{R}$.

Solution 23.   We find

$$|x, \phi\rangle = \frac{1}{\sqrt[4]{\pi}}e^{-x^2/2}\sum_{n=0}^\infty \frac{e^{in\phi}}{\sqrt{2^n n!}}H_n(x)|n\rangle$$

$$= \frac{1}{\sqrt[4]{\pi}}e^{-x^2/2}\exp\left(-\frac{1}{2}(e^{i\phi}b^\dagger)^2 + \sqrt{2}xe^{i\phi}b^\dagger\right)|0\rangle$$

where $H_n(x)$ are the *Hermite polynomials* $(n = 0, 1, \ldots)$.

Problem 24. Let b, $b^\dagger$ be Bose annihilation and creation operators. We have

$$[b, b^\dagger] = cI$$

where $c = 1$. Let $m \geq 1$. Then we have

$$(b + b^\dagger)^m = \sum_{k=0}^{\lfloor m/2 \rfloor} \sum_{j=0}^{m-2k} \frac{c^k m! (b^\dagger)^j b^{m-2k-j}}{2^k k! j! (m - 2k - j)!}$$

where $\lfloor m/2 \rfloor$ is the integer less than or equal to $m/2$.
(i) Give a SymbolicC++ implementation of this sum.
(ii) Let b, $b^\dagger$ be Bose annihilation and creation operators. Let n be an integer number with $n \geq 1$. We set

$$(b^\dagger b)^n = \sum_{k=1}^{n} S(n, k)(b^\dagger)^k b^k.$$

Give a SymbolicC++ implementation that finds the coefficients $S(n, k)$ for a given n. $S(n, k)$ are the *Stirling numbers* of the second kind.
(iii) Give a SymbolicC++ implementation that find the normal ordering of

$$(b^\dagger + b)^4 \quad \text{and} \quad (b^\dagger - b)^4.$$

Solution 24. (i) For $m = 2$ we have $b^2 + 2b^\dagger b + (b^\dagger)^2 + 1$. The SymbolicC++ program is

```
// sum.cpp

#include <iostream>
#include "symbolicc++.h"
using namespace std;

Symbolic fact(int n)
{
 Symbolic f = 1;
 while(n > 1) f *= (n--);
 return f;
}

int main(void)
{
 int m = 2, m2 = m/2, k, kf, k2, j;
 Symbolic b = ~Symbolic("b"), bd = ~Symbolic("bd");
 Symbolic sum = 0, mf = fact(m);
 for(k=0, k2=1;k<=m2;k++, k2*=2)
   for(j=0,kf=fact(k);j<=m-2*k;j++)
     sum += mf*(bd^j)*(b^(m-2*k-j))/(k2*kf*fact(j)*fact(m-2*k-j));
 cout << sum << endl;
 cout << ((b+bd)^m).subst_all(b*bd==1+bd*b)-sum << endl;
 return 0;
}
```

(ii) The SymbolicC++ program is

```
// stirling.cpp

#include <iostream>
#include "symbolicc++.h"
using namespace std;

// expanding integer power
Symbolic pw(const Symbolic &s, int n)
{
 int m = n, sign = (n>=0)?1:-1;
 Symbolic p = 1, x = s;
 m *= sign;
 while(m != 0)
 {
  if(m & 1) p *= x;
  m >>= 1;
  if(m != 0) x *= x;
 }
 return (p^(sign));
}

int main(void)
{
 int n = 5, k;
 Symbolic b = ~Symbolic("b"), bd = ~Symbolic("bd");
 Symbolic S = pw(bd * b,n).subst_all(b*bd==1+bd*b);
 cout << S << endl;
 for(k=1;k<=n;k++)
  cout << "S(" << n << "," << k << ") = " << S.coeff((bd^k)*(b^k)) << endl;
 return 0;
}
```

The output is

```
bd^(5)*b^(5)+10*bd^(4)*b^(4)+25*bd^(3)*b^(3)+15*bd^(2)*b^(2)+bd*b
S(5,1) = 1
S(5,2) = 15
S(5,3) = 25
S(5,4) = 10
S(5,5) = 1
```

(iii) The SymbolicC++ implementation is

```
// normal.cpp

#include <iostream>
#include "symbolicc++.h"
using namespace std;

int main(void)
{
 Symbolic b = ~Symbolic("b"), bd = ~Symbolic("bd");
 cout << ((bd+b)^4).subst_all(b*bd==1+bd*b) << endl;
 cout << ((bd-b)^4).subst_all(b*bd==1+bd*b) << endl;
 return 0;
}
```

The output is

$$(b^\dagger + b)^4 = b^{\dagger^4} + 4b^{\dagger^3}b + 6b^{\dagger^2} + 6b^{\dagger^2}b^2 + 12b^\dagger b + 4b^\dagger b^3 + 6b^2 + b^4 + 3$$

and

$$(b^\dagger - b)^4 = b^{\dagger^4} + 4b^{\dagger^2} - 4b^\dagger b - 2b^{\dagger^2}b^2 - 4b^2 + b^4 + 1.$$

Problem 25. Let $b^\dagger$, b be Bose creation and annihilation operators and I be the identity operator. Let $b_1 = b \otimes I$, $b_2 = b \otimes I$, $b_1^\dagger = b^\dagger \otimes I$, $b_2^\dagger = I \otimes b^\dagger$ (two modes) and

$$\hat{N}_1 = b_1^\dagger b_1, \quad \hat{N}_2 = b_2^\dagger b_2, \quad \hat{N} = \hat{N}_1 + \hat{N}_2 = b_1^\dagger b_1 + b_2^\dagger b_2.$$

(i) Consider the operator

$$Z := b_1 + b_2^\dagger \equiv b \otimes I + I \otimes b.$$

Find the commutators $[Z, Z^\dagger]$ and $[Z, \hat{N}_1 + \hat{N}_2]$.
(ii) Find the commutators $[b_1^\dagger b_2^\dagger, b_1^\dagger b_1 - b_2^\dagger b_2]$, $[b_1 b_2, b_1^\dagger b_1 - b_2^\dagger b_2]$.
(iii) Find the commutator $[i(b_1^\dagger b_2 - b_2^\dagger b_1), (b_1^\dagger)^2 + (b_2^\dagger)^2]$.
(iv) Find the commutator $[b_2 b_1, b_2^\dagger b_1^\dagger]$.
(v) Let $b_1^\dagger$, $b_2^\dagger$, $b_3^\dagger$, b_1, b_2, b_3 be Bose creation and annihilation operators. Find the commutator $[b_3 b_2 b_1, b_3^\dagger b_2^\dagger b_1^\dagger]$.
(vi) Find the commutator

$$[b_j^\dagger b_k^\dagger, b_\ell b_m], \qquad j, k, \ell, m = 1, 2, \dots, N.$$

(vii) Consider the Hamilton operator $\hat{H} = \hat{H}_1 + \hat{H}_2$ with

$$\hat{H}_1 = \hbar\omega_1 b_1^\dagger b_1 + 2\hbar\omega_1 b_2^\dagger b_2, \quad \hat{H}_2 = \hbar\omega_2(b_2^\dagger b_1^2 + b_2(b_1^\dagger)^2).$$

Find the commutator $[\hat{H}_1, \hat{H}_2]$.

Solution 25. (i) Since $Z^\dagger = b_1^\dagger + b_2$ we have

$$[Z, Z^\dagger] = [b_1 + b_2^\dagger, b_1^\dagger + b_2] = [b_1, b_1^\dagger] + [b_2^\dagger, b_2] = 0.$$

(ii) We obtain $[b_1^\dagger b_2^\dagger, b_1^\dagger b_1 - b_2^\dagger b_2] = 0$, $[b_1 b_2, b_1^\dagger b_1 - b_2^\dagger b_2] = 0$.
(iii) We find

$$[i(b_1^\dagger b_2 - b_2^\dagger b_1), (b_1^\dagger)^2 + (b_2^\dagger)^2] = 0.$$

(iv) Applying $b_j b_k^\dagger = \delta_{jk}(I + b_k^\dagger b_j)$ we obtain

$$[b_2 b_1, b_2^\dagger b_1^\dagger] = b_2 b_2^\dagger b_1 b_1^\dagger - b_2^\dagger b_2 b_1^\dagger b_1 = (I + b_2^\dagger b_2)(I + b_1^\dagger b_1) = I + b_1^\dagger b_1 + b_2^\dagger b_2.$$

(v) Applying $b_j b_k^\dagger = \delta_{jk}(I + b_k^\dagger b_j)$ we obtain

$$[b_3 b_2 b_1, b_3^\dagger b_2^\dagger b_1^\dagger] = I + b_1^\dagger b_1 + b_2^\dagger b_2 + b_3^\dagger b_3 + b_2^\dagger b_2 b_1^\dagger b_1 + b_3^\dagger b_3 b_1^\dagger b_1 + b_3^\dagger b_2 b_2^\dagger b_1.$$

(vi) Using repeatedly $[b_j, b_k^\dagger] = \delta_{jk}I$ we obtain

$$[b_j^\dagger b_k^\dagger, b_\ell b_m] = -\delta_{mj}\delta_{k\ell}I - \delta_{mk}\delta_{j\ell}I - \delta_{mj}b_k^\dagger b_\ell - \delta_{mk}b_j^\dagger b_\ell - \delta_{\ell j}b_k^\dagger b_m - \delta_{\ell k}b_j^\dagger b_m.$$

(vii) Note that $[b_1, b_2] = [b_1^\dagger, b_2^\dagger] = [b_1^\dagger, b_2] = [b_1, b_2^\dagger] = 0 \otimes 0$. We obtain $[\hat{H}_1, \hat{H}_2] = 0 \otimes 0$. Consequently both $\hat{H}_1$ and $\hat{H}_2$ are constants of motion.

Problem 26. Consider the operators

$$\hat{A} = b_k e^{-i\phi_k} + b_k^\dagger e^{i\phi_k}, \qquad \hat{B} = b_\ell e^{-i\phi_\ell} + b_\ell^\dagger e^{i\phi_\ell}.$$

Find the commutator $[\hat{A}, \hat{B}]$. What is the condition on the phases ϕ_k and ϕ_ℓ such that $[\hat{A}, \hat{B}] = 0$ for $k = \ell$?

Solution 26. We have

$$\begin{aligned}
[\hat{A}, \hat{B}] &= [b_k e^{-i\phi_k} + b_k^\dagger e^{i\phi_k}, b_\ell e^{-i\phi_\ell} + b_\ell^\dagger e^{i\phi_\ell}] = [b_k e^{-i\phi_k}, b_\ell^\dagger e^{i\phi_\ell}] + [b_k^\dagger e^{i\phi_k}, b_\ell e^{-i\phi_\ell}] \\
&= (e^{i(\phi_\ell - \phi_k)} - e^{i(\phi_k - \phi_\ell)})\delta_{\ell k} \\
&= 2i \sin(\phi_\ell - \phi_k)\delta_{\ell k}.
\end{aligned}$$

Thus the commutator is zero for $k = \ell$ if $\sin(\phi_\ell - \phi_k) = 0$ i.e. $\phi_\ell - \phi_k = n\pi$ and $n \in \mathbb{Z}$.

Problem 27. (i) Find the commutator $[b_1^\dagger b_1 + b_2^\dagger b_2, b_1^\dagger b_2 + b_2^\dagger b_1]$. Discuss.
(ii) Consider the operators

$$A := b_1^\dagger b_2^\dagger, \qquad B := b_1 b_2.$$

Calculate the commutators $[A, B]$, $[A, [A, B]]$, $[B, [A, B]]$.
(iii) Let s be an integer number with $s \geq 2$. We define the *generalized Boson operators* (also called *pseudo-boson operators*)

$$\hat{B}_s := \sum_{j=1}^{s-1} \frac{(-1)^{j-1} b_j}{\sqrt{2^j}} + \frac{(-1)^{s-1} b_s}{\sqrt{2^{s-1}}}.$$

Find the commutator $[\hat{B}_s, \hat{B}_s^\dagger]$.

Solution 27. (i) Using $b_1 b_1^\dagger = I + b_1^\dagger b_1$, $b_2 b_2^\dagger = I + b_2^\dagger b_2$ we obtain

$$[b_1^\dagger b_1 + b_2^\dagger b_2, b_1^\dagger b_2 + b_2^\dagger b_1] = b_1^\dagger b_2 - b_2^\dagger b_1 - b_1^\dagger b_2 + b_2^\dagger b_1 = 0.$$

Note that $b_1^\dagger b_1 + b_2^\dagger b_2$ is the number operator.
(ii) We have $[A, B] = -I - b_1^\dagger b_1 - b_2^\dagger b_2$ and $[A, [A, B]] = 2b_1^\dagger b_2^\dagger$, $[B, [A, B]] = -2b_1 b_2$.
(iii) Using $[b_j, b_k^\dagger] = \delta_{jk} I$ we obtain $[\hat{B}_s, \hat{B}_s^\dagger] = I$.

Problem 28. Consider the number states (two modes)

$$|n_1 n_2\rangle \equiv |n_1\rangle |n_2\rangle \equiv |n_1\rangle \otimes |n_2\rangle$$

where $n_1, n_2 = 0, 1, 2, \ldots$ and

$$|n_1\rangle = \frac{1}{\sqrt{n_1!}} (b_1^\dagger)^{n_1} |0\rangle, \quad |n_2\rangle = \frac{1}{\sqrt{n_2!}} (b_2^\dagger)^{n_2} |0\rangle.$$

The number operators are $\hat{N}_1 = b_1^\dagger b_1$, $\hat{N}_2 = b_2^\dagger b_2$ and the total number operator is $\hat{N} = \hat{N}_1 + \hat{N}_2$. Calculate

$$b_1 |n_1 n_2\rangle, \quad b_2 |n_1 n_2\rangle, \quad b_1^\dagger |n_1 n_2\rangle, \quad b_2^\dagger |n_1 n_2\rangle, \quad \hat{N}_1 |n_1 n_2\rangle, \quad \hat{N}_2 |n_1, n_2\rangle.$$

Solution 28. We obtain

$$b_1|n_1, n_2\rangle = \sqrt{n_1}|n_1 - 1, n_2\rangle, \quad b_2|n_1, n_2\rangle = \sqrt{n_2}|n_1, n_2 - 1\rangle$$

$$b_1^\dagger|n_1, n_2\rangle = \sqrt{n_1 + 1}|n_1 + 1, n_2\rangle, \quad b_2^\dagger|n_1, n_2\rangle = \sqrt{n_2 + 1}|n_1, n_2 + 1\rangle$$

$$\hat{N}_1|n_1, n_2\rangle = n_1|n_1, n_2\rangle, \quad \hat{N}_2|n_1, n_2\rangle = n_2|n_1, n_2\rangle.$$

Problem 29. Find the matrix representation of operator $b_1^\dagger b_2 + b_1 b_2^\dagger$ using the ordering $(0,0)$, $(0,1)$, $(1,0)$, $(0,2)$, $(1,1)$, $(2,0)$, $(0,3)$, etc.

Solution 29. Consider the normalized states $|n\rangle \otimes |m\rangle$ with $n, m = 0, 1, \ldots, \infty$. Since

$$b_1^\dagger b_2 + b_1 b_2^\dagger = b^\dagger \otimes b + b \otimes b^\dagger$$

we have

$$(b^\dagger \otimes b + b \otimes b^\dagger)(|n\rangle \otimes |m\rangle) = b^\dagger|n\rangle \otimes b|m\rangle + b|n\rangle \otimes b^\dagger|m\rangle$$
$$= \sqrt{n + 1}|n + 1\rangle \otimes \sqrt{m}|m - 1\rangle + \sqrt{n}|n - 1\rangle \otimes \sqrt{m + 1}|m + 1\rangle.$$

It follows that

$$\langle k| \otimes \langle \ell|(b^\dagger \otimes b + b \otimes b^\dagger)|n\rangle \otimes |m\rangle = \sqrt{n + 1}\delta_{k,n+1}\sqrt{m}\delta_{\ell,m-1} + \sqrt{n}\delta_{k,n-1}\sqrt{m + 1}\delta_{\ell,m+1}$$

where $k, \ell = 0, 1, \ldots, \infty$.

Problem 30. Let $|0\rangle$ be the vacuum state. Calculate the state

$$\frac{1}{2}(b^\dagger \otimes I - I \otimes b^\dagger)(b^\dagger \otimes I + I \otimes b^\dagger)(|0\rangle \otimes |0\rangle).$$

Solution 30. Since $(b^\dagger \otimes I)(I \otimes b^\dagger) = (I \otimes b^\dagger)(b^\dagger \otimes I)$ we obtain the state

$$\frac{1}{\sqrt{2}}\left(|2\rangle \otimes |0\rangle - |0\rangle \otimes |2\rangle\right).$$

Problem 31. Consider the three operators

$$J_1 = \frac{1}{2}(b_1^\dagger b_2 + b_2^\dagger b_1), \quad J_2 = \frac{1}{2}i(b_1^\dagger b_2 - b_2^\dagger b_1), \quad J_3 = \frac{1}{2}(b_1^\dagger b_1 - b_2^\dagger b_2)$$

and $J_+ := J_1 + iJ_2 = b_1^\dagger b_2$, $J_- := J_1 - iJ_2 = b_2^\dagger b_1$. Let $\phi \in \mathbb{R}$. Find the operators

$$e^{-iJ_3\phi}b_1^\dagger e^{iJ_3\phi}, \quad e^{-iJ_3\phi}b_2^\dagger e^{iJ_3\phi}, \quad e^{-iJ_2\phi}b_1^\dagger e^{iJ_2\phi}, \quad e^{-iJ_2\phi}b_2^\dagger e^{iJ_2\phi}.$$

Solution 31. We obtain

$$e^{-iJ_3\phi}b_1^\dagger e^{iJ_3\phi} = b_1^\dagger e^{-i\phi/2}, \quad e^{-iJ_3\phi}b_2^\dagger e^{iJ_3\phi} = b_2^\dagger e^{-i\phi/2}$$

and

$$e^{-iJ_2\phi}b_1^\dagger e^{iJ_2\phi} = b_1^\dagger\cos(\phi/2) + b_2^\dagger\sin(\phi/2), \quad e^{-iJ_2\phi}b_2^\dagger e^{iJ_2\phi} = b_2^\dagger\cos(\phi/2) - b_1^\dagger\sin(\phi/2).$$

Problem 32. (i) Let $b_j^\dagger$, b_j be Bose creation and annihilation operators, respectively and $j = 1, 2, \ldots, n$. We define

$$B(x) := \sum_{j=1}^n \frac{b_j}{1 - \epsilon_j x}, \qquad f(x) := \sum_{j=1}^n \frac{1}{1 - \epsilon_j x}.$$

Show that

$$[B(x), B^\dagger(y)] = \frac{I}{x - y}(xf(x) - yf(y)), \quad [B(x), B(y)] = 0$$

where $[\,,\,]$ denotes the commutator, I is the identity operator and 0 is the zero operator.
(ii) Consider the operator

$$N(x) := \sum_{j=1}^n \frac{b_j^\dagger b_j}{1 - \epsilon_j x}.$$

Show that

$$[N(x), B^\dagger(y)] = \frac{1}{x - y}(xB^\dagger - yB^\dagger(y)), \quad [N(x), B(y)] = -\frac{1}{x - y}(xB(x) - yB(y)).$$

Solution 32. (i) We have

$$B^\dagger(x) = \sum_{j=1}^n \frac{b_j^\dagger}{1 - \epsilon_j x}.$$

Thus we find

$$
\begin{aligned}
[B(x), B^\dagger(y)] &= \sum_{j=1}^n \frac{b_j}{1 - \epsilon_j x} \sum_{k=1}^n \frac{b_k^\dagger}{1 - \epsilon y} - \sum_{k=1}^n \frac{b_k^\dagger}{1 - \epsilon_k y} \sum_{j=1}^n \frac{b_j}{1 - \epsilon_j x} \\
&= \sum_{j=1}^n \sum_{k=1}^n \frac{b_j b_k^\dagger}{(1 - \epsilon_j x)(1 - \epsilon_k y)} - \sum_{k=1}^n \sum_{j=1}^n \frac{b_k^\dagger b_j}{(1 - \epsilon_k y)(1 - \epsilon_j x)} \\
&= \sum_{j=1}^n \sum_{k=1}^n \frac{b_j b_k^\dagger - b_k^\dagger b_j}{(1 - \epsilon_j x)(1 - \epsilon y)} = \sum_{j=1}^n \sum_{k=1}^n \frac{\delta_{jk} I}{(1 - \epsilon_j x)(1 - \epsilon_k y)} \\
&= I \sum_{j=1}^n \frac{1}{(1 - \epsilon_j x)(1 - \epsilon_j y)} = \frac{I}{x - y} \sum_{j=1}^n \frac{x - y}{(1 - \epsilon_j x)(1 - \epsilon_j y)} \\
&= \frac{I}{x - y} \sum_{j=1}^n \frac{x(1 - \epsilon_j y) - y(1 - \epsilon_j x)}{(1 - \epsilon_j x)(1 - \epsilon_j x)} = \frac{I}{x - y} \sum_{j=1}^n \left(\frac{x}{1 - \epsilon_j x} - \frac{y}{1 - \epsilon_j y} \right) \\
&= \frac{I}{x - y} \left(x \sum_{j=1}^n \frac{1}{1 - \epsilon_j x} - y \sum_{j=1}^n \frac{1}{1 - \epsilon_j y} \right) \\
&= \frac{I}{x - y}(xf(x) - yf(y))
\end{aligned}
$$

where we used that $[b_j, b_k^\dagger] = \delta_{jk} I$. Since $[b_j, b_k] = 0$ the second result is obvious.

(ii) Using the commutators

$$[b_j^\dagger b_j, b_k] = -b_j \delta_{jk}, \qquad [b_j^\dagger b_j, b_k^\dagger] = b_j^\dagger \delta_{jk}$$

and a similar calculation as in (i) we obtain the results.

Problem 33. Assume that the pure bipartite state $|\psi\rangle$ can be written as

$$|\psi\rangle = |\phi_1\rangle \otimes |\phi_2\rangle.$$

Calculate the density matrix $\rho = |\psi\rangle\langle\psi|$ and then the partial traces using the number states $|n\rangle$, where $n = 0, 1, \ldots$.

Solution 33. We have

$$\rho = |\psi\rangle\langle\psi| = (|\phi_1\rangle \otimes |\phi_2\rangle)(\langle\phi_1| \otimes \langle\phi_2|) = (|\phi_1\rangle\langle\phi_1|) \otimes (|\phi_2\rangle\langle\phi_2|).$$

Thus

$$\rho_1 = \sum_{n=0}^{\infty} (I \otimes \langle n|)(|\phi_1\rangle\langle\phi_1| \otimes |\phi_2\rangle\langle\phi_2|)(I \otimes |n\rangle) = \sum_{n=0}^{\infty} (|\phi_1\rangle\langle\phi_1| \otimes \langle n|\phi_2\rangle\langle\phi_2|)(I \otimes |n\rangle)$$

$$= \sum_{n=0}^{\infty} |\phi_1\rangle\langle\phi_1|\langle n|\phi_2\rangle\langle\phi_2|n\rangle = |\phi_1\rangle\langle\phi_1| \sum_{n=0}^{\infty} \langle n|\phi_2\rangle\langle\phi_2|n\rangle$$

$$= |\phi_1\rangle\langle\phi_1| \sum_{n=0}^{\infty} \langle n|\phi_2\rangle\overline{\langle n|\phi_2\rangle} = |\phi_1\rangle\langle\phi_1|\langle\phi_2|\phi_2\rangle$$

$$= |\phi_1\rangle\langle\phi_1|$$

where we used that $\langle\phi_2|\phi_2\rangle = 1$.

Problem 34. Let b_1, b_2 be Bose annihilation operators. Consider the operators

$$d_\pm = \frac{1}{\sqrt{2}}(b_1 \pm b_2 e^{i\theta})$$

where θ is a phase shift. Let

$$I := d_+^\dagger d_+ - d_-^\dagger d_-.$$

Find I in terms of the original operators b_1, b_2.

Solution 34. We have

$$I = d_+^\dagger d_+ - d_-^\dagger d_-$$

$$= \frac{1}{2}(b_1^\dagger + b_2^\dagger e^{-i\theta})(b_1 + b_2 e^{i\theta}) - \frac{1}{2}(b_1^\dagger - b_2^\dagger e^{-i\theta})(b_1 - b_2 e^{i\theta})$$

$$= b_2^\dagger b_1 e^{-i\theta} + b_1^\dagger b_2 e^{i\theta}.$$

This plays a role in *homodyne measurement*, where b_1 describes the signal field and b_2 describes the local oscillator field.

Problem 35.   Let $b_j^\dagger$, b_j be Bose creation and annihilation operators, respectively and $j = 1, 2, \ldots, N$. Define the operators

$$P_j^\dagger := b_j^\dagger \sqrt{\sum_{k=0}^{\infty} \frac{(-2)^k}{(k+1)!} (b_j^\dagger)^k b_j^k}, \qquad P_j = (P_j^\dagger)^\dagger.$$

(i) Find $P_j^\dagger P_j + P_j P_j^\dagger$, $(P_j^\dagger)^2$ and $(P_j)^2$.
(ii) Find

$$[P_j^\dagger, P_k] \text{ for } j \neq k \quad \text{and} \quad [P_j, P_k] \text{ for } j \neq k.$$

Solution 35.   (i) We find $P_j^\dagger P_j + P_j P_j^\dagger = I$, $(P_j^\dagger)^2 = (P_j)^2 = 0$.
(ii) We obtain $[P_j^\dagger, P_k] = [P_j, P_k] = 0$ for $j \neq k$.

Problem 36.   Consider the operator

$$b_1^\dagger b_2 + b_2^\dagger b_1 \equiv b^\dagger \otimes b + b \otimes b^\dagger.$$

Show that 0 is an element of the spectrum of this operator.

Solution 36.   Since $b|0\rangle = 0|0\rangle$ we have

$$(b^\dagger \otimes b + b \otimes b^\dagger)(|n\rangle \otimes |0\rangle \pm |0\rangle \otimes |m\rangle) = 0(|n\rangle \otimes |0\rangle \pm |0\rangle \otimes |m\rangle)$$

where $|n\rangle$, $|m\rangle$ are number states.

Problem 37.   Let $b_1^\dagger$, $b_2^\dagger$, b_1, b_2 be Bose creation and annihilation operators. Consider the operators

$$J_+ = b_1^\dagger b_2, \quad J_- = b_2^\dagger b_1, \quad J_0 = \frac{1}{2}(b_1^\dagger b_1 - b_2^\dagger b_2)$$

and the normalized state

$$|j, m\rangle = \frac{(b_1^\dagger)^{j+m}}{\sqrt{(j+m)!}} \frac{(b_2^\dagger)^{j-m}}{\sqrt{(j-m)!}} |0\rangle \equiv |n_1 = j + m\rangle \otimes |n_2 = j - m\rangle$$

where $j \in \mathbb{N}_0$ and $m = -j, -j+1, \ldots, j$. Find the states $J_0|j, m\rangle$, $J_+|j, m\rangle$, $J_-|j, m\rangle$.

Solution 37.   We obtain the eigenvalue equation $J_0|j, m\rangle = m|j, m\rangle$ and

$$J_+|j, m\rangle = \sqrt{(j-m)(j+m+1)}|j, m+1\rangle, \quad J_-|j, m\rangle = \sqrt{(j+m)(j-m+1)}|j, m-1\rangle.$$

Problem 38.   Consider the unitary operator

$$U = \exp(-i(\omega_1 b_1^\dagger b_1 + \omega_2 b_2^\dagger b_2)t).$$

Find the operator $U^{-1} b_2^\dagger b_1 U$.

Solution 38.   We have $[b_1^\dagger b_1, b_2^\dagger b_2] = 0$. Consequently $U = U_1 U_2$, where $U_1 = \exp(-i(\omega_1 b_1^\dagger b_1)t)$, $U_2 = \exp(-i(\omega_2 b_2^\dagger b_2)t)$ we have

$$U^{-1} b_2^\dagger b_1 U = U^{-1} b_2^\dagger U U^{-1} b_1 U = U_2^{-1} U_1^{-1} b_2^\dagger U_1 U_2 U_2^{-1} U_1^{-1} b_1 U_1 U_2.$$

Thus
$$U^{-1}b_2^\dagger b_1 U = U_2^{-1}b_2 U_2 U_1^\dagger b_1 U_1 = b_2^\dagger b_1 \exp(i(\omega_1 - \omega_2)t).$$

Problem 39. Let
$$U(\theta) = \exp(\theta(b_1^\dagger b_2 - b_1 b_2^\dagger)), \qquad \theta \in \mathbb{R}.$$

Calculate
$$U(\theta)\begin{pmatrix} b_1 \\ b_2 \end{pmatrix} U^\dagger(\theta) \equiv \begin{pmatrix} U(\theta)b_1 U^\dagger(\theta) \\ U(\theta)b_2 U^\dagger(\theta) \end{pmatrix}$$

and show that we can write
$$\begin{pmatrix} U(\theta)b_1 U^\dagger(\theta) \\ U(\theta)b_2 U^\dagger(\theta) \end{pmatrix} = R(\theta)\begin{pmatrix} b_1 \\ b_2 \end{pmatrix}$$

where the 2×2 matrix $R(\theta)$ depends only on θ.

Solution 39. We have $U^\dagger(\theta) = U(-\theta)$. We set
$$f_1(\theta) = U(\theta)b_1 U^\dagger(\theta), \qquad f_2(\theta) = U(\theta)b_2 U^\dagger(\theta)$$

with the initial conditions $f_1(\theta = 0) = b_1$, $f_2(\theta = 0) = b_2$. Differentiation of f_1 with respect to θ provides
$$\frac{df_1}{d\theta} = U(\theta)(b_1^\dagger b_2 - b_1 b_2^\dagger)b_1 U^\dagger(\theta) - U(\theta)b_1(b_1^\dagger b_2 - b_1 b_2^\dagger)U^\dagger(\theta)$$
$$= U(\theta)(b_1^\dagger b_1 - b_1 b_1^\dagger)b_2 U^\dagger(\theta) = -U(\theta)b_2 U^\dagger(\theta)$$
$$= -f_2(\theta).$$

Similarly, we find $df_2/d\theta = f_1$. Thus we have a linear system of differential equations
$$\frac{df_1}{d\theta} = -f_2, \qquad \frac{df_2}{d\theta} = f_1$$

with the initial values $f_1(\theta = 0) = b_1$, $f_2(\theta = 0) = b_2$. The solution is
$$f_1(\theta) = \cos(\theta)b_1 - \sin(\theta)b_2, \qquad f_2(\theta) = \sin(\theta)b_1 + \cos(\theta)b_2$$

or in matrix notation
$$\begin{pmatrix} f_1(\theta) \\ f_2(\theta) \end{pmatrix} = \begin{pmatrix} \cos(\theta) & -\sin(\theta) \\ \sin(\theta) & \cos(\theta) \end{pmatrix}\begin{pmatrix} b_1 \\ b_2 \end{pmatrix}.$$

Thus the matrix $R(\theta)$ is the *rotation matrix*
$$R(\theta) = \begin{pmatrix} \cos(\theta) & -\sin(\theta) \\ \sin(\theta) & \cos(\theta) \end{pmatrix}.$$

Problem 40. Consider the operators
$$J_1 = \frac{1}{2}(b^\dagger \otimes b + b \otimes b^\dagger), \qquad J_2 = \frac{1}{2i}(b^\dagger \otimes b - b \otimes b^\dagger)$$

and
$$J_3 = \frac{1}{2}(b^\dagger b \otimes I - I \otimes b^\dagger b).$$

The *spin ladder operators* are

$$J_+ := J_1 + iJ_2 = b^\dagger \otimes b, \qquad J_- := J_1 - iJ_2 = b \otimes b^\dagger.$$

(i) Find $J^2 := J_1^2 + J_2^2 + J_3^2$.
(ii) Find $J^2(|n_1\rangle \otimes |n_2\rangle)$, $J_3(|n_1\rangle \otimes |n_2\rangle)$, where $n_1, n_2 = 0, 1, 2, \ldots$.

Solution 40.   (i) We obtain

$$J^2 = \frac{1}{2}(b^\dagger b \otimes I + I \otimes b^\dagger b)\left(\frac{1}{2}(b^\dagger b \otimes I + I \otimes b^\dagger b) + I \otimes I\right).$$

(ii) We have

$$J^2(|n_1\rangle \otimes |n_2\rangle) = \frac{1}{2}(n_1 + n_2)\left(\frac{1}{2}(n_1 + n_2) + 1\right)|n_1\rangle \otimes |n_2\rangle$$

$$J_3(|n_1\rangle \otimes |n_2\rangle) = \frac{1}{2}(n_1 - n_2)|n_1\rangle \otimes |n_2\rangle.$$

Thus the two-mode number states are the spin eigenstates $|j, m\rangle$ with

$$j = \frac{1}{2}(n_1 + n_2), \qquad m = \frac{1}{2}(n_1 - n_2).$$

1.2 Coherent States

Let b be a Bose annihilation operator, i.e. $b|0\rangle = 0|0\rangle$. Consider the eigenvalue problem

$$b|\beta\rangle = \beta|\beta\rangle.$$

The eigenstate $|\beta\rangle$ is called a Bose coherent state, where $\beta \in \mathbb{C}$ and $\langle\beta|\beta\rangle = 1$. Thus we have

$$\langle\beta|b^\dagger = \langle\beta|\bar{\beta}.$$

It follows that

$$\langle\beta|b^\dagger b|\beta\rangle = \langle\beta|\bar{\beta}\beta|\beta\rangle = \bar{\beta}\beta\langle\beta|\beta\rangle = \bar{\beta}\beta.$$

Let $|n\rangle$ with $n = 0, 1, 2, \ldots$ be the number states. Using number states a coherent state $|\beta\rangle$ can be written as

$$|\beta\rangle = \exp\left(-\frac{1}{2}|\beta|^2\right) \sum_{n=0}^{\infty} \frac{\beta^n}{\sqrt{n!}}|n\rangle.$$

Let $\beta \in \mathbb{C}$. A coherent state is a minimum uncertainty state. The coherent states are not orthogonal. Starting from the eigenvalue equation $b|\beta\rangle = \beta|\beta\rangle$, the number states $|n\rangle$ and the expansion of $|\beta\rangle$

$$|\beta\rangle = \sum_{n=0}^{\infty} \langle n|\beta\rangle|n\rangle$$

one can show that

$$\langle n|\beta\rangle = \exp(-\frac{1}{2}|\beta|^2)\frac{\beta^n}{\sqrt{n!}}.$$

We have

$$|\psi\rangle = \frac{1}{\pi} \int_{\mathbb{C}} \langle\beta|\psi\rangle|\beta\rangle d^2\beta$$

where the integration over the complex plane $\mathbb{C}$. The *displacement operator* $D(\beta)$ is defined by

$$D(\beta) := e^{\beta b^\dagger - \bar{\beta} b}.$$

Then the coherent states $|\beta\rangle$ can be written as

$$|\beta\rangle := D(\beta)|0\rangle.$$

Since $[\beta b^\dagger, -\bar{\beta} b] = \beta\bar{\beta} I$ we have

$$D(\beta) = e^{\beta b^\dagger - \bar{\beta} b} = e^{-\frac{1}{2}|\beta|^2} e^{\beta b^\dagger} e^{-\bar{\beta} b} = e^{+\frac{1}{2}|\beta|^2} e^{-\bar{\beta} b} e^{\beta b^\dagger}.$$

The displacement operator can also be written as

$$D(q, p) = \exp\left(\frac{i}{\hbar}(p\hat{Q} - q\hat{P})\right)$$

with

$$\exp(i(q\hat{P} - p\hat{Q})/\hbar)\hat{P}\exp(-i(q\hat{P} - p\hat{Q})/\hbar) = \hat{P} + pI$$
$$\exp(i(q\hat{P} - p\hat{Q})/\hbar)\hat{Q}\exp(-i(q\hat{P} - p\hat{Q})/\hbar) = \hat{Q} + qI$$

where I is the identity operator.

Problem 41.   Let $\beta, \gamma \in \mathbb{C}$ and $|\beta\rangle$, $|\gamma\rangle$ be coherent states.
(i) Find the scalar product

$$\langle \beta | \gamma \rangle$$

and thus show that the coherent states are not orthogonal.
(ii) Use this result to find

$$\langle \beta | -\beta \rangle.$$

(iii) Assume that β is real. Calculate $|\langle \beta | -\beta \rangle|^2$. Assume that $\beta \geq 2$. Find $|\langle \beta | -\beta \rangle|^2$. Use the result from (ii).
(iv) Let $|n\rangle$ be the number states. Calculate $\langle n | \beta \rangle$ and $\langle \beta | n \rangle \equiv \overline{\langle n | \beta \rangle}$.
(v) Calculate the square of the distance

$$\| |n\rangle - |\beta\rangle \|^2 = ((\langle n| - \langle \beta|)(|n\rangle - |\beta\rangle).$$

Discuss.

Solution 41.   (i) Utilizing the number state representation $|n\rangle$, $|m\rangle$ $(n, m = 0, 1, \ldots)$ of the coherent states and $\langle n | m \rangle = \delta_{nm}$ we have

$$\langle \beta | \gamma \rangle = \exp(-\frac{1}{2}(|\beta|^2 + |\gamma|^2)) \sum_{n=0}^{\infty} \sum_{m=0}^{\infty} \frac{\beta^{*n}}{\sqrt{n!}} \frac{\gamma^{*m}}{\sqrt{m!}} \langle n | m \rangle = \exp(-\frac{1}{2}(|\beta|^2 + |\gamma|^2)) \sum_{n=0}^{\infty} \frac{\gamma \beta^*}{n!}$$

$$= \exp(-\frac{1}{2}(|\beta|^2 + |\gamma|^2) + \gamma\beta^*).$$

(ii) From (i) with $\gamma = -\beta$ it follows that $\langle \beta | -\beta \rangle = e^{-2\beta\beta^*}$.
(iii) We find $|\langle \beta | -\beta \rangle|^2 = e^{-4\beta^2}$. With $\beta \geq 2$ we have $|\langle \beta | -\beta \rangle|^2 \leq 1.1 \cdot 10^{-7}$.
(iv) Since $\langle n | m \rangle = \delta_{nm}$ we have

$$\langle n | \beta \rangle = \exp\left(-\frac{1}{2}|\beta|^2\right) \frac{\beta^n}{\sqrt{n!}}.$$

It follows that

$$\overline{\langle n | \beta \rangle} = \exp\left(-\frac{1}{2}|\beta|^2\right) \frac{\overline{\beta}^n}{\sqrt{n!}}$$

since $\overline{\beta^n} = \overline{\beta}^n$.
(v) We have

$$\| |n\rangle - |\beta\rangle \|^2 = ((\langle n| - \langle \beta|)(|n\rangle - |\beta\rangle)) = 2 - \langle n | \beta \rangle - \langle \beta | n \rangle = 2 - \langle n | \beta \rangle - \overline{\langle n | \beta \rangle}.$$

Using (i) we obtain

$$\| |n\rangle - |\beta\rangle \|^2 = 2 - \frac{e^{-|\beta|^2/2}}{\sqrt{n!}} \left(\beta^n + \overline{\beta}^n\right).$$

If we set $\beta = re^{i\phi}$ with $r \geq 0$, $\phi \in \mathbb{R}$ and apply the identity

$$2\cos(n\phi) \equiv e^{in\phi} + e^{-in\phi}$$

we obtain

$$\| |n\rangle - |\beta\rangle \|^2 = 2\left(1 - \frac{e^{-|r|^2/2}}{\sqrt{n!}} r^n \cos(n\phi)\right).$$

Problem 42. Let $|\beta\rangle$ be a coherent state, $\hat{N} := b^\dagger b$ be the number operator and $\beta = re^{i\phi}$.
(i) Calculate the expectation value $\langle\beta|\hat{N}|\beta\rangle$.
(ii) Find the expectation value $\langle\beta|\hat{N}^2|\beta\rangle$.
(iii) Find the expectation value $\langle\beta|(b^\dagger + b)|\beta\rangle$.

Solution 42. (i) Since $b|\beta\rangle = \beta|\beta\rangle$, $\langle\beta|b^\dagger = \langle\beta|\beta^*$ we obtain

$$\langle\beta|\hat{N}|\beta\rangle = \beta\beta^* = |\beta|^2 = r^2.$$

(ii) Since $b|\beta\rangle = \beta|\beta\rangle$ and $\langle\beta|b^\dagger = \langle\beta|\beta^*$ we have $\langle\beta|b^\dagger b|\beta\rangle = \beta\beta^*$. Hence

$$\langle\beta|(b^\dagger b)^2|\beta\rangle = \langle\beta|(b^\dagger(I + b^\dagger b)b)|\beta\rangle = \langle\beta|b^\dagger b|\beta\rangle + \langle\beta|b^\dagger b^\dagger bb|\beta\rangle = \beta\beta^* + (\beta\beta^*)^2.$$

Thus we find the real number $\langle\beta|\hat{N}^2|\beta\rangle = (\beta\beta^*)^2 + \beta\beta^*$.
(iii) Since $b|\beta\rangle = \beta|\beta\rangle$ and $\langle\beta|b^\dagger = \langle\beta|\beta^*$ we obtain the real number

$$\langle\beta|(b^\dagger + b)|\beta\rangle = \beta + \beta^* = 2r\cos(\phi).$$

Problem 43. (i) Calculate the expectation value $\langle\beta|(b^\dagger)^j b^k|\beta\rangle$.
(ii) Calculate ($\epsilon \in \mathbb{R}$) the expectation value $\langle\beta|e^{i\epsilon b^\dagger b}|\beta\rangle$.
(iii) Let $\hat{N} = b^\dagger b$ and $\phi \in \mathbb{R}$. Consider the operator $\hat{M}(\phi) = e^{-i\phi\hat{N}}$. Calculate the state

$$\hat{M}(\phi)|\beta\rangle.$$

(iv) Consider the coherent state $|\beta\rangle$ and the Hamilton operator $\hat{H} = \hbar\omega b^\dagger b$. Find the state $\exp(-i\hat{H}t/\hbar)|\beta\rangle$.

Solution 43. (i) Since $b|\beta\rangle = \beta|\beta\rangle$ and $\langle\beta|b^\dagger = \langle\beta|\beta^*$ we find

$$\langle\beta|(b^\dagger)^j b^k|\beta\rangle = (\beta^*)^j \beta^k.$$

(ii) We apply *normal ordering*, i.e.

$$\exp(i\epsilon b^\dagger b) = \sum_{j=0}^{\infty} \frac{1}{j!}(e^{i\epsilon} - 1)^j (b^\dagger)^j b^j.$$

Since $\langle\beta|(b^\dagger)^j b^j)|\beta\rangle = (\beta\beta^*)^j$ we obtain

$$\langle\beta|e^{i\epsilon b^\dagger b}|\beta\rangle = \sum_{j=0}^{\infty} \frac{1}{j!}(e^{i\epsilon} - 1)^j (\beta\beta^*)^j = \sum_{j=0}^{\infty} \frac{1}{j!}((e^{i\epsilon} - 1)\beta\beta^*)^j = e^{(e^{i\epsilon}-1)\beta\beta^*}.$$

(iii) We obtain

$$\hat{M}(\phi) = e^{\beta\beta^*/2} \sum_{n=0}^{\infty} \frac{\beta^n}{n!} e^{-in\phi}(b^\dagger)^n|0\rangle = |\beta e^{-i\theta}\rangle.$$

(iv) Since $b^\dagger b|n\rangle = n|n\rangle$ we obtain

$$\exp(-i\hat{H}t/\hbar)|\beta\rangle = |e^{-i\omega t}\beta\rangle.$$

Problem 44. Let $|0\rangle$, $|1\rangle$, ... be the number states. An arbitrary normalized state $|g\rangle$ can be expanded as

$$|g\rangle = \sum_{j=0}^{\infty} c_j |j\rangle, \qquad \sum_{j=0}^{\infty} c_j^* c_j = 1.$$

Express the state using coherent states $|\beta\rangle$. Consider the special case that $c_j = 0$ for all j except for $c_n = 1$.

Solution 44. We find

$$|g\rangle = \frac{1}{\pi} \int_{\mathbb{C}} e^{-|\beta|^2/2} g(\beta^*) |\beta\rangle d^2\beta$$

where $d^2\beta = d(\Re\beta) d(\Im\beta)$ and

$$g(\beta^*) = \sum_{j=0}^{\infty} \frac{c_j (\beta^*)^j}{\sqrt{j!}}.$$

If all c_j are equal to 0 except for $c_n = 1$ we obtain

$$|g\rangle = |n\rangle = \frac{1}{\pi} \int_{\mathbb{C}} e^{-|\beta|^2/2} \frac{(\beta^*)^n}{\sqrt{n!}} |\beta\rangle d^2\beta.$$

Problem 45. Let $|\beta\rangle$ be a coherent state. Show that

$$\frac{1}{\pi} \int_{\mathbb{C}} |\beta\rangle\langle\beta| d^2\beta = I$$

where I is the identity operator and the integration is over the entire complex plane. Set $\beta = r \exp(i\phi)$ with $0 \leq r < \infty$ and $0 \leq \phi < 2\pi$.

Solution 45. We have

$$\frac{1}{\pi} \int_{\mathbb{C}} |\beta\rangle\langle\beta| d^2\beta = \frac{1}{\pi} \sum_{n=0}^{\infty} \sum_{m=0}^{\infty} \frac{|n\rangle\langle m|}{\sqrt{n!\,m!}} \int_{\mathbb{C}} e^{-|\beta|^2} \beta^{*m} \beta^n d^2\beta.$$

Using $\beta = r \exp(i\phi)$ we arrive at

$$\frac{1}{\pi} \int_{\mathbb{C}} |\beta\rangle\langle\beta| d^2\beta = \frac{1}{\pi} \sum_{n=0}^{\infty} \sum_{m=0}^{\infty} \frac{|n\rangle\langle m|}{\sqrt{n!\,m!}} \int_0^{\infty} r e^{-r^2} r^{n+m} dr \int_0^{2\pi} e^{i(n-m)\phi} d\phi.$$

Since

$$\int_0^{2\pi} e^{i(n-m)\phi} d\phi = 2\pi \delta_{nm}$$

we have

$$\frac{1}{\pi} \int_{\mathbb{C}} |\beta\rangle\langle\beta| d^2\beta = \sum_{n=0}^{\infty} \frac{|n\rangle\langle n|}{n!} \int_0^{\infty} e^{-s} s^n ds$$

where we set $s = r^2$ and therefore $ds = 2r dr$. Thus

$$\frac{1}{\pi} \int_{\mathbb{C}} |\beta\rangle\langle\beta| d^2\beta = \sum_{n=0}^{\infty} |n\rangle\langle n| = I.$$

where we used the completeness relation for the number states.

Problem 46. Let $b^\dagger$, b be Bose creation and annihilation operators and $\beta, \gamma \in \mathbb{C}$. Let $D(\beta)$ be the displacement operator.
(i) Calculate the commutator $[\beta b^\dagger - \bar{\beta}b, \gamma b^\dagger - \bar{\gamma}b]$.
(ii) Show that

$$D(\beta)D(\gamma) \equiv e^{\beta\bar{\gamma} - \bar{\beta}\gamma} D(\gamma)D(\beta), \quad D(\beta + \gamma) \equiv e^{-\frac{1}{2}(z\bar{\gamma} - \bar{\beta}\gamma)} D(\beta)D(\gamma).$$

Solution 46. (i) We have

$$[\beta b^\dagger - \bar{\beta}b, \gamma b^\dagger - \bar{\gamma}b] = -[\bar{\beta}b, \gamma b^\dagger] - [\beta b^\dagger, \bar{\gamma}b] = -\bar{\beta}\gamma[b, b^\dagger] - \beta\bar{\gamma}[b^\dagger, b]$$
$$= (\beta\bar{\gamma} - \bar{\beta}\gamma)I$$

where I is the identity operator.
(ii) Using the result from (i) and that

$$e^{A+B} = e^{-\frac{1}{2}[A,B]}e^A e^B$$

if $[A, B] = cI$ we obtain the two identities.

Problem 47. Let $D(\beta)$ be the displacement operator.
(i) Find the operators $D^\dagger(\beta)$, $D^{-1}(\beta)$, $D^*(\beta)$, $D^T(\beta)$ in terms of $D(\beta)$.
(ii) Find the operators

$$D^{-1}(\beta)bD(\beta), \qquad D^{-1}(\beta)b^\dagger D(\beta).$$

(iii) Let f be an analytic function $f : \mathbb{C}^2 \to \mathbb{C}$. Calculate

$$D^{-1}(\beta)f(b, b^\dagger)D(\beta).$$

Apply it to $D^{-1}(\beta)b^\dagger bD(\beta)$.

Solution 47. (i) We obtain

$$D^\dagger(\beta) = D^{-1}(\beta) = D(-\beta) = \exp(-\beta b^\dagger + \beta^* b)$$

and $D^*(\beta) = D(\beta^*)$, $D^T(\beta) = D(-\beta^*)$. Thus $D^\dagger(\beta)D(\beta) = I$.
(ii) We obtain

$$D^{-1}(\beta)bD(\beta) = b + \beta I, \qquad D^{-1}(\beta)b^\dagger D(\beta) = b^\dagger + \beta^* I.$$

(iii) We find

$$D^{-1}(\beta)f(b, b^\dagger)D(\beta) = f(b + \beta I, b^\dagger + \beta^* I).$$

Consequently $D^{-1}(\beta)b^\dagger bD(\beta) = b^\dagger b + \beta b^\dagger + \bar{\beta}b + \beta\bar{\beta}I$.

Problem 48. Let $D(\beta)$, $D(\gamma)$ be displacement operators. Calculate the commutator $[D(\beta), D(\gamma)]$.

Solution 48. For $\beta, \gamma \in \mathbb{C}$ we have the identities

$$D(\beta)D(\gamma) \equiv e^{\beta\gamma^* - \beta^*\gamma} D(\gamma)D(\beta), \quad D(\gamma)D(\beta) \equiv e^{(\beta\gamma^* - \beta^*\gamma)/2} D(\beta + \gamma).$$

Using these identities we find

$$[D(\beta), D(\gamma)] = D(\beta)D(\gamma) - D(\gamma)D(\beta)$$
$$= e^{\beta\gamma^* - \beta^*\gamma} D(\gamma)D(\beta) - D(\gamma)D(\beta) = (e^{\beta\gamma^* - \beta^*\gamma} - 1)D(\gamma)D(\beta)$$
$$= (e^{\beta\gamma^* - \beta^*\gamma} - 1)e^{(\beta\gamma^* - \beta^*\gamma)/2}D(\beta + \gamma).$$

Problem 49. The displacement operator $D(\beta)$ can be written as

$$D(\beta) = e^{\beta b^\dagger} e^{-|\beta|^2/2} e^{-\beta^* b}.$$

(i) Find the state $D(\beta)|0\rangle$. Is the state normalized?
(ii) Calculate the states $D^\dagger(\beta)|0\rangle$, $D^\dagger(\beta)|1\rangle$. Compare $D(\beta)|0\rangle$ and $D^\dagger(\beta)|0\rangle$.

Solution 49. (i) Since $e^{-\beta^* b}|0\rangle = |0\rangle$ we obtain

$$D(\beta)|0\rangle = e^{-|\beta|^2/2} e^{\beta b^\dagger}|0\rangle.$$

Yes the state is normalized.
(ii) We obtain

$$D^\dagger(\beta)|0\rangle = e^{-|\beta|^2/2} \sum_{n=0}^{\infty} (-\beta)^n \frac{1}{\sqrt{n!}} |n\rangle$$

and

$$D^\dagger(\beta)|1\rangle = e^{-|\beta|^2/2} \left(\sum_{n=0}^{\infty} (-\beta)^n \beta^* \frac{1}{\sqrt{n!}} |n\rangle + \sum_{n=0}^{\infty} (-\beta)^n \frac{\sqrt{n+1}}{\sqrt{n!}} |n+1\rangle \right).$$

For $D(\beta)|0\rangle$ we have β^n in the sum and for $D^\dagger(\beta)|0\rangle$ we have $(-\beta)^n$ in the sum.

Problem 50. Let $D(\beta)$ be the displacement operator.
(i) Find the operator $D(\beta)D(\gamma)$.
(ii) Find the operator $D(\beta)D(\gamma)D(-\beta)$.
(iii) Find the operators $D(\beta)bD(-\beta)$ and $D(\beta)b^\dagger D(-\beta)$.
(iv) Let $\beta_1, \beta_2 \in \mathbb{C}$. Is $D(\beta_1)D(\beta_2) = D(\beta_1 + \beta_2)$?

Solution 50. (i) Using the Baker-Campbell-Hausdorff relation we obtain

$$D(\beta)D(\gamma) = e^{i\Im(\beta\gamma^*)}D(\beta + \gamma).$$

The complex parameter β is additive in the sense that the right-hand side is $D(\beta + \gamma)$ multiplied by a factor of unit modulus. Thus $D(\beta)D(\gamma)$ is also a unitary operator.
(ii) From (i) it follows that

$$D(\beta)D(\gamma)D(-\beta) = e^{2i(\Im(\beta\gamma^*))}D(\gamma) = (e^{-(\gamma\beta^* - \gamma^*\beta)})e^{(\gamma b^\dagger - \gamma^* b)}.$$

(iii) We obtain
$$D(\beta)bD(-\beta) = b - \beta I, \quad D(\beta)b^\dagger D(-\beta) = b^\dagger - \beta^* I.$$

(iv) This is only correct if $\beta_1\beta_2^* - \beta_1^*\beta_2 = 0$. In general we have

$$D(\beta_1)D(\beta_2) = D(\beta_1 + \beta_2) \exp\left(\frac{1}{2}(\beta_1\beta_2^* - \beta_1^*\beta_2) \right).$$

Problem 51. Express

$$D^*(\beta)D(z)D(\beta), \quad D^\dagger(\beta)D(z)D(\beta), \quad D(\beta)D(z)D(\beta), \quad D^T(\beta)D(z)D(\beta)$$

in terms of D.

Solution 51. Let $\beta = x + iy$ with $x, y \in \mathbb{R}$. We have the identities

$$D^*(\beta)D(z)D(\beta) = D(z + 2x)\exp(-2iy(x + \Re(z)))$$
$$D^\dagger(\beta)D(z)D(\beta) = D(z)\exp(z\beta^* - z^*\beta)$$
$$D(\beta)D(z)D(\beta) = D(z + 2\beta)$$
$$D^T(\beta)D(z)D(\beta) = D(z + 2iy)\exp(2ix(y + \Im(z))).$$

Problem 52. Let $D(\beta)$ be the displacement operator and $|n\rangle$ be the number states.
(i) Find the *displaced number state* $D(\beta)|n\rangle$.
(ii) Consider the number state $|1\rangle = b^\dagger|0\rangle$. Find the states $D(\beta)|1\rangle$ and $D^\dagger(\beta)|1\rangle$.
(iii) Calculate the matrix elements

$$\langle n|D(\beta)|m\rangle.$$

Solution 52. (i) Since $b|n\rangle = \sqrt{n}|n-1\rangle$ we have

$$D(\beta)|n\rangle = \exp(-|\beta|^2/2)\exp(\beta b^\dagger)\exp(-\bar{\beta}b)|n\rangle$$
$$= e^{-|\beta|^2/2}\sum_{k=0}^{\infty}\frac{\beta^k}{k!}\sum_{j=0}^{n}\frac{(-\bar{\beta})^j}{j!}\left(\frac{(n-j+k)!n!}{(n-j)!(n-j)!}\right)^{1/2}|n-j+k\rangle.$$

(ii) We have

$$D(\beta)|1\rangle = D(\beta)b^\dagger|0\rangle = D(\beta)b^\dagger D^{-1}(\beta)D(\beta)|0\rangle$$
$$= (b^\dagger - \bar{\beta}I)D(\beta)|0\rangle$$
$$= (b^\dagger - \bar{\beta}I)|\beta\rangle.$$

Analogously we find

$$D^\dagger(\beta)|1\rangle = D^{-1}(\beta)|1\rangle = D(-\beta)|1\rangle = (b^\dagger + \bar{\beta}I)|-\beta\rangle.$$

(iii) The matrix elements of the displacement operator $D(\beta)$ are

$$n \leq m \quad \langle n|D(\beta)|m\rangle = e^{-|\beta|^2/2}\sqrt{\frac{n!}{m!}}(-\bar{\beta})^{m-n}L_n^{(m-n)}(|\beta|^2)$$

$$n \geq m \quad \langle n|D(\beta)|m\rangle = e^{-|\beta|^2/2}\sqrt{\frac{m!}{n!}}\beta^{n-m}L_m^{(n-m)}(|\beta|^2)$$

where $L_n^{(\alpha)}$ are the *associated Laguerre polynomials* defined by

$$L_n^{(\alpha)}(x) := \sum_{j=0}^{n}(-1)^j\binom{n+\alpha}{n-j}\frac{x^j}{j!}.$$

In particular $L_n^{(0)} = L_n$ are the usual *Laguerre polynomials*.

Problem 53. Let $D(\beta)$ be the displacement operator. Calculate the *trace* $\mathrm{tr}(D(\beta))$ defined by

$$\mathrm{tr}(D(\beta)) := \sum_{n=0}^{\infty} \langle n|D(\beta)|n\rangle$$

where $\beta = x + iy$ $(x, y \in \mathbb{R})$ and $|n\rangle$ are the number states.

Solution 53. We find

$$\mathrm{tr}(D(\beta)) = \pi\delta^2(z) = \pi\delta(x)\delta(y)$$

where δ is the delta function.

Problem 54. Let $|\beta\rangle$ be a coherent state. Consider the *density operator* $\rho = |\beta\rangle\langle\beta|$. Calculate the *characteristic function*

$$\chi(\beta) := \mathrm{tr}(\rho e^{\beta b^\dagger - \beta^* b}) \equiv \mathrm{tr}(\rho D(\beta))$$

where $D(\beta)$ is the displacement operator.

Solution 54. We apply coherent states $|\beta\rangle$ to calculate the trace. Then

$$
\begin{aligned}
\chi(\beta) &= \mathrm{tr}(|\beta\rangle\langle\beta|D(\beta)) \\
&= \frac{1}{\pi}\int_C d\gamma \langle\gamma|\beta\rangle\langle\beta|D(\beta)|\gamma\rangle = \frac{1}{\pi}\int_C d\gamma \langle\gamma|\beta\rangle\langle\beta|D(\beta)D(\gamma)|0\rangle \\
&= \frac{1}{\pi}\int_C d\gamma \langle\gamma|\beta\rangle\langle\beta|e^{\frac{1}{2}(\beta\gamma^* - \beta^*\gamma)}D(\beta + \gamma)|0\rangle = \frac{1}{\pi}\int_C d\gamma \langle\gamma|\beta\rangle e^{\frac{1}{2}(\beta\gamma^* - \beta^*\gamma)}\langle\beta|\beta + \gamma\rangle \\
&= \frac{1}{\pi}\int_C d\gamma e^{-\frac{1}{2}(|\beta|^2 + |\gamma|^2) + \beta\gamma^*}e^{\frac{1}{2}(\beta\gamma^* - \beta^*\gamma)}e^{-\frac{1}{2}(|\beta + \gamma|^2 + |\beta|^2) + (\beta + \gamma)\beta^*} \\
&= \frac{1}{\pi}\int_C d\gamma e^{-\gamma\gamma^* - \frac{1}{2}\beta\beta^* + \beta\gamma^*} = \frac{1}{\pi}e^{-\frac{1}{2}\beta\beta^*}\int_C d\gamma e^{-\gamma\gamma^* + \beta\gamma^*} \\
&= \frac{1}{\pi}e^{-\frac{1}{2}\beta\beta^*}\int_{r=0}^{\infty} dr e^{-r^2}\int_{\phi=0}^{2\pi} d\phi e^{\beta r e^{-i\phi}} = 2e^{-\frac{1}{2}\beta\beta^*}\int_{r=0}^{\infty} dr e^{-r^2} \\
&= \sqrt{\pi}e^{-\frac{1}{2}\beta\beta^*}.
\end{aligned}
$$

Problem 55. Let $b^\dagger$, b be Bose creation and annihilation operators and $\theta \in \mathbb{R}$. Consider the unitary operator

$$U(\theta) := \exp(i\theta b^\dagger b) \equiv \exp(i\theta\hat{N}).$$

The displacement operator $D(\beta)$ satisfies the identity

$$\mathrm{tr}(D(\beta)D^\dagger(\gamma)) \equiv \pi\delta^2(\beta - \gamma)$$

where tr denotes the trace, $\beta, \gamma \in \mathbb{C}$ and δ is the delta function. Thus the unitary operator $U(\theta)$ can be expanded in terms of the displacement as

$$U(\theta) = \frac{1}{\pi}\int_C u_\theta(\beta)D(\beta)d^2\beta.$$

Find $u_\theta(\beta)$. Discuss the case for small θ.

Solution 55. We obtain

$$u_\theta(\beta) = \text{tr}(U(\theta)D^\dagger(\beta)) = \frac{ie^{i\theta/2}}{2\sin(\theta/2)} \exp\left(-\frac{i}{2}|\beta|^2 \cot(\theta/2)\right).$$

by evaluating the trace in the coherent basis. For small θ we find

$$u_\theta(\beta) \approx \frac{i}{\theta} \exp\left(-\frac{i}{\theta}|\beta|^2\right).$$

Then the coefficient $u_\theta(\beta)$ has a rapidly oscillating phase, and can be regarded as a distribution with support concentrated on values of β such that $|\beta|^2 \approx \theta$. We have in the sense of generalized functions

$$\lim_{\theta \to 0} u_\theta(\beta) = \pi\delta^2(\beta).$$

Problem 56. Let $D(\beta)$ be the displacement operator and I the identity operator. Calculate

$$D(\beta) \begin{pmatrix} b \\ b^\dagger \\ I \end{pmatrix} D^\dagger(\beta) \equiv \begin{pmatrix} D(\beta)bD^\dagger(\beta) \\ D(\beta)b^\dagger D^\dagger(\beta) \\ D(\beta)ID^\dagger(\beta) \end{pmatrix}$$

and show that we can write

$$\begin{pmatrix} D(\beta)bD^\dagger(\beta) \\ D(\beta)b^\dagger D^\dagger(\beta) \\ D(\beta)ID^\dagger(\beta) \end{pmatrix} = T(\beta, \beta^*) \begin{pmatrix} b \\ b^\dagger \\ I \end{pmatrix}$$

where the 3×3 matrix $T(\beta, \beta^*)$ depends only on β and β^*.

Solution 56. Let $\epsilon \in \mathbb{R}$. We have $D^\dagger(\beta) = D(-\beta)$. We set

$$f(\epsilon) = e^{\epsilon(\beta b^\dagger - \beta^* b)} b e^{-\epsilon(\beta b^\dagger - \beta^* b)}$$

with the initial condition $f(\epsilon = 0) = b$. Thus

$$\frac{df}{d\epsilon} = e^{\epsilon(\beta b^\dagger - \beta^* b)}(\beta b^\dagger - \beta^* b) b e^{-\epsilon(\beta b^\dagger - \beta^* b)} - e^{\epsilon(\beta b^\dagger - \beta^* b)} b(\beta b^\dagger - \beta^* b) e^{-\epsilon(\beta b^\dagger - \beta^* b)}$$

$$= e^{\epsilon(\beta b^\dagger - \beta^* b)}(\beta b^\dagger b - \beta b b^\dagger) e^{-\epsilon(\beta b^\dagger - \beta^* b)} = \beta e^{\epsilon(\beta b^\dagger - \beta^* b)}(b^\dagger b - b b^\dagger) e^{-\epsilon(\beta b^\dagger - \beta^* b)}$$

$$= -\beta I.$$

Integration yields $f(\epsilon) = -\beta I\epsilon + C$. Inserting the initial condition $f(\epsilon = 0) = b$ provides $f(\epsilon) = b - \epsilon\beta I$. Thus

$$D(\beta)bD^\dagger(\beta) = b - \beta I.$$

Analogously

$$D(\beta)b^\dagger D^\dagger(\beta) = b^\dagger - \beta^* I.$$

Thus

$$\begin{pmatrix} D(\beta)bD^\dagger(\beta) \\ D(\beta)b^\dagger D^\dagger(\beta) \\ D(\beta)ID^\dagger(\beta) \end{pmatrix} = \begin{pmatrix} 1 & 0 & -\beta \\ 0 & 1 & -\beta^* \\ 0 & 0 & 1 \end{pmatrix} \begin{pmatrix} b \\ b^\dagger \\ I \end{pmatrix}.$$

Note that $T(\beta, \beta^*)$ is an invertible matrix.

Problem 57.   Let $|\beta\rangle$ be a coherent state. We define the density operator ρ of the negative binomial states using the Glauber-Sudarshan P representation as

$$\rho = \frac{1}{\pi} \int_C p(\beta)|\beta\rangle\langle\beta|d^2\beta$$

where

$$p(\beta) = \frac{1}{\Gamma(1/\epsilon)}(\epsilon|\beta_0|^2)^{-1/\epsilon}|\beta|^{-2(1-1/\epsilon)}\exp(-|\beta|^2/(\epsilon|\beta_0|^2)).$$

Here $p(\beta)$ is the P function of the states, ϵ is a parameter with $\epsilon \in [0,1]$. The mean number of photons of the states is identified by $|\beta_0|^2$ and Γ is the gamma function.
(i) Find $p(\beta)$ for $\epsilon = 0$.
(ii) Find $p(\beta)$ for $\epsilon = 1$.
(iii) Calculate the expectation value $p(n,\epsilon) = |\langle n|\rho|n\rangle|^2$, i.e. the probability distribution of the number of photons.

Solution 57.   (i) For $\epsilon = 0$ we obtain the delta function

$$p(\beta) = \delta(|\beta|^2 - |\beta_0|^2).$$

This is the P function for the random-phase coherent state.
(ii) For $\epsilon = 1$ we find

$$p(\beta) = \frac{1}{|\beta_0|^2}\exp(-|\beta|^2/|\beta_0|^2).$$

This is the P function for the thermal noise state, which is a two-dimensional circularly symmetric Gaussian function of the complex value β.
(iii) We find

$$p(n,\epsilon) = \frac{\Gamma(1/\epsilon + n)}{\Gamma(1/\epsilon)n!}(\epsilon|\beta_0|^2)^n(1 + \epsilon|\beta_0|^2)^{-(n+1/\epsilon)}.$$

In the limiting case $\epsilon = 0$ we obtain

$$p(n,0) = \frac{|\beta_0|^{2n}}{n!}e^{-|\beta_0|^2}$$

which is the *Poisson distribution* with mean $|\beta_0|^2$. In the limiting case $\epsilon = 1$ we have with $\Gamma(1) = 1$, $\Gamma(n+1) = n!$

$$p(n,1) = \frac{|\beta_0|^{2n}}{(1 + |\beta_0|^2)^{n+1}}$$

which is the *Bose-Einstein distribution* for a thermal noise state.

Problem 58.   Consider the *density operator*

$$\rho = (1 - e^{\beta\hbar\omega})\sum_{n=0}^{\infty} e^{-n\beta\hbar\omega}|n\rangle\langle n|.$$

Calculate the trace of ρ.

Solution 58.   Utilizing $\sum_{n=0}^{\infty}|n\rangle\langle n| = I$ we have

$$\text{tr}(\rho) = (1 - e^{\beta\hbar\omega})\sum_{m=0}^{\infty}\sum_{n=0}^{\infty} e^{-n\beta\hbar\omega}\langle m|n\rangle\langle n|m\rangle.$$

Since $\langle m|n \rangle = \delta_{mn}$ we end up with

$$\text{tr}(\rho) = (1 - e^{\beta\hbar\omega}) \sum_{n=0}^{\infty} e^{-n\beta\hbar\omega} = 1$$

as expected for a density matrix.

Problem 59. Let $|\beta\rangle$ be a coherent state and $D(\beta)$ be the displacement operator. Let $V(t) = e^{itb^\dagger b}$, where $t \in \mathbb{R}$.
(i) Calculate $V(t)D(\beta)V(t)^{-1}$.
(ii) Calculate $V(t)|\beta\rangle$. Use the result from (i).

Solution 59. (i) Since

$$V(t)bV(t)^{-1} = e^{-it}b, \qquad V(t)b^\dagger V(t)^{-1} = e^{it}b^\dagger$$

we have

$$V(t)D(\beta)V(t)^{-1} = \exp(\beta V(t)b^\dagger V(t)^{-1} - \beta^* V(t)bV(t)^{-1}) = \exp(\beta e^{it}b^\dagger - \beta^* e^{-it}b) = D(e^{it}\beta).$$

(ii) Since $I = V(t)^{-1}V(t)$, $|\beta\rangle = D(\beta)|0\rangle$ we find

$$V(t)|\beta\rangle = V(t)D(\beta)V(t)^{-1}V(t)|0\rangle = D(e^{it}\beta)|0\rangle = |e^{it}\beta\rangle$$

where we have also used $V(t)|0\rangle = |0\rangle$ since $b^\dagger b|0\rangle = 0|0\rangle$.

Problem 60. Let $\hat{N} = b^\dagger b$. Consider the *parity operator*

$$\hat{P} = e^{i\pi b^\dagger b} \equiv e^{i\pi \hat{N}}$$

and the displacement operator $D(\beta)$.
(i) Find the anticommutators $[\hat{P}, b]_+$, $[\hat{P}, b^\dagger]_+$ and $\hat{P}D(\beta)$.
(ii) Let

$$W(\beta) = \frac{1}{\pi\hbar} D(\beta)\hat{P}D^{-1}(\beta).$$

Find $W(\beta)W(\gamma)$ and the commutator $[W(\beta), W(\gamma)]$.
(iii) The *displaced parity operator* is defined as

$$\Pi(\beta) := D(\beta)e^{i\pi\hat{N}}D^\dagger(\beta).$$

Show that this operator can also be written using the number states and the displacement operator.

Solution 60. (i) Looking the matrix representation of $b^\dagger$, b and

$$\hat{P} = \text{diag}(1 \ -1 \ 1 \ -1 \ ...)$$

we find that the anticommutators vanish, i.e. $[\hat{P}, b]_+ = 0$, $[\hat{P}, b^\dagger]_+ = 0$. It follows that $\hat{P}D(\beta) = D(-\beta)\hat{P}$.

(ii) Using the result $\hat{P}D(\beta) = D(-\beta)\hat{P}$ and $\hat{P}\hat{P} = I$ we have

$$W(\beta)W(\gamma) = \frac{1}{(\pi\hbar)^2}D(\beta)\hat{P}D(-\beta)D(\gamma)\hat{P}D(-\gamma) = \frac{1}{(\pi\hbar)^2}D(\beta)D(\beta)\hat{P}\hat{P}D(-\gamma)D(-\gamma)$$

$$= \frac{1}{(\pi\hbar)^2}D(\beta)D(\beta)D(-\beta)D(-\beta).$$

Since

$$D(\beta)D(\gamma) = \exp(\frac{1}{2}(\beta\gamma^* - \beta^*\gamma))D(\beta + \gamma)$$

we obtain

$$W(\beta)W(\gamma) = \frac{1}{(\pi\hbar)^2}e^{\beta^*\gamma - \beta\gamma^*}D(2(\beta - \gamma)).$$

Thus the commutator follows as

$$[W(\beta), W(\gamma)] = \frac{1}{(\pi\hbar)^2}\left(e^{\beta^*\gamma - \beta\gamma^*}D(2(\beta - \gamma)) - e^{\gamma^*\beta - \beta\gamma^*}D(2(\gamma - \beta))\right).$$

The commutator is the zero operator if and only if $\beta = \gamma$.

(iii) We can write

$$\Pi(\beta) = D(\beta)\left(\sum_{k=0}^{\infty}|2k\rangle\langle 2k| - \sum_{k=0}^{\infty}|2k+1\rangle\langle 2k+1|\right)D^{\dagger}(\beta).$$

Problem 61.   Let $D(z)$ be the displacement operator and $|n\rangle$ $(n = 0, 1, 2, \ldots)$ be the number states. Calculate the matrix elements $\langle n|D(z+w)|n\rangle$ using the *Laguerre polynomials* L_n defined by

$$L_n(z) := e^z\frac{d^n}{dz^n}(z^n e^{-z}), \qquad n = 0, 1, \ldots$$

Solution 61.   We find

$$\langle n|D(z+w)|n\rangle = e^{-|z+w|^2/2}L_n(|z+w|^2)$$

$$= e^{-(z\bar{w}+\bar{z}w)/2}e^{-(|z|^2+|w|^2)/2}L_n(|z+w|^2).$$

We could also use the identity

$$D(z+w) \equiv e^{-(z\bar{w}-\bar{z}w)/2}D(z)D(w)$$

to calculate the matrix element.

Problem 62.   Coherent states $|\beta\rangle$ using number states can be written as

$$|\beta\rangle = \sum_{k=0}^{\infty}\frac{(\beta b^{\dagger} - \beta^* b)^k}{k!}|0\rangle = \sum_{k=0}^{\infty}\sum_{j=0}^{[k/2]}\frac{\sqrt{(k-2j)!}}{k!}c_{k,k-2j}(-\beta^*)^j\beta^{k-j}|k-2j\rangle$$

where $[x]$ denotes the integer $\leq x$. Find the coefficients $c_{k,j}$.

Solution 62.   We obtain

$$c_{k,j} = \binom{k}{j}(k-j-1)!!.$$

Problem 63. Consider the operator $\hat{A} = \exp(-\epsilon b^\dagger b)$, where $\epsilon > 0$. Calculate the trace of A using

$$\text{tr}(\hat{A}) = \sum_{n=0}^{\infty} \langle n|\hat{A}|n \rangle$$

and

$$\text{tr}(\hat{A}) = \frac{1}{\pi} \int_{\mathbb{C}} \langle \beta|\hat{A}|\beta \rangle$$

where $|n\rangle$ are number states and $|\beta\rangle$ are coherent states.

Solution 63. Obviously both calculations lead to the same result. Let $\hat{N} := b^\dagger b$. Since

$$e^{-\epsilon \hat{N}}|n\rangle = e^{-\epsilon n}|n\rangle$$

and $\langle n|n \rangle = 1$ we have

$$\text{tr}(\hat{A}) = \sum_{n=0}^{\infty} \langle n|e^{-\epsilon \hat{N}}|n \rangle = \sum_{n=0}^{\infty} \langle n|e^{-\epsilon n}|n \rangle = \sum_{n=0}^{\infty} e^{-\epsilon n} = \frac{1}{1 - e^{-\epsilon}}.$$

Since

$$|\beta\rangle = e^{-|\beta|^2/2} \sum_{n=0}^{\infty} \frac{\beta^n}{\sqrt{n!}}|n\rangle$$

we have

$$e^{-\epsilon \hat{N}}|\beta\rangle = e^{-|\beta|^2/2} \sum_{n=0}^{\infty} \frac{\beta^n}{\sqrt{n!}} e^{-\epsilon n}|n\rangle.$$

Thus using $\langle m|n \rangle = \delta_{mn}$ we have

$$\langle \beta|e^{-\epsilon \hat{N}}|\beta \rangle = \frac{1}{\pi} \int_{\mathbb{C}} e^{-|\beta|^2} \sum_{m=0}^{\infty} \sum_{n=0}^{\infty} \frac{(\beta^*)^m \beta^n e^{-\epsilon n}}{\sqrt{m!}\sqrt{n!}} \delta_{mn} d^2\beta.$$

It follows that

$$\langle \beta|e^{-\epsilon \hat{N}}|\beta \rangle = \frac{1}{\pi} \sum_{n=0}^{\infty} \frac{e^{-\epsilon n}}{n!} \int_{\mathbb{C}} (\beta^*\beta)^n e^{-|\beta|^2} d^2\beta.$$

Using polar coordinates $d^2\beta = rdrd\phi$ $(0 \leq \phi < 2\pi)$ and applying that the integration over ϕ provides 2π we obtain

$$\langle \beta|e^{-\epsilon \hat{N}}|\beta \rangle = 2 \sum_{n=0}^{\infty} \frac{e^{-\epsilon n}}{n!} \int_0^{\infty} e^{-r^2} r^{2n+1} dr = \sum_0^{\infty} e^{-\epsilon n} = \frac{1}{1 - e^{-\epsilon}}.$$

Problem 64. Find the matrix elements of the displacement operator in the number basis, i.e.

$$\langle n+d|D(\beta)|n \rangle, \quad \langle n|D(\beta)|n+d \rangle, \quad \langle n|D(\beta)|n \rangle$$

where $|n\rangle$ are the number states.

Solution 64.   We find

$$\langle n+d|D(\beta)|n\rangle = \sqrt{\frac{n!}{(n+d)!}} \exp(-\frac{1}{2}|\beta|^2)\beta^d L_n^d(|\beta|^2)$$

$$\langle n|D(\beta)|n+d\rangle = \sqrt{\frac{n!}{(n+d)!}} \exp(-\frac{1}{2}|\beta|^2)(-\beta^*)^d L_n^d(|\beta|^2)$$

$$\langle n|D(\beta)|n\rangle = \exp(-\frac{1}{2}|\beta|^2)L_n(|\beta|^2)$$

where the *Laguerre polynomials* are defined by the recursion relation

$$L_{n+1}(x) - (2n+1-x)L_n(x) + n^2 L_{n-1}(x) = 0, \quad n=1,2,\dots$$

with $L_0(x) = 1$, $L_1(x) = 1-x$. The *associated Laguerre polynomials* are defined by the Laguerre polynomials

$$L_n^m(x) := \frac{d^m}{dx^m} L_n(x), \quad m = 0, 1, \dots$$

where $L_n^0(x) = L_n(x)$. We have $L_1^1(x) = -1$, $L_2^1(x) = 2x - 4$, $L_2^2(x) = 2$.

Problem 65.   (i) Express the operator

$$(-1)^{b^\dagger b} \equiv e^{i\pi b^\dagger b}$$

using the displacement operator $D(\beta)$.
(ii) Express $|0\rangle\langle 0|$ using the displacement operator.

Solution 65.   (i) We have

$$e^{i\pi b^\dagger b} = \frac{1}{2\pi} \int_{\mathbb{C}} D(\beta)d^2\beta.$$

Note that in general we have

$$x^{b^\dagger b} \equiv e^{\ln(x)b^\dagger b} = \frac{1}{\pi(1-x)} \int_{\mathbb{C}} \exp\left(-\frac{1}{2}\frac{1+x}{1-x}|\beta|^2\right)D(\beta)d^2\beta.$$

(ii) We find

$$|0\rangle\langle 0| = \frac{1}{\pi} \int_{\mathbb{C}} \exp(-|\beta|^2/2)D(\beta)d^2\beta.$$

Problem 66.   For the displacement operator $D(\beta) = \exp(\beta b^\dagger - \beta^* b)$ we have

$$D(\beta) = \exp\left(-\frac{|\beta|^2}{2}\right)\exp(\beta b^\dagger)\exp(-\beta^* b) == \exp\left(\frac{|\beta|^2}{2}\right)\exp(-\beta^* b)\exp(\beta b^\dagger).$$

(i) Find the *normal ordering*.
(ii) Find the *antinormal ordering*.

Solution 66.   (i) We have

$$D(\beta) = \exp\left(-\frac{|\beta|^2}{2}\right)\sum_{n,m=0}^{\infty}\frac{(\beta b^\dagger)^n}{n!}\frac{(-\beta^* b)^m}{m!}.$$

(ii) We have

$$D(\beta) = \exp\left(\frac{|\beta|^2}{2}\right) \sum_{n,m=0}^{\infty} \frac{(-\beta^* b)^m}{m!} \frac{(\beta b^\dagger)^n}{n!}.$$

Problem 67. Let $|\beta\rangle$ be a coherent state and $|n\rangle$ be a number state. Then

$$|\langle n|\beta\rangle| = \exp(-|\beta|^2) \frac{(|\beta|^2)^n}{n!}.$$

Coherent states $|\mu\rangle$ of angular momentum are defined as

$$|\mu\rangle := \frac{1}{(1+|\mu|^2)^j} \sum_{p=0}^{2j} \binom{2j}{p}^{1/2} \mu^p |p\rangle$$

where μ is a complex number and $|p\rangle$ are the projections of a single angular momentum j. Find the probability $|\langle p|\mu\rangle|^2$. Discuss.

Solution 67. We find

$$|\langle p|\mu\rangle|^2 = \binom{2j}{p}(|\mu|^2)^p(1+|\mu|^2)^{-2j}.$$

This a *binomial distribution*.

Problem 68. The *thermal mixture of states* with the mean number of photon equal to $\bar{n}$ is represented by the density operator

$$\rho_{\bar{n}} = \sum_{n=0}^{\infty} \frac{\bar{n}^n}{(\bar{n}+1)^{n+1}} |n\rangle\langle n|$$

where $|n\rangle$ are the number states. Calculate the *Husimi distribution* $\langle \beta|\rho_{\bar{n}}|\beta\rangle$.

Solution 68. Since

$$|\beta\rangle = e^{-|\beta|^2/2} \sum_{k=0}^{\infty} \frac{\beta^k}{\sqrt{k!}} |k\rangle, \quad \langle\beta| = e^{-|\beta|^2/2} \sum_{j=0}^{\infty} \frac{(\beta^*)^j}{\sqrt{j!}} \langle j|$$

we have

$$\langle\beta|\rho_{\bar{n}}|\beta\rangle = e^{-|\beta|^2} \sum_{j=0}^{\infty} \frac{(\beta^*)^j}{\sqrt{j!}} \langle j| \sum_{n=0}^{\infty} \frac{\bar{n}^n}{(\bar{n}+1)^{n+1}} |n\rangle\langle n| \sum_{k=0}^{\infty} \frac{\beta^k}{\sqrt{k!}} |k\rangle$$

$$= e^{-|\beta|^2} \sum_{j=0}^{\infty} \frac{\beta^*}{\sqrt{j!}} \langle j| \sum_{n=0}^{\infty} \frac{\bar{n}^n \beta^n}{\sqrt{n!}(\bar{n}+1)^{n+1}} |n\rangle = e^{-|\beta|^2} \sum_{n=0}^{\infty} \frac{(\beta^*)^n \beta^n \bar{n}^n}{\sqrt{n!}\sqrt{n!}(\bar{n}+1)^{n+1}}$$

$$= \frac{e^{-|\beta|^2}}{(\bar{n}+1)} \sum_{n=0}^{\infty} \frac{1}{n!} \left(\frac{\beta^* \beta \bar{n}}{\bar{n}+1}\right)^n = \frac{e^{-|\beta|^2}}{(\bar{n}+1)} e^{\bar{n}|\beta|^2/(\bar{n}+1)}$$

$$= \frac{1}{\bar{n}+1} e^{-|\beta|^2/(\bar{n}+1)}$$

where we used that $\langle n|k\rangle = \delta_{nk}$ and $\langle j|n\rangle = \delta_{jn}$.

Problem 69. Let $|\beta\rangle$ be a coherent state. Consider the *Husimi function*

$$Q(\beta) := \frac{1}{\pi}|\langle\psi|\beta\rangle|^2$$

which is related to the Wigner function $W(\beta)$ as

$$Q(\beta) = \frac{2}{\pi}\int_{\mathbb{C}} d^2\gamma W(\gamma)\exp(-2|\beta - \gamma|^2).$$

(i) Assume that $Q(\beta_0) = 0$ for at least one β_0. Discuss.
(ii) Show that the only pure states characterized by a strictly positive Husimi function turns out to be Gaussian ones.

Solution 69. (i) If $Q(\beta_0) = 0$ for at least one β_0 then the Wigner function $W(\beta)$ must have negative regions since the convolution involves a Gaussian strictly positive integrand.
(ii) Consider a pure state $|\psi\rangle$ expanded in a number state basis $|n\rangle$ $(n = 0, 1, \ldots)$ as

$$|\psi\rangle = \sum_{n=0}^{\infty} c_n|n\rangle$$

and define the function

$$f(\beta) := e^{\frac{1}{2}|\beta|^2}\langle\psi|\beta\rangle = \sum_{n=0}^{\infty} c_n^* \frac{\beta^n}{\sqrt{n!}}.$$

Obviously the function $f(\beta)$ is an analytic function of growth order less than or equal to 2 which will have zeros if and only if $Q(\beta)$ has zeros. Hence we can apply *Hadamard's theorem* which states that any function that is analytic on the complex plane, has no zeros, and is restricted in growth to be of order 2 or less must be a Gaussian function. Consequently the function $Q(\beta)$ and the function $W(\beta)$ are Gaussian.

Problem 70. Let $D(z)$ be the displacement operator and $\theta \in \mathbb{R}$. A Schrödinger cat state of a single-mode radiation field is the normalized superposition of coherent states

$$|\beta, \theta\rangle := \frac{1}{\sqrt{2 + 2\cos(\theta)e^{-2|\beta|^2}}}(|\beta\rangle + e^{i\theta}|-\beta\rangle).$$

Let $\rho(\beta, \theta) := |\beta, \theta\rangle\langle\beta, \theta|$ be the corresponding density matrix. Find the *characteristic function* defined by

$$\chi(\beta, \theta, z) := \operatorname{tr}(\rho(\beta, \theta, z)D(z)).$$

Solution 70. (i) We obtain

$$\operatorname{tr}(\rho(\beta, \theta)D(z)) = \frac{1}{\pi}\int_{\mathbb{C}}\langle\gamma|\rho(\beta, \theta)|z + \gamma\rangle e^{(z\bar{\gamma} - \bar{z}\gamma)/2}d^2\gamma$$

$$= \frac{1}{2 + 2\cos(\theta)e^{-2|\beta|^2}}e^{-|z|^2/2}\left(2\cosh(\bar{\beta}z - \beta\bar{z}) + 2e^{-2|\beta|^2}\cosh(\bar{\beta}z + \beta\bar{z} + i\theta)\right).$$

The corresponding *Wigner function* is defined by

$$W(\beta, \theta, \gamma) := \frac{1}{\pi^2}\int_{\mathbb{C}} e^{\bar{z}\gamma - z\bar{\gamma}}\chi(\beta, \theta, z)d^2z.$$

Problem 71. Let $|\beta\rangle$ be a coherent state and I be the identity operator.
(i) Calculate

$$\langle\beta| \otimes \langle\beta|(b^\dagger \otimes b)|\beta\rangle \otimes |\beta\rangle.$$

(ii) Let $|\beta\rangle$ and $|\gamma\rangle$ be coherent states. Calculate

$$((\langle\gamma| \otimes \langle\beta|)(b^\dagger \otimes b)(|\gamma\rangle \otimes |\beta\rangle)).$$

Solution 71. (i) Using

$$b|\beta\rangle = \beta|\beta\rangle, \quad \langle\beta|b^\dagger = \langle\beta|\beta^*, \quad \langle\beta|\beta\rangle = 1$$

we find

$$
\begin{aligned}
((\langle\beta| \otimes \langle\beta|)(b^\dagger \otimes b)(|\beta\rangle \otimes |\beta\rangle)) &= ((\langle\beta| \otimes \langle\beta|(b^\dagger \otimes I)(I \otimes b)|\beta\rangle \otimes |\beta\rangle) \\
&= ((\langle\beta|b^\dagger \otimes \langle\beta|)(|\beta\rangle \otimes b|\beta\rangle) = ((\langle\beta|\beta^* \otimes \langle\beta|)(|\beta\rangle \otimes \beta|\beta\rangle)) \\
&= \beta\beta^*.
\end{aligned}
$$

(ii) We have

$$
\begin{aligned}
\langle\gamma| \otimes \langle\beta|(b^\dagger \otimes b)|\gamma\rangle \otimes |\beta\rangle &= \langle\gamma| \otimes \langle\beta|(b^\dagger \otimes I)(I \otimes b)|\gamma\rangle \otimes |\beta\rangle \\
&= ((\langle\gamma|b^\dagger \otimes \langle\beta|)(|\gamma\rangle \otimes b|\beta\rangle) = ((\langle\gamma|\gamma^* \otimes \langle\beta|)(|\gamma\rangle \otimes \beta|\beta\rangle)) \\
&= \gamma^*\beta.
\end{aligned}
$$

Problem 72. Let b be a Bose annihilation operator. We define the quadratic phase operator $\hat{X}_1$, $\hat{X}_2$ by

$$b = X_1 + iX_2.$$

Then $\hat{X}_j$ obeys the usual canonical commutation relation $[\hat{X}_1, \hat{X}_2] = iI/2$ and $\hat{X}_j = \hat{X}_j^\dagger$. We define the *variance* $V(\hat{X}_j)$ by

$$V(\hat{X}_j) := \langle\hat{X}_j^2\rangle - \langle\hat{X}_j\rangle^2.$$

Find the variance of $\hat{X}_1$ and $\hat{X}_2$ for coherent states.

Solution 72. We obtain for coherent states that

$$V(\hat{X}_1) = V(\hat{X}_2) = \frac{1}{4}.$$

Problem 73. Let $|\beta_1\rangle$, $|\beta_2\rangle$ be coherent states. Consider the product state $|\beta_1\rangle \otimes |\beta_2\rangle$. Let $\epsilon \in \mathbb{R}$. Calculate

$$((\langle\beta_1| \otimes \langle\beta_2|)(b_1^\dagger e^{i\epsilon(b_1^\dagger b_1 - b_2^\dagger b_2)} b_2)(|\beta_1\rangle \otimes |\beta_2\rangle))$$

$$((\langle\beta_1| \otimes \langle\beta_2|)(b_2^\dagger e^{-i\epsilon(b_1^\dagger b_1 - b_2^\dagger b_2)} b_1)(|\beta_1\rangle \otimes |\beta_2\rangle))$$

and the sum of these two expectation values using $\beta_1 = r_1 e^{i\phi_1}$, $\beta_2 = r_2 e^{i\phi_2}$. Then consider the special case $\beta_1 = \beta_2 = \beta = re^{i\phi}$.

Solution 73. We have $b_1^\dagger = b^\dagger \otimes I$, $b_2^\dagger = I \otimes b^\dagger$ and

$$b_j|\beta_j\rangle = \beta_j|\beta_j\rangle, \quad \langle\beta_j|b_j^\dagger = \langle\beta_j|\beta_j^*$$

and

$$e^{i\alpha(b_1^\dagger b_1 - b_2^\dagger b_2)} = e^{i\alpha b_1^\dagger b_1} e^{-i\alpha b_2^\dagger b_2}.$$

Since (*normal ordering*)

$$e^{i\epsilon b_1^\dagger b_1} = \sum_{j=0}^{\infty} \frac{1}{j!}(e^{i\epsilon} - 1)(b_1^\dagger)^j b_1^j = \sum_{j=0}^{\infty} \frac{1}{j!}(e^{i\epsilon} - 1)((b^\dagger)^j \otimes I)(b^j \otimes I)$$

$$e^{-i\epsilon b_2^\dagger b_2} = \sum_{j=0}^{\infty} \frac{1}{j!}(e^{-i\epsilon} - 1)(b_2^\dagger)^j b_2^j = \sum_{j=0}^{\infty} \frac{1}{j!}(e^{-i\epsilon} - 1)(I \otimes (b^\dagger)^j)(I \otimes b^j)$$

we obtain

$$(\langle\beta_1| \otimes \langle\beta_2|)(b_1^\dagger e^{i\epsilon(b_1^\dagger b_1 - b_2^\dagger b_2)} b_2)(|\beta_1\rangle \otimes |\beta_2\rangle) = \beta_1^* \beta_2((\langle\beta_1| \otimes \langle\beta_2|)(e^{i\epsilon(b_1^\dagger b_1 - b_2^\dagger b_2)})(|\beta_1\rangle \otimes |\beta_2\rangle)$$
$$= \beta_1^* \beta_2 e^{(e^{i\epsilon}-1)\beta_1^*\beta_1} e^{(e^{-i\epsilon}-1)\beta_2^*\beta_2}.$$

Analogously

$$(\langle\beta_1| \otimes \langle\beta_2|)(b_2^\dagger e^{-i\epsilon(b_1^\dagger b_1 - b_2^\dagger b_2)} b_1)(|\beta_1\rangle \otimes |\beta_2\rangle) = \beta_1 \beta_2^* e^{(e^{-i\epsilon}-1)\beta_1^*\beta_1} e^{(e^{i\epsilon}-1)\beta_2^*\beta_2}.$$

The second term is the complex conjugate of the first. Thus we

$$\beta_1^* \beta_2 e^{(e^{i\epsilon}-1)\beta_1^*\beta_1} e^{(e^{-i\epsilon}-1)\beta_2^*\beta_2} + \beta_1 \beta_2^* e^{(e^{-i\epsilon}-1)\beta_1^*\beta_1} e^{(e^{i\epsilon}-1)\beta_2^*\beta_2} =$$

$$2r_1 r_2 e^{(\cos\epsilon - 1)(r_1^2 + r_2^2)} (\cos(\phi_2 - \phi_1)\cos((r_1^2 - r_2^2)\sin\epsilon) - \sin(\phi_2 - \phi_1)\sin((r_1^2 - r_2^2)\sin\epsilon)).$$

If $\beta_1 = \beta_2$, i.e. $r_1 = r_2 = r$, $\phi_1 = \phi_2 = \phi$ we find

$$2r^2 e^{2(\cos\epsilon - 1)r^2}.$$

If $\epsilon = 0$ we have $2r^2$.

Problem 74. Let $\xi, \eta \in \mathbb{C}$ and $\xi = \xi_1 + i\xi_2$, $\eta = \eta_1 + i\eta_2$. Consider the normalized states

$$|\xi\rangle = \exp(-\frac{1}{2}|\xi|^2 I + \xi b_1^\dagger + \xi^* b_2^\dagger - b_1^\dagger b_2^\dagger)|00\rangle$$

$$|\eta\rangle = \exp(-\frac{1}{2}|\eta|^2 I + \eta b_1^\dagger - \eta^* b_2^\dagger + b_1^\dagger b_2^\dagger)|00\rangle$$

and the operators

$$Q_1 = \frac{1}{\sqrt{2}}(b_1 + b_1^\dagger), \qquad Q_2 = \frac{1}{\sqrt{2}}(b_2 + b_2^\dagger),$$

$$P_1 = \frac{1}{\sqrt{2}i}(b_1 - b_1^\dagger), \qquad P_2 = \frac{1}{\sqrt{2}i}(b_2 - b_2^\dagger).$$

(i) Calculate the scalar products $\langle\xi'|\xi\rangle$, $\langle\eta'|\eta\rangle$ and the operators

$$\frac{1}{\pi}\int_\mathbb{C} |\xi\rangle\langle\xi| d\xi, \qquad \frac{1}{\pi}\int_\mathbb{C} |\eta\rangle\langle\eta| d\eta.$$

(ii) Calculate $(Q_1 - Q_2)|\eta\rangle$, $(P_1 + P_2)|\eta\rangle$, $(Q_1 + Q_2)|\xi\rangle$, $(P_1 - P_2)|\xi\rangle$.

Solution 74. (i) We have

$$\langle\xi'|\xi\rangle = \pi\delta(\xi_1 - \xi_1')\delta(\xi_2 - \xi_2'), \qquad \langle\eta|\eta\rangle = \pi\delta(\eta_1 - \eta_2')\delta(\eta_2 - \eta_2')$$

and

$$\frac{1}{\pi}\int_C |\xi\rangle\langle\xi|d\xi = 1, \qquad \frac{1}{\pi}\int_C |\eta\rangle\langle\eta|d\eta = 1$$

where δ denotes the delta function.

(ii) We find the eigenvalue equations

$$(Q_1 - Q_2)|\eta\rangle = \sqrt{2}\eta_1|\eta\rangle, \quad (P_1 + P_2)|\eta\rangle = \sqrt{2}\eta_2|\eta\rangle,$$
$$(Q_1 + Q_2)|\xi\rangle = \sqrt{2}\xi_1|\xi\rangle, \quad (P_1 - P_2)|\xi\rangle = \sqrt{2}\xi_2|\xi\rangle.$$

Problem 75. Consider the N-qubit *W-state*

$$|\psi\rangle = \frac{1}{\sqrt{N}}(|100\ldots0\rangle + |010\ldots0\rangle + \cdots + |00\ldots01\rangle)$$

where $|100\ldots0\rangle \equiv |1\rangle\otimes|0\rangle\otimes|0\rangle\otimes\cdots\otimes|0\rangle$ and $|0\rangle$, $|1\rangle$ are number states. Find the expectation value

$$\Pi_{b_1,b_2,\ldots,b_N}(\beta_1,\beta_2,\ldots,\beta_N) = \langle\psi|\Pi_{b_1}(\beta_1)\otimes\Pi_{b_2}(\beta_2)\otimes\cdots\otimes\Pi_{b_N}(\beta_N)|\psi\rangle$$

where β_j is the coherent displacement for the mode b_j. The operator $\Pi_{b_j}(\beta_j)$ is the *displaced parity operator* and is defined by

$$\Pi_{b_j}(\beta) := D_{b_j}(\beta_j)\exp(i\pi\hat{N}_{b_j})D_{b_j}^\dagger(\beta_j)$$

and $\hat{N}_{b_j} = b_j^\dagger b_j$.

Solution 75. Since

$$D^\dagger(\beta)|0\rangle = e^{-|\beta|^2/2}\sum_{n=0}^\infty(-\beta)^n\frac{1}{\sqrt{n!}}|n\rangle$$

and

$$D^\dagger(\beta)|1\rangle = e^{-|\beta|^2/2}\left(\sum_{n=0}^\infty(-\beta)^n\beta^*\frac{1}{\sqrt{n!}}|n\rangle + \sum_{n=0}^\infty(-\beta)^n\frac{\sqrt{n+1}}{\sqrt{n!}}|n+1\rangle\right)$$

we obtain

$$\Pi_{b_1,b_2,\ldots,b_N}(\beta_1,\beta_2,\ldots,\beta_N) = \frac{1}{N}\left(4\left|\sum_{j=1}^N\beta_j\right|^2 - N\right)\exp(-2\sum_{j=1}^N|\beta_j|^2).$$

Problem 76. Consider the state

$$|\psi\rangle = \frac{1}{\sqrt{2}}(|1\rangle\otimes|0\rangle - |0\rangle\otimes|1\rangle)$$

where $|0\rangle$, $|1\rangle$ are the number states $|n\rangle$ with $n = 0$ and $n = 1$. Consider the operator

$$\Pi_{12}(\beta_1,\beta_2) := D_1(\beta_1)D_2(\beta_2)e^{i\pi(\hat{N}_1+\hat{N}_2)}D_1^\dagger(\beta_1)D_2^\dagger(\beta_2)$$

where $\hat{N}_1 = b_1^\dagger b_1 = b^\dagger b\otimes I$, $\hat{N}_2 = b_2^\dagger b_2 = I\otimes b^\dagger b$ and $D_j(\beta_j)$ are the displacement operators. Calculate the expectation value

$$\langle\psi|\Pi_{12}(\beta_1,\beta_2)|\psi\rangle.$$

Solution 76.   Using $D^\dagger(\beta)|1\rangle = (b^\dagger + \beta^*I)|-\beta\rangle$ we obtain

$$\langle\psi|\Pi_{12}(\beta_1,\beta_2)|\psi\rangle = (2|\beta_1 - \beta_2|^2 - 1)e^{-2|\beta_1|^2 - 2|\beta_2|^2}.$$

Problem 77.   Let b_1 and b_2 denote Bose annihilation operators for the pump and the signal mode, respectively. Consider the operators

$$J_1 := \frac{1}{2}(b_1^\dagger b_2 + b_2^\dagger b_1), \quad J_2 := -\frac{i}{2}(b_1^\dagger b_2 - b_2^\dagger b_1), \quad J_3 := \frac{1}{2}(b_1^\dagger b_1 - b_2^\dagger b_2).$$

(i) Find the commutators $[J_1, J_2]$, $[J_2, J_3]$, $[J_3, J_1]$.
(ii) Let

$$\hat{N} := b_1^\dagger b_1 + b_2^\dagger b_2$$

be the number operator. Calculate the commutators $[J_1, \hat{N}]$, $[J_2, \hat{N}]$, $[J_3, \hat{N}]$.
(iii) Let $\alpha \in \mathbb{R}$. Find

$$e^{i\alpha J_1}J_2 e^{-i\alpha J_1}, \quad e^{i\alpha J_2}J_3 e^{-i\alpha J_2}, \quad e^{i\alpha J_3}J_1 e^{-i\alpha J_3}.$$

Then simplify to the case $\alpha = \pi/2$.
(iv) Show that

$$e^{-i\alpha J_1^2} \equiv e^{i(\pi/2)J_2}e^{-i\alpha J_3^2}e^{-i(\pi/2)J_2}.$$

Solution 77.   (i) Using $[b_j, b_k^\dagger] = \delta_{jk}I$ we obtain

$$[J_1, J_2] = iJ_3, \qquad [J_2, J_3] = iJ_1, \qquad [J_3, J_1] = iJ_2.$$

(ii) Obviously we find $[J_1, \hat{N}] = 0$, $[J_2, N] = 0$, $[J_3, \hat{N}] = 0$.
(iii) We set

$$g(\alpha) := e^{i\alpha J_1}J_2 e^{-i\alpha J_1}.$$

Taking the derivative with respect to α and using the commutation relation $[J_1, J_2] = iJ_3$ we obtain

$$\frac{dg}{d\alpha} = ie^{i\alpha J_1}[J_1, J_2]e^{-i\alpha J_1} = -ie^{i\alpha J_1}J_3 e^{-i\alpha J_1}.$$

Taking the second derivative yields

$$\frac{d^2g}{d\alpha^2} = ie^{i\alpha J_1}[J_3, J_1]e^{-i\alpha J_1} = -e^{i\alpha J_1}J_2 e^{-i\alpha J_1} = -g.$$

Thus we have a second order linear ordinary differential equation with initial conditions

$$\frac{d^2g}{d\alpha^2} + g = 0, \qquad g(0) = J_2, \qquad \frac{dg(0)}{d\alpha} = -J_3.$$

The solution of this initial value problem is

$$g(\alpha) = e^{i\alpha J_1}J_2 e^{-i\alpha J_1} = \cos(\alpha)J_2 - \sin(\alpha)J_3.$$

Thus for $\alpha = \pi/2$ we have

$$e^{i\alpha J_1}J_2 e^{-i\alpha J_1} = -J_3.$$

Analogously we find

$$e^{i\alpha J_2}J_3 e^{-i\alpha J_2} = -\sin(\alpha)J_1 + \cos(\alpha)J_3$$

with

$$e^{i(\pi/2)J_2}J_3e^{-i(\pi/2)J_2} = -J_1$$

and

$$e^{i\alpha J_3}J_1e^{-i\alpha J_3} = \cos(\alpha)J_1 - \sin(\alpha)J_2$$

with

$$e^{i(\pi/2)J_3}J_1e^{-i(\pi/2)J_3} = -J_2.$$

(iv) Using the result from (iii) we have

$$
\begin{aligned}
e^{i(\pi/2)J_2}e^{-i\alpha J_3^2}e^{-i(\pi/2)J_2} &= \exp(-i\alpha e^{i(\pi/2)J_2}J_3^2e^{-i(\pi/2)J_2}) \\
&= \exp(-i\alpha e^{i(\pi/2)J_2}J_3e^{-i(\pi/2)J_2}e^{i(\pi/2)J_2}J_3e^{-i(\pi/2)J_2}) \\
&= \exp(-i\alpha(-J_1)(-J_1)) \\
&= \exp(-i\alpha J_1^2).
\end{aligned}
$$

Problem 78. Consider continuous variable quantum *teleportation*. We teleport from Alice to Bob an unknown quantum state ρ_{in} (density operator). Let $(b_1, b_1^\dagger)$ and $(b_2, b_2^\dagger)$ be the bosonic annihilation and creation operators of two modes. Let $D(x, p)$ be the displacement operator

$$D(x, p) = \exp(i(p\hat{x} - x\hat{p})) \equiv \exp(\beta b^\dagger - \beta^* b) = D(\beta)$$

with $b = (\hat{x} + i\hat{p})/\sqrt{2}$ and $\beta = (x + ip)/\sqrt{2}$. Consider the density operator

$$\rho_{out} = \int_{-\infty}^{\infty} dx \int_{-\infty}^{\infty} dp P(x, p)D(x, p)\rho_{in}D^\dagger(x, p)$$

with the function $P(x, p)$ is given by

$$P(x, p) = \langle\Phi|(I \otimes D(x, p))^\dagger W^{AB}(I \otimes D(x, p)|\Phi\rangle$$

where $|\Phi\rangle$ is the continuous version of the EPR-Bell state

$$|\Phi\rangle = \frac{1}{\sqrt{2\pi}}\int_{-\infty}^{\infty} dx|x\rangle \otimes |x\rangle$$

and $|x\rangle$ is the eigenstate of the canonical position operator, i.e. $\hat{x}|x\rangle = x|x\rangle$. The bipartite state W^{AB} shared by Alice and Bob can be written in terms of the two-mode Wigner function $W(z_1, z_2)$ of the complex variables z_1 and z_2 as

$$W^{AB} = \frac{4}{\pi^2}\int d^2z_1 \int d^2z_2 W(z_1, z_2)e^{i\pi(b_1^\dagger - z_1^*I)(b_1 - z_1I)} \otimes e^{i\pi(b_2^\dagger - z_2^*I)(b_2 - z_2I)}.$$

Express $P(x, p)$ in terms of the Wigner function $W(z_1, z_2)$.

Solution 78. We find

$$P(x, p) = \frac{1}{2\pi^2}\int d^2z W(z^* - \beta^*, z), \qquad \beta = \frac{1}{\sqrt{2}}(x + ip).$$

Problem 79. Consider the operators

$$J_+ = b_1^\dagger b_2, \quad J_- = b_1 b_2^\dagger, \quad J_3 = \frac{1}{2}\left(b_1^\dagger b_1 - b_2^\dagger b_2\right).$$

Let $\zeta = re^{i\theta}$ with $r > 0$ and $\theta \in \mathbb{R}$. Find the disentangled form of

$$U(\zeta) = \exp(\zeta J_+ - \zeta^* J_-).$$

Solution 79.   We obtain

$$
\begin{aligned}
U(\zeta) &= \exp(\zeta J_+ - \zeta^* J_-) \\
&= \exp\left(\frac{\zeta}{|\zeta|} \tan(|\zeta|) J_+ \right) \exp(\ln(1 + \tan^2 |\zeta|) J_3) \exp\left(-\frac{\zeta^*}{|\zeta|} \tan(|\zeta|) J_- \right) \\
&= \exp\left(e^{i\theta} \tan(r) b_1^\dagger b_2 \right) (\cos^2(r))^{b_2^\dagger b_2 - b_1^\dagger b_1} \exp\left(-e^{-i\theta} \tan(r) b_1 b_2^\dagger \right)
\end{aligned}
$$

where

$$(\cos^2(r))^{b_2^\dagger b_2 - b_1^\dagger b_1} \equiv \exp(-\ln(\cos^2(r))(b_2^\dagger b_2 - b_1^\dagger b_1)).$$

Problem 80.   Consider the coherent states

$$|\beta_j\rangle = D(\beta_j)|0\rangle \equiv \exp(\beta_j b_j^\dagger - \bar{\beta}_j b_j)|0\rangle.$$

The N-mode even and odd coherent states are defined as

$$|\boldsymbol{\beta}_\pm\rangle := N_\pm(|\boldsymbol{\beta}\rangle \pm |-\boldsymbol{\beta}\rangle)$$

where

$$|\boldsymbol{\beta}\rangle = |\beta_1, \beta_2, \ldots, \beta_N\rangle \equiv |\beta\rangle \otimes |\beta\rangle \otimes \cdots \otimes |\beta\rangle.$$

(i) Find the normalization constants N_+ and N_-. Discuss
(ii) Find $b_j|\boldsymbol{\beta}_+\rangle$, $b_j|\boldsymbol{\beta}_-\rangle$.
(iii) Find the expectation values $\langle\beta_\pm|b_j|\beta_\pm\rangle$, $\langle\beta_\pm|b_j^\dagger|\beta_\pm\rangle$.
(iv) Find the expectation value $\langle\beta_\pm|b_j b_k|\beta_\pm\rangle$.
(v) Find the *mean photon numbers* $\langle\beta_+|\hat{N}_j|\beta_+\rangle$, $\langle\beta_-|\hat{N}_j|\beta_-\rangle$, where $\hat{N}_j = b_j^\dagger b_j$.

Solution 80.   (i) We find

$$N_+ = \frac{\exp(|\boldsymbol{\beta}|^2/2)}{2\sqrt{\cosh|\boldsymbol{\beta}|^2}}, \qquad N_- = \frac{\exp(|\boldsymbol{\beta}|^2/2)}{2\sqrt{\sinh|\boldsymbol{\beta}|^2}}.$$

Both N_+ and N_- contain the square the complex vector $\boldsymbol{\beta} = (\beta_1, \ldots, \beta_N)$. Thus one has statistical correlations between the modes.
(ii) We obtain

$$b_j|\boldsymbol{\beta}_+\rangle = \beta_j\sqrt{\tanh|\boldsymbol{\beta}|^2}|\boldsymbol{\beta}_-\rangle, \qquad b_j|\boldsymbol{\beta}_-\rangle = \beta_j\sqrt{\coth|\boldsymbol{\beta}|^2}|\boldsymbol{\beta}_+\rangle.$$

(iii) The expectation values are $\langle\beta_\pm|b_j|\beta_\pm = \langle\beta_\pm|b_j^\dagger|\beta_\pm\rangle = 0$.
(iv) We obtain $\langle\beta_\pm|b_j b_k|\beta_\pm\rangle = \beta_j\beta_k$.
(v) The mean photon numbers are

$$\langle\beta_+|\hat{N}_j|\beta_+\rangle = |\beta_j|^2\tanh|\boldsymbol{\beta}|^2, \qquad \langle\beta_-|\hat{N}_j|\beta_-\rangle = |\beta_j|^2\coth|\boldsymbol{\beta}|^2.$$

1.3 Squeezed States

1.3.1 One-Mode Squeezed States

Let $\zeta \in \mathbb{C}$. The one mode *squeezing operator* is defined by

$$S(\zeta) := \exp\left(-\frac{\zeta}{2}(b^\dagger)^2 + \frac{\zeta^*}{2}b^2\right)$$

where $\zeta \in \mathbb{C}$ and ζ is called the *squeezing parameter*. We set

$$\zeta := se^{i\theta}$$

where $s \geq 0$ and $\theta \in \mathbb{R}$ and $\bar{\zeta} = se^{-i\theta}$. The *single mode squeezed states* $|\zeta\rangle$ are defined by

$$|\zeta\rangle := S(\zeta)|0\rangle.$$

The state $|\zeta\rangle$ is normalized, i.e. $\langle\zeta|\zeta\rangle = 1$. The squeezing operator $S(\zeta)$ is unitary since

$$S^\dagger(\zeta) = \exp\left(\frac{\zeta}{2}(b^\dagger)^2 - \frac{\zeta^*}{2}b^2\right) = S(-\zeta) = S^{-1}(\zeta)$$

with $S^\dagger(\zeta)S(\zeta) = I$. Note that in literature

$$S(-\zeta) = S^\dagger(\zeta) = \exp\left(\frac{\zeta}{2}(b^\dagger)^2 - \frac{\zeta^*}{2}b^2\right)$$

is utilized sometimes as the definition of the squeezing operator.

The squeezed state $|\zeta\rangle$ is a right eigenstate of ζ-dependent linear combinations of the Bose creation and annihilation operators $b^\dagger$, b with eigenvalue 0, i.e.

$$(b\cosh(s) + b^\dagger e^{i\theta}\sinh(s))|\zeta\rangle = 0$$

or

$$S(\zeta)bS^\dagger(\zeta)|\zeta\rangle = 0.$$

Note that $\cosh^2(s) - \sinh^2(s) = 1$ and with

$$B := \cosh(s)b + e^{i\theta}\sinh(s)b^\dagger \Rightarrow B^\dagger := \cosh(s)b^\dagger + e^{-i\theta}\sinh(s)b$$

we obtain

$$[B, B^\dagger] = I.$$

The Hamilton operator

$$\hat{H} = \hbar\omega_1 b^\dagger b + \hbar\omega_2((b^\dagger)^2 e^{-2i\omega t} + b^2 e^{2i\omega t})$$

generates squeezed states.

Problem 81.   Consider the single-mode squeezing operator

$$S(\zeta) = \exp\left(\frac{1}{2}\zeta(b^\dagger)^2 - \frac{1}{2}\zeta^* b^2\right).$$

Let $|0\rangle$ be the vacuum state.
(i) Find the state $S(\zeta)|0\rangle \equiv |\zeta\rangle$ using the number states $|n\rangle$.
(ii) Use this result to find the expectation value $\langle\zeta|b^\dagger b|\zeta\rangle$.

Solution 81.   (i) Expansion over the number state basis contains only even components, i.e.

$$S(\zeta)|0\rangle \equiv |\zeta\rangle = \frac{1}{\sqrt{\mu}} \sum_{k=0}^{\infty} \left(\frac{\nu}{2\mu}\right)^k \frac{\sqrt{(2k)!}}{k!}|2k\rangle$$

where $\zeta = se^{i\theta}$, $\mu = \cosh(s)$, $\nu = e^{i\theta}\sinh(s)$. Thus we can also write

$$|\zeta\rangle = \sqrt{(\mathrm{sech}(s))} \sum_{n=0}^{\infty} \frac{\sqrt{((2n)!)}}{n!}\left(-\frac{1}{2}\exp(i\theta)\tanh(s)\right)^n |2n\rangle.$$

(ii) Using this result and $b^\dagger b|2k\rangle = 2k|2k\rangle$ we obtain

$$\langle\zeta|b^\dagger b|\zeta\rangle = |\nu|^2 = \sinh^2(s).$$

Another derivation is as follows. Since $b|0\rangle = 0|0\rangle$ and $\langle 0|b^\dagger = \langle 0|0$ we have

$$\langle\zeta|b^\dagger b|\zeta\rangle = \langle 0|(b^\dagger \cosh(s) - be^{-i\theta}\sinh(s))(b\cosh(s) - b^\dagger e^{i\theta}\sinh(s)|0\rangle = \sinh^2(s).$$

Problem 82.   Let $S(\zeta)$ be the squeezing operator.
(i) Calculate $\tilde{b} = S^\dagger(\zeta)bS(\zeta)$.
(ii) Calculate $\tilde{b}^\dagger = S^\dagger(\zeta)b^\dagger S(\zeta)$.
(iii) Show that $S(\zeta)bS^\dagger(\zeta)|\zeta\rangle = 0$.

Solution 82.   (i) Let $K := \frac{1}{2}(-\zeta^* b^2 + \zeta(b^\dagger)^2$ and $S^\dagger(\zeta) = e^K$. Utilizing the expansion

$$\tilde{b} = b + \frac{1}{1!}[K, b] + \frac{1}{2!}[K, [K, b]] + \frac{1}{3!}[K, [K, [K, b]]] + \cdots$$

with

$$[b, K] = \frac{1}{2}[b, (b^\dagger)^2] = \zeta b^\dagger$$

we obtain

$$\tilde{b} = \cosh(|\zeta|)b - \frac{\zeta}{|\zeta|}\sinh(|\zeta|)b^\dagger \equiv \cosh(s)b - e^{i\theta}\sinh(s)b^\dagger.$$

(ii) Using the result from (i) we find

$$\tilde{b}^\dagger = \cosh(|\zeta|)b^\dagger - \frac{\zeta^*}{|\zeta|}\sinh(|\zeta|)b \equiv \cosh(s)b - e^{-i\theta}\sinh(s)b^\dagger.$$

We can also write

$$S^\dagger(\zeta)bS(\xi) = \cosh(s)b + e^{i\theta}\sinh(s)b^\dagger, \quad S^\dagger(\zeta)b^\dagger S(\zeta) = \cosh(s)b^\dagger + e^{-i\theta}\sinh(s)b.$$

(v) From $S(\zeta)|0\rangle = |\zeta\rangle$ we obtain

$$S^\dagger(\zeta)|\zeta\rangle = S^\dagger(\zeta)S(\zeta)|0\rangle = |0\rangle.$$

Now $b|0\rangle = 0|0\rangle$ and the result follows.

Problem 83. We set $\zeta = se^{i\theta}$. Find the operator $S^\dagger(\zeta)b^\dagger bS(\zeta)$. Discuss.

Solution 83. Since

$$S^\dagger(\zeta)b^\dagger bS(\zeta) = S^\dagger(\zeta)b^\dagger S(\zeta)S^\dagger(\zeta)bS(\zeta)$$

we can utilize the result from a previous problem and find

$$\widetilde{b^\dagger b} = (\cosh^2(s) + \sinh^2(s))b^\dagger b + \sinh^2(s)I - \cosh(s)\sinh(s)(e^{i\theta}b^\dagger b^\dagger + e^{-i\theta}bb).$$

Thus the number operator $\hat{N} = b^\dagger b$ is not preserved under this unitary transformation.

Problem 84. Let $|\zeta\rangle$ be a squeezed state and $|\beta\rangle$ be a coherent states. We set $\beta = re^{i\phi}$ and $\zeta = se^{i\theta}$. Find the scalar product $\langle\beta|\zeta\rangle$.

Solution 84. We have

$$\langle\beta|\zeta\rangle = \langle\beta|S(\zeta)|0\rangle = \sqrt{(\operatorname{sech}(s))}\langle\beta|\exp(-\frac{1}{2}(b^\dagger)^2 e^{i\theta}\tanh(s))|0\rangle$$
$$= \sqrt{(\operatorname{sech}(s))}\exp(-\frac{1}{2}(\beta^*)^2 e^{i\theta}\tanh(s))\exp(-|\beta|^2/2).$$

Problem 85. Let $D(\beta)$ be the displacement operator and $S(\zeta)$ be the squeezing operator with $\beta, \zeta \in \mathbb{C}$. We set $\beta = re^{i\phi}$ and $\zeta = se^{i\theta}$.
(i) Find
$$S(-\zeta)D(-\beta)bD(\beta)S(\zeta).$$

(ii) Find
$$S(-\zeta)D(-\beta)b^\dagger D(\beta)S(\zeta).$$

Solution 85. (i) We obtain

$$S(-\zeta)D(-\beta)bD(\beta)S(\zeta) = S(-\zeta)(b + \beta I)S(\zeta) = b\cosh(s) - b^\dagger e^{i\theta}\sinh(s) + \beta I.$$

(ii) The result from (i) implies

$$S(-\zeta)D(-\beta)b^\dagger D(\beta)S(\zeta) = b^\dagger\cosh(s) - be^{-i\theta}\sinh(s) + \beta^* I.$$

Problem 86. Consider the three operators

$$K_+ = \frac{1}{2}(b^\dagger)^2, \quad K_- = \frac{1}{2}b^2, \quad K_3 = \frac{1}{2}\left(b^\dagger b + \frac{1}{2}I\right).$$

Let $\zeta = se^{i\theta}$ with $s > 0$ and $\theta \in \mathbb{R}$. Find the *disentangled form* of

$$S(\zeta) = \exp(\zeta K_+ - \zeta^* K_-).$$

Solution 86.   Owing to the Lie algebra property $[K_-, K+] = 2K_3$, $[K_3, K_+] = K_+$, $[K_3, K_-] = K_-$ we obtain

$$S(\zeta) = \exp(\zeta K_+ - \zeta^* K_-)$$
$$= \exp\left(\frac{\zeta}{|\zeta|}K_+\right)\exp(\ln(1 - \tanh^2|\zeta|)K_3)\exp\left(-\frac{\zeta^*}{|\zeta|}K_-\right)$$
$$= \exp\left(\frac{e^{i\theta}\sinh(s)}{2\cosh(s)}(b^\dagger)^2\right)(\cosh(s))^{-(b^\dagger b + I/2)}\exp\left(\frac{e^{-i\theta}\sinh(s)}{2\cosh(s)}b^2\right)$$

where

$$(\cosh(s))^{-(b^\dagger b + I/2)} \equiv \exp(-\ln(\cosh(s))(b^\dagger b + I/2)).$$

We also can write

$$S(\zeta) = \exp\left(-\frac{1}{2}(b^\dagger)^2 e^{i\theta}\tanh(s)\right)\exp\left(-\frac{1}{2}(b^\dagger b + bb^\dagger)\ln(\cosh(s))\right)\exp\left(\frac{1}{2}b^2 e^{-i\theta}\tanh(s)\right).$$

Problem 87.   We set $\zeta = se^{i\theta}$ with $s \geq 0$. Find the *squeezed number state* $S^\dagger(\zeta)|n\rangle$.

Solution 87.   Using the disentangled form for the squeezing operator $S(\zeta)$ from the previous problem we obtain

$$S(\zeta)|n\rangle = \left(\frac{1}{\cosh(s)}\right)^{n+1/2}(n!)^{1/2}\sum_{j=0}^{[n/2]}\frac{(-\bar{z})^j(\cosh(s))^{2j}}{n-2j)!j!}\sum_{k=0}^{\infty}\frac{z^k((n-2j+2k)!)^{1/2}}{k!}|n-2j+2k\rangle.$$

Problem 88.   Consider the one-mode squeezing operator $S(\zeta)$, where $\zeta = s\exp(i\theta)$. The phase θ of the squeezing parameter determines the direction of squeezing. Consider the *rotation operator*

$$R(\theta) = \exp(i\theta b^\dagger b).$$

Find the operator $R(\theta/2)S(s)R^\dagger(\theta/2)$.

Solution 88.   We obtain $R(\theta/2)S(s)R^\dagger(\theta/2) = S(s)$. Thus $S(s)$ is invariant under the rotation operator.

Problem 89.   Let $S(\zeta)$ be the one-mode squeezing operator $S(\zeta)$, where $\zeta = s\exp(i\theta)$. Calculate

$$S(\zeta)\begin{pmatrix} b \\ b^\dagger \end{pmatrix}S^\dagger(\zeta) \equiv \begin{pmatrix} S(\zeta)bS^\dagger(\zeta) \\ S(\zeta)b^\dagger S^\dagger(\zeta) \end{pmatrix}$$

and show that we can write

$$\begin{pmatrix} S(\zeta)bS^\dagger(\zeta) \\ S(\zeta)b^\dagger S^\dagger(\zeta) \end{pmatrix} = T(\zeta, \zeta^*)\begin{pmatrix} b \\ b^\dagger \end{pmatrix}$$

where the 2×2 matrix $T(\zeta, \zeta^*)$ depends only on ζ and ζ^*.

Solution 89.   We have $S^\dagger(\zeta) = S(-\zeta)$. Let $\epsilon \in \mathbb{R}$. We set

$$f_1(\epsilon) = e^{\frac{1}{2}\epsilon(\zeta^* b^2 - \zeta(b^\dagger)^2)}be^{-\frac{1}{2}\epsilon(\zeta^* b^2 - \zeta(b^\dagger)^2)}$$

and
$$f_2(\epsilon) = e^{\frac{1}{2}\epsilon(\zeta^* b^2 - \zeta(b^\dagger)^2)} b^\dagger e^{-\frac{1}{2}\epsilon(\zeta^* b^2 - \zeta(b^\dagger)^2)}$$

with the initial conditions $f_1(\epsilon = 0) = b$ and $f_2(\epsilon = 0) = b^\dagger$. We find the system of differential equations for f_1 and f_2. Using $bb^\dagger = I + b^\dagger b$ we have

$$\frac{df_1}{d\epsilon} = -\frac{1}{2}\zeta e^{\frac{1}{2}\epsilon(\zeta^* b^2 - \zeta(b^\dagger)^2)} ((b^\dagger)^2 b - b(b^\dagger)^2) e^{-\frac{1}{2}\epsilon(\zeta^* b^2 - \zeta(b^\dagger)^2)} = \zeta f_2(\epsilon).$$

Analogously, we obtain
$$\frac{df_2}{d\epsilon} = \zeta^* f_1(\epsilon).$$

In matrix notation we have the system
$$\begin{pmatrix} df_1/d\epsilon \\ df_2/d\epsilon \end{pmatrix} = \begin{pmatrix} 0 & \zeta \\ \zeta^* & 0 \end{pmatrix} \begin{pmatrix} f_1 \\ f_2 \end{pmatrix}.$$

Let A be the hermitian matrix
$$A = \begin{pmatrix} 0 & \zeta \\ \zeta^* & 0 \end{pmatrix}.$$

We find
$$\exp(\epsilon A) = \begin{pmatrix} \cosh(\epsilon s) & \zeta \sinh(\epsilon s)/s \\ \zeta^* \sinh(\epsilon s)/s & \cosh(\epsilon s) \end{pmatrix} = \begin{pmatrix} \cosh(\epsilon s) & e^{i\theta}\sinh(\epsilon s) \\ e^{-i\theta}\sinh(\epsilon s) & \cosh(\epsilon s) \end{pmatrix}$$

where we used $\zeta = s\exp(i\theta)$. Taking into account the initial conditions we obtain
$$\begin{pmatrix} S(\zeta)bS^\dagger(\zeta) \\ S(\zeta)b^\dagger S^\dagger(\zeta) \end{pmatrix} = \begin{pmatrix} \cosh(s) & e^{i\theta}\sinh(s) \\ e^{-i\theta}\sinh(s) & \cosh(s) \end{pmatrix} \begin{pmatrix} b \\ b^\dagger \end{pmatrix}.$$

The matrix on the right-hand side can be decomposed as
$$\begin{pmatrix} \cosh(s) & e^{i\theta}\sinh(s) \\ e^{-i\theta}\sinh(s) & \cosh(s) \end{pmatrix} \equiv \begin{pmatrix} 0 & e^{i\theta/2} \\ e^{-i\theta/2} & 0 \end{pmatrix} \begin{pmatrix} \cosh(s) & \sinh(s) \\ \sinh(s) & \cosh(s) \end{pmatrix} \begin{pmatrix} 0 & e^{i\theta/2} \\ e^{-i\theta/2} & 0 \end{pmatrix}.$$

Problem 90. Let $|n\rangle$, $|\beta\rangle$, $|\zeta\rangle$ be number states, coherent states and squeezed states, respectively. We set
$$\beta = re^{i\phi}$$
and
$$\zeta = se^{i\theta}$$

with $r, s \geq 0$ and $\phi, \theta \in \mathbb{R}$.
(i) Calculate the scalar product $\langle \zeta | \zeta' \rangle$.
(ii) Calculate $\langle n | \zeta \rangle \langle \zeta | n \rangle$. Express first $|\zeta\rangle$ with numbers states.
(iii) Let $|\beta\rangle$ be a coherent state. Calculate the scalar product $\langle \beta | \zeta \rangle$. Express first the states $\langle \beta |$ and $|\zeta\rangle$ with number states.

Solution 90. (i) Applying the squeezing operator $S(\zeta)$ we have
$$\langle \zeta | \zeta' \rangle = \langle 0 | S(-\zeta)S(\zeta') | 0 \rangle$$
$$= \sqrt{\text{sech}(s)\text{sech}(s')} \langle 0 | \exp(-\frac{1}{2}b^2 e^{-i\theta}\tanh(s)) \exp(-\frac{1}{2}(b^\dagger)^2 e^{i\theta'}\tanh(s')) | 0 \rangle$$
$$= \sqrt{\text{sech}(s)\text{sech}(s')} \sum_{n=0}^{\infty} \frac{(2n)!}{(n!)^2 2^{2n}} (\exp(i(\theta' - \theta))\tanh(s)\tanh(s'))^n$$
$$= \left(\frac{\text{sech}(s)\text{sech}(s')}{1 - e^{i(\theta' - \theta)}\tanh(s)\tanh(s')} \right)^{1/2}.$$

(ii) Expansions over the number state basis of the squeezed state $S(\zeta)|0\rangle = |\zeta\rangle$ contains only even components

$$S(\zeta)|0\rangle = |\zeta\rangle = \frac{1}{\sqrt{\mu}} \sum_{k=0}^{\infty} \left(\frac{\nu}{2\mu}\right)^k \frac{\sqrt{(2k)!}}{k!}|2k\rangle$$

where $k = 0, 1, \ldots$ and

$$\zeta = se^{i\theta}, \quad \mu = \cosh(s), \quad \nu = e^{i\theta}\sinh(s).$$

Since

$$\langle n|2k\rangle = \delta_{n,2k}$$

we obtain $(m = 0, 1, 2, \ldots)$

$$\langle 2m+1|(|\zeta\rangle\langle\zeta|)|2m+1\rangle = 0$$

and

$$\langle 2m|(|\zeta\rangle\langle\zeta|)|2m\rangle = \frac{(2m)!}{2^{2m}(m!)^2} \frac{(\tanh(s))^{2m}}{\cosh(s)}.$$

(iii) Utilizing the number state expansion of $|\beta\rangle$ and $\langle\beta|(b^\dagger)^2 = \langle\beta|(\bar{\beta})^2$ we obtain

$$\begin{aligned}
\langle\beta|\zeta\rangle &= \langle\beta|S(\zeta)|0\rangle \\
&= \sqrt{\operatorname{sech}(s)}\langle\beta|\exp(-\frac{1}{2}(b^\dagger)^2\exp(i\theta)\tanh(s))|0\rangle \\
&= \sqrt{\operatorname{sech}(s)}\exp(-\frac{1}{2}\bar{\beta}^2\exp(i\theta)\tanh(s))\exp(-|\beta|^2/2).
\end{aligned}$$

Thus the coherent state $|\beta\rangle$ and the squeezed state $|\zeta\rangle$ are never orthogonal.

1.3.2 Two-Mode Squeezed States

Problem 91. Consider the two-mode squeezed operator

$$S_2(\zeta) = \exp(\zeta b_1^\dagger b_2^\dagger - \zeta^* b_1 b_2)$$

where $\zeta \in \mathbb{C}$.
(i) Find

$$S_{12}^\dagger(\zeta) b_1 S_{12}(\zeta), \qquad S_{12}^\dagger(\zeta) b_2 S_{12}(\zeta).$$

(ii) Can the two-mode squeezing operator be expressed as product of two single-mode squeezing operators?
(iii) Introducing the 2-dimensional vectors of operators

$$\mathbf{b}^T = (b_1, b_2), \qquad (\mathbf{b}^\dagger)^T = (b_1^\dagger, b_2^\dagger)$$

one can write

$$S_2(\zeta) = \exp(\zeta (\mathbf{b}^\dagger)^T \sigma_1 \mathbf{b}^\dagger - \zeta^* \mathbf{b}^T \sigma_1 \mathbf{b})$$

where σ_1 is the first Pauli matrix. Calculate

$$e^{i\pi\sigma_2/4} S_2(z) e^{-i\pi\sigma_2/4}$$

(iv) Show that one can write

$$S_2^\dagger(\zeta) \begin{pmatrix} b_1 \\ b_2^\dagger \end{pmatrix} S_2(\zeta) = S_{2\zeta} \begin{pmatrix} b_1 \\ b_2^\dagger \end{pmatrix}.$$

Find the 2×2 matrix $S_{2\zeta}$.

Solution 91. (i) We obtain

$$S_{12}^\dagger(\zeta) b_1 S_{12}(\zeta) = \cosh(|\zeta|) b_1 + \frac{\zeta}{|\zeta|} \sinh(|\zeta|) b_2^\dagger$$

$$S_{12}^\dagger(\zeta) b_2 S_{12}(\zeta) = \cosh(|\zeta|) b_2 + \frac{\zeta}{|\zeta|} \sinh(|\zeta|) b_1^\dagger.$$

Thus a process leading to two-mode squeezing is a process during which a mode is mixed with the conjugate field of another mode owing to the *Bogoliubov transformation*.
(ii) The two-mode squeezing operator cannot be expressed as product of two single-mode squeezing operators. However we have

$$S_{12}(\zeta) = \exp\left(\frac{1}{2}(\zeta (b_+^\dagger)^2 - \gamma^* b_+^2)\right) \exp\left(-\frac{1}{2}(\zeta (b_-^\dagger)^2 - \zeta^* b_-^2)\right)$$

where

$$b_\pm := \frac{1}{\sqrt{2}}(b_1 \pm b_2).$$

(iii) Using the identity

$$e^{i\pi\sigma_2/4} \sigma_1 e^{-i\pi\sigma_2/4} = \sigma_3$$

we have

$$e^{i\pi\sigma_2/4} S_2(\zeta) e^{-i\pi\sigma_2/4} = \exp(z (\mathbf{b}^\dagger)^T \sigma_3 \mathbf{b}^\dagger \mathbf{b}^\dagger - \zeta^* \mathbf{b}^T \sigma_3 \mathbf{b}) = S^{(1)}(\zeta) S^{(2)}(-\zeta)$$

where $S^{(k)}$ denotes a single-mode (k) squeezed operator

$$S(\zeta) = \exp(\zeta(b^\dagger)^2 - \zeta^* b^2)).$$

(iv) We obtain 2×2 matrix

$$S_{2\zeta} = \begin{pmatrix} \mu & \nu \\ \nu^* & \mu \end{pmatrix}$$

where $\zeta = s e^{i\theta}$, $\mu = \cosh(s)$, and $\nu = e^{i\theta} \sinh(s)$.

Problem 92.   (i) Consider the two-mode squeezed state

$$|\psi\rangle = e^{s(b_1^\dagger b_2^\dagger - b_1 b_2)}|00\rangle$$

where $|00\rangle \equiv |0\rangle \otimes |0\rangle$ and s is the squeezing parameter. This state can also be written as

$$|\psi\rangle = \frac{1}{\cosh(s)} \sum_{n=0}^{\infty} (\tanh(s))^n |n\rangle \otimes |n\rangle.$$

This is the *Schmidt basis* for this state. The density operator ρ is given by $\rho = |\psi\rangle\langle\psi|$. Calculate the partial traces of ρ using the number states.
(ii) Use the reduced density operators ρ_1 and ρ_2 from (i) and calculate the entanglement

$$E(s) := -\mathrm{tr}(\rho_1 \log_2 \rho_1) = -\mathrm{tr}(\rho_2 \log_2 \rho_2).$$

Discuss E as a function of the squeezing parameter s.
(iii) Consider the two-mode squeezing operator

$$S_2(\zeta) = \exp(\zeta b_1^\dagger b_2^\dagger - \zeta^* b_1 b_2).$$

Let $|0\rangle \otimes |0\rangle$ be the two-mode vacuum state. Find the state $S_2(\zeta)(|0\rangle \otimes |0\rangle)$ expressed in number states.

Solution 92.   (i) We have

$$\rho = \frac{1}{(\cosh(s))^2} \sum_{n=0}^{\infty} (\tanh(s))^n |n\rangle \otimes |n\rangle \sum_{m=0}^{\infty} (\tanh(s))^m \langle m| \otimes \langle m|$$

$$= \frac{1}{(\cosh(s))^2} \sum_{m=0}^{\infty} \sum_{n=0}^{\infty} (\tanh(s))^n (\tanh(s))^m |n\rangle\langle m| \otimes |n\rangle\langle m|.$$

Let I be the identity operator. Using that $\langle k|n\rangle = \delta_{kn}$ and $\langle m|k\rangle = \delta_{mk}$ we have

$$\rho_1 = \sum_{k=0}^{\infty} (I \otimes \langle k|) \rho (I \otimes |k\rangle)$$

$$= \frac{1}{(\cosh(s))^2} \sum_{k=0}^{\infty} (I \otimes \langle k|) \sum_{m,n=0}^{\infty} (\tanh(s))^n (\tanh(s))^m |n\rangle\langle m| \otimes |n\rangle\langle m| (I \otimes |k\rangle)$$

$$= \frac{1}{(\cosh(s))^2} \sum_{k=0}^{\infty} \sum_{m,n=0}^{\infty} (\tanh(s))^n (\tanh(s))^m |n\rangle\langle m| \delta_{kn} \delta_{mk}$$

$$= \frac{1}{(\cosh(s))^2} \sum_{k=0}^{\infty} (\tanh(s))^{2k} |k\rangle\langle k|.$$

We obtain the same result for ρ_2.

(ii) We have

$$E(s) = -\text{tr}\left(\left(\frac{1}{\cosh^2(s)}\sum_{k=0}^{\infty}(\tanh^{2k}(s))|k\rangle\langle k|\right)\log_2\left(\frac{1}{\cosh^2(s)}\sum_{\ell=0}^{\infty}(\tanh^{2\ell}(s))|\ell\rangle\langle\ell|\right)\right).$$

The two matrices inside the trace are diagonal matrices and thus the product is again a diagonal matrix. Thus

$$E(s) = -\frac{1}{\cosh^2(s)}\sum_{k=0}^{\infty}\tanh^{2k}(s)\log_2\left(\frac{\tanh^{2k}(s)}{\cosh^2(s)}\right).$$

Thus

$$E(s) = -\frac{1}{\cosh^2(s)}\sum_{k=0}^{\infty}\tanh^{2k}(s)\log_2\tanh^{2k}(s) + \frac{\log_2(\cosh^2(s))}{\cosh^2(s)}\sum_{k=0}^{\infty}\tanh^{2k}(s)$$

or using the property of log

$$E(s) = -\frac{\log_2(\tanh^2(s))}{\cosh^2(s)}\sum_{k=0}^{\infty}k\tanh^{2k}(s) + \frac{\log_2(\cosh^2(s))}{\cosh^2(s)}\sum_{k=0}^{\infty}\tanh^{2k}(s).$$

The identity

$$\sum_{k=0}^{\infty}\tanh^{2k}(s) \equiv \cosh^2(s)$$

follows from a geometric series. The identity

$$\sum_{k=0}^{\infty}k\tanh^{2k}(s) \equiv \sinh^2(s)\cosh^2(s)$$

can be obtained from the first identity by parameter differentiation with respect to r. Using these results we obtain

$$\begin{aligned}E(s) &= -\sinh^2(s)(\log_2(\sinh^2(s)) - \log_2(\cosh^2(s)) + \log_2(\cosh^2(s))\\ &= -\sinh^2(s)\log_2(\sinh^2(s)) + (\sinh^2(s) + 1)\log_2(\cosh^2(s))\\ &= -\sinh^2(s)\log_2(\sinh^2(s)) + \cosh^2(s)\log_2(\cosh^2(s)).\end{aligned}$$

For $s = 0$ we have $E(s = 0) = 0$.

(ii) We obtain

$$S_2(\zeta)(|0\rangle \otimes |0\rangle) = \frac{1}{\sqrt{\mu}}\sum_{k=0}^{\infty}\left(\frac{\nu}{\mu}\right)^k|k\rangle \otimes |k\rangle$$

where $\zeta = se^{i\theta}$, $\mu = \cosh(s)$, $\nu = e^{i\theta}\sinh(s)$. This state is known as *two-mode squeezed vacuum* or *twin-beam state*.

Problem 93. Consider the two-mode state

$$|\psi\rangle = e^{s(b_1^\dagger b_2^\dagger - b_1 b_2)}|00\rangle$$

where $|00\rangle = |0\rangle \otimes |0\rangle$ and s is the squeezing parameter. This state can also be written as

$$|\psi\rangle = \frac{1}{\cosh(s)} \sum_{n=0}^{\infty} (\tanh(s))^n |n\rangle \otimes |n\rangle.$$

This is the *Schmidt basis* for this state. The density operator $\rho(s)$ is given by $\rho = |\psi\rangle\langle\psi|$. Calculate the partial traces using coherent states.

Solution 93.   We have

$$\rho(s) = \frac{1}{\cosh^2(s)} \sum_{n=0}^{\infty} \tanh^n(s)|n\rangle \otimes |n\rangle \sum_{m=0}^{\infty} \tanh^m(s)\langle m| \otimes \langle m|$$

$$= \frac{1}{(\cosh(s))^2} \sum_{m=0}^{\infty} \sum_{n=0}^{\infty} (\tanh^n(s)(\tanh^m(s)|n\rangle\langle m| \otimes |n\rangle\langle m|.$$

Let I be the identity operator. We have for the partial trace

$$\rho_1(s) = \frac{1}{\pi} \int_{\mathbb{C}} d^2\beta (I \otimes \langle\beta|)\rho(I \otimes |\beta\rangle).$$

Inserting $\rho(s)$ provides

$$\rho_1(s) = \frac{1}{\pi \cosh^2(s)} \int_{\mathbb{C}} d^2\beta (I \otimes \langle\beta|) \left(\sum_{m,n=0}^{\infty} \tanh^{n+m}(s)|n\rangle\langle m| \otimes |n\rangle\langle m| \right) (I \otimes |\beta\rangle).$$

Thus

$$\rho_1(s) = \int_{\mathbb{C}} d^2\beta \sum_{m,n=0}^{\infty} \tanh^{n+m}(s)|n\rangle\langle m|\langle\beta|n\rangle\langle m|\beta\rangle.$$

Using

$$\langle\beta|n\rangle = \overline{\langle n|\beta\rangle} = e^{-|\beta|^2/2} \frac{(\beta^*)^n}{\sqrt{n!}}, \qquad \langle m|\beta\rangle = e^{-|\beta|^2/2} \frac{\beta^m}{\sqrt{m!}}$$

and $\beta = re^{i\phi}$ with $r \geq 0$ and $0 \leq \phi < 2\pi$ we have

$$\langle\beta|n\rangle\langle m|\beta\rangle = e^{-|\beta|^2} \frac{\beta^m}{\sqrt{m!}} \frac{(\beta^*)^n}{\sqrt{n!}} = \frac{e^{-r^2} r^{n+m}}{\sqrt{m!}\sqrt{n!}} e^{i(m-n)\phi}$$

and $d^2\beta = rdrd\phi$. Thus

$$\rho_1(s) = \frac{1}{\pi \cosh^2(s)} \sum_{m,n=0}^{\infty} \tanh^{n+m}(s)|n\rangle\langle m| \left(\int_0^{2\pi} e^{i(m-n)\phi}d\phi \right) \left(\int_0^{\infty} \frac{r^{n+m+1}e^{-r^2}dr}{\sqrt{m!}\sqrt{n!}} \right).$$

Since

$$\int_0^{2\pi} e^{i(m-n)\phi}d\phi = 2\pi\delta_{mn}$$

we obtain

$$\rho_1(s) = \frac{2}{\cosh^2(s)} \sum_{n=0}^{\infty} \tanh^{2n}(s)|n\rangle\langle n| \frac{1}{n!} \int_0^{\infty} r^{2n+1}e^{-r^2} dr.$$

Since

$$\int_0^{\infty} r^{2n+1}e^{-r^2} dr = \frac{1}{2}\Gamma(n+1) = \frac{n!}{2}$$

we finally arrive at

$$\rho_1(s) = \frac{1}{\cosh^2(s)} \sum_{n=0}^{\infty} \tanh^{2n}(s)|n\rangle\langle n|.$$

We obtain the same result for ρ_2.

Problem 94. Two-mode squeezing is described by the operator

$$U_S(\zeta) = \exp((\zeta b_1^\dagger b_2^\dagger - \zeta^* b_1 b_2)/2)$$

where $\zeta := s\exp(i2\theta)$, $r \in \mathbb{R}$, $\theta \in [0, 2\pi)$. The representation in phase space of the operator $U_S(\zeta)$ is given by the 4×4 matrix

$$S(s, \theta) = \begin{pmatrix} c - hs & 0 & ks & 0 \\ 0 & c + hs & 0 & -ks \\ ks & 0 & c + hs & 0 \\ 0 & -ks & 0 & c - hs \end{pmatrix}$$

where $c := \cosh(2s)$, $s := \sinh(2s)$, $h = \cos(2\theta)$, $k = \sin(2\theta)$.
(i) Let

$$K := \begin{pmatrix} 0 & 1 \\ -1 & 0 \end{pmatrix} \oplus \begin{pmatrix} 0 & 1 \\ -1 & 0 \end{pmatrix}$$

where $\oplus$ denotes the direct sum. Thus $K^T = -K$. Calculate $S^T(s, \theta)KS(s, \theta)$.
(ii) Let

$$J := \begin{pmatrix} 0_2 & I_2 \\ -I_2 & 0_2 \end{pmatrix}$$

where I_2 is the 2×2 identity matrix. Calculate $S^T(s, \theta)JS(s, \theta)$.

Solution 94. (i) We have $S^T(s, \theta)KS(s, \theta) = K$.
(ii) We have $S^T(s, \theta)JS(s, \theta) = J$.

Problem 95. Consider the unitary operator for two-mode squeezing

$$U(\zeta) := \exp(\zeta b_1^\dagger b_2^\dagger - \zeta^* b_1 b_2).$$

Calculate

$$U(\zeta) \begin{pmatrix} b_1 \\ b_2 \\ b_1^\dagger \\ b_2^\dagger \end{pmatrix} U^\dagger(\zeta) \equiv \begin{pmatrix} U(\zeta)b_1 U^\dagger(\zeta) \\ U(\zeta)b_2 U^\dagger(\zeta) \\ U(\zeta)b_1^\dagger U^\dagger(\zeta) \\ U(\zeta)b_2^\dagger U^\dagger(\zeta) \end{pmatrix}$$

and show that we can write

$$\begin{pmatrix} U(\zeta)b_1 U^\dagger(\zeta) \\ U(\zeta)b_2 U^\dagger(\zeta) \\ U(\zeta)b_1^\dagger U^\dagger(\zeta) \\ U(\xi)b_2^\dagger U^\dagger(\zeta) \end{pmatrix} = T(\zeta, \zeta^*) \begin{pmatrix} b_1 \\ b_2 \\ b_1^\dagger \\ b_2^\dagger \end{pmatrix}$$

where the 4×4 matrix $T(\zeta, \zeta^*)$ depends only on ζ and ζ^*.

Solution 95. We have $U^\dagger(\zeta) = U(-\zeta)$. Let $\epsilon \in \mathbb{R}$. We set

$$f_1(\epsilon) = e^{\epsilon(\zeta b_1^\dagger b_2^\dagger - \zeta^* b_1 b_2)} b_1 e^{-\epsilon(\zeta b_1^\dagger b_2^\dagger - \zeta^* b_1 b_2)}$$

$$f_2(\epsilon) = e^{\epsilon(\zeta b_1^\dagger b_2^\dagger - \zeta^* b_1 b_2)} b_2 e^{-\epsilon(\xi b_1^\dagger b_2^\dagger - \xi^* b_1 b_2)}$$

$$f_3(\epsilon) = e^{\epsilon(\zeta b_1^\dagger b_2^\dagger - \zeta^* b_1 b_2)} b_1^\dagger e^{-\epsilon(\zeta b_1^\dagger b_2^\dagger - \zeta^* b_1 b_2)}$$

$$f_4(\epsilon) = e^{\epsilon(\zeta b_1^\dagger b_2^\dagger - \zeta^* b_1 b_2)} b_2^\dagger e^{-\epsilon(\zeta b_1^\dagger b_2^\dagger - \zeta^* b_1 b_2)}$$

with the initial conditions

$$f_1(\epsilon = 0) = b_1, \qquad f_2(\epsilon = 0) = b_2, \qquad f_3(\epsilon = 0) = b_1^\dagger, \qquad f_4(\epsilon = 0) = b_2^\dagger.$$

We find the system of differential equations for f_1, f_2, f_3 and f_4. Using $b_j b_j^\dagger = I + b_j^\dagger b_j$ and $b_j^\dagger b_k = b_k b_j^\dagger$ if $j \neq k$ we have

$$\frac{df_1}{d\epsilon} = e^{\epsilon(\zeta b_1^\dagger b_2^\dagger - \zeta^* b_1 b_2)}(\zeta b_1^\dagger b_2^\dagger b_1 - \zeta b_1 b_1^\dagger b_2^\dagger)e^{-\epsilon(\zeta b_1^\dagger b_2^\dagger - \zeta^* b_1 b_2)}$$

$$= \xi e^{\epsilon(\zeta b_1^\dagger b_2^\dagger - \zeta^* b_1 b_2)}(-b_2^\dagger)e^{-\epsilon(\zeta b_1^\dagger b_2^\dagger - \zeta^* b_1 b_2)}$$

$$= -\zeta f_4(\epsilon).$$

Analogously, we obtain

$$\frac{df_2}{d\epsilon} = -\zeta f_3(\epsilon), \qquad \frac{df_3}{d\epsilon} = -\zeta^* f_2(\epsilon), \qquad \frac{df_4}{d\epsilon} = -\zeta^* f_1(\epsilon).$$

In matrix notation we have the linear system of differential equations

$$\begin{pmatrix} df_1/d\epsilon \\ df_2/d\epsilon \\ df_3/d\epsilon \\ df_4/d\epsilon \end{pmatrix} = \begin{pmatrix} 0 & 0 & 0 & -\zeta \\ 0 & 0 & -\zeta & 0 \\ 0 & -\zeta^* & 0 & 0 \\ -\zeta^* & 0 & 0 & 0 \end{pmatrix} \begin{pmatrix} f_1 \\ f_2 \\ f_3 \\ f_4 \end{pmatrix}.$$

Let

$$A = \begin{pmatrix} 0 & 0 & 0 & -\zeta \\ 0 & 0 & -\zeta & 0 \\ 0 & -\zeta^* & 0 & 0 \\ -\zeta^* & 0 & 0 & 0 \end{pmatrix}.$$

Using $\zeta = s \exp(i\theta)$ and $\zeta^* \zeta = s^2$ we find

$$\exp(\epsilon A) = \begin{pmatrix} \cosh(\epsilon s) & 0 & 0 & -e^{i\theta}\sinh(\epsilon s) \\ 0 & \cosh(\epsilon s) & -e^{i\theta}\sinh(\epsilon s) & 0 \\ 0 & -e^{-i\theta}\sinh(\epsilon s) & \cosh(\epsilon s) & 0 \\ -e^{-i\theta}\sinh(\epsilon s) & 0 & 0 & \cosh(\epsilon s) \end{pmatrix}.$$

Taking into account the initial conditions and $\epsilon = 1$ we obtain

$$\begin{pmatrix} U(\zeta)b_1 U^\dagger(\zeta) \\ U(\zeta)b_2 U^\dagger(\zeta) \\ U(\zeta)b_1^\dagger U^\dagger(\zeta) \\ U(\zeta)b_2^\dagger U^\dagger(\zeta) \end{pmatrix} = \begin{pmatrix} \cosh(s) & 0 & 0 & -e^{i\theta}\sinh(s) \\ 0 & \cosh(s) & -e^{i\theta}\sinh(s) & 0 \\ 0 & -e^{-i\theta}\sinh(s) & \cosh(s) & 0 \\ -e^{-i\theta}\sinh(s) & 0 & 0 & \cosh(s) \end{pmatrix} \begin{pmatrix} b_1 \\ b_2 \\ b_1^\dagger \\ b_2^\dagger \end{pmatrix}.$$

1.4 Coherent Squeezed States

Let $|0\rangle$ be the vacuum state. Then the coherent state is defined as $|\beta\rangle = D(\beta)|0\rangle$ and the squeezed state as $|\zeta\rangle = S(\zeta)|0\rangle$. One defines the coherent squeezed state as

$$|\beta, \zeta\rangle := D(\beta)S(\zeta)|0\rangle$$

where

$$D(\beta) = \exp(\beta b^\dagger - \beta^* b), \qquad S(\zeta) = \exp(-\zeta (b^\dagger)^2/2 + \zeta^* b^2/2)$$

and $\beta, \zeta \in \mathbb{C}$. Thus such states can be generated by squeezing the vacuum and then by displacing it. A single mode coherent squeezed state $|\beta, \zeta\rangle$ is determined by two complex parameters: the complex amplitude

$$\beta = re^{i\phi}$$

and the complex squeeze factor

$$\zeta = se^{i\theta}.$$

The complex squeeze factor ζ is composed by a real squeeze factor s ($r \geq 0$) and a squeeze phase θ ($0 \leq \phi \leq \pi$). The operators $D(\beta)$ and $S(\zeta)$ do not commute. The coherent squeezed states $|\beta, \zeta\rangle$ also minimize the uncertainty relation. However, the variance of both canonically coupled variables are not equal. The modulus s of the complex number $\zeta = se^{i\theta}$ determines the strength of squeezing while the angle θ orients the squeezing axis.

Problem 96. (i) Show that the state $|\beta, \zeta\rangle$ is normalized.
(ii) Show that

$$S(\zeta)D(\beta)|0\rangle \neq D(\beta)S(\zeta)|0\rangle \equiv |\beta, \zeta\rangle.$$

Solution 96. (i) We have

$$\langle \beta, \zeta | \beta, \zeta \rangle = \langle 0|S(-\zeta)D(-\beta)D(\beta)S(\zeta)|0\rangle = 1.$$

(ii) Since $S(-\zeta)S(\zeta) = I$ we have

$$S(\zeta)D(\beta)|0\rangle = S(\zeta)D(\beta)S(-\zeta)S(\zeta)|0\rangle.$$

Now

$$S(\zeta)D(\beta)S(-\zeta) = \exp(\beta(b^\dagger \cosh(s) + b\sinh(s)) - \beta^*(b\cosh(s) + b^\dagger e^{i\theta}\sinh(s)))$$

and therefore

$$S(\zeta)D(\beta)S(-\zeta) = D(\beta \cosh(s) - \beta^* e^{i\theta}\sinh(s)).$$

It follows that

$$S(\zeta)D(\beta)|0\rangle = D(\beta\cosh(s) - \beta^* e^{i\theta}\sinh(s))S(\zeta)$$
$$= |\beta\cosh(s) - \beta^* e^{i\theta}\sinh(r), \zeta\rangle$$

Problem 97. Let $S(\zeta)$ be the one-mode squeeze operator with

$$\zeta = |\zeta|e^{i\theta} = se^{i\theta}.$$

Let $D(\beta)$ be the displacement operator.

(i) Find
$$S(\zeta)D(\beta)S^{-1}(\zeta).$$

(ii) Let $(s \in \mathbb{R})$
$$S(s) = \exp\left(\frac{1}{2}s(b^2 - b^{\dagger 2})\right), \qquad D(\beta) = \exp(\beta b^\dagger - \beta^* b).$$

Find the operator $S^\dagger(s)D(\beta)S(s)$.

Solution 97.   (i) We obtain
$$S(\zeta)D(\beta)S^{-1}(\zeta) = D(\widetilde{\beta})$$

where
$$\widetilde{\beta} = \cosh(|\zeta|)\beta + e^{i\theta}\sinh(|\zeta|)\beta^*.$$

(ii) We find
$$S^\dagger(s)D(\beta)S(s) = D(\beta_r e^s + i\beta_i e^{-s})$$

where the subscripts r and i, respectively, denote the real and imaginary parts of β.

Problem 98.   Let $|\alpha\rangle$ be a coherent state. The *Husimi distribution* of a quantum wave function $|\psi\rangle$ is given by
$$\rho^H_{|\psi\rangle}(\alpha) = |\langle\psi|\alpha\rangle|^2.$$

Find the Husimi distribution
$$\rho^H_{|\zeta,\beta\rangle}(\alpha) = |\langle\alpha|\zeta,\beta\rangle|^2$$

of a coherent squeezed state $|\zeta,\beta\rangle$.

Solution 98.   Let $\alpha = \alpha_1 + i\alpha_2$, where $\alpha_1, \alpha_2 \in \mathbb{R}$. We obtain
$$\rho^H_{|\gamma,\beta\rangle}(\alpha) = |\langle\alpha|\gamma,\beta\rangle|^2 = \exp(-(\Re(\beta) - \alpha_1)^2/(s+1)^2 - (\Im(\beta) - \alpha_2)^2(s+1)^2).$$

Problem 99.   Let $D(\beta) = \exp(\beta b^\dagger - \beta^* b)$ be the displacement operator and
$$S(s,\theta) = \exp(se^{i\theta}(b^\dagger)^2/2 - se^{-i\theta}b^2/2)$$

be the squeeze operator with the squeeze factor $s \geq 0$ and squeeze angle $\theta \in (-\pi, \pi]$ ($\zeta = se^{i\theta}$).
Let
$$\rho_T := \frac{1}{\overline{n}+1}\sum_{n=0}^{\infty}\left(\frac{\overline{n}}{\overline{n}+1}\right)^n |n\rangle\langle n|$$

be the *Bose-Einstein density operator* with the mean occupancy
$$\overline{n} = \left(\exp\left(\frac{\hbar\omega}{k_BT}\right) - 1\right)^{-1}.$$

Consider the displaced squeezed thermal state
$$\rho(\beta,\zeta) = D(\beta)S(s,\theta)\rho_T S^\dagger(s,\theta)D^\dagger(\beta). \tag{1}$$

The *Weyl expansion* of the density operator

$$\rho = \frac{1}{\pi} \int_C d^2\lambda \chi(\lambda) D(-\lambda)$$

with $d^2\lambda = d\Re\lambda d\Im\lambda$ provides the one-to-one correspondence between the density operator ρ and its characteristic function

$$\chi(\lambda) := \mathrm{tr}(\rho D(\lambda)) \equiv \sum_{n=0}^{\infty} \langle n|\rho D(\lambda)|n\rangle. \tag{2}$$

A Gaussian state has a characteristic function of the form

$$\chi(\lambda) = \exp\left(-\left(A + \frac{1}{2}\right)|\lambda|^2 - \frac{1}{2}B^*\lambda^2 - \frac{1}{2}B(\lambda^*)^2 + C^*\lambda - C\lambda^*\right)$$

with $A > 0$. Find the coefficients A, B, C for the given ρ.

Solution 99. Calculating (2) where ρ is given by (1) and comparing coefficients yields

$$A = \left(\bar{n} + \frac{1}{2}\right)\cosh(2s) - \frac{1}{2}, \quad B = -\left(\bar{n} + \frac{1}{2}\right)e^{i\theta}\sinh(2s), \quad C = \beta.$$

Problem 100. Show that the coherent squeezed states $|\beta, \zeta\rangle$ form a overcomplete set in that the identity operator I can be resolved as

$$\frac{1}{\pi} \int_C d^2\beta |\beta, \zeta\rangle\langle\beta, \zeta| = I.$$

Solution 100. We can write

$$S(\zeta)\frac{1}{\pi}\int_C d^2\beta D(\gamma)|0\rangle\langle 0|D(-\gamma)S(-\zeta) = S(\zeta)\left(\frac{1}{\pi}\int_C d^2\beta|\gamma\rangle\langle\gamma|\right)S(-\zeta)$$

where

$$\gamma = \beta\cosh(s) + \bar{\beta}e^{i\theta}\sinh(s).$$

The Jacobian determinant for the change of variable from the complex number β to the complex number γ is unity. Thus it follows that

$$\frac{1}{\pi}\int_C d^2\beta|\beta, \zeta\rangle\langle\beta, \zeta| = S(\zeta)\left(\frac{1}{\pi}\int_C d^2\gamma|\gamma\rangle\langle\gamma|\right)S(-\zeta) = I.$$

1.5 Hamilton Operators

Problem 101. Consider the Hamilton operator

$$\hat{H} = b^\dagger b + \beta^* b + \beta b^\dagger$$

where $\beta \in \mathbb{C}$. Thus $\hat{H}$ represents the Hamilton operator of a displaced harmonic oscillator of unit frequency. Let $\Re(z) > 0$. Calculate

$$\text{tr}(e^{-z\hat{H}}).$$

Solution 101. We have $b^\dagger b |n\rangle = n|n\rangle$. Thus $\hat{H}|\psi_n\rangle = E_n|\psi_n\rangle$ $(n = 0, 1, \ldots)$ with

$$E_n = n - |\beta|^2, \qquad |\psi_n\rangle = D^\dagger(\beta)|n\rangle.$$

Here $D(\beta)$ is the displacement operators. We have

$$D(\beta)\hat{H}D^\dagger(\beta) = b^\dagger b - |\beta|^2 I.$$

Thus

$$\text{tr}(e^{-z\hat{H}}) = \frac{e^{z|\beta|^2}}{1 - e^{-z}}.$$

Problem 102. For generating a squeezed state of one mode we start from the Hamilton operator

$$\hat{H} = \hbar\omega b^\dagger b + i\hbar\Lambda(b^2 e^{2i\omega t} - (b^\dagger)^2 e^{-2i\omega t}).$$

A photon of the driven mode, with frequency 2ω, splits into two photons of the mode of interest, each with frequency ω. Solve the Heisenberg equation of motion for b and $b^\dagger$.

Solution 102. The Heisenberg equation of motion yields

$$\frac{db}{dt} = \frac{1}{i\hbar}[b, \hat{H}](t) = -i\omega b(t) - 2\Lambda b^\dagger(t)e^{-2i\omega t}$$

$$\frac{db^\dagger}{dt} = \frac{1}{i\hbar}[b^\dagger, \hat{H}](t) = i\omega b^\dagger - 2\Lambda b e^{2i\omega t}.$$

Using $b = \tilde{b}\exp(-i\omega t)$ we obtain the system of differential equations

$$\frac{d\tilde{b}}{dt} = -2\Lambda\tilde{b}^\dagger(t), \qquad \frac{d\tilde{b}^\dagger}{dt} = -2\Lambda\tilde{b}(t).$$

Introducing the operators

$$\tilde{b}_P := \frac{1}{2}(\tilde{b} + \tilde{b}^\dagger), \qquad \tilde{b}_Q := \frac{1}{2i}(\tilde{b} - \tilde{b}^\dagger)$$

i.e.

$$\tilde{b} = \tilde{b}_P + i\tilde{b}_Q, \qquad \tilde{b}^\dagger = \tilde{b}_P - i\tilde{b}_Q$$

we finally arrive at the system of differential equations

$$\frac{db_P}{dt} = -2\Lambda b_P(t), \qquad \frac{db_Q}{dt} = 2\Lambda b_Q(t)$$

with the solution of the initial value problem

$$b_P(t) = b_P(0)e^{-2\Lambda t}, \qquad b_Q(t) = b_Q(0)e^{2\Lambda t}.$$

If the state is the vacuum state $|0\rangle$ we obtain a squeezed state. For the *electric field* $E(\mathbf{r}, t)$ we have

$$E(\mathbf{r}, t) = i\mathcal{E}(b(t)e^{i\mathbf{k}\cdot\mathbf{r}} - b^\dagger(t)e^{-i\mathbf{k}\cdot\mathbf{r}}) = i\mathcal{E}(\tilde{b}(t)e^{i\mathbf{k}\cdot\mathbf{r}-i\omega t} - \tilde{b}^\dagger(t)e^{-i\mathbf{k}\cdot\mathbf{r}-i\omega t})$$
$$= -2\mathcal{E}(b_P(t)\sin(\mathbf{k}\cdot\mathbf{r} - \omega t) + b_Q(t)\cos(\mathbf{k}\cdot\mathbf{r} - \omega t)).$$

Thus b_P and b_Q are the amplitudes of two quadrature components of the electric field. They are measurable by phase sensitive detection.

Problem 103. Let $\hat{N} = b^\dagger b$ be the number operator. Consider the Hamilton operator

$$\hat{H} = \hbar\omega b^\dagger b + \hbar\mu(b^\dagger b)^2 \equiv \hbar\omega\hat{N} + \hbar\mu(\hat{N})^2, \quad \mu > 0.$$

(i) Find the *Heisenberg equation of motion*

$$i\hbar\frac{dA}{dt} = [A, \hat{H}](t)$$

for b and $b^\dagger$.
(ii) Solve the Heisenberg equation of motion.
(iii) Consider the coherent state ($\beta_0 \in \mathbb{C}$)

$$|\beta_0\rangle = \exp\left(-\frac{1}{2}|\beta_0|^2\right)\sum_{n=0}^{\infty}\frac{\beta_0^n}{\sqrt{n!}}|n\rangle.$$

Find the functions

$$\beta(t) = \langle\beta_0|b(t)|\beta_0\rangle, \qquad \beta^*(t) = \langle\beta_0|b^\dagger(t)|\beta_0\rangle.$$

Solution 103. (i) Since $[b, b^\dagger b] = b$, $[b^\dagger, b^\dagger b] = -b^\dagger$ and

$$[b, \hat{N}^2] = (I + 2b^\dagger b)b, \qquad [b^\dagger, \hat{N}^2] = -b^\dagger(I + 2b^\dagger b)$$

we obtain

$$i\hbar\frac{db}{dt} = \hbar\omega b + \mu\hbar(I + 2b^\dagger b)b \implies \frac{db}{dt} = -i(\omega I + \mu I + 2\mu b^\dagger b)b.$$

Analogously

$$\frac{db^\dagger}{dt} = ib^\dagger(\omega I + \mu I + 2\mu b^\dagger b).$$

(ii) The Hamilton operator $\hat{H}$ commutes with $\hat{N}$. We set

$$\Omega(b^\dagger b) = \omega I + \mu I + 2\mu b^\dagger b.$$

Thus we have

$$\Omega(b^\dagger(t)b(t)) := \Omega(b^\dagger(0)b(0)).$$

The solution of the Heisenberg equation of motion is given by

$$b^\dagger(t) = b^\dagger(0)e^{i\Omega(b^\dagger b)t}, \qquad b(t) = e^{-i\Omega(b^\dagger b)t}b(0).$$

(iii) Since (normal ordering)

$$\exp(-\epsilon b^\dagger b) = \sum_{j=0}^{\infty} \frac{1}{j!}(e^{-\epsilon} - 1)^j (b^\dagger)^j b^j$$

we obtain

$$\beta(t) = e^{-i(\omega+\mu)t} \exp\left((e^{-2i\mu t} - 1)|\beta_0|^2\right)\beta_0.$$

Consequently

$$\beta^*(t) = e^{i(\omega+\mu)t} \exp\left((e^{2i\mu t} - 1)|\beta_0|^2\right)\beta_0^*.$$

Problem 104.   Consider the Hamilton operator

$$\hat{H} = \hat{H}_0 + V(t)$$

where $\hat{H}_0 = \hbar\omega b^\dagger b$ is the Hamilton operator of a one-dimensional harmonic oscillator, and $V(t) = -xF(t)$ is a time-dependent potential. Here, $F(t)$ is a real c-number function of time and corresponds to a spatially uniform force, and $x = x_0(b + b^\dagger)$, where $x_0 := (\hbar/(2m\omega))^{1/2}$. If the particle has charge q and a uniform electric field $E(t)$ in x-direction is applied, then $F(t) = qE(t)$. In the interaction picture the state vector $|\psi(t)\rangle^I$ is related to the corresponding Schrödinger picture state $|\psi(t)\rangle^S$ by

$$|\psi(t)\rangle^I = e^{i\hat{H}_0 t/\hbar}|\psi(t)\rangle^S$$

and an interaction-picture operator $\hat{O}^I(t)$ is related to the corresponding Schrödinger picture operator $\hat{O}^S(t)$ by

$$\hat{O}^I(t) = e^{i\hat{H}_0 t/\hbar} O^S(t) e^{-i\hat{H}_0 t/\hbar}.$$

The equation of motion for an interaction-picture operator is

$$\frac{d}{dt}\hat{O}^I(t) = -\frac{i}{\hbar}[\hat{O}^I(t), \hat{H}_0]$$

Solve the Schrödinger equation in the interaction picture for the Hamilton operator $\hat{H}$. Assume that $|\psi(0)\rangle^I$ is a coherent state. We can make then the assumption that $|\psi(t)\rangle^I$ is also a coherent state

$$|\psi(t)\rangle^I = e^{i\theta(t)}|\beta(t)\rangle \equiv e^{i\theta(t)}e^{-\frac{1}{2}|\beta(t)|^2}\sum_{n=0}^{\infty}\frac{\beta^n(t)}{\sqrt{n!}}|n\rangle$$

where the number states $|n\rangle$ $(n = 0, 1, \ldots)$ are the eigenstates of $b^\dagger b$ with eigenvalue n. The function $\theta(t)$ is a real function such that $^I\langle\psi(t)|\psi(t)\rangle^I = 1$. Verify this assumption.

Solution 104.   We have

$$\frac{d}{dt}b^I(t) = -\frac{i}{\hbar}[b^I(t), \hat{H}_0] = -i\omega b^I(t).$$

Thus the solution is

$$b^I(t) = b^I(0)e^{-i\omega t}, \qquad b^{\dagger I}(t) = b^{\dagger I}(0)e^{i\omega t}.$$

The Schrödinger equation in the interaction picture is

$$\frac{\partial}{\partial t}|\psi(t)\rangle^I = -\frac{i}{\hbar}V^I(t)|\psi(t)\rangle^I = if(t)(be^{-i\omega t} + b^\dagger e^{i\omega t})|\psi(t)\rangle^I$$

where $f(t) = x_0 F(t)/\hbar$. If $f(t) = 0$, then $|\psi(t)\rangle^I$ is constant. Inserting the ansatz for the coherent state into the right-hand side of the Schrödinger equation yields using $b|n\rangle = \sqrt{n}|n-1\rangle$, $b^\dagger|n\rangle = \sqrt{n+1}|n+1\rangle$

$$-\frac{i}{\hbar}V^I(t)|\psi(t)\rangle^I = if(t)(be^{-i\omega t} + b^\dagger e^{i\omega t})e^{i\theta(t)}e^{-\frac{1}{2}|\beta(t)|^2}\sum_{n=0}^\infty \frac{\beta^n(t)}{\sqrt{n!}}|n\rangle$$

$$= if(t)e^{i\theta(t)}e^{-\frac{1}{2}|\beta(t)|^2}\left(e^{-i\omega t}\sum_{n=0}^\infty \frac{\beta^{n+1}(t)}{\sqrt{n!}}|n\rangle + e^{i\omega t}\sum_{n=1}^\infty \frac{\beta^{n-1}(t)\sqrt{n}}{\sqrt{(n-1)!}}|n\rangle\right).$$

The left-hand side of the Schrödinger equation provides

$$\frac{\partial}{\partial t}|\psi(t)\rangle^I = i\frac{d\theta}{dt}e^{i\theta(t)}e^{-\frac{1}{2}|\beta(t)|^2}\sum_{n=0}^\infty \frac{\beta^n(t)}{\sqrt{n!}}|n\rangle$$

$$-\frac{e^{i\theta(t)}}{2}\left(\frac{d\beta}{dt}\beta^*(t) + \beta(t)\frac{d\beta(t)^*}{dt}\right)e^{-\frac{1}{2}|\beta(t)|^2}\sum_{n=0}^\infty \frac{\beta^n(t)}{\sqrt{n!}}|n\rangle$$

$$+ie^{i\theta(t)}e^{-\frac{1}{2}|\beta(t)|^2}\sum_{n=0}^\infty \frac{n\beta^{n-1}(t)}{\sqrt{n!}}\frac{d\beta}{dt}|n\rangle.$$

We equate the coefficients of $|n\rangle$ and divide by a common factor $e^{i\theta(t)}e^{-|\beta(t)|^2/2}$ to obtain

$$n\left(\frac{d\beta}{dt} - if(t)e^{i\omega t}\right) + i\beta(t)\frac{d\theta}{dt} - \frac{1}{2}\beta(t)\beta^*(t)\frac{d\beta}{dt} - \frac{1}{2}\beta^2(t)\frac{d\beta^*}{dt} - i\beta^2(t)f(t)e^{-i\omega t} = 0.$$

For this equation to be true for all n, the expression in the parenthesis must be equal to 0. This provides a first order differential equation for $\beta(t)$,

$$\frac{d\beta}{dt} = if(t)e^{i\omega t}$$

with the solution of the initial value problem

$$\beta(t) = \beta(0) + i\int_0^t f(t')e^{i\omega t'}\,dt'.$$

Thus we also obtain a differential equation for $\theta(t)$

$$\frac{d\theta}{dt} + \frac{i}{2}\beta^*(t)\frac{d\beta}{dt} + \frac{i}{2}\beta(t)\frac{d\beta^*}{dt} - \beta(t)f(t)e^{-i\omega t} = 0.$$

The imaginary part of this equation is satisfied as long as the ordinary differential equation for $\beta(t)$ is satisfied. The real part of the differential equation is given by

$$\frac{d\theta}{dt} = f(t)(\Re\beta(t)\cos(\omega t) + \Im\beta(t)\sin(\omega t))$$

$$= f(t)(\Re\beta(0)\cos(\omega t) + \Im\beta(t)\sin(\omega t) + \int_0^t \sin(\omega(t-t'))dt')$$

where $\Re$ and $\Im$ stands for the real and imaginary parts. Integrating this differential equation with the initial condition yields

$$\theta(t) = \theta(0) + \Re\beta(0)\int_0^t f(t')\cos(\omega t')dt' + \Im\beta(0)\int_0^t f(t')\sin(\omega t')dt'$$

$$+ \int_0^{t'} dt'f(t')\int_0^{t'} dt''f(t'')\sin(\omega(t'-t'')).$$

Substitution back into the Schrödinger equation verifies the solution and justifies the coherent state ansatz.

Problem 105. (i) Consider the Hamilton function (time-dependent harmonic oscillator)

$$H(p, q, t) = f_1(t)\frac{1}{2}p^2 + f_2(t)pq + f_3(t)\frac{1}{2}q^2$$

where f_1, f_2 and f_3 are smooth functions of t. Find the Hamilton equations of motion and integrate it.
(ii) Consider the functions

$$R_1 = \frac{1}{2}p^2, \quad R_2 = pq, \quad R_3 = \frac{1}{2}q^2$$

and the Poisson bracket $\{\,,\,\}$. Find $\{R_1, R_2\}$, $\{R_1, R_3\}$, $\{R_2, R_3\}$ and thus show they form a basis of a Lie algebra.
(iii) Consider the Hamilton operator for the quantum harmonic oscillator

$$\hat{H}(\hat{p}, \hat{q}, t) = f_1(t)\frac{1}{2}\hat{p}^2 + f_2(t)\frac{1}{2}(\hat{p}\hat{q} + \hat{q}\hat{p}) + f_3(t)\frac{1}{2}\hat{q}^2$$

and the operators

$$\hat{R}_1 = \frac{1}{2}\hat{p}^2, \quad \hat{R}_2 = \frac{1}{2}(\hat{p}\hat{q} + \hat{q}\hat{p}), \quad \hat{R}_3 = \frac{1}{2}\hat{q}^2.$$

Find the commutators $[\hat{R}_1, \hat{R}_2]$, $[\hat{R}_1, \hat{R}_3]$, $[\hat{R}_2, \hat{R}_3]$. Using $\hat{R}_1$, $\hat{R}_2$, $\hat{R}_3$ the Hamilton operator can be written as

$$\hat{H}(p, q, t) = f_1(t)\hat{R}_1 + f_2(t)\hat{R}_2 + f_3(t)\hat{R}_3.$$

Find the quantum evolution operator $U(t)$.

Solution 105. (i) The Hamilton equations of motion can be written in matrix form

$$\frac{d}{dt}\begin{pmatrix} p(t) \\ q(t) \end{pmatrix} = \begin{pmatrix} -f_2(t) & -f_3(t) \\ f_1(t) & f_2(t) \end{pmatrix}\begin{pmatrix} p(t) \\ q(t) \end{pmatrix}.$$

Integrating yields

$$\begin{pmatrix} p(t) \\ q(t) \end{pmatrix} = T\exp\left(\int_0^t \begin{pmatrix} -f_2(\tau) & -f_3(\tau) \\ f_1(\tau) & f_2(\tau) \end{pmatrix} d\tau\right)\begin{pmatrix} p(0) \\ q(0) \end{pmatrix}$$

where T denotes the time-ordered integration. The Poisson bracket provides

$$\{R_1, R_2\} = -2R_1, \quad \{R_1, R_3\} = -R_2, \quad \{R_2, R_3\} = -2R_3.$$

Thus we have a basis of a semi-simple Lie algebra.
(ii) The commutators yield $[\hat{R}_1, \hat{R}_2] = -2i\hat{R}_1$, $[\hat{R}_1, \hat{R}_3] = -i\hat{R}_2$, $[\hat{R}_2, \hat{R}_3] = -2i\hat{R}_3$. The quantum evolution operator is given by

$$U(t) = T\exp\left(-\frac{i}{\hbar}\int_0^t (\sum_{j=1}^3 f_j(\tau)R_j)d\tau\right).$$

Problem 106. Let b_1, b_2 Bose annihilation operators for photons. Describe the Hamilton operator

$$\hat{H} = i\zeta(b_1^\dagger b_2^\dagger - b_1 b_2).$$

Solution 106. The Hamilton operator implements that two photons are simultaneously created or annihilated. The pump process of the amplifier, accounted for the real parameter ζ, must provide the energy source of the photon pair production or the reservoir for annihilation. The parameter ζ can be written as $\zeta = \gamma t$ where t denotes the amplification time and γ the differential gain that depends on the performance of the pump.

Problem 107. The ideal nondegenerate parametric oscillator is described by an interaction Hamilton operator

$$\hat{H} = i\hbar\omega(b_3^\dagger b_1 b_2 - b_3 b_1^\dagger b_2^\dagger)$$

where b_1, b_2 and b_3 are the Boson annihilation operators for the signal, idler, and pump modes, respectively and ω is the parametric coupling constant. Then

$$U = \exp(-i\hat{H}t/\hbar) = e^{\omega t(b_3^\dagger b_1 b_2 - b_3 b_1^\dagger b_2^\dagger)}$$

and therefore

$$U^\dagger = e^{-\omega t(b_3^\dagger b_1 b_2 - b_3 b_1^\dagger b_2^\dagger)}.$$

Calculate

$$\tilde{b}_1 = U b_1 U^\dagger, \quad \tilde{b}_2 = U b_2 U^\dagger, \quad \tilde{b}_3 = U b_3 U^\dagger, \quad \tilde{b}_1^\dagger = U b_1^\dagger U^\dagger, \quad \tilde{b}_2^\dagger = U b_2^\dagger U^\dagger, \quad \tilde{b}_3^\dagger = U b_3^\dagger U^\dagger.$$

Hint. We set ($\epsilon \in \mathbb{R}$)

$$U(\epsilon) := e^{\epsilon\omega t(b_3^\dagger b_1 b_2 - b_3 b_1^\dagger b_2^\dagger)}.$$

$$f_1(\epsilon) = U(\epsilon)b_1 U^\dagger(\epsilon), \quad f_2(\epsilon) = U(\epsilon)b_2 U^\dagger(\epsilon), \quad f_3(\epsilon) = U(\epsilon)b_3 U^\dagger(\epsilon)$$

$$g_1(\epsilon) = U(\epsilon)b_1^\dagger U^\dagger(\epsilon), \quad g_2(\epsilon) = U(\epsilon)b_2^\dagger U^\dagger(\epsilon), \quad g_3(\epsilon) = U(\epsilon)b_3^\dagger U^\dagger(\epsilon)$$

and find the system of ordinary differential equations for f_j, g_j with $j = 1, 2, 3$. Then we solve the system of differential equations together with the initial conditions $f_j(\epsilon = 0) = b_j$ and $g_j(\epsilon = 0) = b_j^\dagger$.

Solution 107. We set $\chi = \omega t$. Using the commutation relation $b_j b_j^\dagger = I + b_j^\dagger b_j$ we find

$$\frac{df_1}{d\epsilon} = \chi U(\epsilon)(b_3 b_2^\dagger)U^\dagger(\epsilon) = \chi U(\epsilon)b_3 U^\dagger(\epsilon)U(\epsilon)b_2^\dagger U^\dagger(\epsilon) = \chi f_3 g_2.$$

Analogously we find

$$\frac{df_2}{d\epsilon} = \chi f_3 g_1, \qquad \frac{df_3}{d\epsilon} = -\chi f_1 f_2.$$

For the functions $g_j(\epsilon)$ we obtain

$$\frac{dg_1}{d\epsilon} = \chi f_2 g_3, \qquad \frac{dg_2}{d\epsilon} = \chi f_1 g_3, \qquad \frac{dg_3}{d\epsilon} = -\chi g_1 g_2.$$

Thus we have a system of six coupled nonlinear ordinary differential equations.

Problem 108. Consider the Hamilton operator $\hat{H} = \hat{H}_0 + \hat{H}_1$ with

$$\hat{H}_0 = \hbar\omega b_1^\dagger b_1 + 2\hbar\omega b_2^\dagger b_2, \quad \hat{H}_1 = \hbar g(b_2^\dagger b_1^2 + b_2(b_1^\dagger)^2)$$

where b_1, b_2 are the Bose annihilation operators of the fundamental and second-harmonic mode, respectively, and the g is the coupling constant between the modes. We set $\hat{N}_1 = b_1^\dagger b_1$ and $\hat{N}_2 = b_2^\dagger b_2$. Calculate the commutator $[\hat{H}_0, \hat{H}_1]$. Discuss.

Solution 108.   We find

$$[\hat{H}_0, \hat{H}_1] = \hbar^2 \omega g[b_1^\dagger b_1 + 2b_2^\dagger b_2, b_2^\dagger b_1^2 + b_2(b_1^\dagger)^2]$$
$$= \hbar^2 \omega g(-2b_1^2 b_2^\dagger + 2b_1^2 b_2^\dagger + 2b_2(b_1^\dagger)^2 - 2b_2(b_1^\dagger)^2)$$
$$= 0.$$

Consequently $\hat{H}_1$ is a constant of motion of $\hat{H}_0$ (or $\hat{N} = \hat{N}_1 + 2\hat{N}_2$ with $n_1, n_2 = 0, 1, 2, \ldots$) or vice versa. Thus we can write

$$\exp(-i\hat{H}t/\hbar) = \exp(-i\hat{H}_0 t/\hbar)\exp(-i\hat{H}_1 t/\hbar).$$

The common eigenstates of the Hamilton operators $\hat{H}_0$ and $\hat{H}_1$ with eigenvalues $E_0 = \hbar\omega N$ and $E_1 = \hbar g M$, respectively, can be given as

$$|N, M\rangle = \sum_{\substack{n_1=0, n_2=0 \\ n_1+2n_2=N}} \langle n_1, n_2 | N, M \rangle |n_1, n_2\rangle$$

where the normalized state $|n_1, n_2\rangle$ is given by

$$|n_1, n_2\rangle = \frac{(b_2^\dagger)^{n_1}(b_1^\dagger)^{n_2}}{\sqrt{n_1! n_2!}}|0, 0\rangle$$

with $|0, 0\rangle$ the vacuum state. Since $\hat{N} = \hat{N}_1 + 2\hat{N}_2$ the Hamilton operator $\hat{H}_0$ splits the number state space into orthogonal vector spaces with $[N/2] + 1$ components, where $[N/2]$ denotes the integer part of $N/2$. Thus we can write

$$|n_1, n_2\rangle = |N - 2k, k\rangle$$

which form a complete set. Using this basis the Hamilton operator H_1 for a given N can be written as a tridiagonal matrix of order $([N/2] + 1) \times ([N/2] + 1)$.

Problem 109.   A nonlinear four-wave mixing device can be described by the Hamilton operator

$$\hat{H} = \frac{1}{4}\hbar\omega(b_1^\dagger b_2 + b_1 b_2^\dagger)^2.$$

(i) Let $|n\rangle$ be a number state. Consider the input state $|n\rangle \otimes |0\rangle$. Find

$$\exp(-i\hat{H}t/\hbar)(|n\rangle \otimes |0\rangle)$$

for the interaction time $t = \pi/\omega$.
(ii) Consider the input state $|n\rangle \otimes |0\rangle$. Find

$$\exp(-i\hat{H}t/\hbar)(|n\rangle \otimes |0\rangle)$$

for the interaction time $t = \pi/(2\omega)$.

Solution 109.   (i) For the interaction time $t = \pi/\omega$ the output state is $i|0\rangle \otimes |n\rangle$ (for n even) and $\exp(-i\pi/4)|n\rangle \otimes |0\rangle$ for n odd. Thus the device acts as an even-odd filter, switching the even numbers from one mode to the other. Under these operating conditions it can be used as a device to measure parity without counting the photon number. It is sufficient to detect any photons in either of the output channels.

(ii) If the interaction time is $t = \pi/(2\omega)$ the output state will have the form

$$\frac{1}{\sqrt{2}}(|n\rangle \otimes |0\rangle + e^{-i(n+1)\pi/2}|0\rangle \otimes |n\rangle)$$

which is a maximally entangled state for n photons.

Problem 110. *Cross phase modulation* is described by the Hamilton operator

$$\hat{H} = -\hbar\omega b_1^\dagger b_1 b_2^\dagger b_2$$

where ω is a function of the third order susceptibility $\chi^{(3)}$. Consider the two-mode number state $|m\rangle \otimes |n\rangle$. Find $\exp(-i\hat{H}t/\hbar)(|m\rangle \otimes |n\rangle)$.

Solution 110. Since $b_1 = b \otimes I$ and $b_2 = I \otimes b$ we have

$$b_1^\dagger b_1|m\rangle \otimes |n\rangle = m|m\rangle \otimes |n\rangle, \quad b_2^\dagger b_2|m\rangle \otimes |n\rangle = n|m\rangle \otimes |n\rangle.$$

Thus

$$\exp(-i\hat{H}t/\hbar)|m\rangle \otimes |n\rangle = e^{i\omega tmn}|m\rangle \otimes |n\rangle.$$

Problem 111. Consider two Bose-Einstein condensates which both occupy the ground-state of their respective traps. They are described by the atom Bose annihilation (creation) operators b_1 ($b_1^\dagger$) and b_2 ($b_2^\dagger$). Atoms are released from each trap with momenta $\mathbf{k}_1$ and $\mathbf{k}_2$, respectively, producing an interference pattern which enables a relative phase to be measured. The intensity $I(\mathbf{x},t)$ of the atomic field is given by

$$I(\mathbf{x},t) = I_0\langle\psi|(b_1^\dagger(t)e^{i\mathbf{k}_1\cdot\mathbf{x}} + b_2^\dagger(t)e^{i\mathbf{k}_2\cdot\mathbf{x}})(b_1(t)e^{-i\mathbf{k}_1\cdot\mathbf{x}} + b_2(t)e^{-i\mathbf{k}_2\cdot\mathbf{x}})|\psi\rangle$$

where I_0 is the single atom intensity. Atoms within each condensate collide. This can be described using the Hamilton operator

$$\hat{H} = \frac{1}{2}\hbar\chi((b_1^\dagger b_1)^2 + (b_2^\dagger b_2)^2)$$

where χ is the collision rate between the atoms within each condensate. Cross-collisions between the two condensates, described by the term $b_1^\dagger b_1 b_2^\dagger b_2$ could also be included. Using the Hamilton operator the intensity $I(\mathbf{x},t)$ is given by

$$I(\mathbf{x},t) = I_0(\langle\psi|b_1^\dagger b_1|\psi\rangle + \langle\psi|b_2^\dagger b_2|\psi\rangle + \langle\psi|b_1^\dagger \exp(i\chi t(b_1^\dagger b_1 - b_2^\dagger b_2))b_2|\psi\rangle e^{-i\phi(\mathbf{x})}$$
$$+ \langle\psi|b_2^\dagger \exp(-i\chi t(b_1^\dagger b_1 - b_2^\dagger b_2))b_1|\psi\rangle e^{i\phi(\mathbf{x})})$$

where $\phi(\mathbf{x}) := (\mathbf{k}_2 - \mathbf{k}_1) \cdot \mathbf{x}$. Calculate $I(\mathbf{x},t)$ for the state $|\psi\rangle = |\beta_1\rangle \otimes |\beta_2\rangle$, where $|\beta_1\rangle$ and $|\beta_2\rangle$ are coherent states.

Solution 111. Since $b|\beta\rangle = \beta|\beta\rangle$, $\langle\beta|b^\dagger = \langle\beta|\beta^*$, and

$$e^{i\chi t b^\dagger b}|\beta\rangle = \sum_{j=0}^\infty \frac{1}{j!}(e^{i\chi t} - 1)(b^\dagger)^j b^j|\beta\rangle = \sum_{j=0}^\infty \frac{1}{j!}(e^{i\chi t} - 1)(b^\dagger)^j \beta^j$$

we have $\langle\beta|e^{i\chi t b^\dagger b}|\beta\rangle = \exp((e^{i\chi t} - 1)\beta^*\beta)$. We also have

$$\langle\beta|e^{i\chi t b^\dagger b}b|\beta\rangle = \beta\exp((e^{i\chi t} - 1)\beta^*\beta), \quad \langle\beta|b^\dagger e^{i\chi t b^\dagger b}|\beta\rangle = \beta^*\exp((e^{i\chi t} - 1)\beta^*\beta).$$

Since $b_1^\dagger b_1 = b^\dagger b \otimes I$ and $b_2^\dagger b_2 = I \otimes b^\dagger b$ we have

$$e^{i\chi t(b_1^\dagger b_1 - b_2^\dagger b_2)} = e^{i\chi t b_1^\dagger b_1} e^{-i\chi t b_2^\dagger b_2} = e^{i\chi t(b^\dagger b \otimes I)} e^{-i\chi t(I \otimes b^\dagger b)}$$
$$= (e^{i\chi t b^\dagger b} \otimes I)(I \otimes e^{-i\chi t b^\dagger b})$$
$$= e^{i\chi t b^\dagger b} \otimes e^{-i\chi t b^\dagger b}.$$

Thus

$$\langle \psi | b_1^\dagger b_1 | \psi \rangle = \beta_1^* \beta_1, \qquad \langle \psi | b_2^\dagger b_2 | \psi \rangle = \beta_2^* \beta_2$$

and

$$\langle \psi | b_1^\dagger \exp(i\chi t(b_1^\dagger b_1 - b_2^\dagger b_2)) b_2 | \psi \rangle e^{-i\phi(\mathbf{x})} = \beta_1^* \beta_2 \exp((e^{i\chi t} - 1)\beta_1^* \beta_1) \exp((e^{-i\chi t} - 1)\beta_2^* \beta_2) e^{-i\phi(\mathbf{x})}$$

$$\langle \psi | b_2^\dagger \exp(-i\chi t(b_1^\dagger b_1 - b_2^\dagger b_2)) b_1 | \psi \rangle e^{i\phi(\mathbf{x})} = \beta_1 \beta_2^* \exp((e^{-i\chi t} - 1)\beta_1^* \beta_1) \exp((e^{i\chi t} - 1)\beta_2^* \beta_2) e^{i\phi(\mathbf{x})}.$$

Problem 112. Let $a_k^\dagger$, $b_k^\dagger$ be Bose creation operator with $k = 1, 2, 3$. Consider the Hamilton operator

$$\hat{H} = c_{12}^{(1)} a_1 a_2^\dagger + c_{13}^{(1)} a_1 a_3^\dagger + c_{23}^{(1)} a_2 a_3^\dagger + c_{12}^{(2)} b_1 b_2^\dagger + c_{13}^{(2)} b_1 b_3^\dagger + c_{23}^{(2)} b_2 b_3^\dagger + \sum_{j=1}^{3} \sum_{k=1}^{3} c_{jk}^{(3)} a_j b_k + \text{h.c.}$$

Does the Hamilton operator $\hat{H}$ commute with the "difference" operator

$$\hat{D} := \sum_{k=1}^{3} a_k^\dagger a_k - \sum_{k=1}^{3} b_k^\dagger b_k ?$$

Solution 112. Yes we find $[\hat{H}, \hat{D}] = 0$.

Problem 113. Consider a particle of mass m and charge q in a harmonic oscillator potential with the Hamilton operator $\hat{H} = \hat{H}_0 + V(t)$, where

$$\hat{H}_0 = \frac{1}{2m}(\hat{p}_1^2 + \hat{p}_2^2 + \hat{p}_3^2) + \frac{1}{2}k(\hat{q}_1^2 + \hat{q}_2^2 + \hat{q}_3^2)$$

and

$$V(t) = qAe^{-(t/\tau)^2} \hat{q}_3$$

At time $t = -\infty$ the harmonic oscillator is in its ground state. Then a perturbation is applied by a spatially uniform time-dependent electric field

$$\mathbf{E}(t) = \begin{pmatrix} 0 \\ 0 \\ Ae^{-(t/\tau)^2} \end{pmatrix}$$

where A and τ are constants. This implies the potential $V(t)$ given above. Find in lowest-order perturbation theory the probability $P(\tau)$ that the oscillator is in an excited state at time $t = +\infty$.

Introduce $\omega = \sqrt{k/m}$ and the Bose operators

$$b_1^\dagger = \left(\frac{m\omega}{2\hbar}\right)^{1/2}\left(q_1 - \frac{i}{m\omega}\hat{p}_1\right), \qquad b_1 = \left(\frac{m\omega}{2\hbar}\right)^{1/2}\left(q_1 + \frac{i}{m\omega}\hat{p}_1\right).$$

Analogously for $b_2^\dagger$, b_2, $b_3^\dagger$, b_3.

Solution 113. The Hamilton operator $\hat{H}_0$ takes the form

$$\hat{H}_0 = \hbar\omega\left(\hat{N}_1 + \hat{N}_2 + \hat{N}_3 + \frac{3}{2}I\right)$$

where $\hat{N}_1 = b_1^\dagger b_1$ etc. The eigenstates of $\hat{H}_0$ are $|n_1n_2n_3\rangle \equiv |n_1\rangle \otimes |n_2\rangle \otimes |n_3\rangle$, where $n_1, n_2, n_3 = 0, 1, 2, \ldots$, i.e.

$$\hat{H}_0|n_1n_2n_3\rangle = \hbar\omega\left(n_1 + n_2 + n_3 + \frac{3}{2}\right)|n_1n_2n_3\rangle.$$

We also have

$$\left(\frac{m\omega}{2\hbar}\right)^{1/2} q_3 = \frac{1}{2}(b_3^\dagger + b_3).$$

Applying the eigenstates $|n_1n_2n_3\rangle$ of $\hat{H}_0$ we can write an arbitrary wave function $|\psi(t)\rangle$ as

$$|\psi(t)\rangle = \sum_{n_1,n_2,n_3=0}^{\infty} c_{n_1n_2n_3}(t)e^{E_{n_1n_2n_3}t/\hbar}$$

where $c_{n_1n_2n_3}(t)$ are time-dependent complex coefficients. Applying time-dependent perturbation theory at $t = +\infty$ up to first order we obtain

$$|c_{n_1n_2n_3}(t)|^2 = \frac{1}{\hbar^2}\left|\int_{-\infty}^{\infty} V_{n_1n_2n_3m_1m_2m_3}(t')e^{i(E_{n_1n_2n_3}-E_{m_1m_2m_3})t'/\hbar}dt'\right|^2$$

where

$$V_{n_1n_2n_3m_1m_2m_3}(t) = \langle n_1n_2n_3|V(t)|m_1m_2m_3\rangle.$$

Let $|m_1m_2m_3\rangle = |000\rangle$. Thus the probability that the system is in any excited state at $t = +\infty$ is given by the sum

$$P(\tau) = \sum_{(n_1n_2n_3)\neq(m_1m_2m_3)} |c_{n_1n_2n_3}(+\infty)|^2.$$

To find this sum we calculate

$$\langle n_1n_2n_3|V(t)|000\rangle = qAe^{-(t/\tau)^2}\langle n_1n_2n_3|q_3|000\rangle.$$

Using $\hat{q}_3$ expressed above with Bose operators we obtain applying $b|0\rangle = 0|0\rangle$

$$\langle n_1n_2n_3|V(t)|000\rangle = qAe^{-(t/\tau)^2}\langle n_1n_2n_3|q_3|000\rangle = qAe^{-(t/\tau)^2}\left(\frac{\hbar}{2m\omega}\right)^{1/2}\delta_{n_1,0}\delta_{n_2,0}\delta_{n_3,1}.$$

Thus only one term in the sum is nonzero. It follows that

$$P(\tau) = \frac{1}{2\hbar m\omega}q^2A^2|I(\tau)|^2 \quad \text{with} I(\tau) = \int_{-\infty}^{\infty} e^{-(t/\tau)^2}e^{i\omega t}dt.$$

This integral can be evaluated providing

$$P(\tau) = \frac{q^2 A^2 \tau^2 \pi}{2m\omega^2 \hbar} e^{-\omega \tau^2/2}.$$

Problem 114. Let $b_1^\dagger$, $b_2^\dagger$ be Bose creation operators. Consider the operators

$$K_+ = b_1^\dagger b_2^\dagger, \quad K_- = b_1 b_2, \quad K_3 = \frac{1}{2}(b_1^\dagger b_1 + b_2^\dagger b_2 + I).$$

(i) Calculate the commutators and thus show that we have a basis of the Lie algebra $su(1,1)$.
(ii) Consider the Hamilton operator

$$\hat{H} = \hbar\omega_1(b_1^\dagger b_1 + b_2^\dagger b_2 + I) + \hbar\omega_2(b_1 b_2 + b_1^\dagger b_2^\dagger).$$

Show that $\hat{H}$ can be expressed using K_+, K_-, K_3.

Solution 114. (i) We have $[K_+, K_-] = -2K_3$, $[K_3, K_\pm] = \pm K_\pm$.
(ii) The Hamilton operator can be written as the linear combination of K_+, K_-, K_3

$$\hat{H} = 2\hbar\omega_1 K_3 + \hbar\omega_2(K_+ + K_-).$$

The initial value problem of the *Schrödinger equation*

$$i\hbar \frac{\partial \psi}{\partial t} = \hat{H}\psi$$

with

$$\psi(t) = e^{-i\hat{H}t/\hbar}\psi(0)$$

can now be solved using the disentangled form

$$e^{-it\hat{H}/\hbar} = e^{a(t)K_3} e^{b(t)K_-} e^{c(t)K_+}.$$

The solution of the Schrödinger equation can be expressed with Hermite polynomials.

Problem 115. Let $b_1^\dagger, \ldots, b_N^\dagger$ be Bose creation operators. Let δ be the delta function. Consider the Hamilton operator $\hat{H} = \hat{H}_0 + \hat{H}_1$, where

$$\hat{H}_0 = c\hat{p} + \epsilon \sum_{j=1}^{N} b_j^\dagger b_j, \quad \hat{H}_1 = g\sum_{j=1}^{N} \delta(\hat{x} - j\ell)b_j^\dagger + c.c.$$

where $\hat{p}$ and $\hat{x}$, respectively, denote the momentum and position operators of the particle. Furthermore, ϵ and g, respectively, denote the energy and the coupling constant and $j\ell$ are the position of the scatterers for the site length ℓ. Find

$$U(t,t') = \exp\left(-i \int_{t'}^{t} \hat{H}_{int}(t'') dt''/\hbar\right)$$

where

$$\hat{H}_{int}(t) = \exp(i\hat{H}_0 t/\hbar)\hat{H}_1 \exp(-i\hat{H}_0 t/\hbar).$$

Find the S matrix

$$S = \lim_{t \to \infty, t' \to -\infty} U(t, t').$$

Solution 115. Since $\exp(\alpha d/dx)f(x) = f(x + \alpha)$ we obtain

$$\hat{H}_{int}(t) = \exp(iH_0 t/\hbar)\hat{H}_1 \exp(-i\hat{H}_0 t/\hbar) = g \sum_{j=1}^{N} \delta(\hat{x} + ct - j\ell)b_j^\dagger(t) + c.c.$$

where $b_j^\dagger(t) = b_j^\dagger \exp(i\epsilon t/\hbar)$. For the S matrix we obtain

$$S = \lim_{t \to \infty, t' \to -\infty} U(t, t') = \exp\left(-i\left(\frac{g}{\hbar c} \sum_{j=1}^{N} b_j^\dagger((j\ell - \hat{x})/c) + c.c.\right)\right).$$

1.6   Linear Optics

Problem 116.   The building blocks of linear optics are phase shifters, half- and quarter-wave plates, beam splitter etc.
(i) The single mode phase shift changes the phase of the electromagnetic field in a given mode. Calculate ($\phi \in \mathbb{R}$)

$$\widetilde{b^\dagger} = e^{i\phi b^\dagger b} b^\dagger e^{-i\phi b^\dagger b}, \qquad \widetilde{b} = e^{i\phi b^\dagger b} b e^{-i\phi b^\dagger b}.$$

(ii) For the *beam splitter* we have the transformation

$$\begin{pmatrix} \widetilde{b_1^\dagger} \\ \widetilde{b_2^\dagger} \end{pmatrix} = \begin{pmatrix} \cos\theta & ie^{-i\phi}\sin\theta \\ ie^{i\phi}\sin\theta & \cos\theta \end{pmatrix} \begin{pmatrix} b_1^\dagger \\ b_2^\dagger \end{pmatrix}.$$

Show that the 2×2 matrix on the right-hand side is unitary.

Solution 116.   (i) We set

$$f(\epsilon) = e^{i\epsilon\phi b^\dagger b} b^\dagger e^{-i\epsilon\phi b^\dagger b}$$

with the initial condition $f(\epsilon = 0) = b^\dagger$. Differentiating f with respect to ϵ and using $bb^\dagger = I + b^\dagger b$ we obtain the linear differential equation

$$\frac{df}{d\epsilon} = i\phi f(\epsilon).$$

The solution is $f(\epsilon) = e^{i\epsilon\phi} b^\dagger$. Thus $\widetilde{b^\dagger} = e^{i\phi} b^\dagger$. Analogously, we have $\widetilde{b} = e^{-i\phi} b$. This can be written in matrix form ($U = \exp(i\phi b^\dagger b)$)

$$U \begin{pmatrix} b \\ b^\dagger \end{pmatrix} U^* \equiv \begin{pmatrix} UbU^* \\ Ub^\dagger U^* \end{pmatrix} = \begin{pmatrix} e^{-i\phi} b \\ e^{i\phi} b^\dagger \end{pmatrix} = \begin{pmatrix} e^{-i\phi} & 0 \\ 0 & e^{i\phi} \end{pmatrix} \begin{pmatrix} b \\ b^\dagger \end{pmatrix}.$$

(ii) We have

$$U = \begin{pmatrix} \cos\theta & ie^{-i\phi}\sin\theta \\ ie^{i\phi}\sin\theta & \cos\theta \end{pmatrix} \Rightarrow U^* = \begin{pmatrix} \cos\theta & -ie^{-i\phi}\sin\theta \\ -ie^{i\phi}\sin\theta & \cos\theta \end{pmatrix}$$

and $U^*U = I_n$. Thus U is unitary.

Problem 117.   The Hilbert space of two modes of an electromagnetic field has the basis $|n_1\rangle \otimes |n_2\rangle$ of joint eigenstates of the photon number operators

$$\hat{N}_1 := b_1^\dagger b_1 \equiv b^\dagger b \otimes I, \qquad \hat{N}_2 := b_2^\dagger b_2 \equiv I \otimes b^\dagger b$$

where $n_1, n_2 = 0, 1, 2, \ldots$. We set

$$j := (n_1 + n_2)/2, \qquad m := (n_1 - n_2)/2$$

and $|jm\rangle \equiv |n_1\rangle \otimes |n_2\rangle$. The state $|jm\rangle$ is the number states with photon numbers $j \pm m$ at modes 1 and 2, respectively. The value of j can be any non-negative half integer and m takes values $-j, -j+1, \ldots, j$ for a given j. Thus for a given m and j we have $n_1 = j + m$ and $n_2 = j - m$.
(i) The beam splitter action on a two-mode state of an electromagnetic field is given by the $SU(2)$ operator

$$\hat{B}(\varphi_1, \varphi_2, \varphi_3) := e^{-i\varphi_1 \hat{J}_z} e^{-i\varphi_2 \hat{J}_y} e^{-i\varphi_3 \hat{J}_z}$$

where the $SU(2)$ generators $\hat{J}_y$ and $\hat{J}_z$ are expressed in the *Schwinger boson representation* as

$$\hat{J}_y = -\frac{i}{2}(b_1^\dagger b_2 - b_2^\dagger b_1), \qquad \hat{J}_z = \frac{1}{2}(b_1^\dagger b_1 - b_2^\dagger b_2).$$

Calculate $\hat{B}(\varphi_1, \varphi_2, \varphi_3)|jm\rangle$.

(ii) Let

$$|\beta_k\rangle = e^{-|\beta_k|^2/2} \sum_{n=0}^{\infty} \frac{\beta_k^n}{\sqrt{n!}} |n\rangle$$

where $k = 1, 2$. Calculate $\hat{B}(\varphi_1, \varphi_2, \varphi_3)(|\beta_1\rangle \otimes |\beta_2\rangle)$.

Solution 117. (i) We find

$$\hat{B}(\varphi_1, \varphi_2, \varphi_3)|jm\rangle = \sum_{m'=-j}^{j} e^{-i(m\varphi_3 + m'\varphi_1)} d_{m'm}^j(\varphi_2)|jm'\rangle$$

where m' in the sum runs from $-j$ to j with unit steps and

$$d_{m'm}^j(\varphi_2) := \langle jm'|e^{-i\varphi_2 \hat{J}_y}|jm\rangle$$

are the $SU(2)$ *Wigner functions*.

(ii) We find $\hat{B}(\varphi_1, \varphi_2, \varphi_3)(|\beta_1\rangle \otimes |\beta_2\rangle) = |\beta_1'\rangle \otimes |\beta_2'\rangle$ where

$$e^{\theta b^\dagger b}|\beta\rangle = |e^\theta \beta\rangle$$
$$\beta_1' = e^{-i\varphi_1/2}(\beta_1 e^{-i\varphi_3/2}\cos(\varphi_2) - \beta_2 e^{i\varphi_3/2}\sin(\varphi_2))$$
$$\beta_2' = e^{i\varphi_1/2}(\beta_1 e^{-i\varphi_3/2}\sin(\varphi_2) + \beta_2 e^{i\varphi_3/2}\cos(\varphi_2)).$$

Problem 118. A *beam splitter* is described by the operator ($\theta \in [0, 2\pi)$)

$$U_{BS}(\theta) = \exp(\theta(b_1^\dagger b_2 - b_1 b_2^\dagger)).$$

It corresponds to a symplectic rotation $B(\theta)$ in phase space given by the 4×4 matrix ($\theta \in [0, 2\pi)$)

$$B(\theta) = \begin{pmatrix} \cos\theta & 0 & -\sin\theta & 0 \\ 0 & \cos\theta & 0 & -\sin\theta \\ \sin\theta & 0 & \cos\theta & 0 \\ 0 & \sin\theta & 0 & \cos\theta \end{pmatrix}.$$

(i) Let K be the 4×4 matrix

$$K := \begin{pmatrix} 0 & 1 \\ -1 & 0 \end{pmatrix} \oplus \begin{pmatrix} 0 & 1 \\ -1 & 0 \end{pmatrix}$$

where $\oplus$ denotes the direct sum. Thus $K^T = -K$. Calculate $B^T(\theta)KB(\theta)$.

(ii) Let J be the 4×4 matrix

$$J := \begin{pmatrix} 0_2 & I_2 \\ -I_2 & 0_2 \end{pmatrix}$$

where I_2 is the 2×2 identity matrix and 0_2 is the 2×2 zero matrix. Calculate $B^T(\theta)JB(\theta)$.

(iii) Can one find a 4×4 permutation matrix P such that $K = P^T J P$?

Solution 118.   (i) We have $B(\theta)^T K B(\theta) = K$.
(ii) We have $B(\theta)^T J B(\theta) = J$.
(iii) We find the permutation matrix

$$
P = \begin{pmatrix} 1 & 0 & 0 & 0 \\ 0 & 0 & 1 & 0 \\ 0 & 1 & 0 & 0 \\ 0 & 0 & 0 & 1 \end{pmatrix}
$$

with $P^T = P$.

Problem 119.   Consider a $50 - 50$ beam splitter. Let $b_1^\dagger$, $b_2^\dagger$ be the incoming modes and $\widetilde{b}_1^\dagger$ and $b_2^\dagger$ denote the outgoing modes. Then the evolution is

$$
b_1^\dagger \rightarrow \frac{1}{\sqrt{2}}(\widetilde{b}_1^\dagger + \widetilde{b}_2^\dagger)|0\rangle.
$$

When a single photon in mode $b_2^\dagger$ impinges on the $50 - 50$ beam splitter, the evolution is

$$
b_2^\dagger \rightarrow \frac{1}{\sqrt{2}}(\widetilde{b}_1^\dagger - b_2^\dagger)|0\rangle.
$$

Find the evolution when two photons impinge together on the beam splitter. Discuss.

Solution 119.   We have

$$
b_2^\dagger b_1^\dagger |0\rangle = \frac{1}{\sqrt{2}}(\widetilde{b}_1^\dagger - b_2^\dagger)\frac{1}{\sqrt{2}}(\widetilde{b}_1^\dagger + b_2^\dagger)|0\rangle = \frac{1}{2}((\widetilde{b}_1^\dagger)^2 - (b_2^\dagger)^2)|0\rangle.
$$

Thus the two photons emerging from the beam splitter are correlated, in a superposition of both being in mode $\widetilde{b}_1^\dagger$ or both being in mode $\widetilde{b}_2^\dagger$. Therefore both photons leave the beam splitter in the same direction. In quantum optics this effect is known as *Hong-Ou-Mandel dip*. The incoming particles must overlap within their coherence length (give definition of coherence length). The coherence length depends on the particle generation and is for photons typically several orders of magnitude larger than their wavelength. For Fermi operators $c_1^\dagger$, $c_2^\dagger$ we have the state

$$
\frac{1}{2}(c_1^\dagger + c_2^\dagger)(c_1^\dagger - c_2^\dagger)|0\rangle = c_2^\dagger c_1^\dagger |0\rangle
$$

since $(c_1^\dagger)^2 = (c_2^\dagger)^2 = 0$. Thus Fermi particles always arrive in separate output port and never bunch together.

Problem 120.   Consider the unitary evolution operator for the *beam splitter*

$$
U_{BS} = \exp(\theta(b_1^\dagger b_2 e^{i\phi} - b_1 b_2^\dagger e^{-i\phi}))
$$

where the real angular parameter θ determines the transmission and reflection coefficients via $T = t^2 = \cos^2(\theta)$ and $R = r^2 = \sin^2(\theta)$. The internal phase shift ϕ between the reflected and transmitted modes is given by the beamsplitter itself. To control ϕ we can place a phase shifter in one of the output channels.
(i) Let $|\psi_{in}\rangle = |0\rangle \otimes |1\rangle$, i.e. the one input is a one-photon state and the other the vacuum state. Calculate $|\psi_{out}\rangle = U_{BS}|\psi_{in}\rangle$.

(ii) To test quantum nonlocality of the state $|\psi_{out}\rangle$ we apply the *displaced parity operator* based on joint parity measurements

$$\hat{\Pi}_{12}(\beta_1, \beta_2) := D_1(\beta_1)D_2(\beta_2) \exp(i\pi(\hat{N}_1 + \hat{N}_2))D_1^\dagger(\beta_1)D_2^\dagger(\beta_2)$$

where $D_1(\beta_1)$ and $D_2(\beta_2)$ are the unitary displacement operators. Calculate

$$\Pi_{12}(\beta_1, \beta_2) := \langle\psi_{out}|\hat{\Pi}_{12}(\beta_1, \beta_2)|\psi_{out}\rangle.$$

(iii) The *two-mode Bell function* $B(\beta_1, \beta_2)$ can be written as

$$B(\beta_1, \beta_2) = \Pi_{12}(0, 0) + \Pi_{12}(\beta_1, 0) + \Pi_{12}(0, \beta_2) - \Pi_{12}(\beta_1, \beta_2).$$

For local realistic theory $B(\beta_1, \beta_2)$ should satisfy the *Bell-CHSH inequality*

$$|B(\beta_1, \beta_2)| \leq 2.$$

The violation of this inequality indicates quantum nonlocality of the single photon entangled state. Calculate $B(\beta_1, \beta_2)$ and discuss the case where $|\beta_1|^2 = |\beta_2|^2$.

(iv) Consider the beam-splitter interaction given by the unitary transformation

$$U = \exp\left(i\frac{\theta}{2}(b_1 b_2^\dagger + b_1^\dagger b_2)\right).$$

Let $|\gamma\rangle$ and $|\beta\rangle$ be coherent states. Calculate the state $U(|\gamma\rangle \otimes |\beta\rangle)$.

(v) Let B be the beam splitter operator

$$B = \exp\left(\frac{\theta}{2}(b_1^\dagger b_2 e^{i\phi} - b_1 b_2^\dagger e^{-i\phi})\right).$$

Let $D(\beta)$ be the displacement operator. Find $BD_1(\beta_1)D_2(\beta_2)|00\rangle$.

Solution 120. (i) We find $|\psi_{out}\rangle = t|0\rangle \otimes |1\rangle + re^{i\phi}|1\rangle \otimes |0\rangle$.
(ii) We obtain

$$\Pi(\beta_1, \beta_2) = \langle\psi_{out}|\hat{\Pi}(\beta_1, \beta_2)|\psi_{out}\rangle = (4|re^{-i\phi}\beta_1 + t\beta_2|^2 - 1)\exp(-2(|\beta_1|^2 + |\beta_2|^2)).$$

(iii) Let $|\beta_1|^2 = |\beta_2|^2 = J$ and let γ_{12} be an arbitrary phase space difference between the two coherent displacements β_1 and β_2. Then we can write $\beta_2 = \beta_1 e^{i\gamma_{12}}$. Thus the two-mode Bell function is given by

$$B(\beta_1, \beta_2) = -1 + (4J - 2)e^{-2J} - (4J - 1)e^{-4J} - 8rtJe^{-4J}\cos(\Delta)$$

where $\Delta := \gamma_{12} + \phi$. When $\gamma_{12} = -\phi$ we obtain the maximal value $|B|_{max}$ of the two mode Bell function

$$|B|_{max} = 1 + (4J - 1)e^{-4J} + 8rtJe^{-4J} - (4J - 2)e^{-2J}.$$

(iv) We find

$$U(|\gamma\rangle \otimes |\beta\rangle) = |\cos(\theta/2)\gamma + i\sin(\theta/2)\beta\rangle \otimes |\cos(\theta/2)\beta + i\sin(\theta/2)\gamma\rangle$$

where $\cos^2(\theta/2)$ $(\sin^2(\theta/2))$ is the reflectivity (transmissivity) of the beamsplitter.
(v) We obtain

$$BD_1(\beta_1)D_2(\beta_2)|00\rangle = D_1(t\beta_1 + re^{i\phi}\beta_2)D_2(t\beta_2 - re^{-i\phi}\beta_1)|00\rangle.$$

1.7   Classical Dynamical Systems

Problem 121.   Consider the nonlinear ordinary differential equation

$$\frac{du}{dt} = -u + u^2.$$

(i) Set $u_n := u^n$ $(n = 1, 2, \ldots)$ and show that

$$\frac{du_n}{dt} = -nu_n + nu_{n+1}.$$

(ii) Set $v_n := u_n/\sqrt{n!}$ and show that

$$\frac{dv_n}{dt} = -nv_n + n\sqrt{n+1}v_{n+1}.$$

We define $v_0 = 0$. Then show that the differential equation can be written as

$$\begin{pmatrix} dv_0/dt \\ dv_1/dt \\ dv_2/dt \\ dv_3/dt \\ \vdots \end{pmatrix} = \begin{pmatrix} 0 & 0 & 0 & 0 & \cdots \\ 0 & -1 & 1\sqrt{2} & 0 & \cdots \\ 0 & 0 & -2 & 2\sqrt{3} & \cdots \\ 0 & 0 & 0 & -3 & \cdots \\ & & \cdots & & \end{pmatrix} \begin{pmatrix} v_0 \\ v_1 \\ v_2 \\ v_3 \\ \vdots \end{pmatrix}.$$

(iii) Consider the infinite dimensional matrix representation of Bose creation and annihilation operators

$$b^\dagger = \begin{pmatrix} 0 & 0 & 0 & 0 & \cdots \\ 1 & 0 & 0 & 0 & \cdots \\ 0 & \sqrt{2} & 0 & 0 & \cdots \\ 0 & 0 & \sqrt{3} & 0 & \cdots \\ & & \cdots & & \end{pmatrix} \Rightarrow b = \begin{pmatrix} 0 & 1 & 0 & 0 & \cdots \\ 0 & 0 & \sqrt{2} & 0 & \cdots \\ 0 & 0 & 0 & \sqrt{3} & \cdots \\ & & \cdots & & \end{pmatrix}.$$

Show that the infinite-dimensional dynamical system can be expressed with the Bose operators.

Solution 121.   (i) Differentiation yields

$$\frac{du_n}{dt} = \frac{d}{dt}u^n = nu^{n-1}\frac{du}{dt} = nu^{n-1}(-u + u^2) = -nu^n + nu^{n+1} = -nu_n + nu_{n+1}.$$

(ii) With $v_n := u_n/\sqrt{n!}$ we obtain

$$\frac{v_n}{dt} = -nv_n + n\sqrt{n+1}v_{n+1}.$$

(iii) We find

$$\frac{dv}{dt} = (b^\dagger bb - b^\dagger b)v.$$

Utilizing the number operator $\hat{N} := b^\dagger b$ we can write $dv/dt = \hat{N}(b - I)v$.

Problem 122.   Consider the differential operators (vector fields)

$$V_1 = \frac{d}{dx}, \qquad V_2 = x\frac{d}{dx}, \qquad V_3 = x^2\frac{d}{dx}$$

corresponding to the differential equations

$$\frac{dx}{d\epsilon} = 1, \quad \frac{dx}{d\epsilon} = x, \quad \frac{dx}{d\epsilon} = x^2$$

with the solutions of the initial value problems

$$x(\epsilon) = x_0 + \epsilon, \quad x(\epsilon) = e^{\epsilon t} x_0, \quad x(\epsilon) = \frac{x_0}{1 - \epsilon x_0}.$$

(i) Show that these differential operators form a Lie algebra under the commutator.

(ii) Consider the Bose operators

$$K_1 = b, \quad K_2 = b^\dagger b, \quad K_3 = b^\dagger b^\dagger b.$$

Show that these operators form a basis of a Lie algebra.

(iii) Show that the two Lie algebras are isomorphic.

Solution 122. (i) We find

$$[V_1, V_2] = V_1, \quad [V_1, V_3] = 2V_2, \quad [V_2, V_3] = V_3.$$

(ii) Since $[b, b^\dagger] = I$, where I is the identity operator we obtain

$$[K_1, K_2] = K_1, \quad [K_1, K_3] = 2K_2, \quad [K_2, K_3] = K_3.$$

(iii) We set

$$V_1 \to K_1, \quad V_2 \to K_2, \quad V_3 \to K_3.$$

Problem 123. Let $f : \mathbb{R} \to \mathbb{R}$ be an analytic function. One dimensional maps

$$x_{t+1} = f(x_t), \quad t = 0, 1, 2, \ldots$$

can be embedded into a linear system

$$|x, t + 1\rangle = \hat{M} |x, t\rangle$$

using Bose operators and coherent states. Here

$$|x, t\rangle := \exp\left(\frac{1}{2}(x_t^2 - x_0^2)\right) |x_t\rangle$$

with

$$|x_t\rangle := \exp\left(-\frac{1}{2}x_t^2\right) \exp(x_t b^\dagger) |0\rangle.$$

The evolution operator $\hat{M}$ is given by

$$M := \sum_{j=0}^{\infty} \frac{b^\dagger}{j!} (f(b) - b)^j.$$

Study *fixed points* of the map f within this approach. The fixed points x^* are defined as solutions of the equation $x^* = f(x^*)$.

Solution 123.   Since for a fixed point of a map f we have $f(x_t) = x_t$ it follows that

$$|x, t\rangle = \hat{M}|x, t\rangle.$$

Thus

$$\exp\left(\frac{1}{2}(x_t^2 - x_0^2)\right)|x_t\rangle = \hat{M}\exp\left(\frac{1}{2}(x_t^2 - x_0^2)\right)|x_t\rangle$$

or

$$|x_t\rangle = \hat{M}|x_t\rangle.$$

It follows that

$$\exp\left(-\frac{1}{2}x_t^2\right)\exp(x_t b^\dagger)|0\rangle = \hat{M}\exp\left(-\frac{1}{2}x_t^2\right)\exp(x_t b^\dagger)|0\rangle$$

or

$$\exp(x_t b^\dagger)|0\rangle = \hat{M}\exp(x_t b^\dagger)|0\rangle.$$

Inserting M yields

$$\exp(x_t b^\dagger)|0\rangle = \sum_{j=0}^{\infty}\frac{(b^\dagger)^j}{j!}(f(b) - b)^j\exp(x_t b^\dagger)|0\rangle.$$

It follows that

$$\exp(x_t b^\dagger)|0\rangle = \sum_{j=0}^{\infty}\frac{(b^\dagger)^j}{j!}(f(x_t) - x_t)^j\exp(x_t b^\dagger)|0\rangle.$$

Thus if $f(x_t) = x_t$ (fixed point) the sum on the right-hand side gives only a contribution at $j = 0$ and with $0^0 = 1$ we obtain the identity

$$\exp(x_t b^\dagger)|0\rangle = \exp(x_t b^\dagger)|0\rangle.$$

Problem 124.   Consider the autonomous system of first order ordinary differential equations

$$\frac{d\mathbf{u}}{dt} = V(\mathbf{u})$$

where $V : \mathbb{C}^n \to \mathbb{C}^n$ is analytic. The analytic vector fields

$$V = V_1\frac{\partial}{\partial u_1} + \cdots + V_n\frac{\partial}{\partial u_n}$$

form a Lie algebra under the commutator. Consider the map

$$V \to M^\dagger := V(\mathbf{b}^\dagger) \cdot \mathbf{b} \equiv \sum_{j=1}^{n}V_j(\mathbf{b}^\dagger)b_j.$$

(i) Show that the operators $M^\dagger$ form a Lie algebra under the commutator.
(ii) Show that the two Lie algebras are isomorphic.

Solution 124.   (i) From

$$V \to M^\dagger = V(\mathbf{b}^\dagger) \cdot \mathbf{b} \equiv \sum_{j=1}^{n}V_j(\mathbf{b}^\dagger)b_j$$

and

$$W \to N^{\dagger} = W(\mathbf{b}^{\dagger}) \cdot \mathbf{b} \equiv \sum_{j=1}^{n} W_j(\mathbf{b}^{\dagger}) b_j$$

we obtain

$$[V, W] = \sum_{k=1}^{n} \sum_{j=1}^{n} \left(V_j \frac{\partial W_k}{\partial u_j} - W_j \frac{\partial V_k}{\partial u_j} \right) \frac{\partial}{\partial u_k}$$

and

$$[M^{\dagger}, N^{\dagger}] = \sum_{k=1}^{n} \sum_{j=1}^{n} \left(V_j(\mathbf{b}^{\dagger}) \frac{\partial W_k(\mathbf{b}^{\dagger})}{\partial b_j^{\dagger}} - W_j(\mathbf{b}^{\dagger}) \frac{\partial V_k(\mathbf{b}^{\dagger})}{\partial b_j^{\dagger}} \right) \frac{\partial}{\partial u_k}.$$

Problem 125. The *Lotka-Volterra model* is given by

$$\frac{du_1}{dt} = -u_1 + u_1 u_2, \qquad \frac{du_2}{dt} = u_2 - u_1 u_2.$$

The vector field associated with the Lotka Volterra model is

$$V(u) = (-u_1 + u_1 u_2) \frac{\partial}{\partial u_1} + (u_2 - u_1 u_2) \frac{\partial}{\partial u_2}.$$

(i) Show that Lotka-Volterra admits the first first integral

$$I(u) = u_1 u_2 \exp(-u_1 - u_2).$$

(ii) The operator $M^{\dagger}$ is given by

$$M^{\dagger} = (-b_1^{\dagger} + b_1^{\dagger} b_2^{\dagger}) b_1 + (b_2^{\dagger} - b_1^{\dagger} b_2^{\dagger}) b_2.$$

Show that the first integral $I(u)$ takes the following form using Bose operators

$$|\phi\rangle \equiv b_1^{\dagger} b_2^{\dagger} \exp(-b_1^{\dagger} - b_2^{\dagger}) |0\rangle.$$

Solution 125. (i) We find

$$L_V I = \left((-u_1 + u_1 u_2) \frac{\partial}{\partial u_1} + (u_2 - u_1 u_2) \frac{\partial}{\partial u_2} \right) (u_1 u_2 \exp(-u_1 - u_2)) = 0$$

where $L_V(.)$ denotes the Lie derivative.

(ii) Using the commutation relations and that fact that

$$b \exp(\alpha b^{\dagger}) |0\rangle = \alpha \exp(\alpha b^{\dagger}) |0\rangle$$

we find that $M^{\dagger} |\phi\rangle = 0|0\rangle$.

1.8   Supplementary Problems

Problem 1.   Let $b^\dagger$, b be Bose creation and annihilation operators, $\hat{N} = b^\dagger b$ be the number operator, I be the identity operator and $z \in \mathbb{C}$.
(i) Show that

$$[b, \hat{N}^2] = (I + 2b^\dagger b)b, \qquad [b^\dagger, \hat{N}^2] = -b^\dagger(I + 2b^\dagger b).$$

(ii) Let $k \geq 1$. Show by induction that

$$b^\dagger(b^\dagger b)^k = (b^\dagger b - I)^k b^\dagger.$$

For $k = 1$ we have

$$b^\dagger b^\dagger b = b^\dagger(bb^\dagger - I) = b^\dagger bb^\dagger - b^\dagger = (b^\dagger b - I)b^\dagger.$$

(iii) Let $m \geq 1$ and let $f : \mathbb{R} \to \mathbb{R}$ be an analytic function. Show that

$$(b^\dagger)^m f(\hat{N}) = f(\hat{N} - mI)(b^\dagger)^m, \quad b^m f(\hat{N}) = f(\hat{N} + mI)b^m.$$

(iv) Show that

$$[b, (bb^\dagger)^k] = \sum_{\ell=1}^{k} \binom{k}{\ell}(bb^\dagger)^{k-\ell}b, \qquad [b^\dagger, (bb^\dagger)^k] = \sum_{\ell=1}^{k} \binom{k}{\ell}(-1)^\ell(bb^\dagger)^{k-\ell}b^\dagger.$$

(v) Let k be a positive integer. Show that $[\hat{N}, b^k] = -kb^k$.

Problem 2.   (i) Show that

$$\exp(\beta(b^\dagger + b^k)) = \exp(\beta b^\dagger)\exp\left(\sum_{j=0}^{k} \frac{1}{j+1}\binom{k}{j}\beta^{j+1}b^{k-j}\right).$$

(ii) Find the operator

$$\sinh(zb^\dagger b) - \sinh(zbb^\dagger).$$

(iii) Let $\epsilon \in \mathbb{R}$. Find the operator

$$f(\epsilon) = \cos(-\epsilon b^\dagger b)b\cos(\epsilon b^\dagger b)$$

using *parameter differentiation*. Note that $f(0) = b$.
(iv) Let $f : \mathbb{C} \to \mathbb{C}$ be an analytic function. Show that

$$\exp(zb^\dagger b)(f(b, b^\dagger))\exp(-zb^\dagger b) = f(be^{-z}, b^\dagger e^z).$$

Problem 3.   Let $\mathbb{N}_0$ be the natural numbers including 0. Consider the separable Hilbert space $\ell_2(\mathbb{N}_0)$. Then the number states $|n\rangle$ $(n = 0, 1, 2, \ldots)$ form a basis in this Hilbert space. Consider the two sub Hilbert spaces

$$\mathcal{H}_0 := \{\, |2n\rangle \ : \ n = 0, 1, 2, \ldots \,\}$$
$$\mathcal{H}_1 := \{\, |2n + 1\rangle \ : \ n = 0, 1, 2, \ldots \,\}.$$

The *projection operators* onto these Hilbert spaces are given by

$$\hat{\Pi}_0 = \sum_{n=0}^{\infty} |2n\rangle\langle 2n|, \qquad \hat{\Pi}_1 = \sum_{n=0}^{\infty} |2n + 1\rangle\langle 2n + 1|.$$

One has $\hat{\Pi}_0 + \hat{\Pi}_1 = I$ and $\hat{\Pi}_j \hat{\Pi}_k = \delta_{jk} \hat{\Pi}_j$, where I is the identity matrix.
(i) Show that the *parity operator* is given by

$$\hat{P} = \hat{\Pi}_0 - \hat{\Pi}_1 = \exp(i\pi b^\dagger b).$$

(ii) Show that $\hat{P} = I$, $\hat{P}b = -b\hat{P}$, $\hat{P}b^\dagger = -b^\dagger \hat{P}$.
(iii) Let $|n\rangle$ be the number states ($n = 0, 1, \ldots$). Let $k = 0, 1, \ldots$. Define the operators

$$T_k := \sum_{n=0}^{\infty} |n\rangle\langle 2n + k|.$$

(iv) Show that $T_k T_{k'}^\dagger = \delta_{kk'} I$.
(v) Show that $T_k^\dagger T_k = P_k$ is a projection operator.
(vi) Show that $\sum_{k=0}^{\infty} P_k = I$.
(vii) Is the operator

$$\sum_{k=0}^{\infty} T_k \otimes T_k^\dagger$$

unitary?

Problem 4. Let $|n\rangle$ be the number states ($n = 0, 1, \ldots$). Let $z \in \mathbb{C}$ and $z = x + iy$. Consider the state

$$|z\rangle = M(|z|^2) \sum_{n=0}^{\infty} \frac{z^n}{\sqrt{\rho_n}} |n\rangle$$

where ($u \geq 0$)

$$M^2(u) = \frac{\sqrt{u}}{\sinh(\sqrt{u})}$$

and

$$\rho_n = \int_0^\infty u^n \rho(u) du, \qquad \rho(u) = \frac{1}{2} \exp(-\sqrt{u}).$$

Thus $\rho_n = (2n + 1)!$. Let $g(u)$ be defined by $\rho(u) = M^2(u)g(u)$. Show that

$$\langle z|z\rangle = 1, \qquad I = \frac{1}{\pi} \int_{\mathbb{C}} |z\rangle\langle z| g(|z|^2) dx dy.$$

Problem 5. Let $\hat{Q}$ and $\hat{P}$ be the self-adjoint operators acting in an appropriate subspace of the Hilbert space $L_2(\mathbb{R})$ with the canonical commutation relation

$$[\hat{P}, \hat{Q}] = -i\hbar I$$

where I is the identity operator.
(i) Show that

$$\exp(i(p'\hat{Q} - q'\hat{P})/\hbar) \exp((i(p\hat{Q} - q\hat{P})/\hbar) = \exp(i(p'q - pq')/(2\hbar)) \exp(i((p+p')\hat{Q} - (q+q')\hat{P})/\hbar).$$

(ii) Let n be a positive integer. Show that

$$[\hat{P}^n, \hat{Q}] = -i\hbar n \hat{P}^{n-1}, \qquad [\hat{P}, \hat{Q}^{n-1}] = -i\hbar n \hat{Q}^{n-1}.$$

(iii) Let n, m be positive integers. Show that

$$[\hat{P}^m, \hat{Q}^n] = \sum_{j=1}^{\min(m,n)} \binom{m}{j}\binom{n}{j}(-i\hbar)^j j! \hat{Q}^{n-j} \hat{P}^{m-j}.$$

(iv) Consider the Hamilton operator for the one-dimensional harmonic oscillator

$$\hat{H}(\hat{P}, \hat{Q}) = \frac{1}{2m}(\hat{P}^2 + m^2\omega^2\hat{Q}^2)$$

and

$$\hat{T} = \exp(-m\omega\hat{Q}^2/(2\hbar)).$$

Show that

$$\tilde{H} = \hat{T}^{-1}\hat{H}\hat{T} = \frac{1}{2m}\hat{P}^2 + \frac{1}{2}i\omega(\hat{Q}\hat{P} + \hat{P}\hat{Q}).$$

Problem 6.   Consider the *parity operator*

$$\hat{P} = \exp(i\pi b^\dagger b) \equiv \exp(i\pi \hat{N}).$$

(i) Find the spectrum of $\hat{P}$.
(ii) Find the commutators $[b^\dagger, \hat{P}]$, $[b^\dagger, \hat{P}]$, $[b^\dagger + b, \hat{P}]$.
(iii) Find the spectrum of the operator $b^\dagger + b$. Set

$$b = \frac{1}{\sqrt{2}}(x + \frac{d}{dx}), \quad b^\dagger = \frac{1}{\sqrt{2}}(x - \frac{d}{dx}).$$

Problem 7.   In the Hilbert space $\mathcal{H} = \ell_2(\mathbb{N}_0)$ Bose annihilation and creation operators denoted by b and $b^\dagger$ are defined as follows: They have a common domain

$$\mathcal{D}(b) = \mathcal{D}(b^\dagger) = \left\{ \xi = (x_0, x_1, x_2, \ldots)^T : \sum_{j=0}^{\infty} j|x_j|^2 < \infty \right\}.$$

Then $b\xi$ is given by

$$b(x_0, x_1, x_2, \ldots)^T = (x_1, \sqrt{2}x_2, \sqrt{3}x_3, \ldots)^T$$

and $b^\dagger\xi$ is given by

$$b^\dagger(x_0, x_1, x_2, \ldots) = (0, x_0, \sqrt{2}x_1, \sqrt{3}x_2, \ldots).$$

The infinite dimensional vectors

$$u_n = (0, 0, \ldots, 0, 1, 0, \ldots)^T$$

where the 1 is at the n position ($n = 0, 1, 2, \ldots$) form the standard basis in $\mathcal{H} = \ell_2(\mathbb{N}_0)$. Is

$$\xi = (1, 1/2, 1/3, \ldots, 1/n, \ldots)$$

an element of $\mathcal{D}(a)$? We have to check that

$$\sum_{j=0}^{\infty} j|x_j|^2 = \sum_{j=1}^{\infty} j|x_j|^2 = \sum_{j=1}^{\infty} j\left(\frac{1}{j+1}\right)^2 < \infty.$$

Problem 8. (i) Study the Hamilton operator

$$\hat{H}(t) = \hbar\omega_1 b^\dagger b + \hbar\omega_2(e^{i\omega t}b^\dagger + e^{-i\omega t}b).$$

(ii) Show that the Hamilton operator

$$\hat{H} = \hbar\omega_1 b^\dagger b + \hbar\omega_2((b^\dagger)^2 e^{-2i\omega t} + b^2 e^{2i\omega t})$$

generates squeezed states.

(iii) Let $f : \mathbb{R} \to \mathbb{R}$ be an analytic function. Let $|n\rangle$, $\beta\rangle$, $|\zeta\rangle$ be number states, coherent states and squeezed states, respectively and

$$\hat{H} = \hbar\omega_1 b^\dagger b + \hbar\omega_2(b + b^\dagger) + \hbar\omega_3 f(b^\dagger b).$$

Find $\langle n|\hat{H}|n\rangle$, $\langle \beta|\hat{H}|\beta\rangle$, $\langle \zeta|\hat{H}|\zeta\rangle$.

(iv) Consider the Hamilton operators

$$\hat{H}_0 = \hbar\omega b^\dagger b + \epsilon(b + b^\dagger), \quad \hat{H}_1 = \frac{\hbar\Omega}{2}\cos(\pi b^\dagger b).$$

Let $D(\beta)$ be the displacement operator. Find $D(\beta)\hat{H}_0 D^\dagger(\beta)$ and set $\beta = \epsilon/(\hbar\omega)$. Discuss. Show that $\hat{H}_1 D^\dagger(\beta) = D(\beta)\hat{H}_1$. Hint. Apply $b\cos(\pi b^\dagger b) = -\cos(\pi b^\dagger b)b$.

Problem 9. Let $|0\rangle$ be the vacuum states and $t \in \mathbb{R}$. Find

$$[e^{tb^2/2}, e^{t(b^\dagger)^2}]|0\rangle$$

utilizing the formula

$$[f(b), g(b^\dagger)] = \sum_{j=1}^{\infty} \frac{1}{j!}\frac{\partial^j}{\partial b^{\dagger j}}g(b^\dagger)\frac{\partial^j}{\partial b^j}f(b)$$

where $f : \mathbb{R} \to \mathbb{R}$ and $g : \mathbb{R} \to \mathbb{R}$ are analytic functions and $b|0\rangle = 0|0\rangle$.

Problem 10. The squeezed coherent density operator ρ is defined as

$$\rho = S(\zeta)D(\beta)|0\rangle\langle 0|D^\dagger(\beta)S^\dagger(\zeta)$$

where

$$D(\eta) := \exp(\beta b^\dagger - \beta^* b), \quad S(\zeta) := \exp(\frac{1}{2}(\zeta^* b^2 - \zeta(b^\dagger)^2)).$$

Calculate the expectation value $\langle 0|\rho|0\rangle$.

Problem 11. Let $\beta \in \mathbb{C}$ and $\zeta \in \mathbb{C}$. Consider the displacement operator $D(\beta)$ and squeeze operator $S(\zeta)$, respectively

$$D(\beta) := \exp(\beta b^\dagger - \bar{\beta}b), \quad S(\zeta) := \exp(\frac{1}{2}\bar{\zeta}b^2 - \frac{1}{2}\zeta(b^\dagger)^2).$$

(i) Find the commutator

$$[\beta b^\dagger - \beta^* b, \zeta^* b^2 - \zeta(b^\dagger)^2].$$

(ii) Find the commutator $[D(\beta), S(\zeta)]$.

(iii) Can one find $\beta, \zeta \in \mathbb{C}$ such that the commutator $[D(\beta), S(\zeta)]$ is a unitary operator?

(iv) Show that

$$\langle\beta|b|\zeta\rangle = -\bar{\beta}e^{i\theta}\tanh(s)\langle\beta|\zeta\rangle.$$

Problem 12.   (i) Coherent states $|\beta\rangle$ can be expressed using number states $|n\rangle$, i.e.

$$|\beta\rangle = e^{-|\beta|^2/2}\sum_{n=0}^{\infty}\frac{\beta^n}{\sqrt{n!}}|n\rangle.$$

Express number states $|n\rangle$ using coherent states $|\beta\rangle$.
(ii) Show that the projection on a coherent state $|\beta\rangle$ is given by

$$|\beta\rangle\langle\beta| =: e^{-(b-\beta I)^\dagger(b-\beta I)} :$$

where : : means normal ordering.
(ii) Let $|n\rangle$, $|m\rangle$ $(n, m = 0, 1, 2, \ldots)$ be number states. Consider the squeezed displaced number states

$$|\beta, \zeta, n\rangle := D(\beta)S(\zeta)|n\rangle.$$

Find $\langle m|D(\beta)S(\zeta)|n\rangle$.

Problem 13.   Let $\hat{N} := b^\dagger b$. The *displaced parity operator* is defined by

$$\Pi(\beta) := D(\beta)(-1)^{\hat{N}}D^\dagger(\beta)$$

where

$$(-1)^{\hat{N}} \equiv e^{i\pi\hat{N}}.$$

Consider the number states $|0\rangle$ and $|1\rangle$. Show that

$$D^\dagger(\beta)|0\rangle = e^{-(1/2)|\beta|^2}\sum_{m=0}^{\infty}(-\beta)^m\frac{1}{\sqrt{m!}}|m\rangle$$

and

$$D^\dagger(\beta)|1\rangle = e^{-(1/2)|\beta|^2}\left(\sum_{m=0}^{\infty}(-\beta)^m\beta^*\frac{1}{\sqrt{m!}}|m\rangle + \sum_{m=0}^{\infty}(-\beta)^m\frac{\sqrt{m+1}}{\sqrt{m!}}|m+1\rangle\right).$$

Then calculate $\Pi(\beta)|0\rangle$ and $\Pi(\beta)|1\rangle$.

Problem 14.   Let $b_1^\dagger$, $b_2^\dagger$, b_1, b_2 be Bose creation and annihilation operators.
(i) Let

$$T = \frac{1}{2i}(b_1^\dagger b_2 - b_2^\dagger b_1)$$

and $\alpha \in \mathbb{R}$. Show that

$$e^{i\alpha T}\begin{pmatrix} b_1^\dagger \\ b_2^\dagger \end{pmatrix}e^{-i\alpha T} = \begin{pmatrix} e^{i\alpha T}b_1^\dagger e^{-i\alpha T} \\ e^{i\alpha T}b_2^\dagger e^{-i\alpha T} \end{pmatrix} = \begin{pmatrix} \cos(\alpha/2) & -\sin(\alpha/2) \\ \sin(\alpha/2) & \cos(\alpha/2) \end{pmatrix}\begin{pmatrix} b_1^\dagger \\ b_2^\dagger \end{pmatrix}.$$

(ii) Let $\theta \in \mathbb{R}$. Show that

$$e^{-i\theta(b_1^\dagger b_2 + b_2^\dagger b_1)}b_1 e^{i\theta(b_1^\dagger b_2 + b_2^\dagger b_1)} = \cos(\theta)b_1 + i\sin(\theta)b_2$$
$$e^{-i\theta(b_1^\dagger b_2 + b_2^\dagger b_1)}b_2 e^{i\theta(b_1^\dagger b_2 + b_2^\dagger b_1)} = i\sin(\theta)b_1 + \cos(\theta)b_2.$$

Problem 15. Bose operators obey the commutation relations

$$[b_j, b_k] = [b_j^\dagger, b_k^\dagger] = 0, \quad [b_j, k_k^\dagger] = \delta_{jk} I.$$

We have

$$[b_j, f(b_k^\dagger)] = \delta_{jk} \frac{\partial}{\partial b_j^\dagger} f(b_j^\dagger)$$

where f is an analytic function.

(i) Do the operators

$$b_1^\dagger b_1, \quad b_2^\dagger b_2, \quad b_1^\dagger b_2, \quad b_2^\dagger b_1$$

commute with the operator

$$b_1^\dagger b_2 b_2^\dagger b_1 + b_2^\dagger b_1 b_1^\dagger b_2 ?$$

(ii) Show that

$$[b_j, \exp(b_j^\dagger B)] = B \exp(b_j^\dagger B)$$

where the operator B does not depend on $b_j^\dagger$.

(iii) Show that

$$b_j f(b_k^\dagger)|0\rangle = \left(\delta_{jk} \frac{d}{db_j^\dagger} f(b_j^\dagger + f(b_k^\dagger) b_j \right) |0\rangle = \delta_{jk} \frac{d}{db_j^\dagger} f(b_j^\dagger)|0\rangle.$$

(iv) Show that

$$b_j^\dagger \exp(b_j^\dagger B)|0\rangle = B^k |0\rangle$$

where B does depend on $b_j^\dagger$.

Problem 16. Let b_j, $b_j^\dagger$ $(j = 1, 2)$ be Bose annihilation and creation operators, respectively. Let $\hat{N}_j := b_j^\dagger b_j$ and the Hamilton operator

$$\hat{H} = \frac{1}{2} \sum_{j=1}^{2} [b_j^\dagger, b_j]_+ \equiv \hat{N}_1 + \hat{N}_2 + I.$$

We define (*Schwinger representation*)

$$J_+ := b_1^\dagger b_2, \quad J_- := b_2^\dagger b_1, \quad J_3 := \frac{1}{2}(\hat{N}_1 - \hat{N}_2).$$

(i) Find the commutators $[J_+, J_-]$, $[J_+, J_3]$, $[J_-, J_3]$.
(ii) Express $\hat{H}^2$ using the operators J_-, J_+, J_3.

Problem 17. Consider the Hamilton operator

$$\hat{H} = \hbar\omega_1(b_1^\dagger b_1^\dagger b_1 b_1 + b_2^\dagger b_2^\dagger b_2 b_2 + b_3^\dagger b_3^\dagger b_3 b_3) + \hbar\omega_2(b_1^\dagger b_2 + b_2^\dagger b_1 + b_3^\dagger b_2 + b_2^\dagger b_3).$$

(i) Show that the Hamilton operator commutes with the number operator

$$\hat{N} = b_1^\dagger b_1 + b_2^\dagger b_2 + b_3^\dagger b_3.$$

(ii) Show that the Heisenberg equations of motion are given by

$$i\frac{db_1}{dt} = 2\omega_1 b_1^\dagger b_1 b_1 + \omega_2 b_2$$

$$i\frac{db_2}{dt} = 2\omega_1 b_2^\dagger b_2 b_2 + \omega_2(b_1 + b_3)$$

$$i\frac{db_3}{dt} = 2\omega_1 b_3^\dagger b_3 b_3 + \omega_2 b_2.$$

Problem 18.   Let $n \in \mathbb{N}_0$. Consider the normalized state

$$|\psi\rangle = \frac{1}{\sqrt{2}}(|n\rangle \otimes |0\rangle + e^{in\theta}|0\rangle \otimes |n\rangle)$$

where $|0\rangle$ is the vacuum state and $|n\rangle$ is the number state. Show that the minimum uncertainty achievable by a suitable quantum measurement on these states is proportional to $1/n$.

Problem 19.   Let $|\beta\rangle$ be a coherent state. Consider the four *Bell-cat states*

$$|B_{00}\rangle = \frac{1}{\sqrt{2}}(|-\beta\rangle \otimes |-\beta\rangle + |\beta\rangle \otimes |\beta\rangle)$$

$$|B_{10}\rangle = \frac{1}{\sqrt{2}}(|-\beta\rangle \otimes |-\beta\rangle - |\beta\rangle \otimes |\beta\rangle)$$

$$|B_{01}\rangle = \frac{1}{\sqrt{2}}(|-\beta\rangle \otimes |\beta\rangle + |\beta\rangle \otimes |-\beta\rangle)$$

$$|B_{11}\rangle = \frac{1}{\sqrt{2}}(|-\beta\rangle \otimes |\beta\rangle - |\beta\rangle \otimes |-\beta\rangle).$$

Are the states orthogonal to each other? Note that $\langle\beta|\beta\rangle = 1$.

Problem 20.   Consider the system of n canonical degrees of freedom representing n harmonic oscillators. We arrange these operators in vector form (column vector)

$$R := (X_1, P_1, X_2, P_2, \dots, X_n, P_n)^T.$$

The canonical commutation relations are

$$[R_j, R_k] = i\sigma_{jk}, \qquad j, k = 1, 2, \dots, 2n$$

where the *symplectic matrix* σ is defined by

$$\sigma := \bigoplus_{j=1}^{n} \begin{pmatrix} 0 & 1 \\ -1 & 0 \end{pmatrix}.$$

Here $\oplus$ denotes the direct sum. Density operators ρ can now be characterized by functions that are defined on phase space. Given a column vector $\boldsymbol{\xi}$ the *Weyl operator* (also called *Glauber operator*) is defined by

$$W(\boldsymbol{\xi}) := \exp(i\boldsymbol{\xi}^T \sigma R).$$

These operators generate displacements in phase space. They are used to define the *characteristic function* $\chi_\rho(\boldsymbol{\xi})$ of the density operator ρ

$$\chi_\rho(\boldsymbol{\xi}) := \mathrm{tr}(\rho W(\boldsymbol{\xi})).$$

Show that this can be inverted to express ρ as an integral of $\chi(\boldsymbol{\xi})$, i.e. show that

$$\rho = \frac{1}{(2\pi)^n} \int_{\mathbb{R}^{2n}} \chi_\rho(-\boldsymbol{\xi}) W(\boldsymbol{\xi}) d^{2n}\boldsymbol{\xi}.$$

Problem 21. Consider a quantum mechanical system where the Hamilton operator $\hat{H}(t)$ depends explicitly on t. Then the *Heisenberg equation of motion* for a linear operator $\hat{O}(t)$ takes the form

$$\frac{d}{dt}\hat{O}(t) = \frac{\partial}{\partial t}\hat{O}(t) + \frac{1}{i\hbar}[\hat{O}(t), \hat{H}(t)]$$

where d/dt is the total time derivative whereas $\partial/\partial t$ differentiates only the parametric time derivative. Consider the Hamilton operator

$$\hat{H}(t) = \begin{pmatrix} \hbar\omega_1 & 0 \\ 0 & \hbar\omega_2 \end{pmatrix} + \begin{pmatrix} w_{11} & w_{12} \\ w_{21} & w_{22} \end{pmatrix} \cos(\omega t)$$

where the w_{jk} are real and $w_{12} = w_{21}$ and $d\cos(\omega t)/dt = -\omega\sin(\omega t)$.

Problem 22. (i) Show that two-mode squeezed states can be generated either by entangling two independent single-mode squeezed states via a 50:50 beamsplitter or by employing the non-degenerate operation of a nonlinear medium in the presence of two incoming modes and the unitary operator describing two-mode squeezing is

$$U_{12}(\zeta) = \exp(-i(\zeta b_1 b_2 + \zeta^* b_1^\dagger b_2^\dagger)/2)$$

where $\zeta \in \mathbb{C}$ is the squeezing parameter $(\zeta = se^{i\theta})$.
(ii) Let $s \geq 0$ be the squeezing parameter. Show that

$$|\psi\rangle = e^{s(b_1^\dagger b_2^\dagger - b_1 b_2)}|00\rangle = (1 - \tanh^2(s))^{1/2} e^{\tanh(s) b_1^\dagger b_2^\dagger}|00\rangle.$$

Show that

$$\mathrm{tr}(b_1^\dagger b_1 |\psi\rangle\langle\psi|) = \sinh^2(s).$$

Problem 23. Some nonlinear optical processes can be described by a Hamilton operator $\hat{H}$ with two degrees of freedom and cubic terms in the creation and annihilation Bose operators

$$\hat{H} = \hbar\omega b_1^\dagger b_1 + 2\hbar\omega b_2^\dagger b_2 + g\hbar\omega(b_1^2 b_2^\dagger + (b_1^\dagger)^2 b_2).$$

Examples are harmonic generation, coherent spontaneous emission and down conversion. Consider the number operator

$$\hat{N} = b_1^\dagger b_1 + b_2^\dagger b_2.$$

(i) Find the commutator $[\hat{H}, \hat{N}]$.
(ii) Find the state

$$\hat{H}(|\beta_1\rangle|\beta_2\rangle) \equiv \hat{H}(|\beta\rangle \otimes |\beta\rangle).$$

Problem 24. A *beam splitter* is an optical device that splits a beam of light in two. In general a signal mode (index 1) is mixed with a reference mode (index 2) (the two inputs) at a beam splitter and measurements are performed on the two output modes. The two output modes are entangled. The action of a beam splitter is described by a unitary operator U connecting the input and output states

$$|\Psi_{out}\rangle = U|\Psi_{in}\rangle$$

where

$$U = \exp(i(\phi_T + \phi_R)L_3)\exp(2i\theta L_2)\exp(i(\phi_T - \phi_R)L_3)$$

and the operators L_1, L_2 and L_3 are given by

$$L_1 = \frac{1}{2}(b_1^\dagger b_2 + b_2^\dagger b_1), \quad L_2 = \frac{1}{2i}(b_1^\dagger b_2 - b_2^\dagger b_1), \quad L_3 = \frac{1}{2}(b_1^\dagger b_1 - b_2^\dagger b_2).$$

Show that the operators satisfy $[L_j, L_k] = i\epsilon_{jk\ell}L_\ell$. The Levi-Cività tensor $\epsilon_{jk\ell}$ is equal to $+1$ and -1 for even and odd permutations of its indices, respectively, and zero otherwise. The complex transmittance T and reflectance R of the beam splitter are defined by

$$T := e^{i\phi_T}\cos(\theta), \qquad R := e^{i\phi_R}\sin(\theta).$$

The input state is a product state

$$|\Psi_{in}\rangle = |\Psi_{in1}\rangle \otimes |\Psi_{in2}\rangle.$$

If $\phi_T = \phi_R = 0$, then the unitary operator U takes the form

$$U = \exp(2i\theta L_2) = \exp(\theta(b_1^\dagger b_2 - b_2^\dagger b_1)).$$

For the measurement one assumes that $\Pi(l)$ is the positive operator-value measure that is realized by the measuring device with

$$\Pi(l) \geq 0, \qquad \sum_{l=1}^{n}\Pi(l) = I.$$

Consider the density operators $\Pi(l) := |\Psi_{out2}\rangle\langle\Psi_{out2}|$ and

$$\rho_{out1} = \frac{\text{tr}_2(\rho_{out}\Pi(l))}{p}, \qquad \rho_{out} = U|\Psi_{in}\rangle\langle\Psi_{in}|U^*.$$

Show that

$$p(l) = \langle\Pi(l)\rangle = \text{tr}_1(\text{tr}_2(\rho_{out}\Pi(l)))$$

is the probability of obtaining the result l. Note that tr_1 and tr_2 denote the partial trace.

Problem 25.   Consider n particles on a line, with the coordinates $x_1, x_2, \ldots, x_n$ and the Hamiltonian operator

$$\hat{H} = \hat{H}_0 + \hat{V}$$

where

$$\hat{H}_0 = -\frac{1}{2}\sum_{j=1}^{n}\frac{\partial^2}{\partial x_j^2} + \frac{1}{2n}\sum_{j<k}^{n}(x_j - x_k)^2, \qquad \hat{V} = g^2\sum_{j<k}^{n}(x_j - x_k)^{-2}.$$

(i) Show that if $x_i = \bar{x} + \xi_i$, where

$$\bar{x} := \frac{1}{n}\sum_{i=1}^{n}x_i$$

then the variable $\bar{x}$ can be separated and discarded. This means that the motion of the centre of gravity of the system is trivial.

(ii) Show that these n variables are constrained, $\sum_{j=1}^{n} \xi_j = 0$.

(iii) Show that $\hat{H}_0$ and $\hat{V}$ takes the form

$$\hat{H}_0 = -\frac{1}{2} \sum_{j=1}^{n} \partial_j^2 + \frac{1}{2} \sum_{j=1}^{n} \xi_j^2, \qquad \hat{V} = \frac{g^2}{2} \sum_{j \neq k}^{n} (\xi_j - \xi_k)^{-2}$$

where

$$\partial_j := \frac{\partial}{\partial \xi_j} - \frac{1}{n} \sum_{k=1}^{n} \frac{\partial}{\partial \xi_k}.$$

(iv) Show that one can introduce operators $b_j^\dagger$ and b_j playing the role of creation and annihilation operators,

$$b_j^\dagger := \xi_j - \partial_j, \qquad b_j := \xi_j + \partial_j, \qquad \sum_{j=1}^{n} b_j = \sum_{j=1}^{n} b_j^\dagger = 0$$

where

$$[b_j, b_k] = [b_j^\dagger, b_k^\dagger] = 0, \qquad [b_j, b_k^\dagger] = \left(\delta_{jk} - \frac{1}{n} \right) I.$$

(v) Show that in terms of these operators the Hamilton operator takes the form

$$H_0 = \frac{1}{2} \left(\sum_{j=1}^{N} b_j^\dagger b_j + (n-1)I \right).$$

(vi) Let B and $B^\dagger$

$$B^\dagger := \frac{1}{2} \sum_{j=1}^{n} (b_j^\dagger)^2 - \hat{V}, \qquad B := \frac{1}{2} \sum_{j=1}^{n} (b_j)^2 - \hat{V}.$$

Show that $[\hat{H}, B^\dagger] = 2B^\dagger$, $[\hat{H}, B] = -2B$, $[B, B^\dagger] = 4\hat{H}$.

Problem 26. Consider the Hamilton function

$$H(\mathbf{q}, \mathbf{p}) = \frac{p_1^2}{2} + \frac{p_2^2}{2} + \frac{q_1^2 q_2^2}{2}.$$

Quantize

$$p_1 \rightarrow -i \frac{\partial}{\partial q_1}, \qquad p_2 \rightarrow -i \frac{\partial}{\partial q_2}$$

and introduce

$$\hat{q}_j = \frac{1}{\sqrt{2}} (b_j^\dagger + b_j), \qquad \hat{p}_j = \frac{i}{\sqrt{2}} (b_j^\dagger - b_j)$$

where $\hbar = 1$ and $j = 1, 2$. Introduce the number basis

$$|n_1, n_2\rangle := \prod_{k=1}^{2} \frac{(b_k^\dagger)^{n_k}}{\sqrt{n_k!}} |0\rangle$$

where $n_k = 0, 1, \ldots, \infty$. Recall that

$$b_1|n_1, n_2\rangle = \sqrt{n_1}|n_1 - 1, n_2\rangle, \quad b_1^\dagger|n_1, n_2\rangle = \sqrt{n_1 + 1}|n_1 + 1, n_2\rangle, \quad b_1^\dagger b_1|n_1, n_2\rangle = n_1|n_1, n_2\rangle$$

and analogously for b_2, $b_2^\dagger$. Calculate the Hamilton operator $\hat{H}$ in terms of $b_j^\dagger$, b_j. Then calculate the infinite dimensional matrix representation of $\hat{H}$ using the basis given above. Choose the order of the basis

$$|00\rangle, \quad |01\rangle, \quad |10\rangle, \quad |02\rangle, \quad |11\rangle, \quad |20\rangle, \ldots$$

Note that $\langle m_2, m_1|n_1, n_2\rangle = \delta_{n_1, m_1}\delta_{n_2, m_2}$. Truncate the infinite dimensional matrix and calculate the eigenvalues numerically.

Problem 27.   Let $\ell = 0, 1, 2, \ldots$. Consider the Hamilton operator

$$\hat{H}(\ell) = -\frac{1}{2}\frac{d^2}{dx^2} + \frac{1}{2}x^2 + \frac{\ell(\ell + 1)}{2}\frac{1}{x^2}.$$

Let

$$B_\ell = \frac{1}{\sqrt{2}}\left(\frac{d}{dx} + x - \frac{\ell}{x}\right), \quad B_\ell^\dagger = \frac{1}{\sqrt{2}}\left(-\frac{d}{dx} + x - \frac{\ell}{x}\right).$$

Then $B_\ell + B_{-\ell}^\dagger = \sqrt{2}x$. Show that

$$[B_\ell, B_\ell^\dagger] = I + \frac{\ell}{x}I, \quad [B_\ell, B_{-\ell}] = -\frac{\ell}{x^2}I, \quad [B_{-\ell}, B_\ell^\dagger] = I$$

where I is the identity operator and

$$\hat{H}(\ell) = B_\ell B_\ell^\dagger + \ell I - \frac{1}{2}I.$$

Problem 28.   Consider the operators

$$S_0 = b_H^\dagger b_H + b_V^\dagger b_V, \quad S_1 = b_H^\dagger b_H - b_V^\dagger b_V,$$

$$S_2 = b_H^\dagger b_V e^{i\theta} + b_V^\dagger b_H e^{-i\theta}, \quad S_3 = ib_V^\dagger b_H e^{-i\theta} - ib_H^\dagger b_V e^{i\theta}$$

where the subscripts H and V label the horizontal and vertical polarization modes, respectively; θ is the phase shift between these modes; and $b_{H,V}$ and $b_{H,V}^\dagger$ are the Bose annihilation and creation operators for the electromagnetic field in frequency space. Show that $[S_1, S_2] = 2iS_3$, $[S_2, S_3] = 2iS_1$, $[S_3, S_1] = 2iS_2$ and $[S_0, S_j] = 0$ for $j = 1, 2, 3$.

Problem 29.   Let b_1, b_2 be Bose annihilation operators. Consider the Hamilton operator

$$\hat{H} = \hbar\omega b_1^\dagger b_1 + \hbar\omega_2 b_2^\dagger b_2 + \hbar(Vb_1^\dagger b_2 + V^*b_1 b_2^\dagger)$$

where V is a complex coupling constant.
(i) Show that the Heisenberg equation of motion for b_1 and b_2 is given by

$$i\hbar\frac{db_1}{dt} = [b_1, \hat{H}](t), \quad i\hbar\frac{db_2}{dt} = [b_2, \hat{H}](t).$$

(ii) Show that inserting the Hamilton operator yields

$$i\hbar\frac{db_1}{dt} = \hbar\omega_1 b_1 + \hbar Vb_2, \quad i\hbar\frac{db_2}{dt} = \hbar\omega_2 b_2 + \hbar V^*b_1.$$

(iii) Show that introducing the definitions

$$b_1 = \widetilde{b}_1 e^{-i\omega_1 t}, \qquad b_2 = \widetilde{b}_2 e^{-i\omega_2 t}$$

yields

$$i\frac{d\widetilde{b}_1}{dt} = V\widetilde{b}_2, \qquad i\frac{d\widetilde{b}_2}{dt} = V^*\widetilde{b}_1.$$

Problem 30. Show that the spectrum of the operator $b_1^\dagger b_2 + b_1 b_2^\dagger$ is discrete and coincides with the set $\mathbb{Z}$ of relative integers.

Problem 31. The *Rayleigh-Schrödinger perturbation theory* for the *anharmonic oscillator* with Hamilton operator $(\hat{p} = -i\hbar d/dq)$

$$\hat{H} = \frac{1}{2m}\hat{p}^2 + \frac{1}{2}m\omega^2\hat{q}^2 + \widetilde{\gamma}\hat{q}^4$$

is divergent with the perturbation $\widetilde{\gamma}\hat{q}^4$. Let

$$\hat{q} =: \sqrt{\frac{\hbar}{2m\omega}}(b^\dagger + b), \qquad \hat{p} =: i\sqrt{\frac{\hbar m\omega}{2}}(b^\dagger - b).$$

Since

$$\hat{q}^2 = \frac{\hbar}{2m\omega}((b^\dagger)^2 + b^2 + 2b^\dagger b + I), \quad \hat{p}^2 = -\frac{\hbar m\omega}{2}((b^\dagger)^2 + b^2 - 2b^\dagger b - I)$$

and

$$\hat{q}^4 = \frac{\hbar^2}{4m^2\omega^2}((b^\dagger)^2 + b^2 + 2b^\dagger b + I)^2$$

we arrive for the Hamilton operator

$$\hat{H} = \hbar\omega(b^\dagger b + \frac{1}{2}I) + \frac{\hbar^2\widetilde{\gamma}}{4m^2\omega^2}((b^\dagger)^4 + b^4 + 6((b^\dagger)^2 + b^2) + 4(b^\dagger b^3 + (b^\dagger)^3 b) + 12b^\dagger b + 6(b^\dagger)^2 b^2 + 3I).$$

With $\hat{K} = \hat{H}/(\hbar\omega)$ and the dimensionless quantity $\gamma = (\hbar\widetilde{\gamma})/(4m^2\omega^3)$ we arrive at

$$\hat{K} = b^\dagger b + \frac{1}{2}I + \gamma((b^\dagger)^4 + b^4 + 6((b^\dagger)^2 + b^2) + 4(b^\dagger b^3 + (b^\dagger)^3 b) + 12b^\dagger b + 6(b^\dagger)^2 b^2 + 3I)$$

or

$$\hat{K} = b^\dagger b(1 + 12\gamma) + \left(\frac{1}{2} + 3\gamma\right)I + \gamma(b^4 + (b^\dagger)^4 + 6(b^2 + (b^\dagger)^2) + 4(b^\dagger b^3 + (b^\dagger)^3 b) + 6(b^\dagger)^2 b^2).$$

Let $\hat{K} = \hat{K}_0 + \hat{K}_1$ with

$$\hat{K}_0 = b^\dagger b(1 + 6\gamma) + \left(\frac{1}{2} + 3\gamma\right)I + 6\gamma(b^\dagger b)^2$$

and

$$\hat{K}_1 = \gamma(b^4 + (b^\dagger)^4 + 6(b^2 + (b^\dagger)^2) + 4(b^\dagger b^3 + (b^\dagger)^3 b)).$$

(i) Study the Rayleigh-Schrödinger perturbation theory for this Hamilton operator with $\hat{K}_1$ the perturbation. Take into account the invariance of the Hamilton operator under the transformation $b \mapsto -b$, $b^\dagger \mapsto -b^\dagger$.

(ii) Find the matrix representation of $\hat{K}$ using number states. Truncate the infinite dimensional matrix and find the eigenvalues as a function of γ.

(iii) Consider coherent states $|\beta\rangle$. Calculate the expectation value $\langle\beta|\hat{K}|\beta\rangle$. Discuss.

(iv) Consider squeezed states $|\zeta\rangle$. Calculate the expectation value $\langle\zeta|\hat{K}|\zeta\rangle$. Discuss.

Chapter 2

Spin Systems

2.1 Spin Matrices, Commutators and Anticommutators

Problem 1. Let s be a fixed number (*spin quantum number*) from the set

$$s \in \left\{ \frac{1}{2}, 1, \frac{3}{2}, 2, \frac{5}{2}, \dots \right\}.$$

Given a fixed s, the indices j, k run over $s, s-1, s-2, \dots, -s+1, -s$. Consider the $(2s+1)$ unit vectors (standard basis)

$$\mathbf{e}_{s,s} = \begin{pmatrix} 1 \\ 0 \\ 0 \\ \vdots \\ 0 \end{pmatrix}, \quad \mathbf{e}_{s,s-1} = \begin{pmatrix} 0 \\ 1 \\ 0 \\ \vdots \\ 0 \end{pmatrix}, \quad \dots, \mathbf{e}_{s,-s} = \begin{pmatrix} 0 \\ 0 \\ \vdots \\ 0 \\ 1 \end{pmatrix}.$$

Obviously the vectors have $(2s+1)$ components. The $(2s+1) \times (2s+1)$ matrices S_+ and S_- are defined as

$$S_+ \mathbf{e}_{s,m} := \hbar \sqrt{(s-m)(s+m+1)} \mathbf{e}_{s,m+1}, \quad m = s-1, s-2, \dots, -s$$
$$S_- \mathbf{e}_{s,m} := \hbar \sqrt{(s+m)(s-m+1)} \mathbf{e}_{s,m-1}, \quad m = s, s-1, \dots, -s+1$$

and $S_+ \mathbf{e}_{s,s} = \mathbf{0}$, $S_- \mathbf{e}_{s,-s} = \mathbf{0}$, where $\hbar$ is the Planck constant. We define

$$S_1 = \frac{1}{2}(S_+ + S_-), \quad S_2 = \frac{-i}{2}(S_+ - S_-).$$

The $(2s+1) \times (2s+1)$ matrix S_3 is defined as (eigenvalue equation)

$$S_3 \mathbf{e}_{s,m} := m\hbar \mathbf{e}_{s,m}, \quad m = s, s-1, \dots, -s.$$

Let $\mathbf{S} := (S_1, S_2, S_3)$.

(i) Find the matrix representation of S_+ and S_-.

(ii) Find the $(2s+1) \times (2s+1)$ matrix $\mathbf{S}^2 := S_1^2 + S_2^2 + S_3^2$.

(iii) Calculate the expectation values

$$\mathbf{e}_{s,s}^* S_+ \mathbf{e}_{s,s}, \quad \mathbf{e}_{s,s}^* S_- \mathbf{e}_{s,s}, \quad \mathbf{e}_{s,s}^* S_3 \mathbf{e}_{s,s}.$$

(iv) Find the commutators $[S_1, S_2], [S_2, S_3], [S_3, S_1]$.

Solution 1. (i) We have for $j, k \in \{s, s-1, \ldots, -s\}$

$$(S_+)_{jk} = (S_-)_{kj} = \hbar\sqrt{(s-k)(s+k+1)}\delta_{j,k+1}$$
$$= \hbar\sqrt{(s+j)(s-j+1)}\delta_{j,k+1}$$

and

$$(S_-)_{jk} = (S_+)_{kj} = \hbar\sqrt{(s+k)(s-k+1)}\delta_{j,k-1}$$
$$= \hbar\sqrt{(s-j)(s+j+1)}\delta_{j,k-1}.$$

It follows that

$$S_+ = \hbar \begin{pmatrix} 0 & \sqrt{2s} & 0 & 0 & \cdots & 0 \\ 0 & 0 & \sqrt{2(2s-1)} & 0 & \cdots & 0 \\ 0 & 0 & 0 & \sqrt{3(2s-2)} & \cdots & 0 \\ \vdots & \vdots & \vdots & \vdots & \ddots & \vdots \\ 0 & 0 & 0 & 0 & \cdots & \sqrt{2s} \\ 0 & 0 & 0 & 0 & \cdots & 0 \end{pmatrix}$$

and $S_- = (S_+)^*$.

(ii) We have

$$\mathbf{S}^2 = \frac{1}{2}(S_+ S_- + S_- S_+) + S_3^2$$

where $S_+ S_-$, $S_- S_+$ and S_3^2 are diagonal matrices. Thus

$$(\mathbf{S})_{jk}^2 = s(s+1)\hbar^2 \delta_{jk}.$$

Therefore $\mathbf{S}^2$ is a diagonal matrix and the eigenvalues are $s(s+1)\hbar^2$ with $s = 1/2, 1, \ldots$.

(iii) We find for the expectation values

$$\mathbf{e}_{ss}^* S_+ \mathbf{e}_{ss} = 0, \quad \mathbf{e}_{ss}^* S_- \mathbf{e}_{ss} = 0, \quad \mathbf{e}_{ss}^* S_3 \mathbf{e}_{ss} = \hbar s.$$

(iv) The commutators are $[S_1, S_2] = iS_3$, $[S_2, S_3] = iS_1$, $[S_3, S_1] = iS_2$.

Problem 2. Utilizing the previous problem write down the spin matrices S_1, S_2, S_3 for spin-$\frac{1}{2}$, spin-1, spin-$\frac{3}{2}$ and spin-2. Give the eigenvalues and normalized eigenvectors of these matrices. Can the eigenvectors of S_1 and S_2 for the spin-$\frac{3}{2}$ case be written as a Kronecker product of two vectors from $\mathbb{C}^2$?

Solution 2. For spin-$\frac{1}{2}$ we have the invertible hermitian matrices

$$S_1 = \frac{1}{2}\begin{pmatrix} 0 & 1 \\ 1 & 0 \end{pmatrix}, \quad S_2 = \frac{1}{2}\begin{pmatrix} 0 & -i \\ i & 0 \end{pmatrix}, \quad S_3 = \frac{1}{2}\begin{pmatrix} 1 & 0 \\ 0 & -1 \end{pmatrix}$$

with $S_1^2 + S_2^2 + S_3^2 = 3I_2/4$. The eigenvalues for S_1 are $+1/2$ and $-1/2$ with the corresponding normalized eigenvectors

$$\mathbf{v}_{1,1} = \frac{1}{\sqrt{2}} \begin{pmatrix} 1 \\ 1 \end{pmatrix}, \quad \mathbf{v}_{1,2} = \frac{1}{\sqrt{2}} \begin{pmatrix} 1 \\ -1 \end{pmatrix}.$$

The eigenvalues for S_2 are $+1/2$ and $-1/2$ with the corresponding normalized eigenvectors

$$\mathbf{v}_{2,1} = \frac{1}{\sqrt{2}} \begin{pmatrix} 1 \\ i \end{pmatrix}, \quad \mathbf{v}_{2,2} = \frac{1}{\sqrt{2}} \begin{pmatrix} 1 \\ -i \end{pmatrix}.$$

The eigenvalues for S_3 are $+1/2$ and $-1/2$ with the corresponding normalized eigenvectors

$$\mathbf{v}_{3,1} = \begin{pmatrix} 1 \\ 0 \end{pmatrix}, \quad \mathbf{v}_{3,2} = \begin{pmatrix} 0 \\ 1 \end{pmatrix}.$$

For spin-1 we have the hermitian 3×3 matrices

$$S_1 = \frac{1}{\sqrt{2}} \begin{pmatrix} 0 & 1 & 0 \\ 1 & 0 & 1 \\ 0 & 1 & 0 \end{pmatrix}, \quad S_2 = \frac{1}{\sqrt{2}} \begin{pmatrix} 0 & -i & 0 \\ i & 0 & -i \\ 0 & i & 0 \end{pmatrix}, \quad S_3 = \begin{pmatrix} 1 & 0 & 0 \\ 0 & 0 & 0 \\ 0 & 0 & -1 \end{pmatrix}$$

with $S_1^2 + S_2^2 + S_3^2 = 2I_3$. The eigenvalues of S_1 are given by $+1, 0, -1$ with the corresponding normalized eigenvectors

$$\frac{1}{2} \begin{pmatrix} 1 \\ \sqrt{2} \\ 1 \end{pmatrix}, \quad \frac{1}{\sqrt{2}} \begin{pmatrix} 1 \\ 0 \\ -1 \end{pmatrix}, \quad \frac{1}{2} \begin{pmatrix} 1 \\ -\sqrt{2} \\ 1 \end{pmatrix}.$$

The eigenvalues of S_2 are given by $+1, 0, -1$ with the corresponding normalized eigenvectors

$$\frac{1}{2} \begin{pmatrix} 1 \\ \sqrt{2}i \\ -1 \end{pmatrix}, \quad \frac{1}{\sqrt{2}} \begin{pmatrix} 1 \\ 0 \\ 1 \end{pmatrix}, \quad \frac{1}{2} \begin{pmatrix} 1 \\ -\sqrt{2}i \\ -1 \end{pmatrix}.$$

The eigenvalues of S_3 are given by $+1, 0, -1$ with the corresponding normalized eigenvectors (standard basis)

$$\begin{pmatrix} 1 \\ 0 \\ 0 \end{pmatrix}, \quad \begin{pmatrix} 0 \\ 1 \\ 0 \end{pmatrix}, \quad \begin{pmatrix} 0 \\ 0 \\ 1 \end{pmatrix}.$$

For spin-$\frac{3}{2}$ we have the invertible hermitian 4×4 matrices

$$S_1 = \frac{1}{2} \begin{pmatrix} 0 & \sqrt{3} & 0 & 0 \\ \sqrt{3} & 0 & 2 & 0 \\ 0 & 2 & 0 & \sqrt{3} \\ 0 & 0 & \sqrt{3} & 0 \end{pmatrix}, \quad S_2 = \frac{1}{2} \begin{pmatrix} 0 & -i\sqrt{3} & 0 & 0 \\ i\sqrt{3} & 0 & -2i & 0 \\ 0 & 2i & 0 & -i\sqrt{3} \\ 0 & 0 & i\sqrt{3} & 0 \end{pmatrix},$$

$$S_3 = \begin{pmatrix} 3/2 & 0 & 0 & 0 \\ 0 & 1/2 & 0 & 0 \\ 0 & 0 & -1/2 & 0 \\ 0 & 0 & 0 & -3/2 \end{pmatrix}$$

with $S_1^2 + S_2^2 + S_3^2 = \frac{15}{4} I_4$. The eigenvalues of S_1 are given by $+3/2, +1/2, -1/2, -3/2$ with the corresponding normalized eigenvectors

$$\frac{1}{\sqrt{8}}\begin{pmatrix} 1 \\ \sqrt{3} \\ \sqrt{3} \\ 1 \end{pmatrix}, \quad \frac{1}{\sqrt{8}}\begin{pmatrix} \sqrt{3} \\ 1 \\ -1 \\ -\sqrt{3} \end{pmatrix}, \quad \frac{1}{\sqrt{8}}\begin{pmatrix} \sqrt{3} \\ -1 \\ -1 \\ \sqrt{3} \end{pmatrix}, \quad \frac{1}{\sqrt{8}}\begin{pmatrix} 1 \\ -\sqrt{3} \\ \sqrt{3} \\ -1 \end{pmatrix}.$$

The eigenvectors cannot be written as a Kronecker product of two vectors of $\mathbb{C}^2$. They are *entangled*. The eigenvalues of S_2 are given by $+3/2, +1/2, -1/2, -3/2$ with the corresponding normalized eigenvectors

$$\frac{1}{\sqrt{8}}\begin{pmatrix} 1 \\ \sqrt{3}i \\ -\sqrt{3} \\ -i \end{pmatrix}, \quad \frac{1}{\sqrt{8}}\begin{pmatrix} \sqrt{3} \\ i \\ 1 \\ \sqrt{3}i \end{pmatrix}, \quad \frac{1}{\sqrt{8}}\begin{pmatrix} \sqrt{3} \\ -i \\ 1 \\ -\sqrt{3}i \end{pmatrix}, \quad \frac{1}{\sqrt{8}}\begin{pmatrix} 1 \\ -\sqrt{3}i \\ -\sqrt{3} \\ i \end{pmatrix}.$$

The eigenvectors cannot be written as a Kronecker product of two vectors in of $\mathbb{C}^2$. They are entangled. The eigenvalues of S_3 are given by $+1, 0, -1$ with the corresponding normalized eigenvectors (standard basis)

$$\begin{pmatrix} 1 \\ 0 \\ 0 \\ 0 \end{pmatrix}, \quad \begin{pmatrix} 0 \\ 1 \\ 0 \\ 0 \end{pmatrix}, \quad \begin{pmatrix} 0 \\ 0 \\ 1 \\ 0 \end{pmatrix}, \quad \begin{pmatrix} 0 \\ 0 \\ 0 \\ 1 \end{pmatrix}.$$

For spin-2 we have the hermitian 5×5 matrices

$$S_1 = \begin{pmatrix} 0 & 1 & 0 & 0 & 0 \\ 1 & 0 & \sqrt{6}/2 & 0 & 0 \\ 0 & \sqrt{6}/2 & 0 & \sqrt{6}/2 & 0 \\ 0 & 0 & \sqrt{6}/2 & 0 & 1 \\ 0 & 0 & 0 & 1 & 0 \end{pmatrix},$$

$$S_2 = \begin{pmatrix} 0 & -i & 0 & 0 & 0 \\ i & 0 & -i\sqrt{6}/2 & 0 & 0 \\ 0 & i\sqrt{6}/2 & 0 & -i\sqrt{6}/2 & 0 \\ 0 & 0 & i\sqrt{6}/2 & 0 & -i \\ 0 & 0 & 0 & i & 0 \end{pmatrix},$$

$$S_3 = \begin{pmatrix} 2 & 0 & 0 & 0 & 0 \\ 0 & 1 & 0 & 0 & 0 \\ 0 & 0 & 0 & 0 & 0 \\ 0 & 0 & 0 & -1 & 0 \\ 0 & 0 & 0 & 0 & -2 \end{pmatrix}$$

with $S_1^2 + S_2^2 + S_3^2 = 6I_5$. The eigenvalues of S_1 are given by $+2, +1, 0, -1, -2$ with the corresponding normalized eigenvectors

$$\frac{1}{4}\begin{pmatrix} 1 \\ 2 \\ \sqrt{6} \\ 2 \\ 1 \end{pmatrix}, \quad \frac{1}{2}\begin{pmatrix} 1 \\ 1 \\ 0 \\ -1 \\ -1 \end{pmatrix}, \quad \frac{1}{\sqrt{8}}\begin{pmatrix} \sqrt{3} \\ 0 \\ -\sqrt{2} \\ 0 \\ \sqrt{3} \end{pmatrix}, \quad \frac{1}{2}\begin{pmatrix} 1 \\ -1 \\ 0 \\ 1 \\ -1 \end{pmatrix}, \quad \frac{1}{4}\begin{pmatrix} 1 \\ -2 \\ \sqrt{6} \\ -2 \\ 1 \end{pmatrix}.$$

The eigenvalues of S_2 are given by $+2$, $+1$, 0, -1, -2 with the corresponding normalized eigenvectors

$$\frac{1}{4}\begin{pmatrix} 1 \\ 2i \\ -\sqrt{6} \\ -2i \\ 1 \end{pmatrix}, \quad \frac{1}{2}\begin{pmatrix} 1 \\ i \\ 0 \\ i \\ -1 \end{pmatrix}, \quad \frac{1}{\sqrt{8}}\begin{pmatrix} \sqrt{3} \\ 0 \\ \sqrt{2} \\ 0 \\ \sqrt{3} \end{pmatrix}, \quad \frac{1}{2}\begin{pmatrix} 1 \\ -i \\ 0 \\ -i \\ -1 \end{pmatrix}, \quad \frac{1}{4}\begin{pmatrix} 1 \\ -2i \\ -\sqrt{6} \\ 2i \\ 1 \end{pmatrix}.$$

The eigenvalues of S_3 are given by $+2$, $+1$, 0, -1, -2 with the corresponding normalized eigenvectors (standard basis)

$$\begin{pmatrix} 1 \\ 0 \\ 0 \\ 0 \\ 0 \end{pmatrix}, \quad \begin{pmatrix} 0 \\ 1 \\ 0 \\ 0 \\ 0 \end{pmatrix}, \quad \begin{pmatrix} 0 \\ 0 \\ 1 \\ 0 \\ 0 \end{pmatrix}, \quad \begin{pmatrix} 0 \\ 0 \\ 0 \\ 1 \\ 0 \end{pmatrix}, \quad \begin{pmatrix} 0 \\ 0 \\ 0 \\ 0 \\ 1 \end{pmatrix}.$$

Problem 3.   Consider the normalized vector in $\mathbb{R}^3$ $(\theta, \phi \in \mathbb{R})$

$$\mathbf{n} = \begin{pmatrix} \sin(\theta)\cos(\phi) \\ \sin(\theta)\sin(\phi) \\ \cos(\theta) \end{pmatrix}$$

i.e. $n_1^2 + n_2^2 + n_3^2 = 1$.
(i) Let σ_1, σ_2, σ_3 be the *Pauli spin matrices*. Calculate the 2×2 matrix

$$U(\theta, \phi) = \mathbf{n} \cdot \boldsymbol{\sigma} \equiv n_1\sigma_1 + n_2\sigma_2 + n_3\sigma_3.$$

Is the matrix $U(\theta, \phi)$ unitary? Find the trace and the determinant. Is the matrix $U(\theta, \phi)$ hermitian? Find the eigenvalues and normalized eigenvectors of $U(\theta, \phi)$.
(ii) For $s = 1$ we have the 3×3 matrices

$$S_1 = \frac{1}{\sqrt{2}}\begin{pmatrix} 0 & 1 & 0 \\ 1 & 0 & 1 \\ 0 & 1 & 0 \end{pmatrix}, \quad S_2 = \frac{1}{\sqrt{2}}\begin{pmatrix} 0 & -i & 0 \\ i & 0 & -i \\ 0 & i & 0 \end{pmatrix}, \quad S_3 = \begin{pmatrix} 1 & 0 & 0 \\ 0 & 0 & 0 \\ 0 & 0 & -1 \end{pmatrix}.$$

Let $\mathbf{n} \in \mathbb{R}^3$ given above. Calculate the eigenvalues and eigenvectors of matrix

$$n_1 S_1 + n_2 S_2 + n_3 S_3.$$

Solution 3.   (i) We find

$$U(\theta, \phi) = \begin{pmatrix} \cos\theta & e^{-i\phi}\sin\theta \\ e^{i\phi}\sin\theta & -\cos\theta \end{pmatrix}.$$

The matrix is unitary since $U(\theta, \phi)U^*(\theta, \phi) = I_2$ and hermitian. The trace is zero and the determinant is -1. The matrix is also hermitian. Since the matrix is unitary, hermitian and the trace is equal to 0, we obtain the eigenvalues $\lambda_1 = +1$ and $\lambda_2 = -1$. Using the identities

$$\sin(\theta) \equiv 2\sin(\theta/2)\cos(\theta/2), \qquad \cos(\theta) \equiv \cos^2(\theta/2) - \sin^2(\theta/2)$$

we obtain the normalized eigenvectors

$$\begin{pmatrix} e^{-i\phi/2}\cos(\theta/2) \\ e^{i\phi/2}\sin(\theta/2) \end{pmatrix}, \qquad \begin{pmatrix} -e^{-i\phi/2}\sin(\theta/2) \\ e^{i\phi/2}\cos(\theta/2) \end{pmatrix}.$$

(ii) We have

$$n_1 S_1 + n_2 S_2 + n_3 S_3 = \begin{pmatrix} \cos(\theta) & \sin(\theta)e^{-i\phi}/\sqrt{2} & 0 \\ \sin(\theta)e^{i\phi}/\sqrt{2} & 0 & \sin(\theta)e^{-i\phi}/\sqrt{2} \\ 0 & \sin(\theta)e^{i\phi}/\sqrt{2} & -\cos(\theta) \end{pmatrix}.$$

The eigenvalues are $+1, 0, -1$ with the corresponding normalized eigenvectors

$$\frac{1}{2}\begin{pmatrix} e^{-i\phi}(1+\cos(\theta)) \\ \sqrt{2}\sin(\theta) \\ e^{-i\phi}(1-\cos(\theta)) \end{pmatrix}, \quad \frac{1}{\sqrt{2}}\begin{pmatrix} e^{-i\phi}\sin(\theta) \\ -\sqrt{2}\cos(\theta) \\ -e^{i\phi}\sin(\theta) \end{pmatrix}, \quad \frac{1}{2}\begin{pmatrix} e^{-i\phi}(1-\cos(\theta)) \\ -\sqrt{2}\sin(\theta) \\ e^{i\phi}(1+\cos(\theta)) \end{pmatrix}.$$

The normalized eigenvectors form an orthonormal basis in the Hilbert space $\mathbb{C}^3$.

Problem 4. Let $\sigma_1, \sigma_2, \sigma_3$ be the Pauli spin matrices, I_2 the 2×2 identity matrix and

$$U_H = \frac{1}{\sqrt{2}}\begin{pmatrix} 1 & 1 \\ 1 & -1 \end{pmatrix} \equiv \frac{1}{\sqrt{2}}(\sigma_3 + \sigma_1)$$

be the *Hadamard matrix*. U_H is unitary with $\det(U_H) = -1$ and $U_H = U_H^*$.
(i) Find the matrices

$$U_H \sigma_1 U_H^*, \qquad U_H \sigma_2 U_H^*, \qquad U_H \sigma_3 U_H^*.$$

(ii) Show that one can find a 2×2 unitary matrix U such that $U\sigma_1 U^* = \sigma_3$.
(iii) Show that one can find a 4×4 unitary matrix V such that

$$V(\sigma_1 \otimes \sigma_1)V^* = \sigma_3 \otimes \sigma_3.$$

Hint. Use the result from (ii).
(iv) Is the 4×4 matrix

$$U = \frac{1}{2}\left(I_2 \otimes I_2 + \sum_{j=1}^{3} \sigma_j \otimes \sigma_j \right)$$

unitary?

Solution 4. (i) We obtain $U_H \sigma_1 U_H^* = \sigma_3$, $U_H \sigma_2 U_H^* = -\sigma_2$, $U_H \sigma_3 U_H^* = \sigma_1$.
(ii) Note that σ_1 is hermitian and unitary. Thus by calculating the eigenvalues and normalized eigenvectors of σ_1 we can construct a unitary matrix such that $U\sigma_1 U^*$ is a diagonal, hermitian and unitary matrix. Now σ_3 is a diagonal, hermitian and unitary matrix. Thus the diagonal elements of σ_3 are the eigenvalues of σ_1. Using the proper ordering of the normalized eigenvectors of σ_1 we obtain the unitary matrix

$$U = \frac{1}{\sqrt{2}}\begin{pmatrix} 1 & 1 \\ 1 & -1 \end{pmatrix} = U^* = U^{-1}$$

with

$$U\begin{pmatrix} 0 & 1 \\ 1 & 0 \end{pmatrix}U^* = \begin{pmatrix} 1 & 0 \\ 0 & -1 \end{pmatrix} = \sigma_3.$$

(iii) Using the result from (ii) we have $V = U \otimes U$, since

$$(U \otimes U)(\sigma_1 \otimes \sigma_1)(U^* \otimes U^*) = ((U\sigma_1) \otimes (U\sigma_1))(U^* \otimes U^*) = (U\sigma_1 U^*) \otimes (U\sigma_1 U^*)$$
$$= \sigma_3 \otimes \sigma_3.$$

(iv) We have $U^* = U$. We find

$$U^*U = \frac{1}{4}\left(I_2 \otimes I_2 + \sum_{j=1}^{3} \sigma_j \otimes \sigma_j \right)\left(I_2 \otimes I_2 + \sum_{k=1}^{3} \sigma_k \otimes \sigma_k \right)$$

$$= \frac{1}{4}(I_4 + 2\sum_{j=1}^{3} \sigma_j \otimes \sigma_j + \sum_{j,k=1}^{3} (\sigma_j\sigma_k) \otimes (\sigma_j\sigma_k))$$

$$= I_4.$$

Problem 5. Let $\mathbf{a} = (a_1, a_2, a_3)^T$ be an element in $\mathbb{R}^3$ and $\boldsymbol{\sigma} = (\sigma_1, \sigma_2, \sigma_3)^T$ be the Pauli spin matrices written as a column vector. We define the *vector product*

$$\mathbf{a} \times \boldsymbol{\sigma} := \begin{pmatrix} a_2\sigma_3 - a_3\sigma_2 \\ a_3\sigma_1 - a_1\sigma_3 \\ a_1\sigma_2 - a_2\sigma_1 \end{pmatrix}.$$

Find the eigenvalues of the 2×2 matrix $(\mathbf{a} \times \boldsymbol{\sigma})^\dagger(\mathbf{b} \times \boldsymbol{\sigma})$, where $\dagger$ denotes transpose and complex conjugate and $\mathbf{b} = (b_1, b_2, b_3)^T$ is an element in $\mathbb{R}^3$.

Solution 5. The matrix $(\mathbf{a} \times \boldsymbol{\sigma})^\dagger(\mathbf{b} \times \boldsymbol{\sigma})$ is given by

$$\begin{pmatrix} 2(a_1b_1 + a_2b_2 + a_3b_3) + (a_1b_2 - a_2b_1)i & (a_2b_3 - a_3b_2)i + (a_3b_1 - a_1b_3) \\ (a_2b_3 - a_3b_2)i - (a_3b_1 - a_1b_3) & 2(a_1b_1 + a_2b_2 + a_3b_3) - (a_1b_2 - a_2b_1)i \end{pmatrix}.$$

Note that $\mathbf{a}^T\mathbf{b} = a_1b_1 + a_2b_2 + a_3b_3$ (scalar product of $\mathbf{a}$ and $\mathbf{b}$) and

$$(\mathbf{a} \times \mathbf{b})^T(\mathbf{a} \times \mathbf{b}) = a_1^2b_2^2 + a_1^2b_3^2 + a_2^2b_1^2 + a_2^2b_3^2 + a_3^2b_1^2 + a_3^2b_2^2 - 2(a_1a_2b_1b_2 + a_1a_3b_1b_3 + a_2a_3b_2b_3).$$

Then the two eigenvalues are

$$\lambda_+ = \sqrt{-(\mathbf{a} \times \mathbf{b})^T(\mathbf{a} \times \mathbf{b}) + 2(\mathbf{a}^T\mathbf{b})} \equiv \sqrt{(\mathbf{a}^T\mathbf{b})^2 - (\mathbf{a}^T\mathbf{a})(\mathbf{b}^T\mathbf{b})} + 2(\mathbf{a}^T\mathbf{b})$$
$$\lambda_- = -\sqrt{-(\mathbf{a} \times \mathbf{b})^T(\mathbf{a} \times \mathbf{b}) + 2(\mathbf{a}^T\mathbf{b})} \equiv -\sqrt{(\mathbf{a}^T\mathbf{b})^2 - (\mathbf{a}^T\mathbf{a})(\mathbf{b}^T\mathbf{b})} + 2(\mathbf{a}^T\mathbf{b}).$$

Study the eigenvalue problem for the 2×2 matrix $(\mathbf{a} \times \boldsymbol{\sigma})^T(\mathbf{b} \times \boldsymbol{\sigma})$. Compare to the result from above.

Problem 6. Let $\sigma_1, \sigma_2, \sigma_3$ be the Pauli spin matrices. Let $a_j \in \mathbb{R}$ with $j = 0, 1, 2, 3$ and

$$a_0^2 + a_1^2 + a_2^2 + a_3^2 = 1.$$

(i) Show that

$$U = e^{i\phi}(a_0 I_2 + a_1 i\sigma_1 + a_2 i\sigma_2 + a_3 i\sigma_3)$$

is a unitary matrix, where $\phi \in \mathbb{R}$.

(ii) Find the cosine-sine decomposition of the Pauli matrix σ_2.

Solution 6. (i) Since $\sigma_1^2 = \sigma_2^2 = \sigma_3^2 = I_2$, $\sigma_1^* = \sigma_1$, $\sigma_2^* = \sigma_2$, $\sigma_3^* = \sigma_3$ and

$$\sigma_1\sigma_2 = i\sigma_3, \quad \sigma_2\sigma_3 = i\sigma_1, \quad \sigma_3\sigma_1 = i\sigma_2$$

$$\sigma_2\sigma_1 = -i\sigma_3, \quad \sigma_3\sigma_2 = -i\sigma_1, \quad \sigma_1\sigma_3 = -i\sigma_2$$

we obtain

$$\begin{aligned}
UU^* &= e^{i\phi}(a_0 I_2 + a_1 i\sigma_1 + a_2 i\sigma_2 + a_3 i\sigma_3)e^{-i\phi}(a_0 I_2 + a_1 i\sigma_1 + a_2 i\sigma_2 + a_3 i\sigma_3) \\
&= (a_0 I_2 + a_1 i\sigma_1 + a_2 i\sigma_2 + a_3 i\sigma_3)(a_0 I_2 + a_1 i\sigma_1 + a_2 i\sigma_2 + a_3 i\sigma_3) \\
&= (a_0^2 + a_1^2 + a_2^2 + a_3^2)I_2 \\
&= I_2.
\end{aligned}$$

(ii) We obtain

$$\sigma_2 = \begin{pmatrix} 0 & -i \\ i & 0 \end{pmatrix} = \begin{pmatrix} i & 0 \\ 0 & i \end{pmatrix} \begin{pmatrix} \cos\alpha & \sin\alpha \\ -\sin\alpha & \cos\alpha \end{pmatrix} \begin{pmatrix} i & 0 \\ 0 & i \end{pmatrix}$$

with $\alpha = \pi/2$.

Problem 7. Let σ_1, σ_2, σ_3 be the Pauli spin matrices and $\sigma_0 = I_2$, where I_2 is the 2×2 unit matrix.
(i) Find the unitary matrices $\sigma_1\sigma_1\sigma_1$, $\sigma_1\sigma_2\sigma_1$, $\sigma_1\sigma_3\sigma_1$.
(ii) Find the unitary matrices $\sigma_j\sigma_k\sigma_j$.
(iii) Find the 4×4 unitary matrices

$$(\sigma_1 \otimes \sigma_1)(\sigma_1 \otimes \sigma_1)(\sigma_1 \otimes \sigma_1), \quad (\sigma_1 \otimes \sigma_1)(\sigma_2 \otimes \sigma_2)(\sigma_1 \otimes \sigma_1), \quad (\sigma_1 \otimes \sigma_1)(\sigma_3 \otimes \sigma_3)(\sigma_1 \otimes \sigma_1).$$

(iv) Find the normalized vector in $\mathbb{C}^4$

$$(\sigma_1 \otimes \sigma_1)\left(\begin{pmatrix} 1 \\ 0 \end{pmatrix} \otimes \begin{pmatrix} 0 \\ 1 \end{pmatrix}\right).$$

(v) Consider the Pauli spin matrix σ_1 and the vectors $\mathbf{u}$, $\mathbf{v}$ in $\mathbb{C}^2$

$$\mathbf{u} = \begin{pmatrix} 1 \\ 0 \end{pmatrix}, \quad \mathbf{v} = \begin{pmatrix} 0 \\ 1 \end{pmatrix}.$$

We identify the vector $\mathbf{u}$ with spin up and the vector $\mathbf{v}$ with spin down. Calculate the vectors in $\mathbb{C}^{2^n}$

$$(\sigma_1 \otimes \sigma_1 \otimes \cdots \otimes \sigma_1)(\mathbf{u} \otimes \mathbf{u} \otimes \cdots \otimes \mathbf{u}), \quad (\sigma_1 \otimes \sigma_1 \otimes \cdots \otimes \sigma_1)(\mathbf{v} \otimes \mathbf{v} \otimes \cdots \otimes \mathbf{v}).$$

Discuss.

Solution 7. (i) Note that $\sigma_j^2 = I_2$ for $j = 0, 1, 2, 3$. Thus we obtain

$$\sigma_1\sigma_1\sigma_1 = \sigma_1, \quad \sigma_1\sigma_2\sigma_1 = -\sigma_2, \quad \sigma_1\sigma_3\sigma_1 = -\sigma_3.$$

(ii) If $j = k$ we have $\sigma_j\sigma_k\sigma_j = \sigma_k$. If $j \neq k$ we obtain $\sigma_j\sigma_k\sigma_j = -\sigma_k$. In general we have

$$\sigma_j\sigma_k\sigma_j = (-1)^{1+\delta_{jk}+\delta_{0k}+\delta_{0j}}\sigma_k, \quad j = 0, 1, 2, 3.$$

(iii) We obtain the unitary and hermitian matrices

$$\begin{aligned}
(\sigma_1 \otimes \sigma_1)(\sigma_1 \otimes \sigma_1)(\sigma_1 \otimes \sigma_1) &= \sigma_1 \otimes \sigma_1, \\
(\sigma_1 \otimes \sigma_1)(\sigma_2 \otimes \sigma_2)(\sigma_1 \otimes \sigma_1) &= \sigma_2 \otimes \sigma_2, \\
(\sigma_1 \otimes \sigma_1)(\sigma_3 \otimes \sigma_3)(\sigma_1 \otimes \sigma_1) &= \sigma_3 \otimes \sigma_3.
\end{aligned}$$

(iv) We obtain the normalized vector

$$(\sigma_1 \otimes \sigma_1)\left(\begin{pmatrix} 1 \\ 0 \end{pmatrix} \otimes \begin{pmatrix} 0 \\ 1 \end{pmatrix}\right) = \begin{pmatrix} 0 \\ 1 \end{pmatrix} \otimes \begin{pmatrix} 1 \\ 0 \end{pmatrix}.$$

Thus the spin's are flipped around.

(v) We find the normalized vectors

$$(\sigma_1 \otimes \sigma_1 \otimes \cdots \otimes \sigma_1)(\mathbf{u} \otimes \mathbf{u} \otimes \cdots \otimes \mathbf{u}) = \mathbf{v} \otimes \mathbf{v} \otimes \cdots \otimes \mathbf{v}$$
$$(\sigma_1 \otimes \sigma_1 \otimes \cdots \otimes \sigma_1)(\mathbf{v} \otimes \mathbf{v} \otimes \cdots \otimes \mathbf{v}) = \mathbf{u} \otimes \mathbf{u} \otimes \cdots \otimes \mathbf{u}.$$

Thus all the spins are flipped around.

Problem 8.   Consider the Pauli spin matrices σ_1, σ_2, σ_3 and the *Hadamard basis* in $\mathbb{C}^2$

$$\frac{1}{\sqrt{2}}\begin{pmatrix} 1 \\ 1 \end{pmatrix}, \qquad \frac{1}{\sqrt{2}}\begin{pmatrix} 1 \\ -1 \end{pmatrix}.$$

(i) Find the three normalized states in $\mathbb{C}^4$

$$(\sigma_1 \otimes \sigma_1)\left(\frac{1}{\sqrt{2}}\begin{pmatrix} 1 \\ 1 \end{pmatrix} \otimes \frac{1}{\sqrt{2}}\begin{pmatrix} 1 \\ -1 \end{pmatrix}\right),$$

$$(\sigma_2 \otimes \sigma_2)\left(\frac{1}{\sqrt{2}}\begin{pmatrix} 1 \\ 1 \end{pmatrix} \otimes \frac{1}{\sqrt{2}}\begin{pmatrix} 1 \\ -1 \end{pmatrix}\right),$$

$$(\sigma_3 \otimes \sigma_3)\left(\frac{1}{\sqrt{2}}\begin{pmatrix} 1 \\ 1 \end{pmatrix} \otimes \frac{1}{\sqrt{2}}\begin{pmatrix} 1 \\ -1 \end{pmatrix}\right).$$

(ii) Consider the Pauli spin matrices σ_3, σ_1, σ_2. The eigenvalues are given by $+1$ and -1 with the corresponding normalized eigenvectors

$$\begin{pmatrix} 1 \\ 0 \end{pmatrix}, \begin{pmatrix} 0 \\ 1 \end{pmatrix}, \quad \frac{1}{\sqrt{2}}\begin{pmatrix} 1 \\ 1 \end{pmatrix}, \frac{1}{\sqrt{2}}\begin{pmatrix} 1 \\ -1 \end{pmatrix}, \quad \frac{1}{\sqrt{2}}\begin{pmatrix} i \\ 1 \end{pmatrix}, \frac{1}{\sqrt{2}}\begin{pmatrix} -i \\ 1 \end{pmatrix}.$$

Consider the three 4×4 matrices $\sigma_1 \otimes \sigma_1$, $\sigma_2 \otimes \sigma_2$, $\sigma_3 \otimes \sigma_3$. Find the eigenvalues. Show that the eigenvectors can be given as product states (unentangled states), but also as entangled states (i.e. they cannot be written as product states). Explain.

Solution 8.   (i) We find

$$(\sigma_1 \otimes \sigma_1)\left(\frac{1}{\sqrt{2}}\begin{pmatrix} 1 \\ 1 \end{pmatrix} \otimes \frac{1}{\sqrt{2}}\begin{pmatrix} 1 \\ -1 \end{pmatrix}\right) = -\frac{1}{\sqrt{2}}\begin{pmatrix} 1 \\ 1 \end{pmatrix} \otimes \frac{1}{\sqrt{2}}\begin{pmatrix} 1 \\ -1 \end{pmatrix}.$$

Thus we have an eigenvalue equation with eigenvalue -1. We find

$$(\sigma_2 \otimes \sigma_2)\left(\frac{1}{\sqrt{2}}\begin{pmatrix} 1 \\ 1 \end{pmatrix} \otimes \frac{1}{\sqrt{2}}\begin{pmatrix} 1 \\ -1 \end{pmatrix}\right) = \frac{1}{\sqrt{2}}\begin{pmatrix} 1 \\ -1 \end{pmatrix} \otimes \frac{1}{\sqrt{2}}\begin{pmatrix} 1 \\ 1 \end{pmatrix}$$

and

$$(\sigma_3 \otimes \sigma_3)\left(\frac{1}{\sqrt{2}}\begin{pmatrix} 1 \\ 1 \end{pmatrix} \otimes \frac{1}{\sqrt{2}}\begin{pmatrix} 1 \\ -1 \end{pmatrix}\right) = \frac{1}{\sqrt{2}}\begin{pmatrix} 1 \\ -1 \end{pmatrix} \otimes \frac{1}{\sqrt{2}}\begin{pmatrix} 1 \\ 1 \end{pmatrix}.$$

(ii) Since the eigenvalues of the Pauli matrices are given by $+1$ and -1, the eigenvalues of the 4×4 matrices $\sigma_1 \otimes \sigma_1$, $\sigma_2 \otimes \sigma_2$, $\sigma_3 \otimes \sigma_3$ are $+1$ (twice) and -1 (twice). Obviously

$$\begin{pmatrix} 1 \\ 0 \end{pmatrix} \otimes \begin{pmatrix} 1 \\ 0 \end{pmatrix}, \quad \begin{pmatrix} 1 \\ 0 \end{pmatrix} \otimes \begin{pmatrix} 0 \\ 1 \end{pmatrix}, \quad \begin{pmatrix} 0 \\ 1 \end{pmatrix} \otimes \begin{pmatrix} 1 \\ 0 \end{pmatrix}, \quad \begin{pmatrix} 0 \\ 1 \end{pmatrix} \otimes \begin{pmatrix} 0 \\ 1 \end{pmatrix}$$

are four product eigenstates of $\sigma_3 \otimes \sigma_3$. Product eigenstates of $\sigma_1 \otimes \sigma_1$ are

$$\frac{1}{2}\begin{pmatrix} 1 \\ 1 \end{pmatrix} \otimes \begin{pmatrix} 1 \\ 1 \end{pmatrix}, \quad \frac{1}{2}\begin{pmatrix} 1 \\ 1 \end{pmatrix} \otimes \begin{pmatrix} 1 \\ -1 \end{pmatrix}, \quad \frac{1}{2}\begin{pmatrix} 1 \\ -1 \end{pmatrix} \otimes \begin{pmatrix} 1 \\ 1 \end{pmatrix}, \quad \frac{1}{2}\begin{pmatrix} 1 \\ -1 \end{pmatrix} \otimes \begin{pmatrix} 1 \\ -1 \end{pmatrix}.$$

Product eigenstates of $\sigma_2 \otimes \sigma_2$ are

$$\frac{1}{2}\begin{pmatrix} i \\ 1 \end{pmatrix} \otimes \begin{pmatrix} i \\ 1 \end{pmatrix}, \quad \frac{1}{2}\begin{pmatrix} i \\ 1 \end{pmatrix} \otimes \begin{pmatrix} -i \\ 1 \end{pmatrix}, \quad \frac{1}{2}\begin{pmatrix} -i \\ 1 \end{pmatrix} \otimes \begin{pmatrix} i \\ 1 \end{pmatrix}, \quad \frac{1}{2}\begin{pmatrix} -i \\ 1 \end{pmatrix} \otimes \begin{pmatrix} -i \\ 1 \end{pmatrix}.$$

All three 4×4 matrices also admit the *Bell basis*

$$\frac{1}{\sqrt{2}}\begin{pmatrix} 1 \\ 0 \\ 0 \\ 1 \end{pmatrix}, \quad \frac{1}{\sqrt{2}}\begin{pmatrix} 0 \\ 1 \\ 1 \\ 0 \end{pmatrix}, \quad \frac{1}{\sqrt{2}}\begin{pmatrix} 1 \\ 0 \\ 0 \\ -1 \end{pmatrix}, \quad \frac{1}{\sqrt{2}}\begin{pmatrix} 0 \\ 1 \\ -1 \\ 0 \end{pmatrix}$$

as eigenvectors which are maximally entangled.

Problem 9. Consider the Pauli spin matrices $\sigma_0 = I_2$, σ_1, σ_2, σ_3. These matrices are unitary and hermitian.
(i) Is the 4×4 matrix

$$A = \frac{1}{\sqrt{2}}\begin{pmatrix} \sigma_0 & \sigma_1 \\ \sigma_2 & \sigma_3 \end{pmatrix}$$

unitary?
(ii) Is the 4×4 matrix

$$B = \frac{1}{\sqrt{2}}\begin{pmatrix} \sigma_0 & \sigma_1 \\ -i\sigma_2 & \sigma_3 \end{pmatrix}$$

unitary?
(iii) Let I_n be the $n \times n$ unitary matrix. Is the $(2n \times 2n)$ matrix

$$C = \frac{1}{\sqrt{2}}\begin{pmatrix} I_n & I_n \\ iI_n & -iI_n \end{pmatrix}$$

unitary?

Solution 9. (i) No. We find $AA^* \neq I_4$.
(ii) Yes. We find $BB^* = I_4$.
(iii) Yes. We find $CC^* = I_{2n}$.

Problem 10. Consider the *Hilbert space* $M_2(\mathbb{C})$ of the 2×2 matrices over the complex numbers with the scalar product

$$\langle A, B \rangle := \text{tr}(AB^*), \qquad A, B \in M_2(\mathbb{C}).$$

Show that the rescaled Pauli matrices $\mu_j = \frac{1}{\sqrt{2}}\sigma_j$, $j = 1, 2, 3$

$$\mu_1 = \frac{1}{\sqrt{2}}\begin{pmatrix} 0 & 1 \\ 1 & 0 \end{pmatrix}, \quad \mu_2 = \frac{1}{\sqrt{2}}\begin{pmatrix} 0 & -i \\ i & 0 \end{pmatrix}, \quad \mu_3 = \frac{1}{\sqrt{2}}\begin{pmatrix} 1 & 0 \\ 0 & -1 \end{pmatrix}$$

plus the rescaled 2×2 identity matrix

$$\mu_0 = \frac{1}{\sqrt{2}}\begin{pmatrix} 1 & 0 \\ 0 & 1 \end{pmatrix}$$

form an orthonormal basis in the Hilbert space $M_2(\mathbb{C})$.

Solution 10.   The Hilbert space $\mathcal{H}$ is four dimensional. Since for the scalar products we find

$$\langle \mu_j, \mu_k \rangle = \delta_{jk}$$

and μ_0, μ_1, μ_2, μ_3 are nonzero matrices we have an orthonormal basis.

Problem 11.   Show that the four *gamma matrices*

$$\gamma_1 = \begin{pmatrix} 0 & 0 & 0 & -i \\ 0 & 0 & -i & 0 \\ 0 & i & 0 & 0 \\ i & 0 & 0 & 0 \end{pmatrix}, \quad \gamma_2 = \begin{pmatrix} 0 & 0 & 0 & -1 \\ 0 & 0 & 1 & 0 \\ 0 & 1 & 0 & 0 \\ -1 & 0 & 0 & 0 \end{pmatrix},$$

$$\gamma_3 = \begin{pmatrix} 0 & 0 & -i & 0 \\ 0 & 0 & 0 & i \\ i & 0 & 0 & 0 \\ 0 & -i & 0 & 0 \end{pmatrix}, \quad \gamma_4 = \begin{pmatrix} 1 & 0 & 0 & 0 \\ 0 & 1 & 0 & 0 \\ 0 & 0 & -1 & 0 \\ 0 & 0 & 0 & -1 \end{pmatrix}$$

can be written as Kronecker products of the Pauli spin matrices σ_1, σ_2, σ_3 including $\sigma_0 = I_2$.

Solution 11.   We have

$$\gamma_1 = \sigma_2 \otimes \sigma_1, \quad \gamma_2 = \sigma_1 \otimes (-i\sigma_2), \quad \gamma_3 = \sigma_2 \otimes \sigma_3, \quad \gamma_4 = \sigma_3 \otimes \sigma_0.$$

Problem 12.   Consider the hermitian matrices

$$K_1 = \sigma_1 \otimes \sigma_2 \otimes I_2 \otimes I_2 + I_2 \otimes \sigma_1 \otimes \sigma_2 \otimes I_2 + I_2 \otimes I_2 \otimes \sigma_1 \otimes \sigma_2 + \sigma_2 \otimes I_2 \otimes I_2 \otimes \sigma_1$$

and

$$K_2 = \sigma_1 \otimes \sigma_2 \otimes I_2 \otimes I_2 + I_2 \otimes \sigma_1 \otimes \sigma_2 \otimes I_2 + I_2 \otimes I_2 \otimes \sigma_1 \otimes \sigma_2.$$

For K_1 we have cyclic boundary conditions and for K_2 we have open end boundary conditions.
(i) Consider the unitary, hermitian and diagonal matrix

$$S = \sigma_3 \otimes \sigma_3 \otimes \sigma_3 \otimes \sigma_3.$$

Find the commutators $[K_1, S]$, $[K_2, S]$.
(ii) Find the commutator $[S, [K_1, K_2]]$.
(iii) Find the commutators

$$[\sigma_3 \otimes \sigma_3, \sigma_1 \otimes \sigma_1]$$
$$[\sigma_3 \otimes \sigma_3 \otimes \sigma_3, \sigma_1 \otimes \sigma_1 \otimes I_2 + I_2 \otimes \sigma_1 \otimes \sigma_1]$$
$$[\sigma_3 \otimes \sigma_3 \otimes \sigma_3 \otimes \sigma_3, \sigma_1 \otimes \sigma_1 \otimes I_2 \otimes I_2 + I_2 \otimes \sigma_1 \otimes \sigma_1 \otimes I_2 + I_2 \otimes I_2 \otimes \sigma_1 \otimes \sigma_1].$$

Solution 12.   (i) Straightforward calculation yields $[K_1, S] = [K_2, S] = 0_{16}$, i.e. K_1 and K_2 commute with S.
(ii) Applying the *Jacobi identity*

$$[K_1, [K_2, S]] + [S, [K_1, K_2]] + [K_2, [S, K_1]] = 0_{16}$$

provides $[S, [K_1, K_2]] = 0_{16}$. Note that $[K_1, K_2] \neq 0_{16}$.

(iii) For all three the commutator is the appropriate zero matrix. We find the same results when we replace σ_1 by σ_2. The operator

$$X = \sigma_3 \otimes \sigma_3 \otimes \cdots \otimes \sigma_3$$

is sometimes called the *parity operator*. X is diagonal matrix.

Problem 13.　Let σ_1, σ_2, σ_3 be the Pauli spin matrices and $\sigma_0 = I_2$ the 2×2 identity matrix.

(i) Show that any 2×2 matrix can be written as linear combination

$$\sum_{j=0}^{3} c_j \sigma_j, \quad c_j \in \mathbb{C}.$$

(ii) Show that any 4×4 matrix can be written as linear combination

$$\sum_{j_1=0}^{3} \sum_{j_2=0}^{3} c_{j_1 j_2} \sigma_{j_1} \otimes \sigma_{j_2}, \quad c_{j_1 j_2} \in \mathbb{C}.$$

Solution 13.　(i) The three Pauli matrices together with the 2×2 identity matrix form a orthogonal basis in Hilbert space of 2×2 matrices.

(ii) Since the sixteen 4×4 matrices $\sigma_j \otimes \sigma_k$ $(j, k = 0, 1, 2, 3)$ form an orthogonal basis in the Hilbert space $M_4(\mathbb{C})$ any 4×4 matrix over $\mathbb{C}$ can be written as a linear combination of these sixteen matrices.

Show that any $2^n \times 2^n$ matrix can be written as linear combination

$$\sum_{j_1=0}^{3} \sum_{j_2=0}^{3} \cdots \sum_{j_n=0}^{3} c_{j_1 j_2 \ldots j_n} \sigma_{j_1} \otimes \sigma_{j_2} \otimes \cdots \otimes \sigma_{j_n}, \quad c_{j_1 j_2 \ldots j_n} \in \mathbb{C}.$$

Problem 14.　(i) Consider the Pauli spin matrices σ_0, σ_1, σ_2, σ_3, where we included the 2×2 identity matrix σ_0. We know that

$$\sigma_k \sigma_j \sigma_k = (-1)^{(1-\delta_{k0})(1-\delta_{kj})} \sigma_j, \quad j, k = 0, 1, 2, 3$$

where δ_{jk} denotes the Kronecker delta. Use this result to calculate the 8×8 unitary matrices

$$(\sigma_j \otimes \sigma_k \otimes \sigma_\ell)(\sigma_1 \otimes \sigma_2 \otimes \sigma_3)(\sigma_j \otimes \sigma_k \otimes \sigma_\ell), \quad j, k, \ell = 0, 1, 2, 3.$$

Solution 14.　We obtain

$$(\sigma_j \otimes \sigma_k \otimes \sigma_\ell)(\sigma_1 \otimes \sigma_2 \otimes \sigma_3)(\sigma_j \otimes \sigma_k \otimes \sigma_\ell) = (\sigma_j \sigma_1 \sigma_j) \otimes (\sigma_k \sigma_2 \sigma_k) \otimes (\sigma_\ell \sigma_3 \sigma_\ell) =$$
$$(-1)^{(1-\delta_{j0})(1-\delta_{j1})}(-1)^{(1-\delta_{k0})(1-\delta_{k2})}(-1)^{(1-\delta_{\ell0})(1-\delta_{\ell3})}(\sigma_1 \otimes \sigma_2 \otimes \sigma_3).$$

Problem 15.   Let H be an hermitian $n \times n$ matrix and U be a unitary matrix such that $UHU^* = H$. We call H invariant under U. From $UHU^* = H$ it follows that $[U, H] = 0_n$. If $\mathbf{v}$ is an eigenvector of H, i.e. $H\mathbf{v} = \lambda\mathbf{v}$, then $U\mathbf{v}$ is also an eigenvector of H since

$$H(U\mathbf{v}) = (HU)\mathbf{v} = (UH)\mathbf{v} = U(H\mathbf{v}) = \lambda(U\mathbf{v}).$$

The set of unitary matrices U_j $(j = 1, \ldots, m)$ that leave a given hermitian matrix H invariant, i.e. $U_j H U_j^* = H$ $(j = 1, \ldots, m)$ form a group under matrix multiplication.
(i) Find all 2×2 hermitian matrices H such that $[H, U] = 0_2$, where $U = \sigma_1$.
(ii) Find all 4×4 matrices H such that $[H, U] = 0_4$, where $U = \sigma_1 \otimes \sigma_1$.

Solution 15.   (i) From the condition $[H, \sigma_1] = 0_2$ we arrive at $h_{11} = h_{22}$ and $h_{12} = h_{21}$. Thus

$$H = \begin{pmatrix} h_{11} & h_{12} \\ h_{12} & h_{11} \end{pmatrix}$$

with h_{11} real and h_{12} real.
(ii) From the condition $[H, \sigma_1 \otimes \sigma_1] = 0_2 \otimes 0_2$ we obtain

$$h_{11} = h_{44}, \qquad h_{22} = h_{22} = h_{33}$$

with

$$h_{12} = h_{43}, \ \ h_{13} = h_{42}, \ \ h_{14} = h_{41}, \ \ h_{23} = h_{32}, \ \ h_{24} = h_{31}, \ \ h_{21} = h_{34}$$

i.e

$$H = \begin{pmatrix} h_{11} & h_{12} & h_{13} & h_{14} \\ h_{21} & h_{22} & h_{23} & h_{24} \\ h_{24} & h_{23} & h_{22} & h_{21} \\ h_{14} & h_{13} & h_{12} & h_{11} \end{pmatrix}$$

with $h_{11}, h_{22}, h_{14}, h_{23}$ real and $h_{12} = \bar{h}_{21}, h_{13} = \bar{h}_{24}$.

Problem 16.   Let A be an $m \times m$ matrix and B be an $n \times n$ matrix. Let λ be an eigenvalue of A with normalized eigenvector $|\mathbf{u}\rangle$ and μ be an eigenvalue of B with normalized eigenvector $|\mathbf{v}\rangle$. Then the matrix

$$A \otimes I_n + I_m \otimes B$$

has the eigenvalue $\lambda + \mu$ with the corresponding normalized eigenvector $|\mathbf{u}\rangle \otimes |\mathbf{v}\rangle$.
(i) Find the eigenvalues and eigenvectors of the $(mn) \times (mn)$ matrix

$$(A \otimes I_n + I_m \otimes B)^2.$$

(ii) Apply the result to the spin matrices for spin-1 and spin-$\frac{1}{2}$, respectively

$$A = \frac{1}{\sqrt{2}} \begin{pmatrix} 0 & 1 & 0 \\ 1 & 0 & 1 \\ 0 & 1 & 0 \end{pmatrix}, \qquad B = \frac{1}{2} \begin{pmatrix} 0 & 1 \\ 1 & 0 \end{pmatrix}.$$

Solution 16.   (i) We have

$$(A \otimes I_n + I_m \otimes B)^2 = A^2 \otimes I_n + I_m \otimes B^2 + 2(A \otimes B).$$

Thus $|\mathbf{u}\rangle \otimes |\mathbf{v}\rangle$ is a normalized eigenvector since

$$(A^2 \otimes I_n + I_m \otimes B^2 + 2(A \otimes B))(|\mathbf{u}\rangle \otimes |\mathbf{v}\rangle) = (\lambda^2 + \mu^2 + 2\lambda\mu)(|\mathbf{u}\rangle \otimes |\mathbf{v}\rangle)$$

and the eigenvalue is given by $\lambda^2 + \mu^2 + 2\lambda\mu$.

(ii) The spin-1 matrix A admits the eigenvalues $\lambda_1 = +1$, $\lambda_2 = 0$, $\lambda_3 = -1$ with the corresponding normalized eigenvectors

$$\mathbf{u}_1 = \frac{1}{2}\begin{pmatrix} 1 \\ \sqrt{2} \\ 1 \end{pmatrix}, \quad \mathbf{u}_2 = \frac{1}{\sqrt{2}}\begin{pmatrix} 1 \\ 0 \\ -1 \end{pmatrix}, \quad \mathbf{u}_3 = \frac{1}{2}\begin{pmatrix} 1 \\ -\sqrt{2} \\ 1 \end{pmatrix}.$$

The spin matrix B admits the eigenvalues $1/2$, $-1/2$ with the corresponding normalized eigenvectors

$$\mathbf{v}_1 = \frac{1}{\sqrt{2}}\begin{pmatrix} 1 \\ 1 \end{pmatrix}, \quad \mathbf{v}_2 = \frac{1}{\sqrt{2}}\begin{pmatrix} 1 \\ -1 \end{pmatrix}.$$

Thus the six eigenvalues are

$$\lambda_j^2 + \mu_k^2 + 2\lambda_j\mu_k, \quad j = 1, 2, 3 \quad k = 1, 2$$

with the corresponding six normalized eigenvectors

$$\mathbf{u}_j \otimes \mathbf{v}_k, \quad j = 1, 2, 3 \quad k = 1, 2.$$

Problem 17. Let $s = 0, 1/2, 1, 3/2, 2, \dots$ (spin) and $n = 2s + 1$. For example, for spin $s = 1/2$ we have $n = 2$. Consider the $n \times n$ matrix V with the entries

$$V_{jk} := \exp(\beta J(s - j + 1)(s - k + 1)), \quad j, k = 1, \dots, n$$

where β, J are positive constants.

(i) Find the matrix V for $s = 1/2$ and show that the matrix can be diagonalized by RVR^{-1}, where R is the *Hadamard matrix*

$$R = \frac{1}{\sqrt{2}}\begin{pmatrix} 1 & 1 \\ 1 & -1 \end{pmatrix}.$$

(ii) Consider the case with spin equal to 1, i.e. $n = 3$. Show that the matrix V can be block-diagonalized.

Solution 17. (i) We obtain the symmetric matrix

$$V = \begin{pmatrix} e^{\beta J/4} & e^{-\beta J/4} \\ e^{-\beta J/4} & e^{\beta J/4} \end{pmatrix}.$$

Then

$$RVR^{-1} = \begin{pmatrix} 2\cosh(\beta J) & 0 \\ 0 & 2\sinh(\beta J) \end{pmatrix}$$

with $R = R^{-1}$. Thus the eigenvalues of V are $2\cosh(\beta J)$ and $2\sinh(\beta J)$.

(ii) We obtain the symmetric matrix

$$V = \begin{pmatrix} e^{\beta J} & 1 & e^{-\beta J} \\ 1 & 1 & 1 \\ e^{-\beta J} & 1 & e^{\beta J} \end{pmatrix}.$$

The matrix R which block diagonalize V is the unitary matrix

$$R = \frac{1}{\sqrt{2}}\begin{pmatrix} 1 & 0 & 1 \\ 0 & \sqrt{2} & 0 \\ 1 & 0 & -1 \end{pmatrix}.$$

with $R = R^{-1}$ and

$$RVR^{-1} = \begin{pmatrix} e^{\beta J} + e^{-\beta J} & \sqrt{2} & 0 \\ \sqrt{2} & 1 & 0 \\ 0 & 0 & e^{\beta J} - e^{-\beta J} \end{pmatrix}.$$

Thus $e^{\beta J} - e^{-\beta J}$ is an eigenvalue of V.

Problem 18. Let $s = 1/2, 1, 3/2, 2, \ldots$ be the spin. Let $n = 2s + 1$, i.e. for $s = 1/2$ we have $n = 2$, for $s = 1$ we have $n = 3$ etc. Consider the $n \times n$ matrix $V_s = (V_{jk})$ with

$$V_{jk} := \exp(c(s - j + 1)(s - k + 1))$$

where $j, k = 1, 2, \ldots, n$ and c is a positive constant.
(i) Let s be positive integer (i.e. $1, 2, 3, \ldots$) with $n = 2s + 1$ and the $n \times n$ matrix

$$R_s = \frac{1}{\sqrt{2}} \begin{pmatrix} 1 & & & & & & 1 \\ & \ddots & & & & \iddots & \\ & & 1 & 0 & 1 & & \\ & & 0 & \sqrt{2} & 0 & & \\ & & 1 & 0 & -1 & & \\ & \iddots & & & & \ddots & \\ 1 & & & & & & -1 \end{pmatrix}.$$

Note that $R_s = R_s^{-1}$ and R_s is unitary. Let $s = 2$, i.e. $n = 5$. Find $R_s V_s R_s^{-1}$.
(iii) Let s be $1/2, 3/2, \ldots$ with $n = 2s + 1$ and the $n \times n$ matrix

$$R_s = \frac{1}{\sqrt{2}} \begin{pmatrix} 1 & & & & & 1 \\ & \ddots & & & \iddots & \\ & & 1 & 1 & & \\ & & 1 & -1 & & \\ & \iddots & & & \ddots & \\ 1 & & & & & -1 \end{pmatrix}.$$

Note that $R_s^{-1} = R_s$. Let $s = 3/2$, i.e. $n = 4$. Find $R_s V_s R_s^{-1}$.

Solution 18. (i) We have

$$R_2 = \frac{1}{\sqrt{2}} \begin{pmatrix} 1 & 0 & 0 & 0 & 1 \\ 0 & 1 & 0 & 1 & 0 \\ 0 & 0 & \sqrt{2} & 0 & 0 \\ 0 & 1 & 0 & -1 & 0 \\ 1 & 0 & 0 & 0 & -1 \end{pmatrix}$$

and

$$V_2 = \begin{pmatrix} e^{4c} & e^{2c} & 1 & e^{-2c} & e^{-4c} \\ e^{2c} & e^{c} & 1 & e^{-c} & e^{-2c} \\ 1 & 1 & 1 & 1 & 1 \\ e^{-2c} & e^{-c} & 1 & e^{c} & e^{2c} \\ e^{-4c} & e^{-2c} & 1 & e^{2c} & e^{4c} \end{pmatrix}.$$

Thus we obtain the direct sum of a 3×3 matrix and a 2×2 matrix

$$R_2 V_2 R_2^{-1} = \begin{pmatrix} 2\cosh(4c) & 2\cosh(2c) & \sqrt{2} \\ 2\cosh(2c) & 2\cosh(c) & \sqrt{2} \\ \sqrt{2} & \sqrt{2} & 1 \end{pmatrix} \oplus \begin{pmatrix} 2\sinh(c) & 2\sinh(2c) \\ 2\sinh(2c) & 2\sinh(4c) \end{pmatrix}.$$

(ii) We have

$$R_{3/2} = \frac{1}{\sqrt{2}} \begin{pmatrix} 1 & 0 & 0 & 1 \\ 0 & 1 & 1 & 0 \\ 0 & 1 & -1 & 0 \\ 1 & 0 & 0 & -1 \end{pmatrix} = R_{3/2}^{-1}$$

and

$$V_{3/2} = \begin{pmatrix} e^{9c/4} & e^{3c/4} & e^{-3c/4} & e^{-9c/4} \\ e^{3c/4} & e^{c/4} & e^{-c/4} & e^{-3c/4} \\ e^{-3c/4} & e^{-c/4} & e^{c/4} & e^{3c/4} \\ e^{-9c/4} & e^{-3c/4} & e^{3c/4} & e^{9c/4} \end{pmatrix}.$$

Thus we obtain the direct sum of two 2×2 matrices

$$R_{3/2} V_{3/2} R_{3/2}^{-1} = \begin{pmatrix} 2\cosh(9c/4) & 2\cosh(3c/4) \\ 2\cosh(3c/4) & 2\cosh(c/4) \end{pmatrix} \oplus \begin{pmatrix} 2\sinh(c/4) & 2\sinh(3c/4) \\ 2\sinh(3c/4) & 2\sinh(9c/4) \end{pmatrix}.$$

Problem 19. Let $\|X\|_F$ be the *Frobenius norm* of an $n \times n$ matrix X over $\mathbb{C}$, i.e.

$$\|X\|_F^2 := \mathrm{tr}(XX^*).$$

Then the inequality

$$\|[X, Y]\|_F \le \sqrt{2} \|X\|_F \cdot \|Y\|_F$$

holds, where $[,]$ denotes the commutator. Let σ_1, σ_2, σ_3 be the Pauli spin matrices. Consider the cases $X = \sigma_1$, $Y = \sigma_2$; $X = \sigma_2$, $Y = \sigma_3$; $X = \sigma_3$, $Y = \sigma_1$. Find the left and right-hand sides of the inequality for the three cases. Discuss.

Solution 19. Since the Pauli matrices hermitian and $\sigma_j^2 = I_2$ $(j = 1, 2, 3)$ we obtain

$$\|\sigma_j\|_F^2 = 2$$

for $j = 1, 2, 3$. Thus the right-hand side is given by $2\sqrt{2}$. Since

$$[\sigma_1, \sigma_2] = 2i\sigma_3, \quad [\sigma_2, \sigma_3] = 2i\sigma_1, \quad [\sigma_3, \sigma_1] = 2i\sigma_2$$

the left-hand side in all three cases is also given by $2\sqrt{2}$. i.e. the inequality becomes an equality. Study the case $X = \sigma_1 \otimes \sigma_1$ and $Y = \sigma_2 \otimes \sigma_2$.

Problem 20. The spin-1 matrices can also be represented by the hermitian matrices

$$S_1 = \begin{pmatrix} 0 & 0 & 0 \\ 0 & 0 & -i \\ 0 & i & 0 \end{pmatrix}, \quad S_2 = \begin{pmatrix} 0 & 0 & i \\ 0 & 0 & 0 \\ -i & 0 & 0 \end{pmatrix}, \quad S_3 = \begin{pmatrix} 0 & -i & 0 \\ i & 0 & 0 \\ 0 & 0 & 0 \end{pmatrix}$$

with the eigenvalues $+1$, 0, -1. They are utilized in Maxwell's equation.
(i) Find the corresponding normalized eigenvectors for the three matrices.
(ii) Find the commutators $[S_1, S_2]$, $[S_2, S_3]$, $[S_3, S_1]$.

Solution 20.   (i) The normalized eigenvectors for S_1 are

$$\frac{1}{\sqrt{2}}\begin{pmatrix} 0 \\ 1 \\ i \end{pmatrix}, \quad \begin{pmatrix} 1 \\ 0 \\ 0 \end{pmatrix}, \quad \frac{1}{\sqrt{2}}\begin{pmatrix} 0 \\ 1 \\ -i \end{pmatrix}.$$

The eigenvectors for S_2 are

$$\frac{1}{\sqrt{2}}\begin{pmatrix} 1 \\ 0 \\ -i \end{pmatrix}, \quad \begin{pmatrix} 0 \\ 1 \\ 0 \end{pmatrix}, \quad \frac{1}{\sqrt{2}}\begin{pmatrix} 1 \\ 0 \\ i \end{pmatrix}.$$

The eigenvectors for S_3 are

$$\begin{pmatrix} 1 \\ 0 \\ 0 \end{pmatrix}, \quad \begin{pmatrix} 0 \\ 1 \\ 0 \end{pmatrix}, \quad \begin{pmatrix} 0 \\ 0 \\ 1 \end{pmatrix}.$$

(ii) For the commutators we find $[S_1, S_2] = iS_3$, $[S_2, S_3] = iS_1$, $[S_3, S_1] = iS_2$.

Problem 21.   Consider the two *spin-1 matrices*

$$S = \frac{1}{\sqrt{2}}\begin{pmatrix} 0 & 1 & 0 \\ 1 & 0 & 1 \\ 0 & 1 & 0 \end{pmatrix}, \quad \widetilde{S} = \begin{pmatrix} 0 & i & 0 \\ -i & 0 & 0 \\ 0 & 0 & 0 \end{pmatrix}.$$

Both matrices are hermitian and admit the eigenvalues $+1$, 0, -1. Describe a method to construct a 3×3 unitary matrix U such that $U^*SU = \widetilde{S}$. Hint. Calculate the normalized eigenvectors of S and $\widetilde{S}$.

Solution 21.   Method. From the normalized eigenvectors we construct unitary matrices V and W such that

$$V^*\widetilde{S}V = \mathrm{diag}(1, 0, -1), \quad W^*SW = \mathrm{diag}(1, 0, -1).$$

Now $V^* = V^{-1}$, $W^* = W^{-1}$ and we learn that $V^{-1}\widetilde{S}V = W^{-1}SW$. Consequently $U = WV^{-1} = WV^*$. The normalized eigenvectors of $\widetilde{S}$ for the eigenvalues $+1$, 0, -1 are

$$\frac{1}{\sqrt{2}}\begin{pmatrix} 1 \\ -i \\ 0 \end{pmatrix}, \quad \begin{pmatrix} 0 \\ 0 \\ 1 \end{pmatrix}, \quad \frac{1}{\sqrt{2}}\begin{pmatrix} 1 \\ i \\ 0 \end{pmatrix}.$$

This provides the unitary matrix

$$V = \begin{pmatrix} 1/\sqrt{2} & 0 & 1/\sqrt{2} \\ -i/\sqrt{2} & 0 & i/\sqrt{2} \\ 0 & 1 & 0 \end{pmatrix} \Rightarrow V^* = \begin{pmatrix} 1/\sqrt{2} & i/\sqrt{2} & 0 \\ 0 & 0 & 1 \\ 1/\sqrt{2} & -i/\sqrt{2} & 0 \end{pmatrix}.$$

The normalized eigenvectors of S for the eigenvalues $+1$, 0, -1 are

$$\frac{1}{2}\begin{pmatrix} 1 \\ \sqrt{2} \\ 1 \end{pmatrix}, \quad \frac{1}{\sqrt{2}}\begin{pmatrix} 1 \\ 0 \\ -1 \end{pmatrix}, \quad \frac{1}{2}\begin{pmatrix} 1 \\ i \\ 0 \end{pmatrix}.$$

This provides the unitary matrix

$$W = \begin{pmatrix} 1/\sqrt{2} & 1/\sqrt{2} & 1/2 \\ 1/\sqrt{2} & 0 & -1/\sqrt{2} \\ 1/2 & -1/\sqrt{2} & 1/2 \end{pmatrix} \Rightarrow W^* = \begin{pmatrix} 1/2 & 1/\sqrt{2} & 1/2 \\ 1/\sqrt{2} & 0 & -1/\sqrt{2} \\ 1/2 & -1/\sqrt{2} & 1/2 \end{pmatrix}.$$

Problem 22. Let S_1, S_2, S_3 be the *spin-1 matrices*

$$S_1 = \frac{1}{\sqrt{2}} \begin{pmatrix} 0 & 1 & 0 \\ 1 & 0 & 1 \\ 0 & 1 & 0 \end{pmatrix}, \quad S_2 = \frac{1}{\sqrt{2}} \begin{pmatrix} 0 & -i & 0 \\ i & 0 & -i \\ 0 & i & 0 \end{pmatrix}, \quad S_3 = \begin{pmatrix} 1 & 0 & 0 \\ 0 & 0 & 0 \\ 0 & 0 & -1 \end{pmatrix}$$

with eigenvalues $+1$, 0, -1 and commutators $[S_1, S_2] = iS_3$, $[S_2, S_3] = iS_1$, $[S_3, S_1] = iS_2$. The matrices are hermitian. Consider the hermitian 3×3 matrices

$$T_1 = \begin{pmatrix} 0 & 0 & 0 \\ 0 & 0 & -i \\ 0 & i & 0 \end{pmatrix}, \quad T_2 = \begin{pmatrix} 0 & 0 & i \\ 0 & 0 & 0 \\ -i & 0 & 0 \end{pmatrix}, \quad T_3 = \begin{pmatrix} 0 & -i & 0 \\ i & 0 & 0 \\ 0 & 0 & 0 \end{pmatrix}$$

with eigenvalues $+1$, 0, -1 and commutation relations $[T_1, T_2] = iT_3$, $[T_2, T_3] = iT_1$, $[T_3, T_1] = iT_2$. Note that the matrices T_1, T_2, T_3 are used to represent Maxwell's equations of the photon which is a spin-1 particle. Show that one can find a unitary 3×3 matrix U such that

$$U^{-1}T_1 U = S_1, \qquad U^{-1}T_2 U = S_2, \qquad U^{-1}T_3 U = S_3.$$

Solution 22. Consider the permutation matrix (cyclic matrix)

$$P = \begin{pmatrix} 0 & 1 & 0 \\ 0 & 0 & 1 \\ 1 & 0 & 0 \end{pmatrix} \Rightarrow P^* = \begin{pmatrix} 0 & 0 & 1 \\ 1 & 0 & 0 \\ 0 & 1 & 0 \end{pmatrix}.$$

Then $P^*T_1 P = T_2$, $P^*T_2 P = T_3$. Thus the equations can now be rewritten as

$$U^*T_1 U = S_1, \qquad (PU)^*T_1(PU) = S_2, \qquad (P^*U)^*T_1(P^*U) = S_3.$$

An orthonormal set of eigenvectors of T_1 is given by

$$\left\{ \frac{1}{\sqrt{2}} \begin{pmatrix} 0 \\ 1 \\ i \end{pmatrix}, \begin{pmatrix} 1 \\ 0 \\ 0 \end{pmatrix}, \frac{1}{\sqrt{2}} \begin{pmatrix} 0 \\ 1 \\ -i \end{pmatrix} \right\}$$

corresponding to the eigenvalues 1, 0 and -1 respectively. Since P^*U diagonalizes T_1 to yield S_3 we have

$$P^*U = \frac{1}{\sqrt{2}} \begin{pmatrix} 0 & \sqrt{2} & 0 \\ 1 & 0 & 1 \\ i & 0 & -i \end{pmatrix} \begin{pmatrix} e^{ia} & 0 & 0 \\ 0 & e^{ib} & 0 \\ 0 & 0 & e^{ic} \end{pmatrix}$$

for some $a, b, c \in \mathbb{R}$. It follows that

$$U = P \frac{1}{\sqrt{2}} \begin{pmatrix} 0 & \sqrt{2} & 0 \\ 1 & 0 & 1 \\ i & 0 & -i \end{pmatrix} \begin{pmatrix} e^{ia} & 0 & 0 \\ 0 & e^{ib} & 0 \\ 0 & 0 & e^{ic} \end{pmatrix}.$$

Inserting this expression into $U^*T_1U = S_1$ and $(PU)^*T_1(PU) = S_2$ yields

$$\exp(i(a - b)) = -1 \quad \text{and} \quad \exp(i(b - c)) = 1.$$

Thus $b = c + 2k_1\pi$ and $a = c + (2k_2 + 1)\pi$ for some $k_1, k_2 \in \mathbb{Z}$. For example, setting $c = k_1 = k_2 = 0$ yields

$$U = \frac{1}{\sqrt{2}}\begin{pmatrix} -1 & 0 & 1 \\ -i & 0 & -i \\ 0 & \sqrt{2} & 0 \end{pmatrix} \Rightarrow U^* = U^{-1} = \frac{1}{\sqrt{2}}\begin{pmatrix} -1 & i & 0 \\ 0 & 0 & \sqrt{2} \\ 1 & i & 0 \end{pmatrix}.$$

A Maxima program that can do the job is given by

```
/* unitarytrans.mac */

T1: matrix([0,0,0],[0,0,-%i],[0,%i,0]);
P: matrix([0,1,0],[0,0,1],[1,0,0]);
V: (matrix([0,sqrt(2),0],[1,0,1],[%i,0,-%i])/sqrt(2))
   . diag_matrix(%e^(%i*a),%e^(%i*b),%e^(%i*c));
V.ctranspose(V);
U: P.V;
ctranspose(U).T1.U;
ctranspose(P.U).T1.P.U;
ctranspose(transpose(P).U).T1.transpose(P).U;
V: subst([a=%pi,b=0,c=0],V);
U: P.V;
ctranspose(U).T1.U;
ctranspose(P.U).T1.P.U;
ctranspose(transpose(P).U).T1.transpose(P).U;
```

Problem 23. An $n \times n$ matrix Π over $\mathbb{C}$ is called a *projection matrix* if $\Pi = \Pi^*$ and $\Pi^2 = \Pi$.
(i) Consider the Hilbert space $\mathbb{C}^4$. Show that the matrices

$$\Pi_1 = \frac{1}{2}(I_2 \otimes I_2 + \sigma_1 \otimes \sigma_1), \qquad \Pi_2 = \frac{1}{2}(I_2 \otimes I_2 - \sigma_1 \otimes \sigma_1)$$

are projection matrices in $\mathbb{C}^4$.
(ii) Find $\Pi_1\Pi_2$. Discuss.
(iii) Let e_1, e_2, e_3, e_4 be the standard basis in $\mathbb{C}^4$. Calculate the vectors

$$\Pi_1 e_j, \qquad \Pi_2 e_j, \qquad j = 1, 2, 3, 4$$

and show that we obtain 2 two-dimensional Hilbert spaces under these projections.
(iv) Do the matrices $I_2 \otimes I_2$, $\sigma_1 \otimes \sigma_1$ form a group under matrix multiplication? Prove or disprove?
(v) Show that the matrices

$$\Pi_j = \frac{1}{2}(I_2 \otimes I_2 + \sigma_j \otimes \sigma_j), \qquad j = 1, 2, 3$$

are projection matrices.

Solution 23. (i) We have $\Pi_1^* = \Pi_1$, $\Pi_2^* = \Pi_2$ and $\Pi_1^2 = \Pi_1$, $\Pi_2^2 = \Pi_2$. Thus Π_1 and Π_2 are projection matrices.

(ii) We obtain $\Pi_1 \Pi_2 = 0$. Let **u** be an arbitrary vector in $\mathbb{C}^4$. Then the vectors $\Pi_1 \mathbf{u}$ and $\Pi_2 \mathbf{u}$ are perpendicular.

(iii) We obtain

$$\Pi_1 e_1 = \frac{1}{2} \begin{pmatrix} 1 \\ 0 \\ 0 \\ 1 \end{pmatrix}, \quad \Pi_1 e_2 = \frac{1}{2} \begin{pmatrix} 0 \\ 1 \\ 1 \\ 0 \end{pmatrix}, \quad \Pi_1 e_3 = \frac{1}{2} \begin{pmatrix} 0 \\ 1 \\ 1 \\ 0 \end{pmatrix}, \quad \Pi_1 e_4 = \frac{1}{2} \begin{pmatrix} 1 \\ 0 \\ 0 \\ 1 \end{pmatrix}$$

$$\Pi_2 e_1 = \frac{1}{2} \begin{pmatrix} 1 \\ 0 \\ 0 \\ -1 \end{pmatrix}, \quad \Pi_2 e_2 = \frac{1}{2} \begin{pmatrix} 0 \\ 1 \\ -1 \\ 0 \end{pmatrix}, \quad \Pi_2 e_3 = \frac{1}{2} \begin{pmatrix} 0 \\ -1 \\ 1 \\ 0 \end{pmatrix}, \quad \Pi_2 e_4 = \frac{1}{2} \begin{pmatrix} -1 \\ 0 \\ 0 \\ 1 \end{pmatrix}.$$

Thus Π_1 projects into a two-dimensional Hilbert space spanned by the normalized vectors

$$\left\{ \frac{1}{\sqrt{2}} \begin{pmatrix} 1 \\ 0 \\ 0 \\ 1 \end{pmatrix}, \quad \frac{1}{\sqrt{2}} \begin{pmatrix} 0 \\ 1 \\ 1 \\ 0 \end{pmatrix} \right\}.$$

The projection matrix Π_2 projects into a two-dimensional Hilbert space spanned by the normalized vectors

$$\left\{ \frac{1}{\sqrt{2}} \begin{pmatrix} 1 \\ 0 \\ 0 \\ -1 \end{pmatrix}, \quad \frac{1}{\sqrt{2}} \begin{pmatrix} 0 \\ 1 \\ -1 \\ 0 \end{pmatrix} \right\}.$$

These four vectors are the *Bell basis*.

(iv) Yes. Matrix multiplication yields

$$(I_2 \otimes I_2)(I_2 \otimes I_2) = I_2 \otimes I_2, \quad (I_2 \otimes I_2)(\sigma_1 \otimes \sigma_1) = \sigma_1 \otimes \sigma_1,$$

$$(\sigma_1 \otimes \sigma_1)(I_2 \otimes I_2) = \sigma_1 \otimes \sigma_1, \quad (\sigma_1 \otimes \sigma_1)(\sigma_1 \otimes \sigma_1) = I_2 \otimes I_2.$$

The neutral element is $I_2 \otimes I_2$. The inverse element of $\sigma_1 \otimes \sigma_1$ is $\sigma_1 \otimes \sigma_1$.

(v) Since $\sigma_j^2 = I_2$ we obtain $\Pi_j^2 = \Pi_j$ and $\Pi_j^* = \Pi_j$ for $j = 1, 2, 3$. Thus the Π_j's are projection matrices.

Problem 24. Let T be an $n \times n$ hermitian matrix with $T^2 = I_n$.

(i) Show that

$$\Pi_+ := \frac{1}{2}(I_n + T), \qquad \Pi_- := \frac{1}{2}(I_n - T)$$

are *projection matrices*

(ii) Let σ_1, σ_2, σ_3 be the Pauli spin matrices and $\sigma_0 = I_2$. These hermitian matrices obey $\sigma_j^2 = I_2$ for $j = 0, 1, 2, 3$. Furthermore all possible Kronecker products of these matrices satisfy

$$(\sigma_{j_1} \otimes \sigma_{j_2} \otimes \cdots \otimes \sigma_{j_n})^2 = I_{2^n}$$

where $j_k \in \{0, 1, 2, 3\}$ and I_{2^n} is the $2^n \times 2^n$ identity matrix. Thus we can construct projection matrices in the Hilbert space $\mathbb{C}^{2^n}$. Let

$$\gamma_5 = \begin{pmatrix} 0_2 & I_2 \\ I_2 & 0_2 \end{pmatrix} \equiv \begin{pmatrix} 0 & 0 & 1 & 0 \\ 0 & 0 & 0 & 1 \\ 1 & 0 & 0 & 0 \\ 0 & 1 & 0 & 0 \end{pmatrix} \equiv \sigma_1 \otimes \sigma_0.$$

Find the projection matrices

$$\Pi_+ = \frac{1}{2}(I_4 + \gamma_5), \qquad \Pi_- = \frac{1}{2}(I_4 - \gamma_5)$$

(the matrices are called the *chirality projection operators*) and calculate

$$\Pi_+ \begin{pmatrix} \psi_1 \\ \psi_2 \\ \psi_3 \\ \psi_4 \end{pmatrix}, \qquad \Pi_- \begin{pmatrix} \psi_1 \\ \psi_2 \\ \psi_3 \\ \psi_4 \end{pmatrix}.$$

Discuss.

Solution 24.   (i) We obtain $\Pi_+^2 = \Pi_+$, $\Pi_-^2 = \Pi_-$, $\Pi_+ = \Pi_+^*$, $\Pi_- = \Pi_-^*$. Thus these matrices are projection matrices.
(ii) We find

$$\Pi_+ = \frac{1}{2}\begin{pmatrix} 1 & 0 & 1 & 0 \\ 0 & 1 & 0 & 1 \\ 1 & 0 & 1 & 0 \\ 0 & 1 & 0 & 1 \end{pmatrix}, \qquad \Pi_- = \frac{1}{2}\begin{pmatrix} 1 & 0 & -1 & 0 \\ 0 & 1 & 0 & -1 \\ -1 & 0 & 1 & 0 \\ 0 & -1 & 0 & 1 \end{pmatrix}$$

and

$$\Pi_+ \begin{pmatrix} \psi_1 \\ \psi_2 \\ \psi_3 \\ \psi_4 \end{pmatrix} = \frac{1}{2}\begin{pmatrix} \psi_1 + \psi_3 \\ \psi_2 + \psi_4 \\ \psi_1 + \psi_3 \\ \psi_2 + \psi_4 \end{pmatrix}, \qquad \Pi_- \begin{pmatrix} \psi_1 \\ \psi_2 \\ \psi_3 \\ \psi_4 \end{pmatrix} = \frac{1}{2}\begin{pmatrix} \psi_1 - \psi_3 \\ \psi_2 - \psi_4 \\ -\psi_1 + \psi_3 \\ -\psi_2 + \psi_4 \end{pmatrix}.$$

Problem 25.   Let

$$S_+ := \begin{pmatrix} 0 & 1 \\ 0 & 0 \end{pmatrix}, \qquad S_- := \begin{pmatrix} 0 & 0 \\ 1 & 0 \end{pmatrix}$$

and

$$S_{j,+} = I_2 \otimes \cdots \otimes I_2 \otimes S_+ \otimes I_2 \otimes \cdots \otimes I_2$$

where S_+ is at the j-th position and there are n Kronecker products. Thus $j = 1, 2, \ldots, n$. Analogously we define $S_{j,-}$. Let Π_j be the $2^n \times 2^n$ matrix

$$\Pi_j := S_{j,+}S_{j,-}.$$

Show that Π_j is a projection matrix.

Solution 25.   We have

$$S_+S_- = \begin{pmatrix} 0 & 1 \\ 0 & 0 \end{pmatrix}\begin{pmatrix} 0 & 0 \\ 1 & 0 \end{pmatrix} = \begin{pmatrix} 1 & 0 \\ 0 & 0 \end{pmatrix}$$

which is a projection matrix. Thus

$$\Pi_j = I_2 \otimes \cdots \otimes I_2 \otimes \begin{pmatrix} 1 & 0 \\ 0 & 0 \end{pmatrix} \otimes I_2 \otimes \cdots \otimes I_2.$$

Since $(S_+S_-)^2 = S_+S_-$ and $I_2^2 = I_2$ it follows that $\Pi_j^2 = \Pi_j$. Obviously we have $\Pi_j^* = \Pi_j$. Thus Π_j is a $2^n \times 2^n$ projection matrix.

Problem 26. An $n \times n$ matrix is a *permutation matrix* if it contains exactly one 1 in each row and column and 0's otherwise. Consider the Pauli spin matrices $\sigma_0 = I_2$, σ_1, σ_2, σ_3.
(i) Show that

$$P = \frac{1}{2} \sum_{j=0}^{3} \sigma_j \otimes \sigma_j$$

is a permutation matrix acting on the Hilbert space $\mathbb{C}^4$.
(ii) Are the 8×8 matrices

$$P_{12} = P \otimes I_2, \quad P_{13} = \frac{1}{2} \sum_{j=0}^{3} \sigma_j \otimes I_2 \otimes \sigma_j, \quad P_{23} = I_2 \otimes P$$

permutation matrices?

Solution 26. (i) We find the permutation matrix

$$P = \begin{pmatrix} 1 & 0 & 0 & 0 \\ 0 & 0 & 1 & 0 \\ 0 & 1 & 0 & 0 \\ 0 & 0 & 0 & 1 \end{pmatrix}.$$

(ii) Since P and I_2 are permutation matrices we find that P_{12} and P_{23} are permutation matrices. For P_{13} we find the permutation matrix

$$P_{13} = \begin{pmatrix} 1 & 0 & 0 & 0 & 0 & 0 & 0 & 0 \\ 0 & 0 & 0 & 0 & 1 & 0 & 0 & 0 \\ 0 & 0 & 1 & 0 & 0 & 0 & 0 & 0 \\ 0 & 0 & 0 & 0 & 0 & 0 & 1 & 0 \\ 0 & 1 & 0 & 0 & 0 & 0 & 0 & 0 \\ 0 & 0 & 0 & 0 & 0 & 1 & 0 & 0 \\ 0 & 0 & 0 & 1 & 0 & 0 & 0 & 0 \\ 0 & 0 & 0 & 0 & 0 & 0 & 0 & 1 \end{pmatrix}.$$

Problem 27. Consider the *Hilbert space* $M_4(\mathbb{C})$ of all 4×4 matrices over $\mathbb{C}$ with the scalar product

$$\langle A, B \rangle := \text{tr}(AB^*)$$

where $A, B \in M_4(\mathbb{C})$. The five γ matrices are given by

$$\gamma_1 = \begin{pmatrix} 0 & 0 & 0 & -i \\ 0 & 0 & -i & 0 \\ 0 & i & 0 & 0 \\ i & 0 & 0 & 0 \end{pmatrix}, \quad \gamma_2 = \begin{pmatrix} 0 & 0 & 0 & -1 \\ 0 & 0 & 1 & 0 \\ 0 & 1 & 0 & 0 \\ -1 & 0 & 0 & 0 \end{pmatrix}$$

$$\gamma_3 = \begin{pmatrix} 0 & 0 & -i & 0 \\ 0 & 0 & 0 & i \\ i & 0 & 0 & 0 \\ 0 & -i & 0 & 0 \end{pmatrix}, \quad \gamma_4 = \begin{pmatrix} 1 & 0 & 0 & 0 \\ 0 & 1 & 0 & 0 \\ 0 & 0 & -1 & 0 \\ 0 & 0 & 0 & -1 \end{pmatrix}$$

and

$$\gamma_5 := \gamma_1 \gamma_2 \gamma_3 \gamma_4 = \begin{pmatrix} 0 & 0 & -1 & 0 \\ 0 & 0 & 0 & -1 \\ -1 & 0 & 0 & 0 \\ 0 & -1 & 0 & 0 \end{pmatrix}.$$

We define the 4×4 matrices

$$\sigma_{jk} := \frac{i}{2}[\gamma_j, \gamma_k], \qquad j < k$$

where $j = 1, 2, 3$, $k = 2, 3, 4$ and $[\,,\,]$ denotes the commutator.
(i) Calculate σ_{12}, σ_{13}, σ_{14}, σ_{23}, σ_{24}, σ_{34}.
(ii) Do the sixteen 4×4 matrices

$$I_4, \gamma_1, \gamma_2, \gamma_3, \gamma_4, \gamma_5, \gamma_5\gamma_1, \gamma_5\gamma_2, \gamma_5\gamma_3, \gamma_5\gamma_4, \sigma_{12}, \sigma_{13}, \sigma_{14}, \sigma_{23}, \sigma_{24}, \sigma_{34}$$

form a basis in the Hilbert space $M_4(\mathbb{C})$? If so is the basis orthogonal?

Solution 27.   (i) We have

$$\sigma_{12} = \frac{i}{2}[\gamma_1, \gamma_2] = \text{diag}(-1, +1, -1, +1)$$

$$\sigma_{13} = \frac{i}{2}[\gamma_1, \gamma_3] = \begin{pmatrix} 0 & -i & 0 & 0 \\ i & 0 & 0 & 0 \\ 0 & 0 & 0 & -i \\ 0 & 0 & i & 0 \end{pmatrix}$$

$$\sigma_{14} = \frac{i}{2}[\gamma_1, \gamma_4] = \begin{pmatrix} 0 & 0 & 0 & -1 \\ 0 & 0 & -1 & 0 \\ 0 & -1 & 0 & 0 \\ -1 & 0 & 0 & 0 \end{pmatrix}$$

$$\sigma_{23} = \frac{i}{2}[\gamma_2, \gamma_3] = \begin{pmatrix} 0 & -1 & 0 & 0 \\ -1 & 0 & 0 & 0 \\ 0 & 0 & 0 & -1 \\ 0 & 0 & -1 & 0 \end{pmatrix}$$

$$\sigma_{24} = \frac{i}{2}[\gamma_2, \gamma_4] = \begin{pmatrix} 0 & 0 & 0 & i \\ 0 & 0 & -i & 0 \\ 0 & i & 0 & 0 \\ -i & 0 & 0 & 0 \end{pmatrix}$$

$$\sigma_{34} = \frac{i}{2}[\gamma_3, \gamma_4] = \begin{pmatrix} 0 & 0 & -1 & 0 \\ 0 & 0 & 0 & 1 \\ -1 & 0 & 0 & 0 \\ 0 & 1 & 0 & 0 \end{pmatrix}.$$

(ii) The 16 matrices are linearly independent since from the equation

$$c_0 I_4 + \sum_{j=1}^{5} c_j \gamma_j + \sum_{j=1}^{4} d_j \gamma_5 \gamma_j + \sum_{j<k}^{4} e_{jk} \sigma_{jk} = 0_4$$

it follows that all the coefficients c_j, d_j, e_{jk} are equal to 0. Calculating the scalar product of all possible pairs of matrices we find 0. For example

$$\langle \gamma_1, \gamma_2 \rangle = \text{tr}(\gamma_1\gamma_2^*) = \text{tr} \begin{pmatrix} i & 0 & 0 & 0 \\ 0 & -i & 0 & 0 \\ 0 & 0 & i & 0 \\ 0 & 0 & 0 & -i \end{pmatrix} = 0.$$

Problem 28. Let $j = 1/2, 1, 3/2, 2, \ldots$ and $\phi \in \mathbb{R}$. Consider the $(2j+1) \times (2j+1)$ matrices

$$
H = \begin{pmatrix}
0 & 1 & & & & \\
0 & 0 & 1 & & & \\
0 & 0 & 0 & 1 & & \\
& & & & \ddots & \\
& & & & & 1 \\
e^{i\phi} & & 0 & & & 0
\end{pmatrix}, \qquad D = \mathrm{diag}(1, \omega, \omega^2, \ldots, \omega^{2j})
$$

where $\omega := \exp(i2\pi/(2j+1))$. Is H unitary? Find $\omega DH - HD$.

Solution 28. H is unitary, i.e. we have $H^{-1} = H^*$. We obtain $\omega DH - HD = 0$.

Problem 29. (i) Find the eigenvalues and eigenvectors of the 3×3 matrix

$$
A(\alpha) = \begin{pmatrix}
e^\alpha & 1 & 1 \\
1 & e^\alpha & 1 \\
1 & 1 & e^\alpha
\end{pmatrix}.
$$

For which values of α is the matrix $A(\alpha)$ not invertible?
(ii) Extend (i) to the $n \times n$ matrix

$$
B(\alpha) = \begin{pmatrix}
e^\alpha & 1 & \cdots & 1 \\
1 & e^\alpha & \cdots & 1 \\
\vdots & \vdots & \ddots & \vdots \\
1 & 1 & \cdots & e^\alpha
\end{pmatrix}
$$

where $\alpha \in \mathbb{R}$. This matrix plays a role for the *Potts model*.

Solution 29. (i) The characteristic equation is

$$
(e^\alpha - \lambda)^3 - 3(e^\alpha - \lambda) + 2 = 0.
$$

The eigenvalues are $\lambda_1 = e^\alpha + 2$, $\lambda_2 = \lambda_3 = e^\alpha - 1$. Thus the matrix $A(\alpha)$ is not invertible if $\alpha = 0$. The normalized eigenvectors are

$$
\frac{1}{\sqrt{3}} \begin{pmatrix} 1 \\ 1 \\ 1 \end{pmatrix}, \quad \frac{1}{\sqrt{2}} \begin{pmatrix} 1 \\ 0 \\ -1 \end{pmatrix}, \quad \frac{1}{\sqrt{2}} \begin{pmatrix} 0 \\ 1 \\ -1 \end{pmatrix}.
$$

They form an orthonormal basis in the Hilbert space $\mathbb{C}^3$.
(ii) We find the eigenvalues

$$
\lambda_1 = e^\alpha + n - 1, \qquad \lambda_2 = \cdots = \lambda_n = e^\alpha - 1.
$$

Obviously a normalized eigenvector of $B(\alpha)$ is

$$
\mathbf{v} = \frac{1}{\sqrt{n}} (1 \quad 1 \quad \cdots \quad 1)^T
$$

with eigenvalue λ_1.

Problem 30.   Consider the Hilbert space $\mathbb{C}^n$. Let A, B, C be $n \times n$ matrices acting in $\mathbb{C}^n$. We consider the nonlinear eigenvalue problem

$$A\mathbf{u} = \lambda B\mathbf{u} + \lambda^2 C\mathbf{u}$$

where $\mathbf{u} \in \mathbb{C}^n$ and $\mathbf{u} \neq \mathbf{0}$.
(i) Let σ_1, σ_2, σ_3 be the Pauli spin matrices. Find the solutions of the nonlinear eigenvalue problem

$$\sigma_1\mathbf{u} = \lambda\sigma_2\mathbf{u} + \lambda^2\sigma_3\mathbf{u}$$

where $\mathbf{u} \in \mathbb{C}^2$ and $\mathbf{u} \neq \mathbf{0}$.
(ii) Consider the basis of the simple Lie algebra $s\ell(2, \mathbb{R})$

$$H = \begin{pmatrix} 1 & 0 \\ 0 & -1 \end{pmatrix}, \quad E = \begin{pmatrix} 0 & 1 \\ 0 & 0 \end{pmatrix}, \quad F = \begin{pmatrix} 0 & 0 \\ 1 & 0 \end{pmatrix}.$$

Solve the nonlinear eigenvalue problem

$$H\mathbf{u} = \lambda E\mathbf{u} + \lambda^2 F\mathbf{u}$$

where $\mathbf{u} \in \mathbb{C}^2$ and $\mathbf{u} \neq \mathbf{0}$.
(iii) Consider the basis of the simple Lie algebra $so(3, \mathbb{R})$

$$A = \begin{pmatrix} 0 & 0 & 0 \\ 0 & 0 & 1 \\ 0 & -1 & 0 \end{pmatrix}, \quad B = \begin{pmatrix} 0 & 0 & -1 \\ 0 & 0 & 0 \\ 1 & 0 & 0 \end{pmatrix}, \quad C = \begin{pmatrix} 0 & 1 & 0 \\ -1 & 0 & 0 \\ 0 & 0 & 0 \end{pmatrix}.$$

Solve the nonlinear eigenvalue problem.

Solution 30.   (i) We obtain $\lambda^4 - \lambda^2 + 1 = 0$ which provides the four eigenvalues

$$\lambda_1 = \sqrt{\frac{1 + \sqrt{5}}{2}}, \quad \lambda_2 = -\lambda_1, \quad \lambda_3 = \sqrt{\frac{1 - \sqrt{5}}{2}}, \quad \lambda_4 = -\lambda_3.$$

The corresponding eigenvectors are

$$\mathbf{u}_1 = \begin{pmatrix} 1 + i\lambda_1 \\ (1 + \sqrt{5})/2 \end{pmatrix}, \quad \mathbf{u}_2 = \begin{pmatrix} 1 - i\lambda_1 \\ (1 + \sqrt{5})/2 \end{pmatrix},$$

$$\mathbf{u}_3 = \begin{pmatrix} 1 + i\lambda_3 \\ (1 - \sqrt{5})/2 \end{pmatrix}, \quad \mathbf{u}_4 = \begin{pmatrix} 1 - i\lambda_3 \\ (1 - \sqrt{5})/2 \end{pmatrix}.$$

(ii) From $\lambda^3 + 1 = 0$ we find the eigenvalues $\lambda_1 = -1$, $\lambda_2 = e^{i\pi/3}$, $\lambda_3 = e^{-i\pi/3}$. The corresponding eigenvectors are

$$\mathbf{u}_1 = \begin{pmatrix} 1 \\ -1 \end{pmatrix}, \quad \mathbf{u}_2 = \begin{pmatrix} e^{i\pi/3} \\ 1 \end{pmatrix}, \quad \mathbf{u}_3 = \begin{pmatrix} e^{-i\pi/3} \\ 1 \end{pmatrix}.$$

(iii) The eigenvalues are $\lambda \in \mathbb{C}$ with the eigenvector

$$\begin{pmatrix} 1 \\ -\lambda \\ -\lambda^2 \end{pmatrix}.$$

Problem 31. An $n \times n$ matrix A over $\mathbb{C}$ is called *normal* if $AA^* = A^*A$. Consider the Pauli spin matrices σ_1, σ_2, σ_3 and the matrix

$$\sigma_3 + i\sigma_1 = \begin{pmatrix} 1 & i \\ i & -1 \end{pmatrix}.$$

(i) Is the matrix normal?
(ii) Find the eigenvalues and eigenvectors of the matrix. Discuss.
(iii) Find the eigenvalues the 4×4 matrix $\sigma_3 \otimes \sigma_3 + i\sigma_1 \otimes \sigma_1$. Is the matrix normal?

Solution 31. (i) The matrix is nonnormal since

$$(\sigma_3 + i\sigma_1)(\sigma_3 - i\sigma_1) \neq (\sigma_3 - i\sigma_1)(\sigma_3 + i\sigma_1).$$

(ii) The eigenvalues are 0 (twice). There is only one eigenvector

$$\frac{1}{\sqrt{2}} \begin{pmatrix} 1 \\ i \end{pmatrix}.$$

(iii) The eigenvalues of the normal, invertible and non-hermitian matrix

$$\sigma_3 \otimes \sigma_3 + i\sigma_1 \otimes \sigma_1 = \begin{pmatrix} 1 & 0 & 0 & i \\ 0 & -1 & i & 0 \\ 0 & i & -1 & 0 \\ i & 0 & 0 & 1 \end{pmatrix}$$

are

$$(-1)^{1/4}\sqrt{2}, \quad -(-1)^{1/4}\sqrt{2}, \quad (-1)^{1/4}i\sqrt{2}, \quad -(-1)^{1/4}i\sqrt{2}.$$

Problem 32. (i) Let σ_1, σ_2, σ_3 be the Pauli spin matrices. Consider the three non-normal matrices

$$A = \sigma_1 + i\sigma_2, \quad B = \sigma_2 + i\sigma_3, \quad C = \sigma_3 + i\sigma_1.$$

Find the commutators and anticommutators. Discuss.
(ii) Consider the three matrices

$$X = \sigma_1 \otimes \sigma_1 + i\sigma_2 \otimes \sigma_2, \quad Y = \sigma_2 \otimes \sigma_2 + i\sigma_3 \otimes \sigma_3, \quad Z = \sigma_3 \otimes \sigma_3 + i\sigma_1 \otimes \sigma_1.$$

Find the commutators.

Solution 32. (i) The commutators are

$$[\sigma_1 + i\sigma_2, \sigma_2 + i\sigma_3] = \begin{pmatrix} 2i & -4i \\ 0 & -2i \end{pmatrix}$$

$$[\sigma_2 + i\sigma_3, \sigma_3 + i\sigma_1] = \begin{pmatrix} 2 & 2i - 2 \\ 2i + 2 & -2 \end{pmatrix}$$

$$[\sigma_3 + i\sigma_1, \sigma_1 + i\sigma_2] = \begin{pmatrix} -2i & 4 \\ 0 & 2i \end{pmatrix}.$$

Are these matrices normal? The anticommutators are the same diagonal matrix

$$[A, B]_+ = [B, C]_+ = [C, A]_+ = 2iI_2.$$

(ii) The commutators are $[X, Y] = [Y, Z] = [Z, X] = 0_4$ i.e. the matrices commute. Are the matrices X, Y, Z normal?

Problem 33. Consider the spin matrices for describing a spin-1 system

$$S_1 = \frac{1}{\sqrt{2}} \begin{pmatrix} 0 & 1 & 0 \\ 1 & 0 & 1 \\ 0 & 1 & 0 \end{pmatrix}, \quad S_2 = \frac{1}{\sqrt{2}} \begin{pmatrix} 0 & -i & 0 \\ i & 0 & -i \\ 0 & i & 0 \end{pmatrix}, \quad S_3 = \begin{pmatrix} 1 & 0 & 0 \\ 0 & 0 & 0 \\ 0 & 0 & -1 \end{pmatrix}$$

and the matrix $S_3 + iS_1$.
(i) Is the matrix $S_3 + iS_1$ normal?
(ii) Find the eigenvalues and eigenvectors of the matrix. Discuss.
(iii) Find the eigenvalues of $S_3 \otimes S_3 + iS_1 \otimes S_1$. Is the matrix normal?

Solution 33. (i) The matrix is nonnormal since

$$(S_3 + iS_1)(S_3 - iS_1) - (S_3 - iS_1)(S_3 + iS_1) = \begin{pmatrix} 0 & -i\sqrt{2} & 0 \\ i\sqrt{2} & 0 & -i\sqrt{2} \\ 0 & i\sqrt{2} & 0 \end{pmatrix}.$$

(ii) The eigenvalues are 0 (three times) with the only normalized eigenvector

$$\frac{1}{2} \begin{pmatrix} -1 \\ -i\sqrt{2} \\ 1 \end{pmatrix}.$$

(iii) The matrix $S_3 \otimes S_3 + iS_1 \otimes S_1$ is nonnormal and the eigenvalues are

$$-i, \quad i, \quad 0 \, (5 \text{ times}) \quad -1, \quad 1.$$

Problem 34. Consider the Pauli spin matrices σ_1, σ_2, σ_3 and $\sigma_0 = I_2$. How many different $2^n \times 2^n$ matrices can we form using the Kronecker product of n such matrices? Are the matrices linearly independent? For example, for $n = 2$, we have 16 elements

$$\sigma_0 \otimes \sigma_0, \quad \sigma_0 \otimes \sigma_1, \quad \sigma_0 \otimes \sigma_2, \quad \sigma_0 \otimes \sigma_3$$

$$\sigma_1 \otimes \sigma_0, \quad \sigma_1 \otimes \sigma_1, \quad \sigma_1 \otimes \sigma_2, \quad \sigma_1 \otimes \sigma_3$$

$$\sigma_2 \otimes \sigma_0, \quad \sigma_2 \otimes \sigma_1, \quad \sigma_2 \otimes \sigma_2, \quad \sigma_2 \otimes \sigma_3$$

$$\sigma_3 \otimes \sigma_0, \quad \sigma_3 \otimes \sigma_1, \quad \sigma_3 \otimes \sigma_2, \quad \sigma_3 \otimes \sigma_3.$$

Solution 34. The elements can be numbered as

$$000...00, \ 000...01, \ 000...02, \ 000...03, \ 000...10, \ 000...11, \ldots, 333...33.$$

Thus we find 4^n different Kronecker products for n matrices given by the Pauli spin matrices and the 2×2 identity matrix. The 4^n matrices are linearly independent and form an orthonormal basis in the Hilbert space of the $2^n \times 2^n$ matrices with the scalar product

$$\langle X, Y \rangle := \text{tr}(XY^*).$$

Problem 35. Let I_2 be the 2×2 identity matrix, and let σ_1, σ_2 and σ_3 be the Pauli spin matrices.

(i) Let A be an $m \times m$ matrix over $\mathbb{C}$ and B be an $n \times n$ matrix over $\mathbb{C}$. Find all solutions for m, n, A and B satisfying

$$\frac{1}{2}I_2 \otimes I_2 + \frac{1}{2}\sigma_1 \otimes \sigma_1 + \frac{1}{2}\sigma_2 \otimes \sigma_2 + \frac{1}{2}\sigma_3 \otimes \sigma_3 = A \otimes B.$$

(ii) Let C, D, E, and F be 2×2 matrices over $\mathbb{C}$. Find all solutions for C, D, E and F satisfying

$$\frac{1}{2}I_2 \otimes I_2 + \frac{1}{2}\sigma_1 \otimes \sigma_1 + \frac{1}{2}\sigma_2 \otimes \sigma_2 + \frac{1}{2}\sigma_3 \otimes \sigma_3 = C \otimes D + E \otimes F.$$

(iii) Find the set stabilized by

$$S = \{\sigma_1 \otimes I_2, I_2 \otimes \sigma_1\},$$

i.e. find $\{\mathbf{u} \in \mathbb{C}^4 : \forall A \in S : A\mathbf{u} = \mathbf{u}\}$.

Solution 35. (i) Straightforward calculation yields

$$\frac{1}{2}I_2 \otimes I_2 + \frac{1}{2}\sigma_1 \otimes \sigma_1 + \frac{1}{2}\sigma_2 \otimes \sigma_2 + \frac{1}{2}\sigma_3 \otimes \sigma_3 = \begin{pmatrix} 1 & 0 & 0 & 0 \\ 0 & 0 & 1 & 0 \\ 0 & 1 & 0 & 0 \\ 0 & 0 & 0 & 1 \end{pmatrix} = A \otimes B.$$

Since $A \otimes B$ is an $mn \times mn$ matrix we have $mn = 4$. For $m = n = 2$ consider

$$A = \begin{pmatrix} a_1 & a_2 \\ a_3 & a_4 \end{pmatrix}, \qquad B = \begin{pmatrix} b_1 & b_2 \\ b_3 & b_4 \end{pmatrix}.$$

Thus we have $a_1 b_1 = 1$, $a_1 b_4 = 0$ and $a_4 b_4 = 1$. Consequently $a_1 \neq 0$ and $b_4 \neq 0$, so that $a_1 b_4 \neq 0$. Thus $m = n = 2$ does not yield a solution.

For $m = 1$ and $n = 4$ we find the solution

$$A = (a), \qquad B = \frac{1}{a}\begin{pmatrix} 1 & 0 & 0 & 0 \\ 0 & 0 & 1 & 0 \\ 0 & 1 & 0 & 0 \\ 0 & 0 & 0 & 1 \end{pmatrix}, \qquad a \in \mathbb{C}/\{0\}.$$

For $m = 4$ and $n = 1$ we find the solution

$$A = \frac{1}{b}\begin{pmatrix} 1 & 0 & 0 & 0 \\ 0 & 0 & 1 & 0 \\ 0 & 1 & 0 & 0 \\ 0 & 0 & 0 & 1 \end{pmatrix}, \qquad B = (b), \qquad b \in \mathbb{C}/\{0\}.$$

(ii) Clearly C and E, and D and F must be linearly independent, otherwise we would have the case $m = n = 2$ discussed in (i) which has no solution. Let

$$C = \begin{pmatrix} c_1 & c_2 \\ c_3 & c_4 \end{pmatrix}, \qquad E = \begin{pmatrix} e_1 & e_2 \\ e_3 & e_4 \end{pmatrix}.$$

We have to satisfy the equations

$$c_1 D + e_1 F = \begin{pmatrix} 1 & 0 \\ 0 & 0 \end{pmatrix}, \qquad c_2 D + e_2 F = \begin{pmatrix} 0 & 0 \\ 1 & 0 \end{pmatrix},$$

$$c_3 D + e_3 F = \begin{pmatrix} 0 & 1 \\ 0 & 0 \end{pmatrix}, \qquad c_4 D + e_4 F = \begin{pmatrix} 0 & 0 \\ 0 & 1 \end{pmatrix}.$$

It follows that $\{D, F\}$ should be a basis for the 2×2 matrices over $\mathbb{C}$, however the span of $\{D, F\}$ is a 2 dimensional space whereas the 2×2 matrices over $\mathbb{C}$ forms a 4 dimensional space. Thus we have a contradiction. There are no solutions.

(iii) The vector $\mathbf{u}$ must simultaneously be an eigenvector of $\sigma_1 \otimes I_2$ and $I_2 \otimes \sigma_1$ with eigenvalue 1. The eigenspace corresponding to the eigenvalue 1 from $\sigma_1 \otimes I_2$ yields

$$\mathbf{u} = \alpha \begin{pmatrix} 1 \\ 1 \end{pmatrix} \otimes \begin{pmatrix} 1 \\ 0 \end{pmatrix} + \beta \begin{pmatrix} 1 \\ 1 \end{pmatrix} \otimes \begin{pmatrix} 0 \\ 1 \end{pmatrix}$$

where $\alpha, \beta \in \mathbb{C}$. We must now satisfy

$$(I_2 \otimes \sigma_1)\mathbf{u} = \alpha(I_2 \otimes \sigma_1) \begin{pmatrix} 1 \\ 1 \end{pmatrix} \otimes \begin{pmatrix} 1 \\ 0 \end{pmatrix} + \beta(I_2 \otimes \sigma_1) \begin{pmatrix} 1 \\ 1 \end{pmatrix} \otimes \begin{pmatrix} 0 \\ 1 \end{pmatrix}$$

$$= \alpha \begin{pmatrix} 1 \\ 1 \end{pmatrix} \otimes \begin{pmatrix} 0 \\ 1 \end{pmatrix} + \beta \begin{pmatrix} 1 \\ 1 \end{pmatrix} \otimes \begin{pmatrix} 1 \\ 0 \end{pmatrix}$$

$$= \mathbf{u}$$

so that $\alpha = \beta$. Thus

$$\mathbf{u} = \alpha \begin{pmatrix} 1 \\ 1 \end{pmatrix} \otimes \begin{pmatrix} 1 \\ 0 \end{pmatrix} + \alpha \begin{pmatrix} 1 \\ 1 \end{pmatrix} \otimes \begin{pmatrix} 0 \\ 1 \end{pmatrix}, \qquad \alpha \in \mathbb{C}.$$

Problem 36.   Let σ_1, σ_2, σ_3 be the Pauli spin matrices which are hermitian and unitary and have the eigenvalues $+1$ and -1. Find all invertible 2×2 matrices X such that

$$X^{-1}\sigma_1 X = \sigma_3.$$

Proceed as follows: Multiply the equation with X and after rearrangements we have

$$\sigma_1 X - X\sigma_3 = \begin{pmatrix} 0 & 0 \\ 0 & 0 \end{pmatrix}.$$

This is a special case of the *Sylvester equation*. The *vec-operation* on a matrix stacks the columns of the matrix one under the other to form a single column. Applying the operation the Sylvester equation can be written as

$$(I_2 \otimes \sigma_1 - \sigma_3^T \otimes I_2)\mathrm{vec}(X) = \begin{pmatrix} 0 \\ 0 \\ 0 \\ 0 \end{pmatrix}, \qquad \mathrm{vec}(X) := \begin{pmatrix} x_{11} \\ x_{21} \\ x_{12} \\ x_{22} \end{pmatrix}.$$

Solve this linear equation to find the matrix X, but one still has to take into account that the matrix X is invertible.

Solution 36.   We have

$$I_2 \otimes \begin{pmatrix} 0 & 1 \\ 1 & 0 \end{pmatrix} + \begin{pmatrix} -1 & 0 \\ 0 & 1 \end{pmatrix} \otimes I_2 = \begin{pmatrix} -1 & 1 & 0 & 0 \\ 1 & -1 & 0 & 0 \\ 0 & 0 & 1 & 1 \\ 0 & 0 & 1 & 1 \end{pmatrix}.$$

Thus we arrive at the system of linear equations

$$\begin{pmatrix} -1 & 1 & 0 & 0 \\ 1 & -1 & 0 & 0 \\ 0 & 0 & 1 & 1 \\ 0 & 0 & 1 & 1 \end{pmatrix} \begin{pmatrix} x_{11} \\ x_{21} \\ x_{12} \\ x_{22} \end{pmatrix} = \begin{pmatrix} 0 \\ 0 \\ 0 \\ 0 \end{pmatrix}$$

with the solution $x_{21} = x_{11}$, $x_{12} = -x_{22}$. Therefore the matrix X is

$$X = \begin{pmatrix} x_{11} & -x_{22} \\ x_{11} & x_{22} \end{pmatrix}$$

with determinant $\det(X) = 2x_{11}x_{22}$. This implies that $x_{11} \neq 0$ and $x_{22} \neq 0$. Then the inverse of X is

$$X^{-1} = \frac{1}{2x_{11}x_{22}} \begin{pmatrix} x_{22} & x_{22} \\ -x_{11} & x_{11} \end{pmatrix}.$$

Problem 37. Consider the spin matrices for spin-$\frac{1}{2}$

$$S_1 = \frac{1}{2} \begin{pmatrix} 0 & 1 \\ 1 & 0 \end{pmatrix}, \quad S_2 = \frac{1}{2} \begin{pmatrix} 0 & -i \\ i & 0 \end{pmatrix}, \quad S_3 = \frac{1}{2} \begin{pmatrix} 1 & 0 \\ 0 & -1 \end{pmatrix}.$$

(i) Calculate the vector

$$(S_1 \otimes S_1 + S_2 \otimes S_2 + S_3 \otimes S_3) \left(\begin{pmatrix} 1 \\ 0 \end{pmatrix} \otimes \begin{pmatrix} 1 \\ 0 \end{pmatrix} \right)$$

in the Hilbert space $\mathbb{C}^4$ and thus show that one has an eigenvalue equation. Is the vector normalized?

(ii) Consider the spin matrices for spin-1

$$S_1 = \frac{1}{\sqrt{2}} \begin{pmatrix} 0 & 1 & 0 \\ 1 & 0 & 1 \\ 0 & 1 & 0 \end{pmatrix}, \quad S_2 = \frac{1}{\sqrt{2}} \begin{pmatrix} 0 & -i & 0 \\ i & 0 & -i \\ 0 & i & 0 \end{pmatrix}, \quad S_3 = \begin{pmatrix} 1 & 0 & 0 \\ 0 & 0 & 0 \\ 0 & 0 & -1 \end{pmatrix}.$$

Calculate the vector

$$(S_1 \otimes S_1 + S_2 \otimes S_2 + S_3 \otimes S_3) \left(\begin{pmatrix} 1 \\ 0 \\ 0 \end{pmatrix} \otimes \begin{pmatrix} 1 \\ 0 \\ 0 \end{pmatrix} \right)$$

in the Hilbert space $\mathbb{C}^9$ and thus show that one has an eigenvalue equation. Is the vector normalized?

Solution 37. (i) We obtain the eigenvalue equation

$$(S_1 \otimes S_1 + S_2 \otimes S_2 + S_3 \otimes S_3) \left(\begin{pmatrix} 1 \\ 0 \end{pmatrix} \otimes \begin{pmatrix} 1 \\ 0 \end{pmatrix} \right) = \frac{1}{2} \begin{pmatrix} 1 \\ 0 \end{pmatrix} \otimes \begin{pmatrix} 1 \\ 0 \end{pmatrix}$$

with the eigenvalue $+\frac{1}{2}$.

(ii) We obtain the eigenvalue equation

$$(S_1 \otimes S_1 + S_2 \otimes S_2 + S_3 \otimes S_3) \left(\begin{pmatrix} 1 \\ 0 \\ 0 \end{pmatrix} \otimes \begin{pmatrix} 1 \\ 0 \\ 0 \end{pmatrix} \right) = \begin{pmatrix} 1 \\ 0 \\ 0 \end{pmatrix} \otimes \begin{pmatrix} 1 \\ 0 \\ 0 \end{pmatrix}$$

with the eigenvalue $+1$.

Problem 38. For spin-3/2 we have the hermitian matrices

$$S_1 = \frac{1}{2} \begin{pmatrix} 0 & \sqrt{3} & 0 & 0 \\ \sqrt{3} & 0 & 2 & 0 \\ 0 & 2 & 0 & \sqrt{3} \\ 0 & 0 & \sqrt{3} & 0 \end{pmatrix}, \quad S_2 = \frac{1}{2} \begin{pmatrix} 0 & -i\sqrt{3} & 0 & 0 \\ i\sqrt{3} & 0 & -2i & 0 \\ 0 & 2i & 0 & -i\sqrt{3} \\ 0 & 0 & i\sqrt{3} & 0 \end{pmatrix},$$

$$S_3 = \begin{pmatrix} 3/2 & 0 & 0 & 0 \\ 0 & 1/2 & 0 & 0 \\ 0 & 0 & -1/2 & 0 \\ 0 & 0 & 0 & -3/2 \end{pmatrix}$$

with $S_1^2 + S_2^2 + S_3^2 = \frac{15}{4} I_4$ and the commutation relations

$$[S_1, S_2] = iS_3, \quad [S_2, S_3] = iS_1, \quad [S_3, S_1] = iS_2.$$

Can these three 4×4 matrices be written as

$$S_1 = A_1 \otimes A_2, \quad S_2 = B_1 \otimes B_2, \quad S_3 = C_1 \otimes C_2$$

where A_j, B_j, C_j $(j = 1, 2)$ are 2×2 matrices?

Solution 38. No for all three cases.

Problem 39. Consider the Pauli spin matrices $\sigma_0 = I_2$, σ_1, σ_2, σ_3.
(i) Calculate the 4×4 matrix

$$R = \frac{1}{2}(\sigma_0 \otimes \sigma_0 + \sigma_0 \otimes \sigma_1 + \sigma_1 \otimes \sigma_0 - \sigma_1 \otimes \sigma_1).$$

Is the matrix unitary?
(ii) Consider the permutation matrix (*flip matrix*)

$$F = \begin{pmatrix} 1 & 0 & 0 & 0 \\ 0 & 0 & 1 & 0 \\ 0 & 1 & 0 & 0 \\ 0 & 0 & 0 & 1 \end{pmatrix}.$$

Find $\tilde{R} = FR$. Let I_2 be the 2×2 identity matrix. Show that (*braid relation*)

$$(I_2 \otimes \tilde{R})(\tilde{R} \otimes I_2)(I_2 \otimes \tilde{R}) = (\tilde{R} \otimes I_2)(I_2 \otimes \tilde{R})(\tilde{R} \otimes I_2).$$

(iii) Apply the matrix $\tilde{R}$ to the *Bell basis*, i.e.

$$\tilde{R}\frac{1}{\sqrt{2}}\begin{pmatrix} 1 \\ 0 \\ 0 \\ 1 \end{pmatrix}, \quad \tilde{R}\frac{1}{\sqrt{2}}\begin{pmatrix} 0 \\ 1 \\ 1 \\ 0 \end{pmatrix}, \quad \tilde{R}\frac{1}{\sqrt{2}}\begin{pmatrix} 0 \\ 1 \\ -1 \\ 0 \end{pmatrix}, \quad \tilde{R}\frac{1}{\sqrt{2}}\begin{pmatrix} 1 \\ 0 \\ 0 \\ -1 \end{pmatrix}.$$

Solution 39. (i) We obtain the unitary matrix

$$R = \frac{1}{2} \begin{pmatrix} 1 & 1 & 1 & -1 \\ 1 & 1 & -1 & 1 \\ 1 & -1 & 1 & 1 \\ -1 & 1 & 1 & 1 \end{pmatrix}.$$

(ii) We find the unitary matrix

$$\tilde{R} = FR = \frac{1}{2} \begin{pmatrix} 1 & 1 & 1 & -1 \\ 1 & -1 & 1 & 1 \\ 1 & 1 & -1 & 1 \\ -1 & 1 & 1 & 1 \end{pmatrix}.$$

We check the braid relation with the following Maxima program

```
/* braidrelation.mac */

sig0: matrix([1,0],[0,1]);
sig1: matrix([0,1],[1,0]);
T1: kronecker_product(sig0,sig0);   T2: kronecker_product(sig0,sig1);
T3: kronecker_product(sig1,sig0);   T4: kronecker_product(sig1,sig1);
R: (T1 + T2 + T3 - T4)/2;
F: matrix([1,0,0,0],[0,0,1,0],[0,1,0,0],[0,0,0,1]);
RT: F . R;
V1: kronecker_product(sig0,RT);   V2: kronecker_product(RT,sig0);
/* braid relation */
Z: V1 . V2 . V1 - V2 . V1 . V2;
/* Bell states */
B1: matrix([1],[0],[0],[1])/sqrt(2);   B2: matrix([0],[1],[1],[0])/sqrt(2);
B3: matrix([0],[1],[-1],[0])/sqrt(2);  B4: matrix([1],[0],[0],[-1])/sqrt(2);
/* applying RT to the Bell states */
RT . B1; RT . B2; RT . B3; RT . B4;
```

(iii) The Maxima program also provides

$$\tilde{R}\frac{1}{\sqrt{2}}\begin{pmatrix}1\\0\\0\\1\end{pmatrix} = \frac{1}{\sqrt{2}}\begin{pmatrix}0\\1\\1\\0\end{pmatrix}, \quad \tilde{R}\frac{1}{\sqrt{2}}\begin{pmatrix}0\\1\\1\\0\end{pmatrix} = \frac{1}{\sqrt{2}}\begin{pmatrix}1\\0\\0\\1\end{pmatrix},$$

$$\tilde{R}\frac{1}{\sqrt{2}}\begin{pmatrix}0\\1\\-1\\0\end{pmatrix} = \frac{1}{\sqrt{2}}\begin{pmatrix}0\\-1\\1\\0\end{pmatrix}, \quad \tilde{R}\frac{1}{\sqrt{2}}\begin{pmatrix}1\\0\\0\\-1\end{pmatrix} = -\frac{1}{\sqrt{2}}\begin{pmatrix}1\\0\\0\\-1\end{pmatrix}.$$

Thus the Bell states are preserved under $\tilde{R}$.

Problem 40. Let

$$S_+ := \begin{pmatrix} 0 & 1 \\ 0 & 0 \end{pmatrix}, \quad S_- := \begin{pmatrix} 0 & 0 \\ 1 & 0 \end{pmatrix}.$$

Consider

$$S_{+,j} := I_2 \otimes \cdots \otimes I_2 \otimes S_+ \otimes I_2 \otimes \cdots \otimes I_2, \quad S_{-,j} := I_2 \otimes \cdots \otimes I_2 \otimes S_- \otimes I_2 \otimes \cdots \otimes I_2$$

where $j = 1, 2, \ldots, n$ and S_+ and S_- are at the j-position. Thus $S_{+,j}$ and $S_{-,j}$ are $2^n \times 2^n$ matrices.
(i) Find the matrices $(S_{+,j})^2$, $(S_{-,j})^2$.
(ii) Let $n = 2$ with

$$S_{+,1} = \begin{pmatrix} 0 & 1 \\ 0 & 0 \end{pmatrix} \otimes I_2, \quad S_{+,2} = I_2 \otimes \begin{pmatrix} 0 & 1 \\ 0 & 0 \end{pmatrix},$$

$$S_{-,1} = \begin{pmatrix} 0 & 0 \\ 1 & 0 \end{pmatrix} \otimes I_2, \quad S_{-,2} = I_2 \otimes \begin{pmatrix} 0 & 0 \\ 1 & 0 \end{pmatrix}.$$

Find the matrices

$$(S_{+,1} + S_{+,2})^2, \quad (S_{-,1} + S_{-,2})^2, \quad (S_{+,1} + S_{+,2} + S_{-,1} + S_{-,2})^2.$$

(iii) Consider the normalized vectors in $\mathbb{C}^2$ (standard basis)

$$\mathbf{u} = \begin{pmatrix} 1 \\ 0 \end{pmatrix}, \quad \mathbf{v} = \begin{pmatrix} 0 \\ 1 \end{pmatrix}.$$

Find the two vectors $(S_{+,1} + S_{+,2})(\mathbf{u} \otimes \mathbf{u})$, $(S_{-,1} + S_{-,2})(\mathbf{v} \otimes \mathbf{v})$.

Solution 40. (i) We obtain the zero matrix $(S_{+,j})^2 = 0_{2^n}$, $(S_{-,j})^2 = 0_{2^n}$.
(ii) We find

$$(S_{+,1} + S_{+,2})^2 = \begin{pmatrix} 0 & 0 & 0 & 2 \\ 0 & 0 & 0 & 0 \\ 0 & 0 & 0 & 0 \\ 0 & 0 & 0 & 0 \end{pmatrix}, \quad (S_{-,1} + S_{-,2})^2 = \begin{pmatrix} 0 & 0 & 0 & 0 \\ 0 & 0 & 0 & 0 \\ 0 & 0 & 0 & 0 \\ 2 & 0 & 0 & 0 \end{pmatrix}$$

and

$$(S_{+,1} + S_{+,2} + S_{-,1} + S_{-,2})^2 = \begin{pmatrix} 2 & 0 & 0 & 2 \\ 0 & 2 & 2 & 0 \\ 0 & 2 & 2 & 0 \\ 2 & 0 & 0 & 2 \end{pmatrix}.$$

(iii) We obtain the zero vectors in both cases.

Problem 41. Consider the finite-dimensional Hilbert space $\mathbb{C}^d$. A symmetric informatically complete positive operator valued measure (SIC-POVM) consists of d^2 outcomes that are subnormalized projection matrices Π_j onto pure states

$$\Pi_j = \frac{1}{d}|\psi_j\rangle\langle\psi_j|$$

for $j, k = 1, \dots, d^2$ such that

$$|\langle\psi_k|\psi_j\rangle|^2 = \frac{1 + d\delta_{jk}}{d + 1}.$$

(i) Consider the case $d = 2$. Show that the normalized vectors

$$|\psi_1\rangle = \begin{pmatrix} \sqrt{(3 + \sqrt{3})/6} \\ e^{i\pi/4}\sqrt{(3 - \sqrt{3})/6} \end{pmatrix}$$

$$|\psi_2\rangle = \begin{pmatrix} \sqrt{(3 + \sqrt{3})/6} \\ -e^{i\pi/4}\sqrt{(3 - \sqrt{3})/6} \end{pmatrix}$$

$$|\psi_3\rangle = \begin{pmatrix} e^{i\pi/4}\sqrt{(3 - \sqrt{3})/6} \\ \sqrt{(3 + \sqrt{3})/6} \end{pmatrix}$$

$$|\psi_4\rangle = \begin{pmatrix} -e^{i\pi/4}\sqrt{(3 - \sqrt{3})/6} \\ \sqrt{(3 + \sqrt{3})/6} \end{pmatrix}$$

satisfy this condition.

(ii) Consider the Pauli spin matrices σ_1, $-i\sigma_2$, σ_3. Find the normalized states

$$\sigma_1|\psi_1\rangle, \quad -i\sigma_2|\psi_1\rangle, \quad \sigma_3|\psi_1\rangle.$$

(iii) Let $d = 2$. Let

$$S_d := \sum_{j=1}^{d} |j\rangle \otimes |j\rangle \otimes \langle j| \otimes \langle j| + \sum_{k>j=1} \frac{1}{\sqrt{2}}(|j\rangle \otimes |k\rangle + |k\rangle \otimes |j\rangle) \otimes \frac{1}{\sqrt{2}}(\langle j| \otimes \langle k| + \langle k| \otimes \langle j|)$$

where $|1\rangle$, $|2\rangle$ denotes the standard basis in $\mathbb{C}^2$, i.e.

$$|1\rangle = \begin{pmatrix} 1 \\ 0 \end{pmatrix}, \quad |2\rangle = \begin{pmatrix} 0 \\ 1 \end{pmatrix}.$$

Find S_2.

(iv) Can one find a SIC-POVM in $\mathbb{C}^4$ using the states from (i) and the Kronecker product?

Solution 41. (i) Since

$$(\sqrt{(3+\sqrt{3})/6})^2 = 1/3 + \sqrt{3}/6, \quad (\sqrt{(3-\sqrt{3})/6})^2 = 1/3 - \sqrt{3}/6,$$

$$\sqrt{(3+\sqrt{3})/6}\sqrt{(3-\sqrt{3})/6} = 1/\sqrt{6}$$

we find that the condition is satisfied.

(ii) We obtain $\sigma_1|\psi_1\rangle = |\psi_3\rangle$, $-i\sigma_2|\psi_1\rangle = |\psi_4\rangle$, $\sigma_3|\psi_1\rangle = |\psi_2\rangle$. Thus using the Pauli spin matrices we can generate $|\psi_2\rangle$, $|\psi_3\rangle$, $|\psi_4\rangle$ from $|\psi_1\rangle$.

(iii) We obtain the matrix

$$S_2 = \begin{pmatrix} 1 & 0 & 0 & 0 \\ 0 & 1/2 & 1/2 & 0 \\ 0 & 1/2 & 1/2 & 0 \\ 0 & 0 & 0 & 1 \end{pmatrix}.$$

(iv) Yes.

Problem 42. Let $\hat{A}$, $\hat{B}$ be two self-adjoint operators in a Hilbert space $\mathcal{H}$ and $|\psi\rangle$ be a normalization vector in this Hilbert space. Then we have (*uncertainty relation*)

$$(\Delta\hat{A})(\Delta\hat{B}) \geq \frac{1}{2}|\langle[\hat{A}, \hat{B}]\rangle|$$

where

$$\Delta\hat{A} := \sqrt{\langle\hat{A}^2\rangle - \langle\hat{A}\rangle^2}$$

and $\langle\hat{A}\rangle := \langle\psi|\hat{A}|\psi\rangle$. Consider the Hilbert space $\mathbb{C}^4$, the hermitian and unitary 4×4 matrices

$$\hat{A} = \sigma_1 \otimes \sigma_1, \quad B = \sigma_2 \otimes \sigma_2$$

and the Bell state

$$|\psi\rangle = \frac{1}{\sqrt{2}} \begin{pmatrix} 1 \\ 0 \\ 0 \\ 1 \end{pmatrix}.$$

Find the left-hand side and right-hand side of the inequality.

Solution 42. First we look at the right-hand side. Since the two matrices commute, i.e.

$$[\sigma_1 \otimes \sigma_1, \sigma_2 \otimes \sigma_2] = 0_4$$

we find that the right-hand side of the inequality is 0. For the left-hand side we find

$$\langle \hat{A} \rangle = 1, \quad \langle \hat{B} \rangle = -1, \quad \langle \hat{A}^2 \rangle = 1, \quad \langle \hat{B}^2 \rangle = 1.$$

Thus the left-hand side also vanishes.

Problem 43. Let $s \in \{1/2, 1, 3/2, 2, \ldots\}$ be the spin quantum number. The spin matrices s_1 and s_3 are defined as the $(2s+1) \times (2s+1)$ matrices $s_1 = (s_+ + s_-)/2$ and $s_3 = \text{diag}(s, s - 1, \ldots, -s)$ where the entries of s_+ and s_- are all zero except for the entries given by (here rows and columns are numbered $s, s - 1, s - 2, \ldots, -s$)

$$(s_+)_{m+1,m} = \sqrt{(s - m)(s + m + 1)} \qquad m = s - 1, s - 2, \ldots, -s$$

and $s_- = s_+^T$. Calculate $[s_1, s_3]$ in terms of s. Find the $\|[s_1, s_3]\|^2$ in terms of s, where

$$\|A\| := \sqrt{\text{tr}(AA^*)}$$

is the *Frobenius norm*.

Solution 43. Using $(m = s - 1, \ldots, -s)$

$$(s_+)_{m+1,m} = \sqrt{(s - m)(s + m + 1)}$$
$$(s_+ s_3)_{m+1,m} = m\sqrt{(s - m)(s + m + 1)}$$
$$(s_3 s_+)_{m+1,m} = (m + 1)\sqrt{(s - m)(s + m + 1)}$$

and

$$(s_-)_{m,m+1} = \sqrt{(s - m)(s + m + 1)}$$
$$(s_- s_3)_{m,m+1} = (m + 1)\sqrt{(s - m)(s + m + 1)}$$
$$(s_3 s_-)_{m,m+1} = m\sqrt{(s - m)(s + m + 1)}$$

we find

$$(s_1 s_3 - s_3 s_1)_{m+1,m} = -\sqrt{(s - m)(s + m + 1)}/2$$
$$(s_1 s_3 - s_3 s_1)_{m,m+1} = \sqrt{(s - m)(s + m + 1)}/2.$$

These are the non-zero entries in $[s_1, s_3]$. The matrix $[s_1, s_3]$ is skew-symmetric. We find

$$\|[s_1, s_3]\|^2 = \text{tr}([s_1, s_3][s_1, s_3]^T) = -\text{tr}([s_1, s_3]^2)$$

$$= \sum_{m=s-1}^{-s} (s - m)(s + m + 1)/4 + (s - m)(s + m + 1)/4$$

$$= \sum_{j=0}^{2s-1} (j + 1)(2s - j)/2 = \sum_{j=0}^{2s-1} s + (2s - 1)j/2 - j^2/2$$

$$= 2s^2 + (2s - 1)\frac{(2s - 1)s}{2} - \frac{(4s - 1)s(2s - 1)}{6}$$

$$= \frac{2s^3 + 3s^2 + s}{3}$$

where we substituted $m = s - 1 - j$.

```
/* Frobenius.mac */

load("diag")$
load("zeilberger")$

sum: ratsimp(GosperSum(s+(2*s-1)*j/2,j,0,2*s-1)-GosperSum(j^2,j,0,2*s-1)/2);
s3: diag([1/2,-1/2])$
s1: matrix([0,1],[1,0])/2$
a: s1.s3-s3.s1;
mat_trace(a.transpose(a));
subst(s=1/2,sum);
s3: diag([1,0,-1])$
s1: matrix([0,1,0],[1,0,1],[0,1,0])/sqrt(2)$
a: s1.s3-s3.s1;
ratsimp(mat_trace(a.transpose(a)));
subst(s=1,sum);
s3: diag([3/2,1/2,-1/2,-3/2])$
s1: matrix([0,sqrt(3),0,0],[sqrt(3),0,2,0],[0,2,0,sqrt(3)],[0,0,sqrt(3),0])/2$
a: s1.s3-s3.s1;
ratsimp(mat_trace(a.transpose(a)));
subst(s=3/2,sum);
```

Problem 44. A measure of *nonnormality* of an $n \times n$ matrix A is given by

$$m(A) := \|A^*A - AA^*\|$$

where $\| \cdot \|$ denotes some matrix norm. Let S_3 and S_1 be real valued and symmetric $n \times n$ matrices, for example the spin matrices S_1 and S_3. Find

$$m(S_3 + e^{i\phi}S_1).$$

Solution 44. The matrices are S_1 and S_3 are symmetric and real valued. Thus for $A = S_3 + e^{i\phi}S_1$ we have

$$AA^* - A^*A = (S_3 + e^{i\phi}S_1)(S_3 + e^{-i\phi}S_1) - (S_3 + e^{-i\phi}S_1)(S_3 + e^{i\phi}S_1) = 2i\sin(\phi)[S_1, S_3]$$

so that

$$\|AA^* - A^*A\| = 2|\sin(\phi)|\|[S_1, S_3]\|.$$

Assume that $[S_1, S_3] \neq 0$. Then A is normal when $\sin(\phi) = 0$ and nonnormal otherwise. The commutator $[S_1, S_3]$ determines the extent to which A is nonnormal under the measure m

Problem 45. Let σ_1, σ_2, σ_3 be the Pauli spin matrices. Let

$$K = \sigma_1 \otimes \sigma_1 + \sigma_2 \otimes \sigma_2 + \sigma_3 \otimes \sigma_3, \qquad e_1 = \begin{pmatrix} 1 \\ 0 \end{pmatrix}.$$

(i) Find the vector $K(e_1 \otimes e_1)$ in $\mathbb{C}^4$. Is the vector normalized? Discuss.
(ii) Let

$$K = \sigma_1 \otimes \sigma_1 \otimes I_2 + I_2 \otimes \sigma_1 \otimes \sigma_1 + \sigma_2 \otimes \sigma_2 \otimes I_2 + I_2 \otimes \sigma_2 \otimes \sigma_2 + \sigma_3 \otimes \sigma_3 \otimes I_2 + I_2 \otimes \sigma_3 \otimes \sigma_3.$$

Find the vector $K(\mathbf{e}_1 \otimes \mathbf{e}_1 \otimes \mathbf{e}_1)$ in the Hilbert space $\mathbb{C}^8$. Is the vector normalized? Discuss.
(iii) Consider the *spin-1 matrices*

$$S_1 = \frac{1}{\sqrt{2}} \begin{pmatrix} 0 & 1 & 0 \\ 1 & 0 & 1 \\ 0 & 1 & 0 \end{pmatrix}, \quad S_2 = \frac{1}{\sqrt{2}} \begin{pmatrix} 0 & -i & 0 \\ i & 0 & -i \\ 0 & i & 0 \end{pmatrix}, \quad S_3 = \begin{pmatrix} 1 & 0 & 0 \\ 0 & 0 & 0 \\ 0 & 0 & -1 \end{pmatrix}$$

and

$$K = S_1 \otimes S_1 + S_2 \otimes S_2 + S_3 \otimes S_3, \quad \mathbf{e}_1 = \begin{pmatrix} 1 \\ 0 \\ 0 \end{pmatrix}.$$

Calculate the vector $K(\mathbf{e}_1 \otimes \mathbf{e}_1)$ in the Hilbert space $\mathbb{C}^9$. Is the vector normalized? Discuss.

Solution 45.   (i) Utilizing that $i \cdot i = -1$ we obtain

$$K(\mathbf{e}_1 \otimes \mathbf{e}_1) = \begin{pmatrix} 0 \\ 1 \end{pmatrix} \otimes \begin{pmatrix} 0 \\ 1 \end{pmatrix} + \begin{pmatrix} 0 \\ i \end{pmatrix} \otimes \begin{pmatrix} 0 \\ i \end{pmatrix} + \begin{pmatrix} 1 \\ 0 \end{pmatrix} \otimes \begin{pmatrix} 1 \\ 0 \end{pmatrix}$$

$$= \begin{pmatrix} 1 \\ 0 \end{pmatrix} \otimes \begin{pmatrix} 1 \\ 0 \end{pmatrix} = \begin{pmatrix} 1 \\ 0 \\ 0 \\ 0 \end{pmatrix}.$$

The contribution to the state from $\sigma_1 \otimes \sigma_1$ and $\sigma_2 \otimes \sigma_2$ cancel out. Thus we have an eigenvalue equation with eigenvalue 1.
(ii) Analogously to (i) the contributions to the vector from the first Pauli matrix σ_1 and second Pauli matrix σ_2 cancel out owing to $i \cdot i = -1$ and we have an eigenvalue equation with eigenvalue 2.
(iii) The contribution to the state from $S_1 \otimes S_1$ and $S_2 \otimes S_2$ cancel out and we have an eigenvalue equation with eigenvalue 1.

Problem 46.   Let $\mathbb{H}_n$ be the vector space of $n \times n$ hermitian matrices. The adjoint (conjugate transpose) of a matrix $A \in \mathbb{C}^{n \times n}$ is denoted by A^*. Consider a family $V_1, V_2, \dots, V_m$ of $n \times n$ matrices over $\mathbb{C}$. We associate with this family the completely positive map $\psi : \mathbb{H}_n \to \mathbb{H}_n$ defined by

$$\psi(X) = \sum_{j=1}^{m} V_j X V_j^*.$$

The map ψ is said to be a *Kraus map* if $\psi(I_n) = I_n$, i.e.

$$\sum_{j=1}^{m} V_j V_j^* = I_n$$

and the matrices $V_1, V_2, \dots, V_m$ are called *Kraus operators*.

Let $m = n = 2$ and consider the spin matrices

$$S_+ = \begin{pmatrix} 0 & 1 \\ 0 & 0 \end{pmatrix}, \quad S_- = \begin{pmatrix} 0 & 0 \\ 1 & 0 \end{pmatrix}.$$

Show that S_+ and S_- are Kraus operators and find the associated Kraus map.

Solution 46. Since

$$S_+ S_+^* + S_- S_-^* = \begin{pmatrix} 0 & 1 \\ 0 & 0 \end{pmatrix} \begin{pmatrix} 0 & 0 \\ 1 & 0 \end{pmatrix} + \begin{pmatrix} 0 & 0 \\ 1 & 0 \end{pmatrix} \begin{pmatrix} 0 & 1 \\ 0 & 0 \end{pmatrix} = \begin{pmatrix} 1 & 0 \\ 0 & 1 \end{pmatrix}$$

the matrices S_+ and S_- are Kraus operators. The associated Kraus map is

$$\psi \begin{pmatrix} a & b \\ c & d \end{pmatrix} = \begin{pmatrix} 0 & 1 \\ 0 & 0 \end{pmatrix} \begin{pmatrix} a & b \\ c & d \end{pmatrix} \begin{pmatrix} 0 & 0 \\ 1 & 0 \end{pmatrix} + \begin{pmatrix} 0 & 0 \\ 1 & 0 \end{pmatrix} \begin{pmatrix} a & b \\ c & d \end{pmatrix} \begin{pmatrix} 0 & 1 \\ 0 & 0 \end{pmatrix}$$

$$= \begin{pmatrix} d & 0 \\ 0 & a \end{pmatrix}.$$

Problem 47. Let σ_1, σ_2, σ_3 be the Pauli spin matrices and

$$\sigma_+ := \sigma_1 + i\sigma_2, \qquad \sigma_- := \sigma_1 - i\sigma_2.$$

Show that the *Toda equation*

$$\frac{d^2 u}{dt^2} = e^{-2u}$$

admits the *Lax representation*

$$\frac{dL}{dt} = [L, A](t)$$

with

$$L = -\frac{du}{dt}\sigma_3 - e^{-u}\sigma_1 \equiv \begin{pmatrix} -du/dt & -e^{-u} \\ -e^{-u} & du/dt \end{pmatrix}$$

$$A = \frac{1}{2}\frac{du}{dt}\sigma_3 + e^{-u}\sigma_- = \begin{pmatrix} \frac{1}{2}du/dt & 0 \\ 2e^{-u} & -\frac{1}{2}du/dt \end{pmatrix}.$$

Solution 47. We have

$$\frac{dL}{dt} = \begin{pmatrix} -d^2u/dt^2 & e^{-u}du/dt \\ e^{-u}du/dt & d^2u/dt^2 \end{pmatrix}$$

and

$$[L, A] = LA - AL = \begin{pmatrix} -e^{-2u} & e^{-u}du/dt \\ e^{-u}du/dt & e^{-2u} \end{pmatrix}.$$

Problem 48. Let σ_0, σ_1, σ_2, σ_3 be the Pauli spin matrices with $\sigma_0 = I_2$ be the 2×2 identity matrix. Let

$$\Psi(x_1, x_2, x_3, t) = \begin{pmatrix} \psi_1(x_1, x_2, x_3, t) \\ \psi_2(x_1, x_2, x_3, t) \end{pmatrix}.$$

The *Weyl equation* describing massless spin-$\frac{1}{2}$ particles is given by

$$I_2 \frac{1}{c}\frac{\partial \Psi}{\partial t} + \sigma_1 \frac{\partial \Psi}{\partial x_1} + \sigma_2 \frac{\partial \Psi}{\partial x_2} + \sigma_3 \frac{\partial \Psi}{\partial x_3} = \begin{pmatrix} 0 \\ 0 \end{pmatrix}.$$

Thus we obtain a system of coupled linear partial differential equations

$$\frac{1}{c}\frac{\partial \psi_1}{\partial t} + \frac{\partial \psi_2}{\partial x_1} - i\frac{\partial \psi_2}{\partial x_2} - \frac{\partial \psi_1}{\partial x_3} = 0$$

$$\frac{1}{c}\frac{\partial \psi_2}{\partial t} + \frac{\partial \psi_1}{\partial x_1} + i\frac{\partial \psi_1}{\partial x_2} - \frac{\partial \psi_2}{\partial x_3} = 0.$$

Note that the Weyl equation does not contain the Planck constant.
(i) Find solutions of the form

$$\Psi(x_1, x_2, x_3, t) = \begin{pmatrix} \psi_1(x_1, x_2, x_3, t) \\ \psi_2(x_1, x_2, x_3, t) \end{pmatrix} = \begin{pmatrix} \chi_1 e^{-i(\mathbf{p}\cdot\mathbf{x} - Et)/\hbar} \\ \chi_2 e^{-i(\mathbf{p}\cdot\mathbf{x} - Et)/\hbar} \end{pmatrix}$$

where $\mathbf{p}\cdot\mathbf{x} := p_1 x_1 + p_2 x_2 + p_3 x_3$. Derive the *dispersion relation*.
(ii) Study the time evolution of the hydrodynamic variables

$$\rho := \Psi^\dagger \Psi, \quad v_j := c\Psi^\dagger \sigma_j \Psi, \quad u_j := \frac{c}{2i}\Psi^\dagger \overset{\leftrightarrow}{\frac{\partial}{\partial x_j}} \Psi$$

where $j = 1, 2, 3$, $\Psi^\dagger = (\bar\psi_1 \quad \bar\psi_2)$ and $\overset{\leftrightarrow}{\nabla} := \overset{\rightarrow}{\nabla} - \overset{\leftarrow}{\nabla}$.

Solution 48.   (i) Inserting the ansatz for Ψ we arrive at

$$\begin{pmatrix} iE/c - ip_3 & -ip_1 - p_2 \\ -ip_1 + p_2 & iE/c + ip_3 \end{pmatrix} \begin{pmatrix} \chi_1 \\ \chi_2 \end{pmatrix} = \begin{pmatrix} 0 \\ 0 \end{pmatrix}.$$

To find nonzero solutions we have

$$\det \begin{pmatrix} iE/c - ip_3 & -ip_1 - p_2 \\ -ip_1 + p_2 & iE/c + ip_3 \end{pmatrix} = 0.$$

Thus the dispersion relation follows

$$\frac{E^2}{c^2} = p_1^2 + p_2^2 + p_3^2.$$

(ii) We obtain for the time evolution of ρ

$$\frac{\partial\rho}{\partial t} = \frac{\partial}{\partial t}(\Psi^\dagger\Psi) = \frac{\partial}{\partial t}(\bar\psi_1\psi_1 + \bar\psi_2\psi_2) = -c\sum_{k=1}^{3}\frac{\partial}{\partial x_k}(\Psi^\dagger\sigma_k\Psi).$$

Show that the Weyl equation can also be written as

$$i\hbar\frac{\partial}{\partial t}\Psi = c(\boldsymbol\sigma\cdot\hat{\mathbf{p}})\Psi$$

where

$$\hat{\mathbf{p}} \equiv (\hat p_1, \hat p_2, \hat p_3) = \left(-i\hbar\frac{\partial}{\partial x_1}, -i\hbar\frac{\partial}{\partial x_2}, -i\hbar\frac{\partial}{\partial x_3}\right)$$

and

$$\boldsymbol\sigma\cdot\hat{\mathbf{p}} = \sigma_1\hat p_1 + \sigma_2\hat p_2 + \sigma_3\hat p_3.$$

Problem 49.   Consider the spin-1 matrices in the form

$$S_1 = \begin{pmatrix} 0 & 0 & 0 \\ 0 & 0 & -i \\ 0 & i & 0 \end{pmatrix}, \quad S_2 = \begin{pmatrix} 0 & 0 & i \\ 0 & 0 & 0 \\ -i & 0 & 0 \end{pmatrix}, \quad S_3 = \begin{pmatrix} 0 & -i & 0 \\ i & 0 & 0 \\ 0 & 0 & 0 \end{pmatrix}.$$

Let $\hat p_k = -i\partial/\partial x_k$ ($k = 1, 2, 3$) and

$$\hat{\mathbf{p}}\cdot\mathbf{S} := p_1 S_1 + p_2 S_2 + p_3 S_3.$$

Consider the system of partial differential equations

$$(\hat{\mathbf{p}} \cdot \mathbf{S})\mathbf{F} = \frac{i}{c}\frac{\partial \mathbf{F}}{\partial t}$$

where $\mathbf{F}(\mathbf{x}, t) = \mathbf{E}(\mathbf{x}, t) + ic\mathbf{B}(\mathbf{x}, t)$, where $\mathbf{F}$ is called *Kramer's vector*, $\mathbf{E}$ and $\mathbf{B}$ are real valued vector fields and c is the speed of light. Find the system of partial differential equations for $\mathbf{E}$ and $\mathbf{B}$.

Solution 49.   Since $\mathbf{F}^* = \mathbf{E} - ic\mathbf{B}$ we obtain

$$\mathrm{curl}\mathbf{F} = \frac{i}{c}\frac{\partial \mathbf{F}}{\partial t}, \qquad \mathrm{curl}\mathbf{F}^* = -\frac{i}{c}\frac{\partial \mathbf{F}^*}{\partial t}.$$

Looking at the real and imaginary part of these equations leads to *Maxwell's equations*

$$\mathrm{curl}\mathbf{B} = \frac{1}{c^2}\frac{\partial \mathbf{E}}{\partial t}, \qquad \mathrm{curl}\mathbf{E} = -\frac{\partial \mathbf{B}}{\partial t}.$$

Problem 50.   Consider the *Dirac-Hamilton operator*

$$\hat{H} = mc^2\beta + c\alpha \cdot \mathbf{p} \equiv mc^2\beta + c(\alpha_1 p_1 + \alpha_2 p_2 + \alpha_3 p_3)$$

where m is the rest mass, c is the speed of light and

$$\beta := \begin{pmatrix} 1 & 0 & 0 & 0 \\ 0 & 1 & 0 & 0 \\ 0 & 0 & -1 & 0 \\ 0 & 0 & 0 & -1 \end{pmatrix}, \quad \alpha_1 := \begin{pmatrix} 0 & 0 & 0 & 1 \\ 0 & 0 & 1 & 0 \\ 0 & 1 & 0 & 0 \\ 1 & 0 & 0 & 0 \end{pmatrix},$$

$$\alpha_2 := \begin{pmatrix} 0 & 0 & 0 & -i \\ 0 & 0 & i & 0 \\ 0 & -i & 0 & 0 \\ i & 0 & 0 & 0 \end{pmatrix}, \quad \alpha_3 := \begin{pmatrix} 0 & 0 & 1 & 0 \\ 0 & 0 & 0 & -1 \\ 1 & 0 & 0 & 0 \\ 0 & -1 & 0 & 0 \end{pmatrix}$$

$$\hat{p}_1 := -i\hbar\frac{\partial}{\partial x_1}, \quad \hat{p}_2 := -i\hbar\frac{\partial}{\partial x_2}, \quad \hat{p}_3 := -i\hbar\frac{\partial}{\partial x_3}.$$

Thus the Dirac-Hamilton operator takes the form

$$\hat{H} = c\hbar \begin{pmatrix} mc/\hbar & 0 & -i\frac{\partial}{\partial x_3} & -i\frac{\partial}{\partial x_1} - \frac{\partial}{\partial x_2} \\ 0 & mc/\hbar & -i\frac{\partial}{\partial x_1} + \frac{\partial}{\partial x_2} & i\frac{\partial}{\partial x_3} \\ -i\frac{\partial}{\partial x_3} & -i\frac{\partial}{\partial x_1} - \frac{\partial}{\partial x_2} & -mc/\hbar & 0 \\ -i\frac{\partial}{\partial x_1} + \frac{\partial}{\partial x_2} & i\frac{\partial}{\partial x_3} & 0 & -mc/\hbar \end{pmatrix}.$$

Let I_4 be the 4×4 unit matrix. Use the *Heisenberg equation of motion* to find the time evolution of β, α_j and $I_4 p_j$, where $j = 1, 2, 3$. The *Heisenberg equation of motion* is given by

$$i\hbar\frac{d\hat{A}}{dt} = [\hat{A}, \hat{H}](t).$$

Solution 50.   We find for the commutator

$$[\beta, \hat{H}] = 2c\hbar \begin{pmatrix} 0 & 0 & -i\frac{\partial}{\partial x_3} & -i\frac{\partial}{\partial x_1} - \frac{\partial}{\partial x_2} \\ 0 & 0 & -i\frac{\partial}{\partial x_1} + \frac{\partial}{\partial x_2} & i\frac{\partial}{\partial x_3} \\ i\frac{\partial}{\partial x_3} & i\frac{\partial}{\partial x_1} + \frac{\partial}{\partial x_2} & 0 & 0 \\ i\frac{\partial}{\partial x_1} - \frac{\partial}{\partial x_2} & -i\frac{\partial}{\partial x_3} & 0 & 0 \end{pmatrix}.$$

Thus the Heisenberg equation of motion takes the form

$$i\hbar\frac{d\beta}{dt} = 2c\beta\alpha_1(p_1 - ip_2)(t) + 2c\alpha_3 p_3(t).$$

Since

$$[\alpha_1, \hat{H}] = -2\hat{H}\alpha_1 + 2cI_4p_1$$

we obtain

$$i\hbar\frac{d\alpha_1}{dt} = -(2\hat{H}\alpha_1 + 2cI_4p_1)(t).$$

Analogously

$$i\hbar\frac{d\alpha_2}{dt} = -(2\hat{H}\alpha_2 + 2cI_4p_2)(t), \quad i\hbar\frac{d\alpha_3}{dt} = -(2\hat{H}\alpha_3 + 2cI_4p_3)(t).$$

Furthermore we obviously have

$$[I_4p_j, \hat{H}] = 0, \qquad j = 1, 2, 3.$$

Thus

$$i\hbar\frac{d}{dt}I_4p_j = 0, \qquad j = 1, 2, 3.$$

From these equations we obtain

$$i\hbar\frac{d^2\alpha}{dt^2} = 2\frac{d\alpha}{dt}\hat{H} = -2\hat{H}\frac{d\alpha}{dt}.$$

This equation can be integrated once and we find

$$\frac{d\alpha}{dt} = \frac{d\alpha(0)}{dt}\exp(-2i\hat{H}t/\hbar).$$

We also have the identities $[\hat{H}, \dot{\alpha}]_+ = 0$, $[\hat{H}, \dot{\alpha}] = 2\hat{H}\alpha$.

2.2   Spin Matrices and Functions

Problem 51.   (i) Consider the spin matrices (spin-$\frac{1}{2}$)

$$S_+ = \begin{pmatrix} 0 & 1 \\ 0 & 0 \end{pmatrix}, \qquad S_- = \begin{pmatrix} 0 & 0 \\ 1 & 0 \end{pmatrix}.$$

Let $z \in \mathbb{C}$. Calculate e^{zS_+} and e^{zS_-}. Utilize the expansions

$$e^{zS_+} = \sum_{j=0}^{\infty} \frac{(zS_+)^j}{j!}, \qquad e^{zS_-} = \sum_{j=0}^{\infty} \frac{(zS_-)^j}{j!}.$$

Is the matrix unitary? Calculate the vector

$$\exp(zS_+) \begin{pmatrix} 0 \\ 1 \end{pmatrix}.$$

Is the vector normalized? If not normalize the vector. Then calculate the scalar product of this vector with the vector $(0 \quad 1)$.
(ii) Consider the spin matrices (spin-1)

$$S_+ = \begin{pmatrix} 0 & \sqrt{2} & 0 \\ 0 & 0 & \sqrt{2} \\ 0 & 0 & 0 \end{pmatrix}, \qquad S_- = \begin{pmatrix} 0 & 0 & 0 \\ \sqrt{2} & 0 & 0 \\ 0 & \sqrt{2} & 0 \end{pmatrix}.$$

Let $z \in \mathbb{C}$. Calculate e^{zS_+} and e^{zS_-}. Utilize the expansions

$$e^{zS_+} = \sum_{j=0}^{\infty} \frac{(zS_+)^j}{j!}, \qquad e^{zS_-} = \sum_{j=0}^{\infty} \frac{(zS_-)^j}{j!}.$$

Solution 51.   (i) Since $(S_+)^2 = 0_2$ and $(S_-)^2 = 0_2$ we obtain

$$e^{zS_+} = \begin{pmatrix} 1 & z \\ 0 & 1 \end{pmatrix}, \qquad e^{zS_-} = \begin{pmatrix} 1 & 0 \\ z & 1 \end{pmatrix}.$$

The matrices are invertible with determinant equal to 1 for all $z \in \mathbb{C}$, but not unitary if $z \neq 0$.
Now

$$e^{zS_+} \begin{pmatrix} 0 \\ 1 \end{pmatrix} = \begin{pmatrix} z \\ 1 \end{pmatrix}.$$

The normalized vector is

$$\frac{1}{\sqrt{1 + z\bar{z}}} \begin{pmatrix} z \\ 1 \end{pmatrix}.$$

The scalar product is

$$(0 \quad 1) \frac{1}{\sqrt{1 + \bar{z}z}} \begin{pmatrix} z \\ 1 \end{pmatrix} = \frac{1}{\sqrt{1 + \bar{z}z}}.$$

(ii) Since

$$(S_+)^2 = \begin{pmatrix} 0 & 0 & 2 \\ 0 & 0 & 0 \\ 0 & 0 & 0 \end{pmatrix}, \qquad (S_+)^3 = \begin{pmatrix} 0 & 0 & 0 \\ 0 & 0 & 0 \\ 0 & 0 & 0 \end{pmatrix}$$

we obtain

$$e^{zS_+} = \begin{pmatrix} 1 & \sqrt{2}z & z^2 \\ 0 & 1 & \sqrt{2}z \\ 0 & 0 & 1 \end{pmatrix}.$$

The matrix is invertible with determinant equal to 1 for all $z \in \mathbb{C}$, but not unitary if $z \neq 0$. Since

$$(S_-)^2 = \begin{pmatrix} 0 & 0 & 0 \\ 0 & 0 & 0 \\ 2 & 0 & 0 \end{pmatrix}, \quad (S_-)^3 = \begin{pmatrix} 0 & 0 & 0 \\ 0 & 0 & 0 \\ 0 & 0 & 0 \end{pmatrix}$$

we obtain

$$e^{zS_-} = \begin{pmatrix} 1 & 0 & 0 \\ \sqrt{2}z & 1 & 0 \\ z^2 & \sqrt{2}z & 1 \end{pmatrix}.$$

The matrix is invertible with determinant equal to 1 for all $z \in \mathbb{C}$, but not unitary if $z \neq 0$.

Problem 52. Consider the spin-$\frac{1}{2}$ matrices

$$S_1 = \frac{1}{2}\begin{pmatrix} 0 & 1 \\ 1 & 0 \end{pmatrix}, \quad S_2 = \frac{1}{2}\begin{pmatrix} 0 & -i \\ i & 0 \end{pmatrix}, \quad S_3 = \frac{1}{2}\begin{pmatrix} 1 & 0 \\ 0 & -1 \end{pmatrix}.$$

Let $z \in \mathbb{C}$. Find $\exp(zS_1)$, $\exp(zS_2)$, $\exp(zS_3)$. Then set $z = -i\omega t$.

Solution 52. Since $S_1^2 = I_2/4$, $S_2^2 = I_2/4$, $S_3^2 = I_2/4$ we obtain

$$\exp(zS_1) = \cosh(z/2)I_2 + \sinh(z/2)S_1 = \begin{pmatrix} \cosh(z/2) & \sinh(z/2) \\ \sinh(z/2) & \cosh(z/2) \end{pmatrix}$$

$$\exp(zS_2) = \cosh(z/2)I_2 + \sinh(z/2)S_2 = \begin{pmatrix} \cosh(z/2) & -i\sinh(z/2) \\ i\sinh(z) & \cosh(z/2) \end{pmatrix}$$

$$\exp(zS_3) = \cosh(z/2)I_2 + \sinh(z/2)S_3 = \begin{pmatrix} e^{z/2} & 0 \\ 0 & e^{-z/2} \end{pmatrix}.$$

Thus with $z = -i\omega t$ we have $\cosh(-i\omega t/2) \equiv \cos(\omega t/2)$ and $\sinh(-i\omega t/2) \equiv -i\sin(\omega t/2)$. Then the matrices are unitary.

Problem 53. Consider the three (hermitian) spin-1 matrices

$$S_1 = \frac{1}{\sqrt{2}}\begin{pmatrix} 0 & 1 & 0 \\ 1 & 0 & 1 \\ 0 & 1 & 0 \end{pmatrix}, \quad S_2 = \frac{1}{\sqrt{2}}\begin{pmatrix} 0 & -i & 0 \\ i & 0 & -i \\ 0 & i & 0 \end{pmatrix}, \quad S_3 = \begin{pmatrix} 1 & 0 & 0 \\ 0 & 0 & 0 \\ 0 & 0 & -1 \end{pmatrix}$$

all with eigenvalues $+1, 0, -1$.
(i) Find S_j^3 for $j = 1, 2, 3$.
(ii) Let $z \in \mathbb{C}$. Find $\exp(zS_1)$, $\exp(zS_2)$, $\exp(zS_3)$. Then set $z = -i\omega t$.
(iii) Let $\phi \in \mathbb{R}$. Find $\exp(i\phi S_j)$.
(iv) Consider the hermitian matrix $\hat{H} = \hbar\omega S_1$. Find the unitary matrix $\exp(-i\hat{H}t/\hbar)$.
(v) Consider the normalized vector in the Hilbert space $\mathbb{C}^3$

$$|\psi(0)\rangle = \frac{1}{\sqrt{3}}\begin{pmatrix} 1 \\ 1 \\ 1 \end{pmatrix}.$$

Calculate $|\psi(t)\rangle = \exp(-i\hat{H}t/\hbar)|\psi(0)\rangle$.
(vi) Find the probability of finding $|\psi(t)\rangle$ in the initial state $|\psi(0)\rangle$, i.e. $|\langle\psi(t)|\psi(0)\rangle|^2$.

Solution 53. (i) We obtain $S_1^3 = S_1$, $S_2^3 = S_2$, $S_3^3 = S_3$.
(ii) From (i) it follows that

$$\exp(zS_1) = I_3 + \sinh(z)S_1 + (\cosh(z) - 1)S_1^2$$
$$\exp(zS_2) = I_3 + \sinh(z)S_2 + (\cosh(z) - 1)S_2^2$$
$$\exp(zS_3) = I_3 + \sinh(z)S_3 + (\cosh(z) - 1)S_3^2.$$

With $z = -i\omega t$ we have $\cosh(-i\omega t/2) \equiv \cos(\omega t)$, $\sinh(-i\omega t) \equiv -i\sin(\omega t)$. Then the matrices are unitary.
(iii) Since $S_j^3 = S_j$ for $j = 1, 2, 3$ we obtain

$$\exp(i\phi S_j) = I_3 + i\sin(\phi)S_j - (1 - \cos(\phi))S_j^2$$

which is a unitary matrix with determinant equal to $+1$.
(iv) From (ii) we obtain

$$\exp(-i\hat{S}_1 t/\hbar) = \begin{pmatrix} \frac{1}{2}\cos(\omega t) + \frac{1}{2} & -\frac{i}{\sqrt{2}}\sin(\omega t) & \frac{1}{2}\cos(\omega t) - \frac{1}{2} \\ -\frac{i}{\sqrt{2}}\sin(\omega t) & \cos(\omega t) & -\frac{i}{\sqrt{2}}\sin(\omega t) \\ \frac{1}{2}\cos(\omega t) - \frac{1}{2} & -\frac{i}{\sqrt{2}}\sin(\omega t) & \frac{1}{2}\cos(\omega t) + \frac{1}{2} \end{pmatrix}.$$

(v) We obtain

$$\psi(t) = \exp(-i\hat{S}_1 t/\hbar)\psi(0) = \frac{1}{\sqrt{3}}\begin{pmatrix} \cos(\omega t) - \frac{i}{\sqrt{2}}\sin(\omega t) \\ \cos(\omega t) - i\sqrt{2}\sin(\omega t) \\ \cos(\omega t) - \frac{i}{\sqrt{2}}\sin(\omega t) \end{pmatrix}.$$

(vi) We have

$$|\langle\psi(t)|\psi(0)\rangle|^2 = \frac{1}{9}|3\cos(\omega t) - i2\sqrt{2}\sin(\omega t)|^2 = 1 - \frac{1}{9}\sin^2(\omega t).$$

Problem 54. Let S_1, S_2, S_3 be spin matrices with the commutation relations

$$[S_1, S_2] = iS_3, \quad [S_2, S_3] = iS_1, \quad [S_3, S_1] = iS_2.$$

Let $z \in \mathbb{C}$. Calculate $e^{zS_3}S_1e^{-zS_3}$ using the following two methods
(i) Set

$$f(z) = e^{zS_3}S_1e^{-zS_3}$$

with $f(0) = S_1$ and find a differential equation for f by taking the derivative with respect to z. Solve the initial value problem.
(ii) Apply the expansion

$$e^{zS_3}S_1e^{-zS_3} = S_1 + z[S_3, S_1] + \frac{z^2}{2!}[S_3, [S_3, S_1]] + \frac{z^3}{3!}[S_3, [S_3, [S_3, S_1]]] + \cdots$$

(iii) Set $z = i\omega t$ in the result, where ω is the frequency and t the time.

Solution 54.   (i) We obtain

$$\frac{df}{dz} = e^{zS_3}[S_3, S_1]e^{-zS_3} = e^{zS_3}iS_2e^{-zS_3} = ie^{zS_3}S_2e^{-zS_3}$$

with $df(0)/dz = iS_2$ and

$$\frac{d^2 f}{dz^2} = ie^{zS_3}[S_3, S_2]e^{-zS_3} = e^{zS_3}S_1e^{-zS_3} = f(z).$$

Thus we obtain the second order linear matrix differential equation

$$\frac{d^2 f}{dz^2} = f.$$

The solution with the initial conditions $f(0) = S_1$ and $df(0)/dz = iS_2$ is given by

$$f(z) = e^{zS_3}S_1e^{-zS_3} = S_1\cosh(z) + iS_2\sinh(z).$$

(ii) Since $[S_3, S_1] = iS_2$, $[S_3, [S_3, S_1]] = S_1$, $[S_3, [S_3, [S_3, S_1]]] = iS_2$ etc we obtain obviously the same result as above.
(iii) We have $\cosh(i\omega t) \equiv \cos(\omega t)$ and $\sinh(i\omega t) \equiv i\sin(\omega t)$.

Problem 55.   (i) Let A be a 2×2 matrix over $\mathbb{C}$ with two different eigenvalues λ_1 and λ_2. The *Cayley-Hamilton theorem* tells us that

$$e^A = c_1 A + c_0 I_2$$

with

$$e^{\lambda_1} = c_1\lambda_1 + c_0, \qquad e^{\lambda_2} = c_1\lambda_2 + c_0.$$

Thus we have two linear equations with two unknowns c_0, c_1. Let $z \in \mathbb{C}$ and $z \neq 0$. Apply the theorem to calculate $\exp(z\sigma_1)$, $\exp(z\sigma_2)$, $\exp(z\sigma_3)$. Then set $z = -i\omega t$.
(ii) Let A be a 3×3 matrix over $\mathbb{C}$ with three pairwise different eigenvalues λ_1, λ_2 and λ_3. The *Cayley-Hamilton theorem* tells us that

$$e^A = c_2 A^2 + c_1 A + c_0 I_2$$

with

$$e^{\lambda_1} = c_2\lambda_1^2 + c_1\lambda_1 + c_0, \quad e^{\lambda_2} = c_2\lambda_2^2 + c_1\lambda_2 + c_0, \quad e^{\lambda_3} = c_2\lambda_3^2 + c_1\lambda_3 + c_0.$$

Thus we have three linear equations with the unknowns c_0, c_1, c_2. Let $z \in \mathbb{C}$ and $z \neq 0$. Apply the theorem to calculate $\exp(zS_1)$, $\exp(zS_2)$, $\exp(zS_3)$, where S_1, S_2, S_3 are the spin-1 matrices. Then set $z = -i\omega t$.

Solution 55.   (i) Solving the two linear equations for c_0 and c_1 we obtain

$$c_1 = \frac{e^{\lambda_1} - e^{\lambda_2}}{\lambda_1 - \lambda_2}, \quad c_0 = \frac{e^{\lambda_2}\lambda_1 - e^{\lambda_1}\lambda_2}{\lambda_1 - \lambda_2}.$$

Consequently

$$e^A = \frac{e^{\lambda_1} - e^{\lambda_2}}{\lambda_1 - \lambda_2}A + \frac{e^{\lambda_2}\lambda_1 - e^{\lambda_1}\lambda_2}{\lambda_1 - \lambda_2}I_2.$$

Since the eigenvalues for all three Pauli spin matrices are given by $\lambda_1 = z$, $\lambda_2 = -z$ it follows that

$$e^{z\sigma_1} = \sinh(z)\sigma_1 + \cosh(z)I_2, \quad e^{z\sigma_2} = \sinh(z)\sigma_2 + \cosh(z)I_2, \quad e^{z\sigma_3} = \sinh(z)\sigma_3 + \cosh(z)I_2$$

with $\sinh(-i\omega t) \equiv -i\sinh(\omega t)$, $\cosh(-i\omega t) \equiv \cos(\omega t)$.

(ii) Solving the three linear equations for the unknowns c_0, c_1, c_2 we obtain

$$c_2 = \frac{1}{D}(\lambda_1(e^{\lambda_3} - e^{\lambda_2}) + \lambda_2(e^{\lambda_1} - e^{\lambda_3}) + \lambda_3(e^{\lambda_2} - e^{\lambda_1}))$$

$$c_1 = \frac{1}{D}(\lambda_1^2(e^{\lambda_2} - e^{\lambda_3}) + \lambda_2^2(e^{\lambda_3} - e^{\lambda_1}) + \lambda_3^2(e^{\lambda_1} - e^{\lambda_2}))$$

$$c_0 = \frac{1}{D}(\lambda_1(e^{\lambda_2}\lambda_1^2 - e^{\lambda_1}\lambda_2^2) + \lambda_2(e^{\lambda}\lambda_2^2 - e^{\lambda}\lambda_3^2) + \lambda_3(e^{\lambda}\lambda_2^2 - e^{\lambda}\lambda_1^2))$$

where
$$D := \lambda_1(\lambda_3^2 - \lambda_2^2) + \lambda_2(\lambda_1^2 - \lambda_3^2) + \lambda_3(\lambda_2^2 - \lambda_1^2).$$

The eigenvalues for all three spin-1 matrices zS_1, zS_2, zS_3 are given by $\lambda_1 = z$, $\lambda_2 = 0$, $\lambda_3 = -z$,

$$S_1^2 = \frac{1}{2}\begin{pmatrix} 1 & 0 & 1 \\ 0 & 2 & 0 \\ 1 & 0 & 1 \end{pmatrix}, \quad S_2^2 = \frac{1}{2}\begin{pmatrix} 1 & 0 & -1 \\ 0 & 2 & 0 \\ -1 & 0 & 1 \end{pmatrix}, \quad S_3^2 = \begin{pmatrix} 1 & 0 & 0 \\ 0 & 0 & 0 \\ 0 & 0 & 1 \end{pmatrix}$$

and $S_1^3 = S_1$, $S_2^3 = S_2$, $S_3^3 = S_3$. Thus we learn that

$$e^{zS_1} = I_3 + \sinh(z)S_1 + (\cosh(z) - 1)S_1^2$$
$$e^{zS_2} = I_3 + \sinh(z)S_2 + (\cosh(z) - 1)S_2^2$$
$$e^{zS_3} = I_3 + \sinh(z)S_3 + (\cosh(z) - 1)S_3^2$$

with $\sinh(-i\omega t) \equiv -i\sinh(\omega t)$, $\cosh(-i\omega t) \equiv \cos(\omega t)$.

Problem 56. Consider the spin-$\frac{3}{2}$ matrix

$$S_1 = \frac{1}{2}\begin{pmatrix} 0 & \sqrt{3} & 0 & 0 \\ \sqrt{3} & 0 & 2 & 0 \\ 0 & 2 & 0 & \sqrt{3} \\ 0 & 0 & \sqrt{3} & 0 \end{pmatrix}$$

with the eigenvalues $\lambda_1 = +3/2$, $\lambda_2 = +1/2$, $\lambda_3 = -1/2$, $\lambda_4 = -3/2$ and the corresponding normalized eigenvectors

$$\mathbf{v}_1 = \frac{1}{\sqrt{8}}\begin{pmatrix} 1 \\ \sqrt{3} \\ \sqrt{3} \\ 1 \end{pmatrix}, \quad \mathbf{v}_2 = \frac{1}{\sqrt{8}}\begin{pmatrix} \sqrt{3} \\ 1 \\ -1 \\ -\sqrt{3} \end{pmatrix}, \quad \mathbf{v}_3 = \frac{1}{\sqrt{8}}\begin{pmatrix} \sqrt{3} \\ -1 \\ -1 \\ \sqrt{3} \end{pmatrix}, \quad \mathbf{v}_4 = \frac{1}{\sqrt{8}}\begin{pmatrix} 1 \\ -\sqrt{3} \\ \sqrt{3} \\ -1 \end{pmatrix}.$$

The eigenvectors form an orthonormal basis in the Hilbert space $\mathbb{C}^4$. Let $z \in \mathbb{C}$ and $z \neq 0$. Then the eigenvalues of zS_1 are given by $3z/2$, $z/2$, $-z/2$, $-3z/2$.

(i) Calculate $\exp(zS_1)$ applying the *Cayley-Hamilton theorem*. Let A be an $n \times n$ matrix over $\mathbb{C}$ and $p(\lambda) = \det(\lambda I_n - A)$ be the characteristic polynomial of A. Then $p(A) = 0_n$, where 0_n is the $n \times n$ unit matrix. Applying the Cayley-Hamilton theorem to calculate $\exp(zS_1)$ one starts from

$$\exp(zS_1) = c_3(zS_1)^3 + c_2(zS_1)^2 + c_1(zS_1) + c_0 I_4$$

and the system of four linear equations

$$e^{\lambda_1} = c_3\lambda_1^3 + c_2\lambda_1^2 + c_1\lambda_1 + c_0, \quad e^{\lambda_2} = c_3\lambda_2^3 + c_2\lambda_2^2 + c_1\lambda_2 + c_0$$
$$e^{\lambda_3} = c_3\lambda_3^3 + c_2\lambda_3^2 + c_1\lambda_3 + c_0, \quad e^{\lambda_4} = c_3\lambda_4^3 + c_2\lambda_4^2 + c_1\lambda_4 + c_0.$$

In matrix form we have

$$\begin{pmatrix} \lambda_1^3 & \lambda_1^2 & \lambda_1 & 1 \\ \lambda_2^3 & \lambda_2^2 & \lambda_2 & 1 \\ \lambda_3^3 & \lambda_3^2 & \lambda_3 & 1 \\ \lambda_4^3 & \lambda_4^2 & \lambda_4 & 1 \end{pmatrix} \begin{pmatrix} c_3 \\ c_2 \\ c_1 \\ c_0 \end{pmatrix} = \begin{pmatrix} e^{\lambda_1} \\ e^{\lambda_2} \\ e^{\lambda_3} \\ e^{\lambda_4} \end{pmatrix}.$$

Solve these linear equations for c_0, c_1, c_2, c_3 and substitute it into the expansion for e^{zS_1}.

(ii) Find $\exp(zS_1)$ applying the *spectral theorem*. For applying the spectral theorem for a normal $n \times n$ matrix one finds first the eigenvalues of A and then computes a basis for each eigenspace of A. Next one utilizes (if necessary) the Gram-Schmidt orthogonalization process to obtain an orthonormal basis for each eigenspace of the normal matrix A. Then combine the orthonormal bases for each eigenspace to obtain an orthonormal basis for eigenvectors of A. For the present case with $A = S_1$ the last two steps are not necessary.

Solution 56. (i) The following Maxima program will do the job to solve the system of linear equations and then substitute it into the matrix

```
/* S132.mac */

I4: matrix([1,0,0,0],[0,1,0,0],[0,0,1,0],[0,0,0,1]);
S1: matrix([0,sqrt(3)/2,0,0],[sqrt(3)/2,0,1,0],
           [0,1,0,sqrt(3)/2],[0,0,sqrt(3)/2,0]);
M: c3*z*z*z*(S1 . S1 . S1) + c2*z*z*(S1 . S1) + c1*z*S1 + c0*I4;
lamb1: z*3/2; lamb2: z*1/2; lamb3: -z*1/2; lamb4: -z*3/2;
sol: solve([exp(lamb1)-c3*lamb1*lamb1*lamb1-c2*lamb1*lamb1-c1*lamb1-c0,
       exp(lamb2)-c3*lamb2*lamb2*lamb2-c2*lamb2*lamb2-c1*lamb2-c0,
       exp(lamb3)-c3*lamb3*lamb3*lamb3-c2*lamb3*lamb3-c1*lamb3-c0,
       exp(lamb4)-c3*lamb4*lamb4*lamb4-c2*lamb4*lamb4-c1*lamb4-c0],
       [c3,c2,c1,c0]);
M1: subst(first(sol),M);
```

We find the solution

$$c_3 = \frac{e^{-3z/2}(e^{3z} - 3e^{2z} + 3e^z - 1)}{6z^3}, \quad c_2 = \frac{e^{-3z/2}(e^{3z} - e^{2z} - e^z + 1)}{4z^2}$$

$$c_1 = \frac{e^{-3z/2}(e^{3z} - 27e^{2z} + 27e^z - 1)}{24z}, \quad c_0 = \frac{e^{-3z/2}(e^{3z} - 9e^{2z} - 9e^z + 1)}{16}$$

and

$$(zS_1)^3 = \frac{z^3}{8}\begin{pmatrix} 0 & 7\sqrt{3} & 0 & 6 \\ 7\sqrt{3} & 0 & 20 & 0 \\ 0 & 20 & 0 & 7\sqrt{3} \\ 6 & 0 & 7\sqrt{3} & 0 \end{pmatrix}, \quad (zS_1)^2 = \frac{z^2}{4}\begin{pmatrix} 3 & 0 & 2\sqrt{3} & 0 \\ 0 & 7 & 0 & 2\sqrt{3} \\ 2\sqrt{3} & 0 & 7 & 0 \\ 0 & 2\sqrt{3} & 0 & 3 \end{pmatrix}.$$

(ii) For the spectral representation of zS_1 we have

$$zS_1 = z\lambda_1\mathbf{v}_1\mathbf{v}_1^* + z\lambda_2\mathbf{v}_2\mathbf{v}_2^* + z\lambda_3\mathbf{v}_3\mathbf{v}_3^* + z\lambda_4\mathbf{v}_4\mathbf{v}_4^*.$$

Since the vectors $\mathbf{v}_1$, $\mathbf{v}_2$, $\mathbf{v}_3$, $\mathbf{v}_4$ form an orthonormal bases we have $\mathbf{v}_j\mathbf{v}_j^*\mathbf{v}_k\mathbf{v}_k^* = \mathbf{v}_j\mathbf{v}_j^*\delta_{jk}$ we obtain

$$e^{zS_1} = e^{z\lambda_1}\mathbf{v}_1\mathbf{v}_1^* + e^{z\lambda_2}\mathbf{v}_2\mathbf{v}_2^* + e^{z\lambda_3}\mathbf{v}_3\mathbf{v}_3^* + e^{z\lambda_4}\mathbf{v}_4\mathbf{v}_4^*$$

with

$$\mathbf{v}_1\mathbf{v}_1^* = \frac{1}{8}\begin{pmatrix} 1 & \sqrt{3} & \sqrt{3} & 1 \\ \sqrt{3} & 3 & 3 & \sqrt{3} \\ \sqrt{3} & 3 & 3 & \sqrt{3} \\ 1 & \sqrt{3} & \sqrt{3} & 1 \end{pmatrix}$$

$$\mathbf{v}_2\mathbf{v}_2^* = \frac{1}{8}\begin{pmatrix} 3 & \sqrt{3} & -\sqrt{3} & -3 \\ \sqrt{3} & 1 & -1 & -\sqrt{3} \\ -\sqrt{3} & -1 & 1 & \sqrt{3} \\ -3 & -\sqrt{3} & \sqrt{3} & 3 \end{pmatrix}$$

$$\mathbf{v}_3\mathbf{v}_3^* = \frac{1}{8}\begin{pmatrix} 1 & \sqrt{3} & \sqrt{3} & 1 \\ \sqrt{3} & 3 & 3 & \sqrt{3} \\ \sqrt{3} & 3 & 3 & \sqrt{3} \\ 1 & \sqrt{3} & \sqrt{3} & 1 \end{pmatrix}$$

$$\mathbf{v}_4\mathbf{v}_4^* = \frac{1}{8}\begin{pmatrix} 1 & -\sqrt{3} & \sqrt{3} & -1 \\ -\sqrt{3} & 3 & -3 & \sqrt{3} \\ \sqrt{3} & -3 & 3 & -\sqrt{3} \\ -1 & \sqrt{3} & -\sqrt{3} & 1 \end{pmatrix}.$$

Problem 57. The *Heisenberg equation of motion* is given by

$$i\hbar\frac{d\hat{A}(t)}{dt} = [\hat{A}, \hat{H}](t)$$

where $\hat{H}$ is the Hamilton operator and $\hat{A}(t=0) = \hat{A}$. The formal solution of the initial value problem is

$$A(t) = e^{i\hat{H}t/\hbar}Ae^{-i\hat{H}t/\hbar}.$$

Consider the *spin-1 matrices*

$$S_1 = \frac{1}{\sqrt{2}}\begin{pmatrix} 0 & 1 & 0 \\ 1 & 0 & 1 \\ 0 & 1 & 0 \end{pmatrix}, \quad S_2 = \frac{1}{\sqrt{2}}\begin{pmatrix} 0 & -i & 0 \\ i & 0 & -i \\ 0 & i & 0 \end{pmatrix}, \quad S_3 = \begin{pmatrix} 1 & 0 & 0 \\ 0 & 0 & 0 \\ 0 & 0 & -1 \end{pmatrix}$$

with the commutation relations $[S_1, S_2] = iS_3$, $[S_2, S_3] = iS_1$, $[S_3, S_1] = iS_2$ and the Hamilton operator

$$\hat{H} = \hbar\omega S_3.$$

Find the time-evolution of S_1, i.e solve the Heisenberg equation of motion.

Solution 57. We have

$$\exp(-i\hat{H}t/\hbar) = \begin{pmatrix} e^{-i\omega t} & 0 & 0 \\ 0 & 1 & 0 \\ 0 & 0 & e^{i\omega t} \end{pmatrix}.$$

Thus we obtain the hermitian matrix

$$S_1(t) = \begin{pmatrix} e^{i\omega t} & 0 & 0 \\ 0 & 1 & 0 \\ 0 & 0 & e^{-i\omega t} \end{pmatrix}\frac{1}{\sqrt{2}}\begin{pmatrix} 0 & 1 & 0 \\ 1 & 0 & 1 \\ 0 & 1 & 0 \end{pmatrix}\begin{pmatrix} e^{-i\omega t} & 0 & 0 \\ 0 & 1 & 0 \\ 0 & 0 & e^{i\omega t} \end{pmatrix} = \frac{1}{\sqrt{2}}\begin{pmatrix} 0 & e^{i\omega t} & 0 \\ e^{-i\omega t} & 0 & e^{i\omega t} \\ 0 & e^{-i\omega t} & 0 \end{pmatrix}.$$

Problem 58. Let A, B be $n \times n$ matrices over $\mathbb{C}$. Assume that

$$A^2 = B^2 = I_n, \qquad [A, B]_+ = 0_n,$$

where $[,]_+$ denotes the anticommutator. Let $z \in \mathbb{C}$.
(i) Calculate

$$\exp(z(A + B)) := \sum_{j=0}^{\infty} \frac{z^j (A + B)^j}{j!}.$$

(ii) Apply the results from (i) or (ii) to the case $A = \sigma_3$, $B = \sigma_1$.
(iii) Let A, B, C be $n \times n$ matrices over $\mathbb{C}$ such that $A^2 = I_n$, $B^2 = I_n$ and $C^2 = I_n$.
Furthermore assume that

$$[A, B]_+ \equiv AB + BA = 0_n, \quad [B, C]_+ \equiv BC + CB = 0_n, \quad [C, A]_+ \equiv CA + AC = 0_n$$

i.e. the anticommutators vanish. Let $\alpha, \beta, \gamma \in \mathbb{C}$. Calculate $e^{\alpha A + \beta B + \gamma C}$ using

$$e^{\alpha A + \beta B + \gamma C} = \sum_{j=0}^{\infty} \frac{(\alpha A + \beta B + \gamma C)^j}{j!}.$$

Solution 58. (i) Using

$$(A + B)^2 = 2I_n, \qquad (A + B)^3 = 2(A + B)$$

and the expansion given above we arrive at

$$e^{z(A+B)} = I_n \left(1 + \frac{2^1 z^2}{2!} + \frac{2^2 z^4}{4!} + \frac{2^3 z^6}{6!} + \cdots \right) + (A + B) \left(z + \frac{2 z^3}{3!} + \frac{2^2 z^5}{5!} + \frac{2^3 z^7}{7!} + \cdots \right).$$

Consequently

$$e^{z(A+B)} = I_n \cosh(\sqrt{2}z) + (A + B) \frac{1}{\sqrt{2}} \sinh(\sqrt{2}z).$$

(ii) We have

$$e^{z(\sigma_3 + \sigma_1)} = I_2 \cosh(\sqrt{2}z) + \begin{pmatrix} 1 & 1 \\ 1 & -1 \end{pmatrix} \sinh(\sqrt{2}z).$$

Try to calculate $e^{z(A+B)}$ applying the *Lie-Trotter formula* given by

$$e^{z(A+B)} \equiv \lim_{p \to \infty} \left(e^{A/p} e^{B/p} \right)^p.$$

Note that

$$e^{zA/p} = I_n \cosh(z/p) + A \sinh(z/p), \qquad e^{zB/p} = I_n \cosh(z/p) + B \sinh(z/p)$$

and $(AB)^2 = -I_n$, $(A + B)^2 = 2I_n$.
(iii) Since the anticommutators vanish we have

$$(\alpha A + \beta B + \gamma C)^2 = (\alpha^2 + \beta^2 + \gamma^2)^2 I_n$$
$$(\alpha A + \beta B + \gamma C)^3 = (\alpha^2 + \beta^2 + \gamma^2)(\alpha A + \beta B + \gamma C).$$

Thus in general we have for positive n

$$(\alpha A + \beta B + \gamma C)^n = (\alpha^2 + \beta^2 + \gamma^2)^{n/2} I_n \quad \text{for} \quad n \text{ even}$$

and
$$(\alpha A + \beta B + \gamma C)^n = (\alpha^2 + \beta^2 + \gamma^2)^{n/2-1} \quad \text{for } n \text{ odd}$$

We have the expansion (we set $\delta^2 := \alpha^2 + \beta^2 + \gamma^2$)

$$e^{\alpha A + \beta B + \gamma C} = I_n \left(1 + \frac{1}{2!}(\delta^2) + \frac{1}{4!}(\delta^2)^2 + \frac{1}{6!}(\delta^2)^3 + \cdots\right)$$
$$+ (\alpha A + \beta B + \gamma C)\left(1 + \frac{1}{3!}(\delta^2) + \frac{1}{5!}(\delta^2)^2 + \cdots\right).$$

This can be summed up to

$$e^{\alpha A + \beta B + \gamma C} = I_n \cosh(\sqrt{\delta^2}) + \frac{\alpha A + \beta B}{\sqrt{\delta^2}} \sinh(\sqrt{\delta^2}).$$

Problem 59. Let K be an $n \times n$ skew-hermitian matrix with eigenvalues $\mu_1, \ldots, \mu_n$ (counted according to multiplicity) and the corresponding normalized eigenvectors $\mathbf{u}_1, \ldots, \mathbf{u}_n$, where $\mathbf{u}_j^* \mathbf{u}_k = 0$ for $k \neq j$. Then K can be written as

$$K = \sum_{j=1}^{n} \mu_j \mathbf{u}_j \mathbf{u}_j^*$$

and $\mathbf{u}_j \mathbf{u}_j^* \mathbf{u}_k \mathbf{u}_k^* = 0$ for $k \neq j$ and $j, k = 1, 2, \ldots, n$. Note that the matrices $\mathbf{u}_j \mathbf{u}_j^*$ are projection matrices and

$$\sum_{j=1}^{n} \mathbf{u}_j \mathbf{u}_j^* = I_n.$$

(i) Calculate $\exp(K)$.
(ii) Every $n \times n$ unitary matrix U can be written as $U = \exp(K)$, where K is a skew-hermitian matrix. Find U from a given K.
(iii) Use the result from (ii) to find for a given U a possible K.
(iv) Apply the result from (ii) and (iii) to the unitary 2×2 matrix

$$U(\theta) = \begin{pmatrix} \cos\theta & \sin\theta \\ -\sin\theta & \cos\theta \end{pmatrix}.$$

(v) Apply the result from (ii) and (iii) to the 2×2 unitary matrix

$$V(\theta, \phi) = \begin{pmatrix} \cos\theta & -e^{i\phi}\sin\theta \\ e^{-i\phi}\sin\theta & \cos\theta \end{pmatrix}.$$

(vi) Every hermitian matrix H can be written as $H = iK$, where K is a skew-hermitian matrix. Find H for the examples given above.

Solution 59. (i) Using the properties of $\mathbf{u}_j \mathbf{u}_j^*$ we find

$$\exp(K) = \exp\left(\sum_{j=1}^{n} \mu_j \mathbf{u}_j \mathbf{u}_j^*\right) = \sum_{j=1}^{n} e^{\mu_j} \mathbf{u}_j \mathbf{u}_j^*.$$

(ii) From $U = \exp(K)$ we find

$$U = \sum_{j=1}^{n} e^{\mu_j} \mathbf{u}_j \mathbf{u}_j^*$$

where $\mathbf{u}_j$ $(j = 1, 2, \ldots, n)$ are the normalized eigenvectors of U.

(iii) The matrix K is given by

$$K = \sum_{j=1}^{n} \ln(\lambda_j)\mathbf{u}_j\mathbf{u}_j^*$$

where λ_j $(j = 1, 2, \ldots, n)$ are the eigenvalues if U and $\mathbf{u}_j$ are the normalized eigenvectors of U. Note that the eigenvalues of U are of the form $\exp(i\alpha)$ with $\alpha \in \mathbb{R}$. Thus we have $\ln(e^{i\alpha}) = i\alpha$.

(iv) The eigenvalues of the matrix $U(\theta)$ are $e^{i\theta}$ and $e^{-i\theta}$ with the corresponding normalized eigenvectors

$$\mathbf{u}_1 = \frac{1}{\sqrt{2}}\begin{pmatrix} 1 \\ i \end{pmatrix}, \qquad \mathbf{u}_2 = \frac{1}{\sqrt{2}}\begin{pmatrix} 1 \\ -i \end{pmatrix}.$$

Thus

$$K(\theta) = \ln(e^{i\theta})\mathbf{u}_1\mathbf{u}_1^* + \ln(e^{i\theta})\mathbf{u}_2\mathbf{u}_2 = \frac{i\theta}{2}\begin{pmatrix} 1 & -i \\ i & 1 \end{pmatrix} - \frac{i\theta}{2}\begin{pmatrix} 1 & i \\ -i & 1 \end{pmatrix} = \theta\begin{pmatrix} 0 & 1 \\ -1 & 0 \end{pmatrix}.$$

(v) For the matrix $V(\theta, \phi)$ the eigenvalues are $e^{-i\theta}$ and $e^{i\theta}$ with the corresponding normalized eigenvectors

$$\frac{1}{\sqrt{2}}\begin{pmatrix} 1 \\ ie^{-i\phi} \end{pmatrix}, \qquad \frac{1}{\sqrt{2}}\begin{pmatrix} 1 \\ -ie^{-i\phi} \end{pmatrix}.$$

Thus

$$K(\theta, \phi) = \ln(e^{-i\theta})\mathbf{u}_1\mathbf{u}_1^* + \ln(e^{i\theta})\mathbf{u}_2\mathbf{u}_2^* = \begin{pmatrix} 0 & -\theta e^{i\phi} \\ \theta e^{-i\phi} & 0 \end{pmatrix}.$$

(vi) For $U(\theta)$ we find

$$H(\theta) = i\theta\begin{pmatrix} 0 & 1 \\ -1 & 0 \end{pmatrix}.$$

For $V(\theta, \phi)$ we find

$$H(\theta, \phi) = \begin{pmatrix} 0 & -i\theta e^{i\phi} \\ i\theta e^{-i\phi} & 0 \end{pmatrix}.$$

Problem 60. Let $f : \mathbb{R} \to \mathbb{R}$ be an analytic function. Let $\theta \in \mathbb{R}$, $\mathbf{n}$ a normalized vector in $\mathbb{R}^3$ and σ_1, σ_2, σ_3 the Pauli spin matrices. We define

$$\mathbf{n} \cdot \boldsymbol{\sigma} := n_1\sigma_1 + n_2\sigma_2 + n_3\sigma_3.$$

Then

$$f(\theta\mathbf{n} \cdot \boldsymbol{\sigma}) \equiv \frac{1}{2}(f(\theta) + f(-\theta))I_2 + \frac{1}{2}(f(\theta) - f(-\theta))(\mathbf{n} \cdot \boldsymbol{\sigma}).$$

(i) Find the matrix $\sin(\theta\mathbf{n} \cdot \boldsymbol{\sigma})$.

(ii) Let $z \in \mathbb{C}$. Find the matrices

$$\sinh(z\sigma_j), \qquad \cosh(z\sigma_j).$$

(iii) Find the matrices

$$\sin(z\sigma_j), \qquad \cos(z\sigma_j).$$

Solution 60. (i) Since $\sin(-x) = -\sin(x)$ we obtain

$$f(\theta\mathbf{n} \cdot \boldsymbol{\sigma}) = \sin(\theta)\mathbf{n} \cdot \boldsymbol{\sigma}.$$

(ii) Since $\sigma_j^2 = I_2$ We have

$$\sinh(z\sigma_j) = \sum_{k=0}^{\infty} \frac{(z\sigma_j)^{2k+1}}{(2k+1)!} = \sinh(z)\sigma_j, \quad \cosh(z\sigma_j) = \sum_{k=0}^{\infty} \frac{(z\sigma_j)^{2k}}{(2k)!} = \cosh(z)I_2.$$

(iii) Since $\sigma_j^2 = I_2$ We have

$$\sin(z\sigma_j) = \sum_{k=0}^{\infty}(-1)^k \frac{(z\sigma_j)^{2k+1}}{(2k+1)!} = \sinh(z)\sigma_j, \quad \cosh(z\sigma_j) = \sum_{k=0}^{\infty}(-1)^k \frac{(z\sigma_j)^{2k}}{(2k)!} = \cos(z)I_2.$$

Problem 61. Let σ_1, σ_2, σ_3 be the Pauli spin matrices. Let $z \in \mathbb{C}$.
(i) Calculate

$$e^{z\sigma_1}\sigma_2 e^{-z\sigma_1}, \quad e^{z\sigma_2}\sigma_3 e^{-z\sigma_2}, \quad e^{z\sigma_3}\sigma_1 e^{-z\sigma_3}.$$

Then consider the case $z = i\alpha$, where $\alpha \in \mathbb{R}$.
(ii) Can one find an $\alpha \in \mathbb{R}$ such that

$$\exp(i\alpha\sigma_3)\sigma_1 \exp(-i\alpha\sigma_3) = \sigma_2 ?$$

Solution 61. (i) Since $\exp(z\sigma_j) = \cosh(z)I_2 + \sinh(z)\sigma_j$ we obtain

$$e^{z\sigma_1}\sigma_2 e^{-z\sigma_1} = \sigma_2\cosh(2z) + i\sigma_3\sinh(2z)$$
$$e^{z\sigma_2}\sigma_3 e^{-z\sigma_2} = \sigma_3\cosh(2z) + i\sigma_1\sinh(2z)$$
$$e^{z\sigma_3}\sigma_1 e^{-z\sigma_3} = \sigma_1\cosh(2z) + i\sigma_2\sinh(2z).$$

(ii) We have

$$\exp(i\alpha\sigma_3)\sigma_1 \exp(-i\alpha\sigma_3) = \begin{pmatrix} 0 & e^{2i\alpha} \\ e^{-2i\alpha} & 0 \end{pmatrix}.$$

Thus we have to solve $\exp(2i\alpha) = -i$, $\exp(-2i\alpha) = i$. For $\alpha \in [0, 2\pi)$ we obtain $\alpha = 3\pi/2$.

Problem 62. Find the *square roots* of the Pauli spin matrices

$$\sigma_1 = \begin{pmatrix} 0 & 1 \\ 1 & 0 \end{pmatrix}, \quad \sigma_2 = \begin{pmatrix} 0 & -i \\ i & 0 \end{pmatrix}, \quad \sigma_3 = \begin{pmatrix} 1 & 0 \\ 0 & -1 \end{pmatrix}.$$

Solution 62. First we note that the square root of 1 is $1 = e^0$ and $-1 = e^{i\pi}$. The square root of -1 is $i = e^{i\pi/2}$ and $-i = e^{i3\pi/2} = e^{-i\pi/2}$. For the identity matrix $\sigma_0 = I_2$ we find

$$R_{01} = \begin{pmatrix} 1 & 0 \\ 0 & 1 \end{pmatrix}, \quad R_{02} = \begin{pmatrix} -1 & 0 \\ 0 & 1 \end{pmatrix}, \quad R_{03} = \begin{pmatrix} 1 & 0 \\ 0 & -1 \end{pmatrix}, \quad R_{04} = \begin{pmatrix} -1 & 0 \\ 0 & -1 \end{pmatrix}.$$

For the Pauli spin matrix σ_3 we find

$$R_{30} = \begin{pmatrix} 1 & 0 \\ 0 & i \end{pmatrix}, \quad R_{31} = \begin{pmatrix} -1 & 0 \\ 0 & i \end{pmatrix}, \quad R_{32} = \begin{pmatrix} 1 & 0 \\ 0 & -i \end{pmatrix}, \quad R_{33} = \begin{pmatrix} -1 & 0 \\ 0 & -i \end{pmatrix}.$$

For the square roots of σ_1 we apply the spectral decomposition theorem. For the pairs $(1, i)$, $(1, -i)$, $(-1, i)$, $(-1, -i)$, respectively, we find

$$R_{10} = \frac{1}{2}\begin{pmatrix} 1+i & 1-i \\ 1-i & 1+i \end{pmatrix}, \quad R_{11} = \frac{1}{2}\begin{pmatrix} 1-i & 1+i \\ 1+i & 1-i \end{pmatrix},$$

$$R_{12} = \frac{1}{2}\begin{pmatrix} -1+i & -1-i \\ -1-i & -1+i \end{pmatrix}, \quad R_{13} = \frac{1}{2}\begin{pmatrix} -1-i & -1+i \\ -1+i & -1-i \end{pmatrix}.$$

For the square roots of σ_2 we apply the spectral decomposition theorem. For the pairs $(1, i)$, $(1, -i)$, $(-1, i)$, $(-1, -i)$, respectively, we find

$$R_{20} = \frac{1}{2}\begin{pmatrix} 1+i & -1-i \\ 1+i & 1+i \end{pmatrix}, \quad R_{21} = \frac{1}{2}\begin{pmatrix} 1-i & 1-i \\ -1+i & 1-i \end{pmatrix},$$

$$R_{22} = \frac{1}{2}\begin{pmatrix} -1+i & -1+i \\ 1-i & -1+i \end{pmatrix}, \quad R_{23} = \frac{1}{2}\begin{pmatrix} -1-i & 1+i \\ -1-i & -1-i \end{pmatrix}.$$

Problem 63. Any 2×2 matrix can be written as a linear combination of the Pauli spin matrices and the 2×2 identity matrix

$$A = aI_2 + b\sigma_1 + c\sigma_2 + d\sigma_3$$

where $a, b, c, d \in \mathbb{C}$.
(i) Find A^2 and A^3.
(ii) Use the result from (i) to find all matrices A such that $A^3 = \sigma_1$, i.e. find the third root of σ_1.

Solution 63. (i) We obtain

$$A^2 = (b^2 + c^2 + d^2 - a^2)I_2 + 2aA$$
$$A^3 = a(a^2 + 3b^2 + 3c^2 + 3d^2)I_2 + b(3a^2 + b^2 + c^2 + d^2)\sigma_1$$
$$+ c(3a^2 + b^2 + c^2 + d^2)\sigma_2 + d(3a^2 + b^2 + c^2 + d^2)\sigma_3.$$

(ii) The Pauli spin matrices σ_1, σ_2, σ_3 together with I_2 form a basis for the vector space of the 2×2 matrices. Setting $A^3 = \sigma_1$ yields the four equations

$$a(a^2 + 3b^2 + 3c^2 + 3d^2) = 0$$
$$b(3a^2 + b^2 + c^2 + d^2) = 1$$
$$c(3a^2 + b^2 + c^2 + d^2) = 0$$
$$d(3a^2 + b^2 + c^2 + d^2) = 0.$$

Thus $b \neq 0$ and $c = d = 0$. Case 1. If $a = 0$, then $b^3 = 1$. Therefore $A = b\sigma_1$, where $b \in \{1, e^{2\pi i/3}, e^{4\pi i/3}\}$. Case 2. If $a \neq 0$, then

$$a^2 + 3b^2 = 0, \quad b(3a^2 + b^2) = 1.$$

It follows that $b^3 = -1/8$, i.e. $b = -1/2$ and $a = \pm\sqrt{3}i/2$. Thus

$$A = \frac{1}{2}\begin{pmatrix} \pm i\sqrt{3} & -1 \\ -1 & \pm i\sqrt{3} \end{pmatrix}.$$

Problem 64. Consider first the Lie algebra $su(2)$. Let σ_1, σ_2, σ_3 be the Pauli spin matrices. A basis for the simple Lie algebra $su(2)$ is given by the skew-hermitian and unitary matrices

$$A_1 = -i\sigma_1 = \begin{pmatrix} 0 & -i \\ -i & 0 \end{pmatrix}, \quad A_2 = -i\sigma_2 = \begin{pmatrix} 0 & -1 \\ 1 & 0 \end{pmatrix}, \quad A_3 = -i\sigma_3 = \begin{pmatrix} -i & 0 \\ 0 & i \end{pmatrix}$$

with the commutators $[A_1, A_2] = 2A_3$, $[A_2, A_3] = 2A_1$, $[A_3, A_1] = 2A_2$. Find the 2×2 matrices K_1, K_2, K_3 such that $A_1 = e^{K_1}$, $A_2 = e^{K_2}$, $A_3 = e^{K_3}$. Find the commutators $[K_1, K_2]$, $[K_2, K_3]$, $[K_3, K_1]$. Apply the *spectral representation* of A_1, A_2, A_3 to find K_1, K_2, K_3.

Solution 64. The eigenvalues of A_1 are $\lambda_1 = -i$, $\lambda_2 = +i$ with the corresponding normalized eigenvectors

$$\mathbf{v}_1 = \frac{1}{\sqrt{2}} \begin{pmatrix} 1 \\ 1 \end{pmatrix}, \qquad \mathbf{v}_2 = \frac{1}{\sqrt{2}} \begin{pmatrix} 1 \\ -1 \end{pmatrix}.$$

Then

$$A_1 = \lambda_1 \mathbf{v}_1 \mathbf{v}_1^* + \lambda_2 \mathbf{v}_2 \mathbf{v}_2^* = -i \frac{1}{\sqrt{2}} \begin{pmatrix} 1 \\ 1 \end{pmatrix} \frac{1}{\sqrt{2}} (1 \ \ 1) + i \frac{1}{\sqrt{2}} \begin{pmatrix} 1 \\ -1 \end{pmatrix} \frac{1}{\sqrt{2}} (1 \ \ -1).$$

Since $-i = e^{-i\pi/2}$, $i = e^{i\pi/2}$ we have $\ln(-i) = -i\pi/2$, $\ln(i) = i\pi/2$. Consequently the spectral representation for K_1 is

$$K_1 = -\frac{i\pi}{2} \frac{1}{\sqrt{2}} \begin{pmatrix} 1 \\ 1 \end{pmatrix} \frac{1}{\sqrt{2}} (1 \ \ 1) + \frac{i\pi}{2} \frac{1}{\sqrt{2}} \begin{pmatrix} 1 \\ -1 \end{pmatrix} \frac{1}{\sqrt{2}} (1 \ \ -1)$$

$$= \frac{\pi}{2} \begin{pmatrix} 0 & -i \\ -i & 0 \end{pmatrix} = \frac{\pi}{2} A_1.$$

Analogously we do the calculation for K_2 and K_3 and obtain

$$K_2 = \frac{\pi}{2} A_2, \qquad K_3 = \frac{\pi}{2} A_3.$$

The commutators are $[K_1, K_2] = \pi K_3$, $[K_2, K_3] = \pi K_1$, $[K_3, K_1] = \pi K_2$.

Problem 65. (i) Let σ_1, σ_2, σ_3 be the Pauli spin matrices and $\alpha \in \mathbb{R}$. Find the unitary 4×4 matrices

$$U_{jk}(\alpha) = \exp(i\alpha(\sigma_j \otimes I_2 + I_2 \otimes \sigma_k))$$

where $j, k = 1, 2, 3$.
(ii) Let $z \in \mathbb{C}$. Calculate the commutator $[\sigma_2 \otimes \sigma_2, \sigma_1 \otimes \sigma_1]$ and

$$\exp(-z\sigma_2 \otimes \sigma_2)(\sigma_1 \otimes \sigma_1) \exp(z\sigma_2 \otimes \sigma_2).$$

Solution 65. (i) Since $[\sigma_j \otimes I_2, I_2 \otimes \sigma_k] = 0_4 = 0_2 \otimes 0_2$ we have

$$U_{jk} = e^{i\alpha\sigma_j \otimes I_2} e^{i\alpha I_2 \otimes \sigma_k}.$$

With $\sigma_j^2 = I_2$, $(\sigma_j \otimes I_2)^2 = I_2 \otimes I_2$, $(I_2 \otimes \sigma_j)^2 = I_2 \otimes I_2$ we obtain

$$U_{jk} = ((I_2 \otimes I_2)\cosh(i\alpha) + (\sigma_j \otimes I_2)\sinh(i\alpha))((I_2 \otimes I_2)\cosh(i\alpha) + (I_2 \otimes \sigma_k)\sinh(i\alpha)).$$

Consequently

$$U_{jk} = (I_2 \otimes I_2)\cosh^2(i\alpha) + (I_2 \otimes \sigma_k)\cosh(i\alpha)\sinh(i\alpha) +$$
$$(\sigma_j \otimes I_2)\cosh(i\alpha)\sinh(i\alpha) + (\sigma_j \otimes \sigma_k)\sinh^2(i\alpha)$$

with $\cosh(i\alpha) \equiv \cos(\alpha)$, $\sinh(i\alpha) \equiv i\sin(\alpha)$.

(ii) Since $[\sigma_2 \otimes \sigma_2, \sigma_1 \otimes \sigma_1] = 0_4 = 0_2 \otimes 0_2$ we obtain

$$\exp(-z\sigma_2 \otimes \sigma_2)(\sigma_1 \otimes \sigma_1)\exp(z\sigma_2 \otimes \sigma_2) = \sigma_1 \otimes \sigma_1.$$

Problem 66.   (i) Consider the Pauli matrix σ_1. Find the matrix

$$\exp\left(-\frac{1}{2}i\pi(\sigma_1 - I_2)\right).$$

(ii) Let B be an $n \times n$ matrix over $\mathbb{C}$ with $B^2 = I_n$ and $k \in \mathbb{Z}$. Find

$$\exp\left(\frac{1}{2}(2k+1)i\pi(B - I_n)\right).$$

Solution 66.   (i) We obtain

$$\exp\left(-\frac{1}{2}i\pi(\sigma_1 - I_2)\right) = \sigma_1.$$

(ii) We find

$$\exp\left(\frac{1}{2}(2k+1)i\pi(B - I_n)\right) = B.$$

Problem 67.   Let A be an $n \times n$ matrix over $\mathbb{C}$ with $A^2 = I_n$. Let $z \in \mathbb{C}$. Then

$$e^{zA} = I_n \cosh(z) + A \sinh(z).$$

Let σ_1, σ_2, σ_3 be the Pauli spin matrices and $\sigma_0 = I_2$. These matrices are unitary and hermitian and satisfy $\sigma_j^2 = I_2$ for $j = 0, 1, 2, 3$.
(i) Consider the $2^n \times 2^n$ matrix

$$X = \sigma_{j_1} \otimes \sigma_{j_2} \otimes \cdots \otimes \sigma_{j_n}$$

with $j_k \in \{0, 1, 2, 3\}$ and $k = 1, \ldots, n$. Calculate $\exp(zX)$.
(ii) Let $z \in \mathbb{C}$. Calculate

$$U(z) = \exp(z(\sigma_1 \otimes \sigma_2)).$$

Then set $z = -i\omega t$.
(iii) Calculate the normalized state

$$U(-i\omega t)\frac{1}{2}\begin{pmatrix} 1 \\ 1 \\ 1 \\ 1 \end{pmatrix} \equiv U(-i\omega t)\frac{1}{\sqrt{2}}\begin{pmatrix} 1 \\ 1 \end{pmatrix} \otimes \frac{1}{\sqrt{2}}\begin{pmatrix} 1 \\ 1 \end{pmatrix}.$$

(iv) Consider the Hamilton operator $\hat{H} = \hbar\omega(\sigma_1 \otimes \sigma_2)$. Find the unitary matrix

$$U(t) = \exp(-i\hat{H}t/\hbar).$$

Set $\omega t = \pi/4$ to find the *Yang-Baxter gate*. Then find the normalized state

$$U(t)\frac{1}{2}\begin{pmatrix} 1 \\ 1 \\ 1 \\ 1 \end{pmatrix}.$$

Solution 67. (i) Since

$$(\sigma_{j_1} \otimes \sigma_{j_2} \otimes \cdots \otimes \sigma_{j_n})^2 = I_2 \otimes I_2 \otimes \cdots \otimes I_2 = I_{2^n}$$

we obtain

$$e^{zX} = I_{2^n} \cosh(z) + X \sinh(z) = I_{2^n} \cosh(z) + (\sigma_{j_1} \otimes \sigma_{j_2} \otimes \cdots \otimes \sigma_{j_n}) \sinh(z).$$

(ii) Since $(\sigma_1 \otimes \sigma_2)^2 = I_2 \otimes I_2$ we obtain

$$U(z) = (I_2 \otimes I_2) \cosh(z) + (A \otimes B) \sinh(z).$$

(iii) With $z = -i\omega t$ it follows that

$$U(-i\omega t) = \begin{pmatrix} \cos(\omega t) & 0 & 0 & -\sin(\omega t) \\ 0 & \cos(\omega t) & \sin(\omega t) & 0 \\ 0 & -\sin(\omega t) & \cos(\omega t) & 0 \\ \sin(\omega t) & 0 & 0 & \cos(\omega t) \end{pmatrix}$$

and

$$U(-i\omega t)\frac{1}{2}\begin{pmatrix} 1 \\ 1 \\ 1 \\ 1 \end{pmatrix} = \frac{1}{2}\begin{pmatrix} \cos(\omega t) - \sin(\omega t) \\ \cos(\omega t) + \sin(\omega t) \\ -\sin(\omega t) + \cos(\omega t) \\ \sin(\omega t) + \cos(\omega t) \end{pmatrix}.$$

Can this state be entangled for any ωt? Of course for $\omega t = 0$ the state is not entangled.
(iv) Since $(\sigma_1 \otimes \sigma_2)^2 = I_2 \otimes I_2$ and

$$\begin{pmatrix} 0 & 1 \\ 1 & 0 \end{pmatrix} \otimes \begin{pmatrix} 0 & -i \\ i & 0 \end{pmatrix} = \begin{pmatrix} 0 & 0 & 0 & -i \\ 0 & 0 & i & 0 \\ 0 & -i & 0 & 0 \\ i & 0 & 0 & 0 \end{pmatrix}$$

we obtain

$$e^{-i\omega t(\sigma_1 \otimes \sigma_2)} = \begin{pmatrix} \cos(\omega t) & 0 & 0 & -\sin(\omega t) \\ 0 & \cos(\omega t) & \sin(\omega t) & 0 \\ 0 & -\sin(\omega t) & \cos(\omega t) & 0 \\ \sin(\omega t) & 0 & 0 & \cos(\omega t) \end{pmatrix}$$

with $\cos(\pi/4) = 1/\sqrt{2}$, $\sin(\pi/4) = 1/\sqrt{2}$. From (iii) it follows that

$$U(t)\frac{1}{2}\begin{pmatrix} 1 \\ 1 \\ 1 \\ 1 \end{pmatrix} = \frac{1}{2}\begin{pmatrix} \cos(\omega t) - \sin(\omega t) \\ \cos(\omega t) + \sin(\omega t) \\ -\sin(\omega t) + \cos(\omega t) \\ \sin(\omega t) + \cos(\omega t) \end{pmatrix}.$$

Problem 68. Let $s = 1/2, 1, 3/2, 2, \ldots$ be the spin and S_1, S_2, S_3 be the spin matrices. The spin matrices are hermitian $(2s + 1) \times (2s + 1)$ matrices satisfying

$$[S_1, S_2] = iS_3, \quad [S_2, S_3] = iS_1, \quad [S_3, S_1] = iS_2.$$

One defines $S_+ := S_1 + iS_2$, $S_- := S_1 - iS_2$. Let $|0\rangle$ be the ground state with $S_3|0\rangle = s|0\rangle$ (eigenvalue equation). Let $z \in \mathbb{C}$. The normalized *coherent spin state* $|z\rangle$ is defined as

$$|z\rangle := \frac{1}{(1 + |z|^2)^s} \exp(zS_-)|0\rangle.$$

Then we have ($z_1, z_2 \in \mathbb{C}$)

$$\langle z_1 | z_2 \rangle = \frac{(1 + \bar{z}_1 z_2)^{2s}}{(1 + |z_1|^2)^s (1 + |z_2|^2)^s}.$$

and (completeness relation)

$$\int_{\mathbb{C}} dz m(|z|^2) |z\rangle \langle z| = I$$

where I is the $(2s + 1) \times (2s + 1)$ identity matrix and

$$m(w) := \frac{2s + 1}{\pi (1 + w)^2}$$

We also have

$$\langle z_1 | S_3 | z_2 \rangle = s \frac{1 - \bar{z}_1 z_2}{1 + \bar{z}_1 z_2} \langle z_1 | z_2 \rangle, \quad \langle z_1 | S_+ | z_2 \rangle = 2s \frac{z_2}{1 + \bar{z}_1 z_2} \langle z_1 | z_2 \rangle, \quad \langle z_1 | S_- | z_2 \rangle = 2s \frac{\bar{z}_1}{1 + \bar{z}_1 z_2} \langle z_1 | z_2 \rangle.$$

(i) Consider the spin 1/2 and the spin matrices

$$S_+ = \begin{pmatrix} 0 & 1 \\ 0 & 0 \end{pmatrix}, \quad S_- = \begin{pmatrix} 0 & 0 \\ 1 & 0 \end{pmatrix}.$$

The ground state is

$$|0\rangle = \begin{pmatrix} 1 \\ 0 \end{pmatrix}.$$

Find $\exp(zS_-)$ and then the state $|z\rangle = \exp(zS_-)|0\rangle$. Check the identities given above for $|z\rangle$.

(ii) Consider the spin-1 matrices

$$S_+ = \begin{pmatrix} 0 & \sqrt{2} & 0 \\ 0 & 0 & \sqrt{2} \\ 0 & 0 & 0 \end{pmatrix}, \quad S_- = \begin{pmatrix} 0 & 0 & 0 \\ \sqrt{2} & 0 & 0 \\ 0 & \sqrt{2} & 0 \end{pmatrix}.$$

The ground state is

$$|0\rangle = \begin{pmatrix} 1 \\ 0 \\ 0 \end{pmatrix}.$$

Find $\exp(zS_-)$ and then the coherent state $|z\rangle$. Check the identities given above for $|z\rangle$.

Solution 68.   (i) We have

$$\exp(zS_-) = \begin{pmatrix} 1 & 0 \\ 0 & 1 \end{pmatrix} + z \begin{pmatrix} 0 & 0 \\ 1 & 0 \end{pmatrix} = \begin{pmatrix} 1 & 0 \\ z & 1 \end{pmatrix}.$$

Thus for spin $\frac{1}{2}$ we have the normalized state

$$|z\rangle = \frac{1}{1 + |z|^2} \begin{pmatrix} 1 & 0 \\ z & 1 \end{pmatrix} \begin{pmatrix} 1 \\ 0 \end{pmatrix} = \frac{1}{1 + |z|^2} \begin{pmatrix} 1 \\ z \end{pmatrix}.$$

(ii) We have

$$\exp(zS_-) = I_3 + zS_- + \frac{z^2}{2} S_-^2 = \begin{pmatrix} 1 & 0 & 0 \\ \sqrt{2}z & 1 & 0 \\ z^2 & \sqrt{2}z & 1 \end{pmatrix}.$$

Thus for spin 1 we have the normalized state

$$|z\rangle = \frac{1}{1+|z|^2} \begin{pmatrix} 1 \\ \sqrt{2}z \\ z^2 \end{pmatrix}.$$

Problem 69. Consider the *GHZ state* in the Hilbert space $\mathbb{C}^8$

$$|GHZ\rangle = \frac{1}{\sqrt{2}} \left(\begin{pmatrix} 1 \\ 0 \end{pmatrix} \otimes \begin{pmatrix} 1 \\ 0 \end{pmatrix} \otimes \begin{pmatrix} 1 \\ 0 \end{pmatrix} + \begin{pmatrix} 0 \\ 1 \end{pmatrix} \otimes \begin{pmatrix} 0 \\ 1 \end{pmatrix} \otimes \begin{pmatrix} 0 \\ 1 \end{pmatrix} \right).$$

(i) Find the state $(\sigma_1 \otimes \sigma_1 \otimes \sigma_1)|GHZ\rangle$. Discuss.
(ii) Consider the unitary 8×8 matrix

$$U(\phi_1, \phi_2) = e^{i\phi_1\sigma_3} \otimes e^{i\phi_2\sigma_3} \otimes e^{-i(\phi_1+\phi_2)\sigma_3}.$$

Find $U(\phi_1, \phi_2)|GHZ\rangle$. Discuss.

Solution 69. (i) We obtain

$$(\sigma_1 \otimes \sigma_1 \otimes \sigma_1)|GHZ\rangle = |GHZ\rangle.$$

This means the $|GHZ\rangle$ state is invariant under $\sigma_1 \otimes \sigma_1 \otimes \sigma_1$. Additionally we have an eigenvalue equation with eigenvalue $+1$.
(ii) We find $U(\phi_1, \phi_2)|GHZ\rangle = |GHZ\rangle$, i.e. the $GHZ\rangle$ is invariant under $U(\phi_1, \phi_2)$. Additionally we have an eigenvalue equation with eigenvalue $+1$.

Problem 70. Given the 2×2 spin matrices $\mathbf{S} = (S_1, S_2, S_3)$

$$S_1 = \frac{1}{2}\begin{pmatrix} 0 & 1 \\ 1 & 0 \end{pmatrix}, \quad S_2 = \frac{1}{2}\begin{pmatrix} 0 & -i \\ i & 0 \end{pmatrix}, \quad S_3 = \frac{1}{2}\begin{pmatrix} 1 & 0 \\ 0 & -1 \end{pmatrix}.$$

Note that $S_1^2 + S_2^2 + S_3^2 = \frac{3}{4}I_2$. Let $\mathbf{n} = (n_1, n_2, n_3)$ be the normalized vector

$$\mathbf{n} = (\sin(\theta)\cos(\phi), \sin(\theta)\sin(\phi), \cos(\theta))$$

and

$$\mathbf{n} \cdot \mathbf{S} := n_1 S_1 + n_2 S_2 + n_3 S_3 = \frac{1}{2}\begin{pmatrix} \cos(\theta) & \sin(\theta)e^{-i\phi} \\ \sin(\theta)e^{i\phi} & -\cos(\theta) \end{pmatrix}.$$

Let $\alpha \in \mathbb{R}$. Calculate the *class operator* defined by

$$C(\alpha) = \int_0^{2\pi} d\phi \int_0^{\pi} d\theta \sin(\theta) \exp(i\alpha(n_1 S_1 + n_2 S_2 + n_3 S_3)).$$

The integration is over the differentiable manifold S^2, i.e.

$$S^2 = \{ (x_1, x_2, x_3) : x_1^2 + x_2^2 + x_3^2 = 1 \}.$$

Apply the *spectral theorem* to $\mathbf{n} \cdot \mathbf{S}$. So calculate first the eigenvalues and normalized eigenvectors of the 2×2 matrices $n_1 S_1 + n_2 S_2 + n_3 S_3$.

Solution 70.   The eigenvalues of the 2×2 matrix $n_1 S_1 + n_2 S_2 + n_3 S_3$ are $\lambda_1 = 1/2$ and $\lambda_2 = -1/2$ with the corresponding normalized eigenvectors

$$\mathbf{v}_1 = \begin{pmatrix} \cos(\theta/2)e^{-i\phi} \\ \sin(\theta/2) \end{pmatrix}, \qquad \mathbf{v}_2 = \begin{pmatrix} \sin(\theta/2)e^{-i\phi} \\ -\cos(\theta/2) \end{pmatrix}.$$

It follows that

$$\mathbf{v}_1 \mathbf{v}_1^* = \begin{pmatrix} \cos^2(\theta/2) & \cos(\theta/2)\sin(\theta/2)e^{-i\phi} \\ \sin(\theta/2)\cos(\theta/2)e^{i\phi} & \sin^2(\theta/2) \end{pmatrix}$$

$$\mathbf{v}_2 \mathbf{v}_2^* = \begin{pmatrix} \sin^2(\theta/2) & -\cos(\theta/2)\sin(\theta/2)e^{-i\phi} \\ -\sin(\theta/2)\cos(\theta/2)e^{i\phi} & \cos^2(\theta/2) \end{pmatrix}.$$

Using these results we obtain

$$\exp(i\alpha(n_1 S_1 + n_2 S_2 + n_3 S_3)) = \exp(i\alpha(\frac{1}{2}\mathbf{v}_1\mathbf{v}_1^* - \frac{1}{2}\mathbf{v}_2\mathbf{v}_2^*))$$

$$= I_2 + (e^{i\alpha/2} - 1)\mathbf{v}_1\mathbf{v}_1^* + (e^{-i\alpha/2} - 1)\mathbf{v}_2\mathbf{v}_2^*$$

$$= e^{i\alpha/2}\mathbf{v}_1\mathbf{v}_1^* + e^{-i\alpha/2}\mathbf{v}_2\mathbf{v}_2^*$$

$$= \cos(\alpha/2)I_2 + i\sin(\alpha/2)(\mathbf{v}_1\mathbf{v}_1^* - \mathbf{v}_2\mathbf{v}_2^*).$$

Since

$$\int_0^{2\pi} e^{i\phi}d\phi = \int_0^{2\pi} e^{-i\phi}d\phi = 0, \qquad \int_0^{2\pi} d\phi = 2\pi,$$

$$\int_0^{\pi} d\theta \sin(\theta)\cos(\theta) = 0, \qquad \int_0^{\pi} d\theta \sin(\theta) = 2$$

we end up with the result

$$C(\alpha) = 4\pi \begin{pmatrix} \cos(\alpha/2) & 0 \\ 0 & \cos(\alpha/2) \end{pmatrix}.$$

Perform the same calculations for the 3×3 spin-1 matrices

$$S_1 = \frac{1}{\sqrt{2}} \begin{pmatrix} 0 & 1 & 0 \\ 1 & 0 & 1 \\ 0 & 1 & 0 \end{pmatrix}, \quad S_2 = \frac{1}{\sqrt{2}} \begin{pmatrix} 0 & -i & 0 \\ i & 0 & -i \\ 0 & i & 0 \end{pmatrix}, \quad S_3 = \begin{pmatrix} 1 & 0 & 0 \\ 0 & 0 & 0 \\ 0 & 0 & -1 \end{pmatrix}.$$

2.3 Spin Hamilton Operators

Problem 71. Let σ_1, σ_2, σ_3 be the Pauli spin matrices and I_2 the 2×2 identity matrix. Find the eigenvalues and normalized eigenvectors of the Hamilton operator

$$\hat{H} = \varepsilon_0 I_2 + \hbar\omega\sigma_3 + \Delta_1\sigma_1 + \Delta_2\sigma_2$$

where $\varepsilon_0 > 0$. Are the normalized eigenvectors orthonormal to each other?

Solution 71. In matrix form we have the (hermitian) 2×2 matrix

$$\hat{H} = \begin{pmatrix} \varepsilon_0 + \hbar\omega & \Delta_1 - i\Delta_2 \\ \Delta_1 + i\Delta_2 & \varepsilon_0 - \hbar\omega \end{pmatrix}.$$

From

$$\det(\hat{H} - EI_2) = (\varepsilon + \hbar\omega - E)(\varepsilon - \hbar\omega - E) - (\Delta_1 + i\Delta_2)(\Delta_1 - i\Delta_2) = 0$$

we obtain the characteristic equation

$$E^2 - 2\varepsilon_0 E + \varepsilon_0^2 = \hbar^2\omega^2 + \Delta_1^2 + \Delta_2^2.$$

Thus the two eigenvalues E_+, E_- are

$$E_\pm = \varepsilon \pm \sqrt{\hbar^2\omega^2 + \Delta_1^2 + \Delta_2^2}.$$

Let $S := \sqrt{\hbar^2\omega^2 + \Delta_1^2 + \Delta_2^2}$. For the eigenvector of E_+ we have to solve

$$\begin{pmatrix} \varepsilon_0 + \hbar\omega & \Delta_1 - i\Delta_2 \\ \Delta_1 + i\Delta_2 & \varepsilon_0 - \hbar\omega \end{pmatrix} \begin{pmatrix} v_1 \\ v_2 \end{pmatrix} = E_+ \begin{pmatrix} v_1 \\ v_2 \end{pmatrix}$$

or

$$(\varepsilon_0 + \hbar\omega)v_1 + (\Delta_1 - i\Delta_2)v_2 = E_+v_1 = (\varepsilon_0 + S)v_1$$
$$(\Delta_1 + i\Delta_2)v_1 + (\varepsilon_0 - \hbar\omega)v_2 = (\varepsilon_0 + S)v_2.$$

We can set $v_1 = S + \hbar\omega$. Thus $v_2 = \Delta_1 + i\Delta_2$. After normalizing we have the eigenvector

$$\frac{1}{\sqrt{(S + \hbar\omega)^2 + \Delta_1^2 + \Delta_2^2}} \begin{pmatrix} S + \hbar\omega \\ \Delta_1 + i\Delta_2 \end{pmatrix}.$$

Analogously for E_- we obtain the normalized eigenvector

$$\frac{1}{\sqrt{(S + \hbar\omega)^2 + \Delta_1^2 + \Delta_2^2}} \begin{pmatrix} \Delta_1 - i\Delta_2 \\ -S - \hbar\omega \end{pmatrix}.$$

Obviously the two eigenvectors are orthonormal to each other, i.e. the scalar product vanishes. Note that the eigenvector does not depend on ε_0.

Problem 72. Consider the spin-1 matrices

$$S_1 = \frac{1}{\sqrt{2}} \begin{pmatrix} 0 & 1 & 0 \\ 1 & 0 & 1 \\ 0 & 1 & 0 \end{pmatrix}, \quad S_2 = \frac{1}{\sqrt{2}} \begin{pmatrix} 0 & -i & 0 \\ i & 0 & -i \\ 0 & i & 0 \end{pmatrix}, \quad S_3 = \begin{pmatrix} 1 & 0 & 0 \\ 0 & 0 & 0 \\ 0 & 0 & -1 \end{pmatrix}$$

and the Hamilton operator

$$\hat{H} = \epsilon_0 I_3 + \hbar\omega S_3 + \Delta_1 S_1 + \Delta_2 S_2.$$

Find the eigenvalues and eigenvectors of $\hat{H}$.

Solution 72. The Hamilton operator $\hat{H}$ takes the form

$$\hat{H} = \begin{pmatrix} \epsilon_0 + \hbar\omega & (\Delta_1 - i\Delta_2)/\sqrt{2} & 0 \\ (\Delta_1 + i\Delta_2)/\sqrt{2} & \epsilon_0 & (\Delta_1 - i\Delta_2)/\sqrt{2} \\ 0 & (\Delta_1 + i\Delta_2)/\sqrt{2} & \epsilon_0 - \hbar\omega \end{pmatrix}$$

We set $E := \sqrt{\hbar\omega^2 + \Delta_1^2 + \Delta_2^2}$. The eigenvalues are ϵ_0, $\epsilon_0 - E$, $\epsilon_0 + E$ with the corresponding eigenvectors

$$\begin{pmatrix} -\Delta_1 + i\Delta_2 \\ \sqrt{2}\hbar\omega \\ \Delta_1 + i\Delta_2 \end{pmatrix}$$

$$\begin{pmatrix} (-\Delta_1 + i\Delta_2)(i\Delta_1 + \Delta_2)^2 \\ \sqrt{2}(E + \hbar\omega)(i\Delta_1 + \Delta_2)^2 \\ -(E + \hbar\omega)^2(-\Delta_1 + i\Delta_2) \end{pmatrix}$$

$$\begin{pmatrix} (-\Delta_1 + i\Delta_2)(i\Delta_1 + \Delta_2)^2 \\ -\sqrt{2}(E + \hbar\omega)(i\Delta_1 + \Delta_2)^2 \\ -(E - \hbar\omega)^2(-\Delta_1 + i\Delta_2) \end{pmatrix}.$$

Problem 73.   Let σ_1, σ_2, σ_3 be the Pauli spin matrices. Consider the Hamilton operator $\hat{H}$ acting in the Hilbert space $\mathbb{C}^4$

$$\hat{H} = J(\sigma_1 \otimes \sigma_1) + B(\sigma_3 \otimes I_2 + I_2 \otimes \sigma_3).$$

The second term represents the interaction of each spin with a uniform magnetic field in the z-direction and the first term describes an interaction between the spins in an orthogonal direction.
(i) Find the eigenvalues and the normalized eigenvectors of $\hat{H}$.
(ii) Show that

$$[\sigma_1 \otimes \sigma_1, \sigma_3 \otimes I_2 + I_2 \otimes \sigma_3] \neq 0_4.$$

(iii) Find the matrix

$$(\sigma_3 \otimes \sigma_3)\hat{H}(\sigma_3 \otimes \sigma_3).$$

Note that $\sigma_3^{-1} = \sigma_3$. Discuss.
(iv) Find the *partition function* $Z(\beta) = \text{tr}(\exp(-\beta\hat{H}))$.

Solution 73.   (i) The Hamilton operator takes the form

$$\hat{H} = \begin{pmatrix} 2B & 0 & 0 & J \\ 0 & 0 & J & 0 \\ 0 & J & 0 & 0 \\ J & 0 & 0 & -2B \end{pmatrix}.$$

The four eigenvalues of $\hat{H}$ are given by

$$-\sqrt{J^2 + 4B^2}, \quad +\sqrt{J^2 + 4B^2}, \quad -J, \quad J$$

with the corresponding non-normalized eigenvectors

$$\begin{pmatrix} J \\ 0 \\ 0 \\ -(\sqrt{J^2+4B^2}+2B) \end{pmatrix}, \begin{pmatrix} J \\ 0 \\ 0 \\ \sqrt{J^2+4B^2}-2B \end{pmatrix}, \begin{pmatrix} 0 \\ 1 \\ -1 \\ 0 \end{pmatrix}, \begin{pmatrix} 0 \\ 1 \\ 1 \\ 0 \end{pmatrix}.$$

(ii) The commutator is given by

$$[\sigma_1 \otimes \sigma_1, \sigma_3 \otimes I_2 + I_2 \otimes \sigma_3] = (-2i\sigma_2) \otimes \sigma_1 + \sigma_1 \otimes (-2i\sigma_2).$$

Thus the commutator is non-zero.

(iii) We find

$$(\sigma_3 \otimes \sigma_3)\hat{H}(\sigma_3 \otimes \sigma_3) = \hat{H}.$$

Thus $\hat{H}$ is invariant under this transformation. This can be used to reduce the Hilbert space $\mathbb{C}^4$ into invariant sub Hilbert spaces.

(iv) With the four eigenvalues

$$-\sqrt{J^2+4B^2}, \quad \sqrt{J^2+B^2}, \quad -J, \quad J$$

we obtain for the partition function

$$Z(\beta) = e^{\beta\sqrt{J^2+4B^2}} + e^{-\beta\sqrt{J^2+4B^2}} + e^{\beta J} + e^{-\beta J} \equiv 2\cosh(\beta\sqrt{J^2+4B^2}) + 2\cosh(\beta J).$$

Problem 74. Let σ_1, σ_2, σ_3 be the Pauli spin matrices. The spin matrices for spin-$\frac{1}{2}$ are given by

$$S_1 := \frac{1}{2}\sigma_1, \qquad S_2 := \frac{1}{2}\sigma_2, \qquad S_3 := \frac{1}{2}\sigma_3.$$

Consider the two-point Heisenberg model

$$\hat{H} = J\sum_{j=1}^{2} \mathbf{S}_j \cdot \mathbf{S}_{j+1}$$

where J is the so-called exchange constant ($J > 0$ or $J < 0$) and $\cdot$ denotes the scalar product. We impose cyclic boundary conditions, i.e. $\mathbf{S}_3 = \mathbf{S}_1$. It follows that

$$\hat{H} = J(\mathbf{S}_1 \cdot \mathbf{S}_2 + \mathbf{S}_2 \cdot \mathbf{S}_3) \equiv J(\mathbf{S}_1 \cdot \mathbf{S}_2 + \mathbf{S}_2 \cdot \mathbf{S}_1).$$

It follows that

$$\hat{H} = J(S_{1,1}S_{1,2} + S_{2,1}S_{2,2} + S_{3,1}S_{3,2} + S_{1,2}S_{1,1} + S_{2,2}S_{2,1} + S_{3,2}S_{3,1}).$$

Since $S_{1,1} := S_1 \otimes I_2$, $S_{1,2} := I_2 \otimes S_1$ etc., where I_2 is the 2×2 unit matrix and $\otimes$ is the Kronecker product, we obtain

$$\hat{H} = J((S_1 \otimes I_2)(I_2 \otimes S_1) + (S_2 \otimes I_2)(I_2 \otimes S_2) + (S_3 \otimes I_2)(I_2 \otimes S_3)$$

$$+(I_2 \otimes S_1)(S_1 \otimes I_2) + (I_2 \otimes S_2)(S_2 \otimes I_2) + (I_2 \otimes S_3)(S_3 \otimes I_2)).$$

Therefore the Hamilton operator takes the form

$$\hat{H} = 2J((S_1 \otimes S_1) + (S_2 \otimes S_2) + (S_3 \otimes S_3)).$$

$$S_1 \otimes S_1 = \frac{1}{4}\begin{pmatrix} 0 & 1 \\ 1 & 0 \end{pmatrix} \otimes \begin{pmatrix} 0 & 1 \\ 1 & 0 \end{pmatrix} = \frac{1}{4}\begin{pmatrix} 0 & 0 & 0 & 1 \\ 0 & 0 & 1 & 0 \\ 0 & 1 & 0 & 0 \\ 1 & 0 & 0 & 0 \end{pmatrix}$$

$$S_2 \otimes S_2 = \frac{1}{4}\begin{pmatrix} 0 & -i \\ i & 0 \end{pmatrix} \otimes \begin{pmatrix} 0 & -i \\ i & 0 \end{pmatrix} = \frac{1}{4}\begin{pmatrix} 0 & 0 & 0 & -1 \\ 0 & 0 & 1 & 0 \\ 0 & 1 & 0 & 0 \\ -1 & 0 & 0 & 0 \end{pmatrix}$$

$$S_3 \otimes S_3 = \frac{1}{4}\begin{pmatrix} 1 & 0 \\ 0 & -1 \end{pmatrix} \otimes \begin{pmatrix} 1 & 0 \\ 0 & -1 \end{pmatrix} = \frac{1}{4}\begin{pmatrix} 1 & 0 & 0 & 0 \\ 0 & -1 & 0 & 0 \\ 0 & 0 & -1 & 0 \\ 0 & 0 & 0 & 1 \end{pmatrix}.$$

Then the Hamilton operator $\hat{H}$ is given by the 4×4 symmetric matrix

$$\hat{H} = \frac{J}{2}\begin{pmatrix} 1 & 0 & 0 & 0 \\ 0 & -1 & 2 & 0 \\ 0 & 2 & -1 & 0 \\ 0 & 0 & 0 & 1 \end{pmatrix} \equiv \frac{J}{2}\left((1) \oplus \begin{pmatrix} -1 & 2 \\ 2 & -1 \end{pmatrix} \oplus (1) \right)$$

where $\oplus$ denotes the *direct sum*. Find the eigenvalues and eigenvectors of $\hat{H}$.

Solution 74.   (i) We find the eigenvalues

$$E_1 = \frac{J}{2} \text{ three times degenerate}, \quad E_2 = -\frac{3J}{2}$$

and the corresponding normalized eigenvectors

$$\begin{pmatrix} 1 \\ 0 \\ 0 \\ 0 \end{pmatrix}, \begin{pmatrix} 0 \\ 0 \\ 0 \\ 1 \end{pmatrix}, \frac{1}{\sqrt{2}}\begin{pmatrix} 0 \\ 1 \\ 1 \\ 0 \end{pmatrix}, \frac{1}{\sqrt{2}}\begin{pmatrix} 0 \\ 1 \\ -1 \\ 0 \end{pmatrix}.$$

Problem 75.   Let σ_1, σ_2, σ_3 be the Pauli spin matrices.
(i) Find the commutators $[\sigma_1 \otimes \sigma_2, \sigma_2 \otimes \sigma_3]$, $[\sigma_2 \otimes \sigma_3, \sigma_3 \otimes \sigma_1]$, $[\sigma_3 \otimes \sigma_1, \sigma_1 \otimes \sigma_2]$.
(ii) Find the spectrum of the Hamilton operators

$$\hat{K} = \frac{\hat{H}}{\hbar\omega} = \sigma_1 \otimes \sigma_2 + \sigma_2 \otimes \sigma_3 + \sigma_3 \otimes \sigma_1.$$

Solution 75.   (i) For all three commutators we find the 4×4 zero matrix.
(ii) The eigenvalues are $-3\hbar\omega$ and $\hbar\omega$ (three times) with the corresponding normalized eigenvectors

$$\frac{1}{2}\begin{pmatrix} 1 \\ -1 \\ -i \\ -i \end{pmatrix}, \frac{1}{\sqrt{2}}\begin{pmatrix} 1 \\ 0 \\ 0 \\ i \end{pmatrix}, \frac{1}{\sqrt{2}}\begin{pmatrix} 0 \\ 1 \\ 0 \\ -i \end{pmatrix}, \frac{1}{\sqrt{2}}\begin{pmatrix} 0 \\ 0 \\ 1 \\ -1 \end{pmatrix}.$$

Problem 76.   Let

$$S_+ := \begin{pmatrix} 0 & 1 \\ 0 & 0 \end{pmatrix}, \quad S_- := \begin{pmatrix} 0 & 0 \\ 1 & 0 \end{pmatrix}, \quad S_3 := \frac{1}{2}\begin{pmatrix} 1 & 0 \\ 0 & -1 \end{pmatrix}$$

with $[S_+, S_-] = 2S_3$. Let $N \geq 2$. Consider the Hamilton operator

$$\hat{H} = \sum_{j=0}^{N-1} (S_{+,j} S_{-,j+1} + S_{-,j} S_{+,j+1} + 2h S_{3,j})$$

with cyclic boundary conditions, i.e. $N \equiv 0$

$$S_{+,N} = S_{+,0}, \quad S_{-,N} = S_{-,0}, \quad S_{3,N} = S_{3,0}.$$

We have

$$S_{+,j} = I_2 \otimes \cdots \otimes I_2 \otimes S_+ \otimes I_2 \otimes \cdots \otimes I_2$$

where S_+ is at the j-th position and $j = 0, 1, \ldots, N-1$. Analogously we define $S_{-,j}$ and $S_{3,j}$. Let $N = 4$ and $h = 0$. Consider the state

$$|\psi\rangle = f(0) \begin{pmatrix} 0 \\ 1 \end{pmatrix} \otimes \begin{pmatrix} 1 \\ 0 \end{pmatrix} \otimes \begin{pmatrix} 1 \\ 0 \end{pmatrix} \otimes \begin{pmatrix} 1 \\ 0 \end{pmatrix} + f(1) \begin{pmatrix} 1 \\ 0 \end{pmatrix} \otimes \begin{pmatrix} 0 \\ 1 \end{pmatrix} \otimes \begin{pmatrix} 1 \\ 0 \end{pmatrix} \otimes \begin{pmatrix} 1 \\ 0 \end{pmatrix}$$

$$+ f(2) \begin{pmatrix} 1 \\ 0 \end{pmatrix} \otimes \begin{pmatrix} 1 \\ 0 \end{pmatrix} \otimes \begin{pmatrix} 0 \\ 1 \end{pmatrix} \otimes \begin{pmatrix} 1 \\ 0 \end{pmatrix} + f(3) \begin{pmatrix} 1 \\ 0 \end{pmatrix} \otimes \begin{pmatrix} 1 \\ 0 \end{pmatrix} \otimes \begin{pmatrix} 1 \\ 0 \end{pmatrix} \otimes \begin{pmatrix} 0 \\ 1 \end{pmatrix}.$$

Find the condition on the expansion coefficients $f(j)$ ($j = 0, 1, 2, 3$) such that $|\psi\rangle$ is an eigenstate of the Hamilton operator $\hat{H}$.

Solution 76. We have

$$\begin{aligned}
S_{+,0} S_{-,1} &= S_+ \otimes S_- \otimes I_2 \otimes I_2, & S_{+,1} S_{-,2} &= I_2 \otimes S_+ \otimes S_- \otimes I_2 \\
S_{+,2} S_{-,3} &= I_2 \otimes I_2 \otimes S_+ \otimes S_-, & S_{+,3} S_{-,0} &= S_- \otimes I_2 \otimes I_2 \otimes S_+ \\
S_{-,0} S_{+,1} &= S_- \otimes S_+ \otimes I_2 \otimes I_2, & S_{-,1} S_{+,2} &= I_2 \otimes S_- \otimes S_+ \otimes I_2 \\
S_{-,2} S_{+,3} &= I_2 \otimes I_2 \otimes S_- \otimes S_+, & S_{-,3} S_{+,0} &= S_+ \otimes I_2 \otimes I_2 \otimes S_-.
\end{aligned}$$

Note that

$$S_+ \begin{pmatrix} 1 \\ 0 \end{pmatrix} = \begin{pmatrix} 0 \\ 0 \end{pmatrix}, \quad S_+ \begin{pmatrix} 0 \\ 1 \end{pmatrix} = \begin{pmatrix} 1 \\ 0 \end{pmatrix}, \quad S_- \begin{pmatrix} 1 \\ 0 \end{pmatrix} = \begin{pmatrix} 0 \\ 1 \end{pmatrix}, \quad S_- \begin{pmatrix} 0 \\ 1 \end{pmatrix} = \begin{pmatrix} 0 \\ 0 \end{pmatrix}.$$

We find

$$\hat{H}|\psi\rangle = (f(1) + f(3)) \begin{pmatrix} 0 \\ 1 \end{pmatrix} \otimes \begin{pmatrix} 1 \\ 0 \end{pmatrix} \otimes \begin{pmatrix} 1 \\ 0 \end{pmatrix} \otimes \begin{pmatrix} 1 \\ 0 \end{pmatrix}$$

$$+ (f(0) + f(2)) \begin{pmatrix} 1 \\ 0 \end{pmatrix} \otimes \begin{pmatrix} 0 \\ 1 \end{pmatrix} \otimes \begin{pmatrix} 1 \\ 0 \end{pmatrix} \otimes \begin{pmatrix} 1 \\ 0 \end{pmatrix}$$

$$+ (f(1) + f(3)) \begin{pmatrix} 1 \\ 0 \end{pmatrix} \otimes \begin{pmatrix} 1 \\ 0 \end{pmatrix} \otimes \begin{pmatrix} 0 \\ 1 \end{pmatrix} \otimes \begin{pmatrix} 1 \\ 0 \end{pmatrix}$$

$$+ (f(2) + f(0)) \begin{pmatrix} 1 \\ 0 \end{pmatrix} \otimes \begin{pmatrix} 1 \\ 0 \end{pmatrix} \otimes \begin{pmatrix} 1 \\ 0 \end{pmatrix} \otimes \begin{pmatrix} 0 \\ 1 \end{pmatrix}.$$

We obtain the four conditions

$$f(1) + f(3) = E f(0), \quad f(0) + f(2) = E f(1), \quad f(1) + f(3) = E f(2), \quad f(2) + f(0) = E f(3).$$

With $f(2) = -f(0)$, $f(3) = -f(1)$ we find the eigenvalue $E = 0$ (twice). With $f(0) = f(1) = f(2) = f(3)$ we find the eigenvalue $E = 2$. With $f(1) = -f(0)$, $f(2) = f(0)$, $f(3) = -f(0)$ we find the eigenvalue $E = -2$.

Problem 77.   Let S_1, S_2, S_3 be the spin-1 matrices

$$S_1 = \frac{1}{\sqrt{2}} \begin{pmatrix} 0 & 1 & 0 \\ 1 & 0 & 1 \\ 0 & 1 & 0 \end{pmatrix}, \quad S_2 = \frac{1}{\sqrt{2}} \begin{pmatrix} 0 & -i & 0 \\ i & 0 & -i \\ 0 & i & 0 \end{pmatrix}, \quad S_3 = \begin{pmatrix} 1 & 0 & 0 \\ 0 & 0 & 0 \\ 0 & 0 & -1 \end{pmatrix}.$$

(i) Consider the Hamilton operator (9×9 hermitian matrix with trace $\text{tr}(\hat{H}) = 0$)

$$\hat{H} = \hbar\omega_1(S_1 \otimes S_1) + \hbar\omega_2(S_3 \otimes I_3 + I_3 \otimes S_3).$$

Find the eigenvalues of $\hat{H}$.

(ii) Find the eigenvalues and eigenvectors of the Hamilton operator

$$\hat{K} = \frac{\hat{H}}{\hbar\omega} = S_1 \otimes S_1 + S_2 \otimes S_2 + S_3 \otimes S_3$$

which is a hermitian 9×9 matrix with $\text{tr}(\hat{K}) = 0$. The Hamilton operator can be written as

$$\frac{1}{2}(\mathbf{S}_1 \cdot \mathbf{S}_2 + \mathbf{S}_2 \cdot \mathbf{S}_1 \equiv S_{1,1}S_{2,x} + S_{1,2}S_{2,y} + S_{1,3}S_{2,3} + S_{2,1}S_{1,1} + S_{2,2}S_{1,2} + S_{2,3}S_{1,3})$$

where $S_{1,1} = S_1 \otimes I_3$, $S_{2,1} = I_3 \otimes S_1$ etc.

Solution 77.   (i) We obtain the 9 eigenvalues

$$\pm\frac{1}{\sqrt{2}}\sqrt{\hbar\omega_1\sqrt{\hbar^2\omega_1^2 + 4\hbar^2\omega_2^2} + \hbar^2\omega_1^2 + 2\hbar^2\omega_2^2}$$

$$\pm\frac{1}{\sqrt{2}}\sqrt{-\hbar\omega_1\sqrt{\hbar^2\omega_1^2 + 4\hbar^2\omega_2^2} + \hbar^2\omega_1^2 + 2\hbar^2\omega_2^2}$$

$$\pm\frac{1}{\sqrt{2}}\sqrt{\sqrt{\hbar^4\omega_1^4 + 16\hbar^4\omega_2^4} + \hbar^2\omega_1^2 + 4\hbar^2\omega_2^2}$$

$$\pm\frac{1}{\sqrt{2}}\sqrt{-\sqrt{\hbar^4\omega_1^4 + 16\hbar^4\omega_2^4} + \hbar^2\omega_1^2 + 4\hbar^2\omega_2^2}$$

$$0$$

which are non-degenerate if $\omega_1 \neq \omega_2$, $\omega_1 \neq 0$, $\omega_2 \neq 0$. The eigenvector for the eigenvalue 0 is given by

$$\frac{1}{\sqrt{2}}(0 \quad 0 \quad 1 \quad 0 \quad 0 \quad 0 \quad -1 \quad 0 \quad 0)^T$$

i.e. the eigenvector of 0 does not depend on ω_1 and ω_2. The other eigenvectors depend on ω_1 and ω_2.

(ii) The following Maxima program will find the matrix $\hat{H}$ and then the eigenvalues and eigenvectors.

```
/* Spin1Heisenberg.mac */

I3: matrix([1,0,0],[0,1,0],[0,0,1]);
S1: matrix([0,1/sqrt(2),0],[1/sqrt(2),0,1/sqrt(2)],[0,1/sqrt(2),0]);
S2: matrix([0,-%i/sqrt(2),0],[%i/sqrt(2),0,-%i/sqrt(2)],[0,%i/sqrt(2),0]);
S3: matrix([1,0,0],[0,0,0],[0,0,-1]);
T11: kronecker_product(S1,I3); T12: kronecker_product(I3,S1);
T21: kronecker_product(S2,I3); T22: kronecker_product(I3,S2);
T31: kronecker_product(S3,I3); T32: kronecker_product(I3,S3);
K: T11 . T12 + T21 . T22 + T31 . T32;
r1: eigenvectors(K);
```

The Hamilton operator can be written as a direct sum of two 1×1 matrices and a 7×7 matrix

$$(1) \oplus \begin{pmatrix} 0 & 0 & 1 & 0 & 0 & 0 & 0 \\ 0 & -1 & 0 & 1 & 0 & 0 & 0 \\ 1 & 0 & 0 & 0 & 0 & 0 & 0 \\ 0 & 1 & 0 & 0 & 0 & 1 & 0 \\ 0 & 0 & 0 & 0 & 0 & 0 & 1 \\ 0 & 0 & 0 & 1 & 0 & -1 & 0 \\ 0 & 0 & 0 & 0 & 1 & 0 & 0 \end{pmatrix} \oplus (1).$$

We obtain the eigenvalues -2 (1 times), -1 (3 times), 1 (5 times) with the corresponding normalized eigenvectors

$$\frac{1}{\sqrt{3}} (0 \ 0 \ 1 \ 0 \ -1 \ 0 \ 1 \ 0 \ 0)^T$$

$$\frac{1}{\sqrt{2}} (0 \ 1 \ 0 \ -1 \ 0 \ 0 \ 0 \ 0 \ 0)^T$$

$$\frac{1}{\sqrt{2}} (0 \ 0 \ 1 \ 0 \ 0 \ 0 \ -1 \ 0 \ 0)^T$$

$$\frac{1}{\sqrt{2}} (0 \ 0 \ 0 \ 0 \ 0 \ 1 \ 0 \ -1 \ 0)^T$$

$$(1 \ 0 \ 0 \ 0 \ 0 \ 0 \ 0 \ 0 \ 0)^T$$

$$\frac{1}{\sqrt{2}} (0 \ 1 \ 0 \ 1 \ 0 \ 0 \ 0 \ 0 \ 0)^T$$

$$\frac{1}{2} (0 \ 0 \ 1 \ 0 \ 2 \ 0 \ 1 \ 0 \ 0)^T$$

$$\frac{1}{\sqrt{2}} (0 \ 0 \ 0 \ 0 \ 0 \ 1 \ 0 \ 1 \ 0)^T$$

$$(0 \ 0 \ 0 \ 0 \ 0 \ 0 \ 0 \ 0 \ 1)^T.$$

Problem 78. Consider the Hamilton operator

$$\hat{K} = \frac{\hat{H}}{\hbar\omega} = \sum_{j=1}^{3} (\sigma_j \otimes \sigma_j \otimes I_2 + I_2 \otimes \sigma_j \otimes \sigma_j + \sigma_j \otimes I_2 \otimes \sigma_j)$$

which is a hermitian 8×8 matrix with trace equal to 0.
(i) Find the eigenvalues and eigenvectors of $\hat{K}$. Discuss.
(ii) Consider the hermitian 8×8 matrices

$$K_1 = \sigma_1 \otimes I_2 \otimes I_2 + I_2 \otimes \sigma_1 \otimes I_2 + I_2 \otimes I_2 \otimes \sigma_1,$$
$$K_2 = \sigma_2 \otimes I_2 \otimes I_2 + I_2 \otimes \sigma_2 \otimes I_2 + I_2 \otimes I_2 \otimes \sigma_2,$$
$$K_3 = \sigma_3 \otimes I_2 \otimes I_2 + I_2 \otimes \sigma_3 \otimes I_2 + I_2 \otimes I_2 \otimes \sigma_3.$$

Find the commutators $[\hat{K}, \hat{K}_1]$, $[\hat{K}, \hat{K}_2]$, $[\hat{K}, \hat{K}_3]$. Discuss.
(iii) Consider the hermitian 8×8 matrices

$$D_1 = \sigma_1 \otimes \sigma_2 \otimes \sigma_1, \quad D_2 = \sigma_2 \otimes \sigma_2 \otimes \sigma_2, \quad D_3 = \sigma_3 \otimes \sigma_3 \otimes \sigma_3.$$

Find the commutators $[\hat{K}, \hat{D}_1]$, $[\hat{K}, \hat{D}_2]$, $[\hat{K}, \hat{D}_3]$. Discuss.

Solution 78.   (i) The eigenvalues are $+3$ (four times) with the normalized eigenvectors

$$(1\ \ 0\ \ 0\ \ 0\ \ 0\ \ 0\ \ 0\ \ 0)^T$$

$$\frac{1}{\sqrt{3}}(0\ \ 1\ \ 1\ \ 0\ \ 1\ \ 0\ \ 0\ \ 0)^T$$

$$\frac{1}{\sqrt{3}}(0\ \ 0\ \ 0\ \ 1\ \ 0\ \ 1\ \ 1\ \ 0)^T$$

$$((0\ \ 0\ \ 0\ \ 0\ \ 0\ \ 0\ \ 0\ \ 1)^T$$

and -3 (four times) with the normalized eigenvectors

$$\frac{1}{\sqrt{2}}(0\ \ 1\ \ 0\ \ 0\ \ -1\ \ 0\ \ 0\ \ 0)^T$$

$$\frac{1}{\sqrt{2}}(0\ \ 0\ \ 1\ \ 0\ \ -1\ \ 0\ \ 0\ \ 0)^T$$

$$\frac{1}{\sqrt{2}}(0\ \ 0\ \ 0\ \ 1\ \ 0\ \ 0\ \ -1\ \ 0)^T$$

$$\frac{1}{\sqrt{2}}(0\ \ 0\ \ 0\ \ 0\ \ 0\ \ 1\ \ -1\ \ 0)^T.$$

(ii) The commutators $[\hat{K}, K_1]$, $[\hat{K}, K_2$, $[\hat{K}, K_3]$ all vanish. Note that commutators $[K_1, K_2]$, $[K_1, K_3]$, $[K_2, K_3]$ are non-zero.

(iii) The commutators $[\hat{K}, D_1]$, $[\hat{K}, D_2$, $[\hat{K}, D_3]$ all vanish. Thus we can construct the projection matrices

$$\Pi_{11} = \frac{1}{2}(I_8 + D_1), \quad \Pi_{12} = \frac{1}{2}(I_8 - D_1)$$

$$\Pi_{21} = \frac{1}{2}(I_8 + D_2), \quad \Pi_{22} = \frac{1}{2}(I_8 - D_2)$$

$$\Pi_{31} = \frac{1}{2}(I_8 + D_3), \quad \Pi_{32} = \frac{1}{2}(I_8 - D_3)$$

with $\Pi_{11}\Pi_{12} = 0_8$, $\Pi_{21}\Pi_{22} = 0_8$, $\Pi_{31}\Pi_{32} = 0_8$ to decompose the Hilbert space $\mathbb{C}^8$ in sub-Hilbert space to simplify the eigenvalue problem. Note that the commutators $[D_1, D_2]$, $[D_1, D_3]$, $[D_2, D_3]$ are non-zero.

Problem 79.   (i) Consider the spin matrices for describing a spin-$\frac{1}{2}$ system

$$S_1 = \frac{1}{2}\begin{pmatrix} 0 & 1 \\ 1 & 0 \end{pmatrix}, \quad S_2 = \frac{1}{2}\begin{pmatrix} 0 & -i \\ i & 0 \end{pmatrix}, \quad S_3 = \frac{1}{2}\begin{pmatrix} 1 & 0 \\ 0 & -1 \end{pmatrix}$$

and the spin matrices for describing a spin-1 system

$$T_1 = \frac{1}{\sqrt{2}}\begin{pmatrix} 0 & 1 & 0 \\ 1 & 0 & 1 \\ 0 & 1 & 0 \end{pmatrix}, \quad T_2 = \frac{1}{\sqrt{2}}\begin{pmatrix} 0 & -i & 0 \\ i & 0 & -i \\ 0 & i & 0 \end{pmatrix}, \quad T_3 = \begin{pmatrix} 1 & 0 & 0 \\ 0 & 0 & 0 \\ 0 & 0 & -1 \end{pmatrix}.$$

Find the spectrum (eigenvalues and eigenvector) of the hermitian 12×12 matrix

$$\hat{K} = \frac{\hat{H}}{\hbar\omega} = S_1 \otimes T_1 \otimes S_1 + S_2 \otimes T_2 \otimes S_2 + S_3 \otimes T_3 \otimes S_3.$$

Note that $\mathrm{tr}(\hat{K}) = 0$.

Solution 79.   We find the eigenvalues

$$\frac{\sqrt{3}}{4} \ (4 \, \mathrm{times}), \quad 0 \ (4 \, \mathrm{times}), \quad -\frac{\sqrt{3}}{4} \ (4 \, \mathrm{times}).$$

Problem 80.   Let σ_1, σ_2, σ_3 be the Pauli spin matrices. Consider the Hamilton operator

$$\hat{K} = \frac{\hat{H}}{\hbar\omega} = \sigma_1 \otimes \sigma_2 \otimes \sigma_3.$$

(i) Let Q be the unitary matrix $Q = \sigma_3 \otimes \sigma_3 \otimes \sigma_3$. Then $Q = Q^*$. Is $Q\hat{K}Q^* = \hat{K}$?

(ii) Find the eigenvalues and eigenvectors of 8×8 hermitian and unitary matrix

$$\sigma_1 \otimes \sigma_2 \otimes \sigma_3.$$

Can one find entangled eigenvectors?

Solution 80.   (i) Since $\sigma_3\sigma_1\sigma_3 = -\sigma_1$, $\sigma_3\sigma_2\sigma_3 = -\sigma_2$ and

$$Q\hat{K}Q^* = (\sigma_3\sigma_1\sigma_3) \otimes (\sigma_3\sigma_2\sigma_3) \otimes (\sigma_3\sigma_3\sigma_3)$$

we find that $Q\hat{K}Q^* = \hat{K}$.

(ii) We find the eigenvalue $+1$ with multiplicity 4 and -1 with multiplicity 4. The eigenvectors for $+1$ are

$$\begin{pmatrix} 1 & i & 1 & -i & -i & 1 & i & 1 \end{pmatrix}^T, \quad \begin{pmatrix} 1 & i & 1 & 0 & -i & 0 & i & 1 \end{pmatrix}^T,$$

$$\begin{pmatrix} 1 & i & 0 & -i & 0 & 1 & i & 1 \end{pmatrix}^T, \quad \begin{pmatrix} 1 & 0 & 1 & -i & -i & 1 & i & 1 \end{pmatrix}^T.$$

The eigenvectors for -1 are

$$\begin{pmatrix} -1 & i & -1 & -i & -i & -1 & i & -1 \end{pmatrix}^T, \quad \begin{pmatrix} 0 & i & -1 & -i & -i & -1 & 0 & -1 \end{pmatrix}^T,$$

$$\begin{pmatrix} -1 & i & 0 & -i & 0 & -1 & i & -1 \end{pmatrix}^T, \quad \begin{pmatrix} -1 & i & -1 & 0 & -i & 0 & i & -1 \end{pmatrix}^T.$$

Problem 81.   Consider the Pauli spin matrices σ_1, σ_2, σ_3 and the spin-1 matrices

$$S_1 = \frac{1}{\sqrt{2}} \begin{pmatrix} 0 & 1 & 0 \\ 1 & 0 & 1 \\ 0 & 1 & 0 \end{pmatrix}, \quad S_2 = \frac{1}{\sqrt{2}} \begin{pmatrix} 0 & -i & 0 \\ i & 0 & -i \\ 0 & i & 0 \end{pmatrix}, \quad S_3 = \begin{pmatrix} 1 & 0 & 0 \\ 0 & 0 & 0 \\ 0 & 0 & -1 \end{pmatrix}.$$

Find the eigenvalues and eigenvectors of the Hamilton operator

$$\hat{K} = \frac{\hat{H}}{\hbar\omega} = \frac{1}{2}\sigma_1 \otimes S_1 + \frac{1}{2}\sigma_2 \otimes S_2 + \frac{1}{2}\sigma_3 \otimes S_3.$$

Solution 81.   The Hamilton operator $\hat{K}$ is the hermitian 6×6 matrix

$$\hat{K} = \begin{pmatrix} 1/2 & 0 & 0 & 0 & 0 & 0 \\ 0 & 0 & 0 & 1/\sqrt{2} & 0 & 0 \\ 0 & 0 & -1/2 & 0 & 1/\sqrt{2} & 0 \\ 0 & 1/\sqrt{2} & 0 & -1/2 & 0 & 0 \\ 0 & 0 & 1/\sqrt{2} & 0 & 0 & 0 \\ 0 & 0 & 0 & 0 & 0 & 1/2 \end{pmatrix}.$$

The 6×6 matrix can be written as direct sum

$$(1/2) \oplus \begin{pmatrix} 0 & 0 & 1/\sqrt{2} & 0 \\ 0 & -1/2 & 0 & 1/\sqrt{2} \\ 1/\sqrt{2} & 0 & -1/2 & 0 \\ 0 & 1/\sqrt{2} & 0 & 0 \end{pmatrix} \oplus (1/2).$$

The eigenvalues are $1/2$ (4 times) and -1 (2 times) with the corresponding eigenvectors for $1/2$

$$\begin{pmatrix} 1 \\ 0 \\ 0 \\ 0 \\ 0 \\ 0 \end{pmatrix}, \quad \begin{pmatrix} 0 \\ \sqrt{2} \\ 0 \\ 1 \\ 0 \\ 0 \end{pmatrix}, \quad \begin{pmatrix} 0 \\ 0 \\ 1 \\ 0 \\ \sqrt{2} \\ 0 \end{pmatrix}, \quad \begin{pmatrix} 0 \\ 0 \\ 0 \\ 0 \\ 0 \\ 1 \end{pmatrix}$$

and for -1

$$\begin{pmatrix} 0 \\ 1 \\ 0 \\ -\sqrt{2} \\ 0 \\ 0 \end{pmatrix}, \quad \begin{pmatrix} 0 \\ 0 \\ \sqrt{2} \\ 0 \\ -1 \\ 0 \end{pmatrix}.$$

Study the eigenvalue problem for

$$S_1 \otimes \frac{1}{2}\sigma_1 + S_2 \otimes \frac{1}{2}\sigma_2 + S_3 \otimes \frac{1}{2}\sigma_3.$$

Problem 82.   Let σ_1, σ_2, σ_3 be the Pauli spin matrices. Consider the Hamilton operator with the *triple spin interaction*

$$\hat{H} = J(\sigma_1 \otimes \sigma_2 \otimes \sigma_3) + B(\sigma_3 \otimes I_2 \otimes I_2 + I_2 \otimes \sigma_3 \otimes I_2 + I_2 \otimes I_2 \otimes \sigma_3).$$

(i) Find the eigenvalues of $\hat{H}$.
(ii) Find the commutator

$$[\sigma_3 \otimes I_2 \otimes I_2 + I_2 \otimes \sigma_3 \otimes I_2 + I_2 \otimes I_2 \otimes \sigma_3, \sigma_1 \otimes \sigma_2 \otimes \sigma_3].$$

Solution 82.   (i) The eight eigenvalues are

$$-\sqrt{J^2 + 4B^2} - B, \quad \sqrt{J^2 + 4B^2} - B, \quad -\sqrt{J^2 + 4B^2} + B, \quad \sqrt{J^2 + 4B^2} + B,$$

$$B - J, \quad J + B, \quad -J - B, \quad J - B.$$

Does energy level crossing occur?
(ii) The commutator we evaluate with the following SymbolicC++ program

```
// kroncommu.cpp

#include <iostream>
#include "symbolicc++.h"
using namespace std;
```

```
int main(void)
{
using SymbolicConstant::i;
Symbolic I2("I2",2,2);
I2 = I2.identity();  // 2 times 2 identity matrix
Symbolic sig1("sig1",2,2);
sig1(0,0) = 0; sig1(0,1) = 1; sig1(1,0) = 1; sig1(1,1) = 0;
Symbolic sig2("sig2",2,2);
sig2(0,0) = 0; sig2(0,1) = -i; sig2(1,0) = i; sig2(1,1) = 0;
Symbolic sig3("sig3",2,2);
sig3(0,0) = 1; sig3(0,1) = 0; sig3(1,0) = 0; sig3(1,1) = -1;
Symbolic H = kron(sig1,kron(sig2,sig3));
Symbolic B = kron(sig3,kron(I2,I2))+kron(I2,kron(sig3,I2))+kron(I2,kron(I2,sig3));
Symbolic C = B*H - H*B;
cout << "C = " << C << endl;
return 0;
}
```

The output is the nonzero skew-hermitian 8×8 matrix

$$
4i \begin{pmatrix}
0 & 0 & 0 & 0 & 0 & 0 & -1 & 0 \\
0 & 0 & 0 & 0 & 0 & 0 & 0 & 1 \\
0 & 0 & 0 & 0 & 0 & 0 & 0 & 0 \\
0 & 0 & 0 & 0 & 0 & 0 & 0 & 0 \\
0 & 0 & 0 & 0 & 0 & 0 & 0 & 0 \\
0 & 0 & 0 & 0 & 0 & 0 & 0 & 0 \\
-1 & 0 & 0 & 0 & 0 & 0 & 0 & 0 \\
0 & 1 & 0 & 0 & 0 & 0 & 0 & 0
\end{pmatrix}.
$$

Problem 83. Consider the two Hamilton operators

$$
\frac{\hat{H}}{\hbar\omega} = \sigma_{1,1}\sigma_{1,2} + \sigma_{1,2}\sigma_{1,3} + \sigma_{2,1}\sigma_{2,2} + \sigma_{2,2}\sigma_{2,3}
$$

and

$$
\frac{\hat{K}}{\hbar\omega} = \sigma_{1,1}\sigma_{1,2} + \sigma_{1,2}\sigma_{1,3} + \sigma_{1,3}\sigma_{1,1} + \sigma_{2,1}\sigma_{2,2} + \sigma_{2,2}\sigma_{2,3} + \sigma_{2,3}\sigma_{2,1}
$$

where for $\hat{H}$ we have open end boundary conditions and for $\hat{K}$ we have cyclic boundary conditions. Note that

$$
\sigma_{1,1} = \sigma_1 \otimes I_2 \otimes I_2, \quad \sigma_{1,2} = I_2 \otimes \sigma_1 \otimes I_2, \quad \sigma_{1,3} = I_2 \otimes I_2 \otimes \sigma_1
$$

etc. Thus the Hamilton operators act in the Hilbert space $\mathbb{C}^8$. Find the spectrum for both Hamilton operators. Discuss.

Solution 83. For

$$
\frac{\hat{H}}{\hbar\omega} = \sigma_1 \otimes \sigma_1 \otimes I_2 + \sigma_2 \otimes \sigma_2 \otimes I_2 + I_2 \otimes \sigma_1 \otimes \sigma_1 + I_2 \otimes \sigma_2 \otimes \sigma_2
$$

we obtain the hermitian 8×8 matrix which can be written as direct sum

$$(0) \oplus \begin{pmatrix} 0 & 2 & 0 & 0 & 0 & 0 \\ 2 & 0 & 0 & 2 & 0 & 0 \\ 0 & 0 & 0 & 0 & 2 & 0 \\ 0 & 2 & 0 & 0 & 0 & 0 \\ 0 & 0 & 2 & 0 & 0 & 2 \\ 0 & 0 & 0 & 0 & 2 & 0 \end{pmatrix} \oplus (0)$$

with the eigenvalues $-2\sqrt{2}$ (2 times), $2\sqrt{2}$ (2 times), 0 (4 times) and the corresponding eigenvectors for $-2\sqrt{2}$

$$(0 \ \ 1 \ \ -\sqrt{2} \ \ 0 \ \ 1 \ \ 0 \ \ 0 \ \ 0)^T, \quad (0 \ \ 0 \ \ 0 \ \ 1 \ \ 0 \ \ -\sqrt{2} \ \ 1 \ \ 0)^T,$$

for $2\sqrt{2}$

$$(0 \ \ 1 \ \ \sqrt{2} \ \ 0 \ \ 1 \ \ 0 \ \ 0 \ \ 0)^T, \quad (0 \ \ 0 \ \ 0 \ \ 1 \ \ 0 \ \ \sqrt{2} \ \ 1 \ \ 0)^T,$$

and for 0

$$(1 \ \ 0 \ \ 0 \ \ 0 \ \ 0 \ \ 0 \ \ 0 \ \ 0)^T, \quad (0 \ \ 1 \ \ 0 \ \ 0 \ \ -1 \ \ 0 \ \ 0 \ \ 0)^T,$$

$$(0 \ \ 0 \ \ 0 \ \ 1 \ \ 0 \ \ 0 \ \ -1 \ \ 0)^T, \quad (0 \ \ 0 \ \ 0 \ \ 0 \ \ 0 \ \ 0 \ \ 0 \ \ 1)^T.$$

For

$$\frac{\hat{K}}{\hbar\omega} = \sigma_1 \otimes \sigma_1 \otimes I_2 + \sigma_2 \otimes \sigma_2 \otimes I_2 + I_2 \otimes \sigma_1 \otimes \sigma_1 + I_2 \otimes \sigma_2 \otimes \sigma_2 + \sigma_1 \otimes I_2 \otimes \sigma_1 + \sigma_2 \otimes I_2 \otimes \sigma_2$$

we obtain the hermitian 8×8 matrix which can be written as direct sum

$$(0) \oplus \begin{pmatrix} 0 & 2 & 0 & 2 & 0 & 0 \\ 2 & 0 & 0 & 2 & 0 & 0 \\ 0 & 0 & 0 & 0 & 2 & 2 \\ 2 & 2 & 0 & 0 & 0 & 0 \\ 0 & 0 & 2 & 0 & 0 & 2 \\ 0 & 0 & 2 & 0 & 2 & 0 \end{pmatrix} \oplus (0)$$

with the eigenvalues 4 (2 times), -2 (4 times), 0 (2 times) and the corresponding eigenvectors for 4

$$(0 \ \ 0 \ \ 0 \ \ 1 \ \ 0 \ \ 1 \ \ 1 \ \ 0)^T, \quad (0 \ \ 1 \ \ 0 \ \ 0 \ \ -1 \ \ 0 \ \ 0 \ \ 0)^T,$$

for -2

$$(0 \ \ 1 \ \ 0 \ \ 0 \ \ -1 \ \ 0 \ \ 0 \ \ 0)^T, \quad (0 \ \ 0 \ \ 1 \ \ 0 \ \ -1 \ \ 0 \ \ 0 \ \ 0)^T,$$

$$(0 \ \ 0 \ \ 0 \ \ 1 \ \ 0 \ \ 0 \ \ -1 \ \ 0)^T, \quad (0 \ \ 0 \ \ 0 \ \ 0 \ \ 0 \ \ 1 \ \ -1 \ \ 0)^T,$$

and for 0

$$(1 \ \ 0 \ \ 0 \ \ 0 \ \ 0 \ \ 0 \ \ 0 \ \ 0)^T, \quad (0 \ \ 0 \ \ 0 \ \ 0 \ \ 0 \ \ 0 \ \ 0 \ \ 1)^T.$$

Problem 84.   (i) Study the eigenvalue problem for the Hamilton operator

$$\hat{H} = \hbar\omega_1 \sigma_1 \otimes \sigma_1 + \hbar\omega_3 \sigma_3 \otimes \sigma_3.$$

(ii) Study the eigenvalue problem for the Hamilton operator

$$\hat{K} = \hbar\omega_1 \sigma_1 \otimes \sigma_1 \otimes \sigma_1 + \hbar\omega_3 \sigma_3 \otimes \sigma_3 \otimes \sigma_3.$$

Solution 84. (i) The Hamilton operator is given by the hermitian 4×4 matrix

$$\hat{H} = \begin{pmatrix} \hbar\omega_3 & 0 & 0 & \hbar\omega_1 \\ 0 & -\hbar\omega_3 & \hbar\omega_1 & 0 \\ 0 & \hbar\omega_1 & -\hbar\omega_3 & 0 \\ \hbar\omega_1 & 0 & 0 & \hbar\omega_3 \end{pmatrix}$$

with $\text{tr}(\hat{H}) = 0$. The eigenvalues are

$$-\hbar(\omega_1 + \omega_3), \quad \hbar(\omega_1 - \omega_3), \quad -\hbar(\omega_1 - \omega_3), \quad \hbar(\omega_1 + \omega_3).$$

Thus if $\omega_1 \neq \omega_3$ and nonzero the spectrum is non-degenerate. The normalized eigenvectors are

$$\frac{1}{\sqrt{2}}\begin{pmatrix} 0 \\ 1 \\ -1 \\ 0 \end{pmatrix}, \quad \frac{1}{\sqrt{2}}\begin{pmatrix} 0 \\ 1 \\ 1 \\ 0 \end{pmatrix}, \quad \frac{1}{\sqrt{2}}\begin{pmatrix} 1 \\ 0 \\ 0 \\ -1 \end{pmatrix}, \quad \frac{1}{\sqrt{2}}\begin{pmatrix} 1 \\ 0 \\ 0 \\ 1 \end{pmatrix}.$$

Note that the eigenvectors do not depend on ω_1 and ω_3. These eigenvectors are the *Bell states*. They form an orthonormal basis in $\mathbb{C}^4$.

(ii) The Hamilton operator is given by the 8×8 hermitian matrix

$$\hat{K} = \hbar \begin{pmatrix} \omega_3 & 0 & 0 & 0 & 0 & 0 & 0 & \omega_1 \\ 0 & -\omega_3 & 0 & 0 & 0 & 0 & \omega_1 & 0 \\ 0 & 0 & -\omega_3 & 0 & 0 & \omega_1 & 0 & 0 \\ 0 & 0 & 0 & \omega_3 & \omega_1 & 0 & 0 & 0 \\ 0 & 0 & 0 & \omega_1 & -\omega_3 & 0 & 0 & 0 \\ 0 & 0 & \omega_1 & 0 & 0 & \omega_3 & 0 & 0 \\ 0 & \omega_1 & 0 & 0 & 0 & 0 & \omega_3 & 0 \\ \omega_1 & 0 & 0 & 0 & 0 & 0 & 0 & -\omega_3 \end{pmatrix}$$

with trace equal to 0. The eigenvalues are highly degenerate

$$-\sqrt{\hbar^2\omega_1^2 + \hbar^2\omega_3^2} \ (4 \text{ times}), \qquad \sqrt{\hbar^2\omega_1^2 + \hbar^2\omega_3^2} \ (4 \text{ times})$$

with the corresponding eigenvectors for $-\sqrt{\hbar^2\omega_1^2 + \hbar^2\omega_3^2}$ (we set $E = \sqrt{\hbar^2\omega_1^2 + \hbar^2\omega_3^2}$)

$$\begin{pmatrix} \hbar\omega_1 \\ 0 \\ 0 \\ 0 \\ 0 \\ 0 \\ 0 \\ -E + \hbar\omega_2 \end{pmatrix}, \quad \begin{pmatrix} 0 \\ \hbar\omega_1 \\ 0 \\ 0 \\ 0 \\ 0 \\ -E - \hbar\omega_2 \\ 0 \end{pmatrix}, \quad \begin{pmatrix} 0 \\ 0 \\ \hbar\omega_1 \\ 0 \\ 0 \\ -E - \hbar\omega_2 \\ 0 \\ 0 \end{pmatrix}, \quad \begin{pmatrix} 0 \\ 0 \\ 0 \\ \hbar\omega_1 \\ -E + \omega_2 \\ 0 \\ 0 \\ 0 \end{pmatrix},$$

and for $\sqrt{\hbar^2\omega_1^2 + \hbar^2\omega_3^2}$ (we set $E = \sqrt{\hbar^2\omega_1^2 + \hbar^2\omega_3^2}$

$$\begin{pmatrix} \hbar\omega_1 \\ 0 \\ 0 \\ 0 \\ 0 \\ 0 \\ 0 \\ E - \hbar\omega_2 \end{pmatrix}, \quad \begin{pmatrix} 0 \\ \hbar\omega_1 \\ 0 \\ 0 \\ 0 \\ 0 \\ E + \hbar\omega_2 \\ 0 \end{pmatrix}, \quad \begin{pmatrix} 0 \\ 0 \\ \hbar\omega_1 \\ 0 \\ 0 \\ E + \hbar\omega_2 \\ 0 \\ 0 \end{pmatrix}, \quad \begin{pmatrix} 0 \\ 0 \\ 0 \\ \hbar\omega_1 \\ E - \omega_2 \\ 0 \\ 0 \\ 0 \end{pmatrix}.$$

Problem 85.   Study the spectrum of the spin Hamilton operator

$$\hat{H} = \hbar\omega_1(\sigma_1 \otimes I_2 \otimes I_2 + I_2 \otimes \sigma_2 \otimes I_2 + I_2 \otimes I_2 \otimes \sigma_3)$$
$$+ \hbar\omega_2(\sigma_1 \otimes \sigma_2 \otimes I_2 + \sigma_1 \otimes I_2 \otimes \sigma_3 + I_2 \otimes \sigma_2 \otimes \sigma_3) + \hbar\omega_3\sigma_1 \otimes \sigma_2 \otimes \sigma_3)$$

which is a hermitian 8×8 matrix with $\mathrm{tr}(\hat{H}) = 0$. Hint. Use the normalized eigenvectors $\mathbf{u}_1$, $\mathbf{u}_2$ of σ_1, the normalized eigenvectors $\mathbf{v}_1$, $\mathbf{v}_2$ of σ_2, the normalized eigenvectors $\mathbf{w}_1$, $\mathbf{w}_2$ of σ_3 and the Kronecker product $\otimes$.

Solution 85.   The eight eigenvectors are given by

$$\mathbf{u}_j \otimes \mathbf{v}_k \otimes \mathbf{w}_\ell, \quad j, k, \ell = 1, 2.$$

With the eigenvalues of σ_1 given by λ_1, λ_2, the eigenvalues of σ_2 given by μ_1, μ_2 and the eigenvalues of σ_3 given by ν_1, ν_2 the eigenvalues of the Hamilton operator follow as

$$\hat{H}(\mathbf{u}_j \otimes \mathbf{v}_k \otimes \mathbf{w}_\ell) = (\hbar\omega_1(\lambda_j + \mu_k + \nu_\ell) + \hbar\omega_2(\lambda_j\mu_k + \lambda_j\nu_\ell + \mu_k\nu_\ell) + \hbar\omega_3\lambda_1\mu_k\nu_\ell)(\mathbf{u}_j \otimes \mathbf{v}_k \otimes \mathbf{w}_\ell).$$

Problem 86.   Let S_1, S_2, S_3 be the spin-$\frac{1}{2}$ matrices

$$S_1 = \frac{1}{2}\begin{pmatrix} 0 & 1 \\ 1 & 0 \end{pmatrix}, \quad S_2 = \frac{1}{2}\begin{pmatrix} 0 & -i \\ i & 0 \end{pmatrix}, \quad S_3 = \frac{1}{2}\begin{pmatrix} 1 & 0 \\ 0 & -1 \end{pmatrix}.$$

Consider the two spin Hamilton operators

$$\hat{H}_1 = \sum_{j=1}^{2} \hbar\omega_j(S_j \otimes S_j \otimes I_2 + I_2 \otimes S_j \otimes S_j)$$

$$\hat{H}_2 = \sum_{j=1}^{2} \hbar\omega_j(S_j \otimes S_j \otimes I_2 + I_2 \otimes S_j \otimes S_j + S_j \otimes I_2 \otimes S_j)$$

with nearest neighbour interaction for the following lattices with three vertices (the first is open end and the second with cyclic boundary conditions)

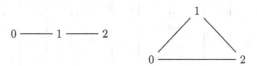

We can write $(\gamma = (\hbar\omega_2)/(\hbar\omega_1))$

$$\hat{K}_1 = S_1 \otimes S_1 \otimes I_2 + I_2 \otimes S_1 \otimes S_1 + \gamma(S_2 \otimes S_2 \otimes I_2 + I_2 \otimes S_2 \otimes S_2)$$

and

$$\hat{K}_2 = S_1 \otimes S_1 \otimes I_2 + I_2 \otimes S_1 \otimes S_1 + S_1 \otimes I_2 \otimes S_1 + \gamma(S_2 \otimes S_2 \otimes I_2 + I_2 \otimes S_2 \otimes S_2 + S_2 \otimes I_2 \otimes S_2)$$

with $\hat{K}_1 = \hat{H}_1/(\hbar\omega_1)$ and $\hat{K}_2 = \hat{H}_2/(\hbar\omega_1)$. Thus the Hamilton operators $\hat{K}_1$ and $\hat{K}_2$ are 8×8 hermitian matrices with trace equal to 0. Applying the following SymbolicC++ program

```
// SpinHamilton.cpp

#include <iostream>
#include "symbolicc++.h"
using namespace std;

int main(void)
{
using SymbolicConstant::i;
Symbolic g("g");
Symbolic half = Symbolic(1)/2;
Symbolic I2("I2",2,2);
I2 = I2.identity();  // 2 times 2 identity matrix
Symbolic sig1("sig1",2,2);
sig1(0,0) = 0; sig1(0,1) = 1; sig1(1,0) = 1; sig1(1,1) = 0;
Symbolic S1 = half*sig1;
Symbolic sig2("sig2",2,2);
sig2(0,0) = 0; sig2(0,1) = -i; sig2(1,0) = i; sig2(1,1) = 0;
Symbolic S2 = half*sig2;
Symbolic K1 = kron(S1,kron(S1,I2))+kron(I2,kron(S1,S1))
            + g*(kron(S2,kron(S2,I2))+kron(I2,kron(S2,S2)));
cout << "K1 = " << K1 << endl;
Symbolic K2 = kron(S1,kron(S1,I2))+kron(I2,kron(S1,S1))+kron(S1,kron(I2,S1))
            + g*(kron(S2,kron(S2,I2))+kron(I2,kron(S2,S2))+kron(S2,kron(I2,S2)));
cout << "K2 = " << K2 << endl;
return 0;
}
```

provides the 8×8 matrices for $\hat{K}_1$ and $\hat{K}_2$

$$
K_1 = \frac{1}{4}
\begin{pmatrix}
0 & 0 & 0 & 1-\gamma & 0 & 0 & 1-\gamma & 0 \\
0 & 0 & 1+\gamma & 0 & 0 & 0 & 0 & 1-\gamma \\
0 & 1+\gamma & 0 & 0 & 1+\gamma & 0 & 0 & 0 \\
1-\gamma & 0 & 0 & 0 & 0 & 1+\gamma & 0 & 0 \\
0 & 0 & 1+\gamma & 0 & 0 & 0 & 0 & 1-\gamma \\
0 & 0 & 0 & 1+\gamma & 0 & 0 & 1+\gamma & 0 \\
1-\gamma & 0 & 0 & 0 & 0 & 1+\gamma & 0 & 0 \\
0 & 1-\gamma & 0 & 0 & 1-\gamma & 0 & 0 & 0
\end{pmatrix}
$$

$$
K_2 = \frac{1}{4}
\begin{pmatrix}
0 & 0 & 0 & 1-\gamma & 0 & 1-\gamma & 1-\gamma & 0 \\
0 & 0 & 1+\gamma & 0 & 1+\gamma & 0 & 0 & 1-\gamma \\
0 & 1+\gamma & 0 & 0 & 1+\gamma & 0 & 0 & 1-\gamma \\
1-\gamma & 0 & 0 & 0 & 0 & 1+\gamma & 1+\gamma & 0 \\
0 & 1+\gamma & 1+\gamma & 0 & 0 & 0 & 0 & 1-\gamma \\
1-\gamma & 0 & 0 & 1+\gamma & 0 & 0 & 1+\gamma & 0 \\
1-\gamma & 0 & 0 & 1+\gamma & 0 & 1+\gamma & 0 & 0 \\
0 & 1-\gamma & 1-\gamma & 0 & 1-\gamma & 0 & 0 & 0
\end{pmatrix}.
$$

The matrices are symmetric over the real numbers. Find find the eigenvalues of $\hat{K}_1$ and $\hat{K}_2$. Study the dependence of the eigenvalues as function of γ.

Solution 86. The eigenvalues of $\hat{K}_1$ (open end boundary conditions) are

$$
-\frac{1}{2}\sqrt{1+\gamma^2} \ (2 \text{ times}), \quad \frac{1}{2}\sqrt{1+\gamma^2} \ (2 \text{ times}), \quad 0 \ (4 \text{ times}).
$$

The eigenvalues for $\hat{K}_2$ (cyclic boundary conditions) are

$$-\frac{1}{2}\sqrt{1-\gamma+\gamma^2}+\frac{1}{4}\gamma+\frac{1}{4} \text{ (2 times)}, \quad \frac{1}{2}\sqrt{1-\gamma+\gamma^2}+\frac{1}{4}\gamma+\frac{1}{4} \text{ (2 times)},$$

and

$$-\frac{1}{4}\gamma-\frac{1}{4} \text{ (4 times)}.$$

Problem 87.   Consider the four 16×16 hermitian and unitary matrices

$$X_{12} = \sigma_1 \otimes \sigma_2 \otimes I_2 \otimes I_2, \quad X_{23} = I_2 \otimes \sigma_1 \otimes \sigma_2 \otimes I_2,$$

$$X_{34} = I_2 \otimes I_2 \otimes \sigma_1 \otimes \sigma_2, \quad X_{41} = \sigma_2 \otimes I_2 \otimes I_2 \otimes \sigma_1$$

and the Hamilton operators

$$\hat{K}_1 = \frac{\hat{H}_1}{\hbar\omega} = X_{12} + X_{23} + X_{34}$$

$$\hat{K}_2 = \frac{\hat{H}_2}{\hbar\omega} = X_{12} + X_{23} + X_{34} + X_{41}.$$

Let

$$C_1 = [X_{12}, X_{23}] + [X_{23}, X_{34}], \quad C_2 = [X_{12}, X_{23}] + [X_{23}, X_{34}] + [X_{34}, X_{41}].$$

Find the commutator $[\hat{K}_1, C_1]$ and anticommutator $[\hat{K}_1, C_1]_+$. Find the commutator $[\hat{K}_2, C_2]$ and anticommutator $[\hat{K}_2, C_2]_+$.

Solution 87.   The following computer algebra program

```
// CommuAnti.cpp

#include <iostream>
#include "symbolicc++.h"
using namespace std;

int main(void)
{
using SymbolicConstant::i;
Symbolic I2("I2",2,2);
I2 = I2.identity();  // 2 times 2 identity matrix
Symbolic sig1("sig1",2,2);
sig1(0,0) = 0; sig1(0,1) = 1; sig1(1,0) = 1; sig1(1,1) = 0;
Symbolic sig2("sig2",2,2);
sig2(0,0) = 0; sig2(0,1) = -i; sig2(1,0) = i; sig2(1,1) = 0;
Symbolic sig3("sig3",2,2);
sig3(0,0) = 1; sig3(0,1) = 0; sig3(1,0) = 0; sig3(1,1) = -1;
Symbolic X12 = kron(sig1,kron(sig2,kron(I2,I2)));
Symbolic X23 = kron(I2,kron(sig1,kron(sig2,I2)));
Symbolic X34 = kron(I2,kron(I2,kron(sig1,sig2)));
Symbolic X41 = kron(sig2,kron(I2,kron(I2,sig1)));
Symbolic K1 = X12 + X23 + X34;
Symbolic K2 = X12 + X23 + X34 + X41;
Symbolic C1 = X12*X23-X23*X12 + X23*X34-X34*X23;
Symbolic C2 = X12*X23-X23*X12 + X23*X34-X34*X23 + X34*X41-X41*X34;
```

```
Symbolic R1 = K1*C1-C1*K1; cout << "R1 = " << R1 << endl;
Symbolic R2 = K1*C1+C1*K1; cout << "R2 = " << R2 << endl;
Symbolic R3 = K2*C2-C2*K1; cout << "R3 = " << R3 << endl;
Symbolic R4 = K2*C2+C2*K2; cout << "R4 = " << R4 << endl;
return 0;
}
```

provides the solutions. The commutator $[\hat{K}_1, C_1]$ is a non-zero matrix and the anticommutator $[\hat{K}_1, C_1]_+$ vanishes, i.e. we obtain the 16×16 zero matrix. The commutator $[\hat{K}_2, C_2]$ is a non-zero matrix and the anticommutator $[\hat{K}_2, C_2]_+$ vanishes, i.e. we obtain the 16×16 zero matrix.

Problem 88. Let $\mathbf{S} = (S_1, S_2, S_3)$ with the spin-1 matrices

$$S_1 = \frac{1}{\sqrt{2}} \begin{pmatrix} 0 & 1 & 0 \\ 1 & 0 & 1 \\ 0 & 1 & 0 \end{pmatrix}, \quad S_2 = \frac{1}{\sqrt{2}} \begin{pmatrix} 0 & -i & 0 \\ i & 0 & -i \\ 0 & i & 0 \end{pmatrix}, \quad S_3 = \begin{pmatrix} 1 & 0 & 0 \\ 0 & 0 & 0 \\ 0 & 0 & -1 \end{pmatrix}$$

with the eigenvalues $+1, 0, -1$. Study the eigenvalue problem for the Hamilton operator

$$\hat{K}(\theta) = \cos(\theta)(\mathbf{S}_1 \cdot \mathbf{S}_2) + \sin(\theta)(\mathbf{S}_1 \cdot \mathbf{S}_2)^2$$

i.e.

$$\hat{K}(\theta) = \cos(\theta)(S_{1,1}S_{2,1} + S_{1,2}S_{2,2} + S_{1,3}S_{2,3}) + \sin(\theta)(S_{1,1}S_{2,1} + S_{1,2}S_{2,2} + S_{1,3}S_{2,3})^2$$

where $S_{1,j} = S_j \otimes I_3$, $S_{2,j} = I_3 \otimes S_j$ $(j = 1, 2, 3)$ and I_3 is the 3×3 identity matrix. Thus $\hat{H}(\theta)$ is hermitian 9×9 matrix.

Solution 88. From

$$\hat{K}(\theta) = \cos(\theta)(S_1 \otimes S_1 + S_2 \otimes S_2 + S_3 \otimes S_3) + \sin(\theta)(S_1 \otimes S_1 + S_2 \otimes S_2 + S_3 \otimes S_3)^2$$

we obtain the 9×9 matrix which can be written as a direct sum

$$(s+c) \oplus \begin{pmatrix} s & 0 & c & 0 & 0 & 0 & 0 \\ 0 & 2s-c & 0 & c-s & 0 & s & 0 \\ c & 0 & s & 0 & 0 & 0 & 0 \\ 0 & c-s & 0 & 2s & 0 & c-s & 0 \\ 0 & 0 & 0 & 0 & s & 0 & c \\ 0 & s & 0 & c-s & 0 & 2s-c & 0 \\ 0 & 0 & 0 & 0 & c & 0 & s \end{pmatrix} \oplus (s+c)$$

where we set $s \equiv \sin(\theta)$ and $c \equiv \cos(\theta)$. The trace of the matrix is $12\sin(\theta)$. The eigenvalues are

$$\sin(\theta) + \cos(\theta) \ (5 \text{ times}), \quad \sin(\theta) - \cos(\theta) \ (3 \text{ times}), \quad 4\sin(\theta) - 2\cos(\theta) \ (1 \text{ time})$$

Problem 89. Consider the spin-$\frac{1}{2}$ matrices

$$S_+ = \begin{pmatrix} 0 & 1 \\ 0 & 0 \end{pmatrix}, \quad S_- = \begin{pmatrix} 0 & 0 \\ 1 & 0 \end{pmatrix} = (S_+)^*, \quad S_3 = \frac{1}{2} \begin{pmatrix} 1 & 0 \\ 0 & -1 \end{pmatrix}$$

with the commutation relations $[S_+, S_-] = 2S_3$, $[S_+, S_3] = -S_+$, $[S_-, S_+] = S_-$ and $S_+^2 = S_-^2 = 0_2$. Consider the 8×8 matrices (counting starts from 0)

$$S_{0,+} = S_+ \otimes I_2 \otimes I_2, \quad S_{1,+} = I_2 \otimes S_+ \otimes I_2, \quad S_{2,+} = I_2 \otimes I_2 \otimes S_+,$$
$$S_{0,-} = S_- \otimes I_2 \otimes I_2, \quad S_{1,-} = I_2 \otimes S_- \otimes I_2, \quad S_{2,-} = I_2 \otimes I_2 \otimes S_-,$$
$$S_{0,3} = S_3 \otimes I_2 \otimes I_2, \quad S_{1,3} = I_2 \otimes S_3 \otimes I_2, \quad S_{2,3} = I_2 \otimes I_2 \otimes S_3.$$

Given the Hamilton operator with cyclic boundary conditions $S_{3,+} \equiv S_{0,+}$, $S_{3,-} \equiv S_{0,-}$

$$\hat{K} = \frac{\hat{H}}{\hbar\omega} = \sum_{j=0}^{2}(S_{j+1,+}S_{j,-} + S_{j+1,-}S_{j,+})$$
$$= S_- \otimes S_+ \otimes I_2 + I_2 \otimes S_- \otimes S_+ + S_+ \otimes I_2 \otimes S_-$$
$$+ S_+ \otimes S_- \otimes I_2 + I_2 \otimes S_+ \otimes S_- + S_- \otimes I_2 \otimes S_+.$$

(i) Find the matrix representation of the Hamilton operator and the eigenvalues and eigenvectors of $\hat{H}$.
(ii) Find the Heisenberg equations of motions for $S_{j,+}$, $S_{j,-}$, $S_{3,j}$ for $j = 0, 1, 2$, i.e.

$$i\hbar\frac{dS_{j,+}}{dt} = [S_{j,+}, \hat{H}](t), \quad i\hbar\frac{dS_{j,-}}{dt} = [S_{j,+}, \hat{H}](t), \quad i\hbar\frac{dS_{j,3}}{dt} = [S_{j,3}, \hat{H}](t).$$

Solution 89. The hermitian 8×8 matrix $\hat{K}$ can be written as a direct sum

$$\hat{K} = (0) \oplus \begin{pmatrix} 0 & 1 & 0 & 1 & 0 & 0 \\ 1 & 0 & 0 & 1 & 0 & 0 \\ 0 & 0 & 0 & 0 & 1 & 1 \\ 1 & 1 & 0 & 0 & 0 & 0 \\ 0 & 0 & 1 & 0 & 0 & 1 \\ 0 & 0 & 1 & 0 & 1 & 0 \end{pmatrix} \oplus (0).$$

The eigenvalues are 2 (2 times), -1 (4 times), 0 (2 times) with the corresponding eigenvectors

$$\begin{pmatrix} 0 \\ 1 \\ 1 \\ 0 \\ 1 \\ 0 \\ 0 \\ 0 \end{pmatrix}, \quad \begin{pmatrix} 0 \\ 0 \\ 0 \\ 1 \\ 0 \\ 1 \\ 1 \\ 0 \end{pmatrix}, \quad \begin{pmatrix} 0 \\ 0 \\ 1 \\ 0 \\ -1 \\ 0 \\ 0 \\ 0 \end{pmatrix}, \quad \begin{pmatrix} 0 \\ 0 \\ 0 \\ 1 \\ 0 \\ -1 \\ 0 \\ 0 \end{pmatrix},$$

$$\begin{pmatrix} 0 \\ 0 \\ 0 \\ 1 \\ 0 \\ 0 \\ -1 \\ 0 \end{pmatrix}, \quad \begin{pmatrix} 0 \\ 0 \\ 0 \\ 0 \\ 0 \\ 1 \\ -1 \\ 0 \end{pmatrix}, \quad \begin{pmatrix} 1 \\ 0 \\ 0 \\ 0 \\ 0 \\ 0 \\ 0 \\ 0 \end{pmatrix}, \quad \begin{pmatrix} 0 \\ 0 \\ 0 \\ 0 \\ 0 \\ 0 \\ 0 \\ 1 \end{pmatrix}.$$

(ii) To find $i\hbar dS_{0,+}/dt = [S_{j,+}, \hat{H}](t)$ we first find the commutator

$$[S_+ \otimes I_2 \otimes I_2, \hat{K}] = (S_+S_- - S_-S_+) \otimes S_+ \otimes I_2 + (S_+S_- - S_-S_+) \otimes I_2 \otimes S_+$$
$$= 2S_3 \otimes S_+ \otimes I_2 + 2S_3 \otimes I_2 \otimes S_+.$$

Using the definitions given above we obtain

$$i\frac{dS_{0,+}}{dt} = 2\omega S_{0,3}(S_{1,+} + S_{2,+})(t).$$

Analogously we find

$$i\frac{dS_{1,+}}{dt} = 2\omega S_{1,3}(S_{0,+} + S_{2,+})(t), \quad i\frac{dS_{2,+}}{dt} = 2\omega S_{2,3}(S_{0,+} + S_{1,+})(t)$$

and

$$i\frac{dS_{0,-}}{dt} = -2\omega S_{0,3}(S_{1,-} + S_{2,-})(t)$$

$$i\frac{dS_{1,-}}{dt} = -2\omega S_{1,3}(S_{0,-} + S_{2,-})(t)$$

$$i\frac{dS_{2,-}}{dt} = -2\omega S_{2,3}(S_{0,-} + S_{1,-})(t).$$

To find the time evolution for $S_{3,0}$ we first have to calculate the commutator

$$[S_3 \otimes I_2 \otimes I_2, \hat{K}] = -S_- \otimes S_+ \otimes I_2 + S_+ \otimes I_2 \otimes S_- + S_+ \otimes S_- \otimes I_2 - S_- \otimes I_2 \otimes S_+.$$

Thus we have

$$i\frac{dS_{0,3}}{dt} = \omega S_{0,+}(S_{1,-} + S_{2,-})(t) - \omega S_{0,-}(S_{2,+} + S_{0,+})(t).$$

Analogously we find

$$i\frac{dS_{1,3}}{dt} = \omega S_{1,+}(S_{2,-} + S_{0,-})(t) - \omega S_{1,-}(S_{2,+} + S_{0,+})(t)$$

$$i\frac{dS_{2,3}}{dt} = \omega S_{2,+}(S_{0,-} + S_{1,-})(t) - \omega S_{2,-}(S_{0,+} + S_{1,+})(t).$$

Thus we have a closed system of matrix differential equations. Solve the system of matrix differential equations.

Problem 90. Let $J, \delta, \gamma \in \mathbb{R}$ and nonzero. Consider the Hamilton operator

$$\hat{H} = \frac{J}{2} \sum_{j=1}^{N}(\gamma\sigma_{1,j} + \sigma_{3,j}\sigma_{3,j+1} - i\delta(I_{2^N} - \sigma_{1,j}))$$

which is a non-hermitian $2^N \times 2^N$ matrix owing to the term $i\delta(I_{2^N} - \sigma_{1,j})$. Periodic boundary conditions are imposed, i.e. $\sigma_{1,N+1} = \sigma_{1,1}$, $\sigma_{2,N+1} = \sigma_{2,1}$, $\sigma_{3,N+1} = \sigma_{2,1}$. Apply the *Jordan-Wigner transformation*

$$\sigma_{1,j} = I - 2c_j^\dagger c_j$$

$$\sigma_{2,j} = i\left(c_j^\dagger - c_j\right)\prod_{\ell < j}\left(I - 2c_\ell^\dagger c_\ell\right)$$

$$\sigma_{3,j} = -\left(c_j + c_j^\dagger\right)\prod_{\ell < j}\left(I - 2c_\ell^\dagger c_\ell\right)$$

to express the Hamilton operator $\hat{H}$ with spinless Fermi creation and annihilation operators $c_j^\dagger$, c_j.

(ii) Apply the discrete Fourier transform

$$c_n = \frac{e^{-i\pi/4}}{\sqrt{N}} \sum_{k \in S} c_k e^{i2\pi kn/N}$$

to this Hamilton operator, where the set S is given by

$$S = \left\{ -\frac{N}{2}, -\frac{N}{2} + 1, \ldots, 0, 1, \ldots, \frac{N}{2} - 1 \right\}$$

Solution 90. (i) We obtain

$$\hat{H} = \frac{J}{2} \sum_{j=1}^{N} (c_j^\dagger c_{j+1} + c_{j+1}^\dagger c_j + c_{j+1} c_j + c_j^\dagger c_{j+1}^\dagger + \gamma(I - 2c_j^\dagger c_j) - 2i\delta c_j^\dagger c_j)$$

with the boundary condition $c_{N+1} = -e^{i\pi \hat{N}} c_1$, where

$$\hat{N} = \sum_{j=1}^{N} c_j^\dagger c_j.$$

(ii) Utilizing $c_k c_k = c_k^\dagger c_k^\dagger = 0$ we find

$$\hat{H} = -\frac{J}{2} \sum_{k \in S} \left(2(\gamma + i\delta - \cos(2\pi k/N))c - k^\dagger c_k - \gamma I + \sin(2\pi k/N)(c_k^\dagger c_{-k}^\dagger + c_{-k} c_k) \right).$$

2.4 Supplementary Problems

Problem 1. Let σ_1, σ_2, σ_3 be the Pauli spin matrices.
(i) Let A, B be two arbitrary 2×2 matrices over $\mathbb{C}$. Is

$$\frac{1}{2}\text{tr}(AB) \equiv \sum_{j=1}^{3} \left(\frac{1}{2}\text{tr}(\sigma_j A)\right)\left(\frac{1}{2}\text{tr}(\sigma_j B)\right)?$$

(ii) Let

$$S_+ := \frac{1}{2}(\sigma_1 + i\sigma_2) = \begin{pmatrix} 0 & 1 \\ 0 & 0 \end{pmatrix}, \qquad S_- := \frac{1}{2}(\sigma_1 - i\sigma_2) = \begin{pmatrix} 0 & 0 \\ 1 & 0 \end{pmatrix}.$$

Find the matrices

$$e^{i\pi\sigma_3/4}S_+e^{-i\pi\sigma_3/4}, \qquad e^{i\pi\sigma_3/4}S_-e^{-i\pi\sigma_3/4}.$$

(iii) Consider the *Hadamard gate*

$$U_H = \frac{1}{\sqrt{2}}(\sigma_1 + \sigma_3) = \frac{1}{\sqrt{2}}\begin{pmatrix} 1 & 1 \\ 1 & -1 \end{pmatrix}.$$

Show that

$$U_H\sigma_1 U_H^{-1} = \sigma_3, \quad U_H\sigma_2 U_H^{-1} = -\sigma_2, \quad U_H\sigma_3 U_H^{-1} = \sigma_1.$$

Note that $U_H^{-1} = U_H$.
(iv) Find all 2×2 hermitian matrices such that $[H, \sigma_1] = 0_2$, $[H, \sigma_2] = 0_2$, $[H, \sigma_3] = 0_2$.

Problem 2. Let σ_1, σ_2, σ_3 be the Pauli spin matrices. Let $z \in \mathbb{C}$.
(i) Calculate

$$\sinh(z\sigma_1), \quad \sinh(z\sigma_2), \quad \sinh(z\sigma_3), \quad \cosh(z\sigma_1), \quad \cosh(z\sigma_2), \quad \cosh(z\sigma_3).$$

(ii) Let S_1, S_2, S_3 be the spin-1 matrices. Calculate

$$\sinh(zS_1), \quad \sinh(zS_2), \quad \sinh(zS_3), \quad \cosh(zS_1), \quad \cosh(zS_2), \quad \cosh(zS_3).$$

Problem 3. (i) Find all 2×2 matrices A such that

$$[A, A^*] = \sigma_3 = \begin{pmatrix} 1 & 0 \\ 0 & -1 \end{pmatrix}.$$

(ii) Find all 3×3 matrices B such that

$$[B, B^*] = \begin{pmatrix} 1 & 0 & 0 \\ 0 & 0 & 0 \\ 0 & 0 & -1 \end{pmatrix}.$$

Problem 4. The *quaternions*

$$\mathbb{H} = \{a1 + bI + cJ + dK : a, b, c, d \in \mathbb{R}\}$$

form a real associative algebra with products specified by Hamilton's formula

$$I^2 = J^2 = K^2 = IJK = -1.$$

The conjugate of a quaternion $x = a1 + bI + cJ + dK$ is defined by $\bar{x} = a1 - bI - cJ - dK$, and its norm $|x|$ is defined by

$$|x|^2 = x\bar{x} = a^2 + b^2 + c^2 + d^2.$$

Show that

$$1 \mapsto \begin{pmatrix} 1 & 0 \\ 0 & 1 \end{pmatrix}, \quad I \mapsto -i \begin{pmatrix} 0 & 1 \\ 1 & 0 \end{pmatrix} = -i\sigma_1,$$

$$J \mapsto -i \begin{pmatrix} 0 & -i \\ i & 0 \end{pmatrix} = -i\sigma_2, \quad K \mapsto -i \begin{pmatrix} 1 & 0 \\ 0 & -1 \end{pmatrix} = -i\sigma_3$$

is a representation.

Problem 5. Let σ_0, σ_1, σ_2, σ_3 be the Pauli spin matrices with $\sigma_0 = I_2$ and 0_2 be the 2×2 zero matrix.
(i) Show that the three *alpha matrices*

$$\alpha_1 = \begin{pmatrix} 0_2 & \sigma_1 \\ \sigma_1 & 0_2 \end{pmatrix}, \quad \alpha_2 = \begin{pmatrix} 0_2 & \sigma_2 \\ \sigma_2 & 0_2 \end{pmatrix}, \quad \alpha_3 = \begin{pmatrix} 0_2 & \sigma_3 \\ \sigma_3 & 0_2 \end{pmatrix}$$

for the Dirac equation cannot be simultaneously diagonalized.
(ii) Consider the 4×4 *gamma matrices*

$$\gamma_1 = \begin{pmatrix} 0_2 & \sigma_1 \\ -\sigma_1 & 0_2 \end{pmatrix}, \quad \gamma_2 = \begin{pmatrix} 0_2 & \sigma_2 \\ -\sigma_2 & 0_2 \end{pmatrix}, \quad \gamma_3 = \begin{pmatrix} 0_2 & \sigma_3 \\ -\sigma_3 & 0_2 \end{pmatrix}$$

and

$$\gamma_0 = \begin{pmatrix} I_2 & 0_2 \\ 0_2 & -I_2 \end{pmatrix}.$$

Find the matrix $\gamma_1\gamma_2\gamma_3\gamma_0$ and $\mathrm{tr}(\gamma_1\gamma_2\gamma_3\gamma_0)$.
(iii) The gamma matrices are the four 4×4 matrices

$$\gamma_k = \begin{pmatrix} 0_2 & \sigma_k \\ -\sigma_k & 0_2 \end{pmatrix}, \quad k = 0, 1, 2, 3$$

where 0_2 is the 2×2 zero matrix. Are the matrices γ_k linearly independent?
(iv) Find the eigenvalues and eigenvectors of the γ_k's.
(v) Are the matrices γ_k invertible. Use the result from (iv). If so, find the inverse.
(vi) Find the commutators $[\gamma_k, \gamma_\ell]$ for $k, \ell = 0, 1, 2, 3$. Find the anticommutators $[\gamma_k, \gamma_\ell]_+$ for $k, \ell = 0, 1, 2, 3$.
(vii) Can the matrices γ_k be written as the Kronecker product of two 2×2 matrices?
(viii) Consider the 4×4 matrix

$$\rho = \frac{1}{4} \sum_{j,k=0}^{3} \alpha_{j,k} \sigma_j \otimes \sigma_k$$

with the real expansion coefficients α_{jk}. What is the conditions on the α_{jk}'s such that ρ is a density matrix, i.e. ρ is a positive semidefinite matrix with $\mathrm{tr}(\rho) = 1$? One can assume that $\alpha_{00} = 1$.

Problem 6. Let σ_1, σ_2, σ_3 be the Pauli spin matrices with the commutators

$$[\sigma_1, \sigma_2] = 2i\sigma_3, \quad [\sigma_2, \sigma_3] = 2i\sigma_1, \quad [\sigma_3, \sigma_1] = 2i\sigma_2$$

and anticommutators $[\sigma_1, \sigma_2]_+ = [\sigma_2, \sigma_3]_+ = [\sigma_3, \sigma_1]_+ = 0_2$. Let A, B be 2×2 matrices. We define the *star operation* as

$$A \star B := \begin{pmatrix} a_{11} & 0 & 0 & a_{12} \\ 0 & b_{11} & b_{12} & 0 \\ 0 & b_{21} & b_{22} & 0 \\ a_{21} & 0 & 0 & a_{22} \end{pmatrix}.$$

(i) Find the commutators $[\sigma_1 \star \sigma_1, \sigma_2 \star \sigma_2]$, $[\sigma_2 \star \sigma_2, \sigma_3 \star \sigma_3]$, $[\sigma_3 \star \sigma_3, \sigma_1 \star \sigma_1]$.
(ii) Find the anticommutators

$$[\sigma_1 \star \sigma_1, \sigma_2 \star \sigma_2]_+, \quad [\sigma_2 \star \sigma_2, \sigma_3 \star \sigma_3]_+, \quad [\sigma_3 \star \sigma_3, \sigma_1 \star \sigma_1]_+.$$

Discuss.

Problem 7. Let σ_1, σ_2, σ_3 be the Pauli spin matrices.
(i) Consider the unitary and hermitian 4×4 matrix

$$R := \sigma_1 \otimes \sigma_2.$$

Apply the *vec-operator* on R, i.e. find $\text{vec}(R)$.
(ii) Calculate the discrete Fourier transform of the vector in $\mathbb{C}^4$. Then apply vec^{-1} to this vector.
(iii) Compare this 4×4 matrix with the 4×4 matrix R. Discuss.
(iv) Let

$$\hat{H} = \hbar\omega(\sigma_1 \otimes \sigma_2 + \sigma_2 \otimes \sigma_1).$$

Calculate $\exp(-i\hat{H}t/\hbar)$.

Problem 8. Let σ_1, σ_2, σ_3 be the Pauli spin matrices.
(i) Find all 2×2 hermitian matrices H such that $\sigma_1 H \sigma_1 = H$, $\sigma_2 H \sigma_2 = H$, $\sigma_3 H \sigma_3 = H$.
(ii) Find all 2×2 hermitian matrices K such that $\sigma_1 K \sigma_1 = K$, $\sigma_2 K \sigma_2 = K$.
(iii) Find all 4×4 hermitian matrices H such that

$$(\sigma_1 \otimes \sigma_1)H(\sigma_1 \otimes \sigma_1) = H, \quad (\sigma_2 \otimes \sigma_2)H(\sigma_2 \otimes \sigma_2) = H, \quad (\sigma_3 \otimes \sigma_3)H(\sigma_3 \otimes \sigma_3) = H.$$

(iv) Find all 4×4 hermitian matrices K such that

$$(\sigma_1 \otimes \sigma_1)K(\sigma_1 \otimes \sigma_1) = K, \quad (\sigma_2 \otimes \sigma_2)K(\sigma_2 \otimes \sigma_2) = K.$$

(v) Let $S_1 = \frac{1}{2}\sigma_1$, $S_2 = \frac{1}{2}\sigma_2$, $S_3 = \frac{1}{2}\sigma_3$ be the spin matrices. Let R be an $n \times n$ matrix over $\mathbb{C}$. Consider

$$V = S_3 \otimes (I_n + R) + iS_2 \otimes (I_n - R).$$

What is the condition on R such that V is unitary? Hint. Calculate VV^*.

Problem 9. Let S_1, S_2, S_3 be the 3×3 spin-1 matrices

$$S_1 = \frac{1}{\sqrt{2}}\begin{pmatrix} 0 & 1 & 0 \\ 1 & 0 & 1 \\ 0 & 1 & 0 \end{pmatrix}, \quad S_2 = \frac{i}{\sqrt{2}}\begin{pmatrix} 0 & -1 & 0 \\ 1 & 0 & -1 \\ 0 & 1 & 0 \end{pmatrix}, \quad S_3 = \begin{pmatrix} 1 & 0 & 0 \\ 0 & 0 & 0 \\ 0 & 0 & -1 \end{pmatrix}.$$

Let $S_+ := S_1 + iS_2$, $S_- := S_1 - iS_2$ with

$$S_+ = \begin{pmatrix} 0 & \sqrt{2} & 0 \\ 0 & 0 & \sqrt{2} \\ 0 & 0 & 0 \end{pmatrix}, \quad S_- = \begin{pmatrix} 0 & 0 & 0 \\ \sqrt{2} & 0 & 0 \\ 0 & \sqrt{2} & 0 \end{pmatrix}.$$

(i) Let $z \in \mathbb{C}$. Find

$$\exp(zS_+ - \bar{z}S_2), \quad \exp(zS_+ - \bar{z}S_-) \begin{pmatrix} 0 \\ 0 \\ 1 \end{pmatrix}.$$

. Is the state normalized?
(ii) Find the commutators $[S_+, S_-]$, $[S_3, S_+]$, $[S_3, S_-]$.
(iii) Show that the commutation relations are preserved under the transformation

$$\tilde{S}_+ = \cos(\theta)S_+ + \sin(\theta)S_- e^{i\pi S_3}, \quad \tilde{S}_- = \cos(\theta)S_- + \sin(\theta)S_+ e^{-i\pi S_3}, \quad \tilde{S}_3 = \frac{1}{2}[\tilde{S}_+, \tilde{S}_-].$$

Problem 10. Let $\sigma_1, \sigma_2, \sigma_3$ be the Pauli spin matrices and $\sigma_0 = I_2$, where I_2 is the 2×2 unit matrix. The generators of the *Clifford algebra* $C\ell(2n, \mathbb{C}) \simeq M(2^n, \mathbb{C})$ in the *Jordan-Wigner representation* are given by the Kronecker product of the Pauli matrices

$$E_{2k-1} = i \underbrace{\sigma_3 \otimes \cdots \otimes \sigma_3}_{k-1} \otimes \sigma_1 \otimes \underbrace{I_2 \otimes \cdots \otimes I_2}_{n-k}$$

$$E_{2k} = i \underbrace{\sigma_3 \otimes \cdots \otimes \sigma_3}_{k-1} \otimes \sigma_2 \otimes \underbrace{I_2 \otimes \cdots \otimes I_2}_{n-k}$$

where $k = 1, \ldots, n$. Thus E_{2k-1} and E_{2k} are elements of the Pauli group.
(i) Find the eigenvalues and eigenvectors of E_{2k-1} and E_{2k}.
(ii) Consider the *gamma matrices*

$$\gamma_0 = -i\sigma_1 \otimes I_2, \quad \gamma_1 = \sigma_2 \otimes \sigma_1, \quad \gamma_2 = \sigma_2 \otimes \sigma_2, \quad \gamma_3 = \sigma_2 \otimes \sigma_3, \quad \gamma_5 = \sigma_3 \otimes I_2.$$

For the construction of the Clifford algebra $C\ell(4n, \mathbb{C}) \simeq M(4^n, \mathbb{C})$ one considers the $4^n \times 4^n$ matrix

$$E_j^{[k]} = \underbrace{\gamma_5 \otimes \cdots \otimes \gamma_5}_{k-1} \otimes \gamma_j \otimes \underbrace{I_4 \otimes \cdots \otimes I_4}_{n-k}$$

where $k = 1, \ldots, n$, $j = 0, \ldots, 3$ and I_4 is the 4×4 unit matrix with $I_4 = I_2 \otimes I_2$. Find the eigenvalues and eigenvectors of $E_j^{[k]}$.
(iii) Show that a representation of a four-dimensional Clifford algebra is given by

$$T_{+1} = \sigma_1 \otimes I_2, \quad T_{+2} = \sigma_3 \otimes \sigma_1, \quad T_{-1} = \sigma_2 \otimes I_2, \quad T_{-2} = \sigma_3 \otimes \sigma_2.$$

Problem 11. Consider the Pauli spin matrices to describe a spin-$\frac{1}{2}$ particle. In the square array of 4×4 matrices

$$I_2 \otimes \sigma_3 \quad \sigma_3 \otimes I_2 \quad \sigma_3 \otimes \sigma_3$$
$$\sigma_1 \otimes I_2 \quad I_2 \otimes \sigma_1 \quad \sigma_1 \otimes \sigma_1$$
$$\sigma_1 \otimes \sigma_3 \quad \sigma_3 \otimes \sigma_1 \quad \sigma_2 \otimes \sigma_2$$

each row and each column is a triad of commuting operators. Consider the hermitian 3×3 matrices to describe a particle with spin-1

$$S_1 := \frac{1}{\sqrt{2}} \begin{pmatrix} 0 & 1 & 0 \\ 1 & 0 & 1 \\ 0 & 1 & 0 \end{pmatrix}, \quad S_2 := \frac{1}{\sqrt{2}} \begin{pmatrix} 0 & -i & 0 \\ i & 0 & -i \\ 0 & i & 0 \end{pmatrix}, \quad S_3 := \begin{pmatrix} 1 & 0 & 0 \\ 0 & 0 & 0 \\ 0 & 0 & -1 \end{pmatrix}.$$

Is in the square array of 9×9 matrices

$$
\begin{array}{ccc}
I_3 \otimes S_3 & S_3 \otimes I_3 & S_3 \otimes S_3 \\
S_1 \otimes I_3 & I_3 \otimes S_1 & S_1 \otimes S_1 \\
S_1 \otimes S_3 & S_3 \otimes S_1 & S_2 \otimes S_2
\end{array}
$$

each row and each column a triad of commuting operators?

Problem 12. Let $|j\rangle$ $(j = 0, 1, \ldots, N)$ be the standard basis in the Hilbert space $\mathbb{C}^{N+1}$, i.e.

$$
|0\rangle = \begin{pmatrix} 1 \\ 0 \\ \vdots \\ 0 \end{pmatrix}, \ldots, \quad |N\rangle = \begin{pmatrix} 0 \\ \vdots \\ 0 \\ 1 \end{pmatrix}.
$$

A *coherent spin state* can be written as

$$
|\theta, \phi\rangle_N = \sum_{j=0}^{N} |j\rangle \binom{N}{j}^{1/2} (\cos(\theta/2))^{N-j} (\sin(\theta/2)e^{i\phi})^j
$$

where θ, ϕ be the two angles in the spherical coordinate system. Write down the spin state for $N+1$, $N = 2$, $N = 3$ and $N = 4$. Find the density matrix $\rho = |\theta, \phi\rangle\langle\phi, \theta|$.

Problem 13. Let σ_1, σ_2, σ_3 be the Pauli spin matrices. Consider the 4×4 matrix

$$
R = a(\lambda, \mu)\sigma_1 \otimes \sigma_1 + b(\lambda, \mu)(\sigma_2 \otimes \sigma_2 + \sigma_3 \otimes \sigma_3)
$$

where

$$
a(\lambda, \mu) = \frac{1}{4}\frac{\lambda^2 + \mu^2}{\lambda^2 - \mu^2}, \qquad b(\lambda, \mu) = \frac{1}{2}\frac{\lambda\mu}{\lambda^2 - \mu^2}.
$$

Does R satisfy the *braid like relation*

$$
(R \otimes I_2)(I_2 \otimes R)(R \otimes I_2) = (I_2 \otimes R)(R \otimes I_2)(I_2 \otimes R)?
$$

Problem 14. Let A, B be $n \times n$ matrices over $\mathbb{C}$. Let $\mathbf{v}$ be a normalized (column) vector in $\mathbb{C}^n$. Let $\langle A \rangle := \mathbf{v}^* A \mathbf{v}$ and $\langle B \rangle := \mathbf{v}^* B \mathbf{v}$. We have the identity

$$
AB \equiv (A - \langle A \rangle I_n)(B - \langle B \rangle I_n) + A\langle B \rangle + B\langle A \rangle - \langle A \rangle\langle B \rangle I_n.
$$

We approximate the $n \times n$ matrix AB as

$$
AB \approx A\langle B \rangle + B\langle A \rangle - \langle A \rangle\langle B \rangle I_n.
$$

Let $n = 2$ and

$$
A = \sigma_1, \quad B = \sigma_2, \quad \mathbf{v} = \frac{1}{\sqrt{2}}\begin{pmatrix} 1 \\ 1 \end{pmatrix}.
$$

Find AB and $A\langle B \rangle + B\langle A \rangle - \langle A \rangle\langle B \rangle I_n$ and the distance (Frobenius norm) between the two matrices. Discuss.

Problem 15. Consider the three spin matrices S_1, S_2, S_3 for spin $s = 1/2$, $s = 1$, $s = 3/2$, $s = 2$, etc. For spin-1/2 we have the 2×2 matrices

$$S_1 = \frac{1}{2} \begin{pmatrix} 0 & 1 \\ 1 & 0 \end{pmatrix}, \quad S_2 = \frac{1}{2} \begin{pmatrix} 0 & -i \\ i & 0 \end{pmatrix}, \quad S_3 = \frac{1}{2} \begin{pmatrix} 1 & 0 \\ 0 & -1 \end{pmatrix}$$

and for spin-1 we have the 4×4 matrices

$$S_1 = \frac{1}{\sqrt{2}} \begin{pmatrix} 0 & 1 & 0 \\ 1 & 0 & 1 \\ 0 & 1 & 0 \end{pmatrix}, \quad S_2 = \frac{1}{\sqrt{2}} \begin{pmatrix} 0 & -i & 0 \\ i & 0 & -i \\ 0 & i & 0 \end{pmatrix}, \quad S_3 = \begin{pmatrix} 1 & 0 & 0 \\ 0 & 0 & 0 \\ 0 & 0 & -1 \end{pmatrix}.$$

The spin-matrices S_1, S_2, S_3 satisfy the commutation relations $[S_1, S_2] = iS_3$, $[S_2, S_3] = iS_1$, $[S_3, S_1] = iS_2$. Consider the hierarchy of spin Hamilton operators

$$\hat{H} = \hbar\omega_{11} S_1 \otimes I_{2s+1} \otimes I_{2s+1} + \hbar\omega_{12} I_{2s+1} \otimes S_2 \otimes I_{2s+1} + \hbar\omega_{13} I_{2s+1} \otimes I_{2s+1} \otimes S_3$$
$$+ \hbar\omega_{21} S_1 \otimes S_2 \otimes I_{2s+1} + \hbar\omega_{22} S_1 \otimes I_{2s+1} \otimes S_3 + \hbar\omega_{23} I_{2s+1} \otimes S_2 \otimes S_3 + \hbar\omega_3 S_1 \otimes S_2 \otimes S_3$$

where I_{2s+1} is the $(2s+1) \times (2s+1)$ identity matrix for the given spin s. Thus the Hamilton operator $\hat{H}$ is a hermitian $(2s+1)^3 \times (2s+1)^3$ matrix with trace equal to 0. Find the eigenvalues and normalized eigenvectors of $\hat{H}$. Then calculate the *partition function*

$$Z(\beta) := \frac{e^{-\beta\hat{H}}}{\text{tr}(e^{-\beta\hat{H}})}.$$

Apply the following: Let A_1, A_2, A_3 be $n \times n$ matrices and I_n the $n \times n$ identity matrix. Let λ_1 be an eigenvalue of A_1 with normalized eigenvector $\mathbf{v}_1$, λ_2 an eigenvalue of A_2 with normalized eigenvector $\mathbf{v}_2$ and λ_3 an eigenvalue of A_3 with normalized eigenvector $\mathbf{v}_3$. Then

$$\mathbf{v} := \mathbf{v}_1 \otimes \mathbf{v}_2 \otimes \mathbf{v}_3$$

is a normalized eigenvector of the $n^3 \times n^3$ matrix

$$\hat{K} = c_{11} A_1 \otimes I_n \otimes I_n + c_{12} I_n \otimes A_2 \otimes I_n + c_{13} I_n \otimes I_n \otimes A_3$$
$$+ c_{21} A_1 \otimes A_2 \otimes I_n + c_{22} A_1 \otimes I_n \otimes A_3 + c_{23} I_n \otimes A_2 \otimes A_3 + c_3 A_1 \otimes A_2 \otimes A_3$$

with

$$\hat{K}\mathbf{v} = (c_{11}\lambda_1 + c_{12}\lambda_2 + c_{13}\lambda_3 + c_{21}\lambda_1\lambda_2 + c_{22}\lambda_1\lambda_3 + c_{23}\lambda_2\lambda_3 + c_3\lambda_1\lambda_2\lambda_3)\mathbf{v}.$$

Problem 16. Consider the vector space $\mathbb{R}^2$ and the standard basis $\{ \mathbf{e}_1, \mathbf{e}_2 \}$.
(i) Show that none of the four vectors

$$\mathbf{e}_1 \otimes \mathbf{e}_1 + \mathbf{e}_2 \otimes \mathbf{e}_2, \quad \mathbf{e}_1 \otimes \mathbf{e}_1 - \mathbf{e}_2 \otimes \mathbf{e}_2, \quad \mathbf{e}_1 \otimes \mathbf{e}_2 + \mathbf{e}_2 \otimes \mathbf{e}_1, \quad \mathbf{e}_1 \otimes \mathbf{e}_2 - \mathbf{e}_2 \otimes \mathbf{e}_1$$

can be written as the Kronecker product of two vectors in $\mathbb{R}^2$.
(ii) Let $\mathbf{e}_1, \ldots, \mathbf{e}_n$ be an orthonormal basis in the vector space $\mathbb{C}^n$. Consider the vectors

$$\mathbf{e}_j \otimes \mathbf{e}_k - \mathbf{e}_k \otimes \mathbf{e}_j, \quad \mathbf{e}_j \otimes \mathbf{e}_k + \mathbf{e}_k \otimes \mathbf{e}_j$$

where $j \neq k$ and $j, k = 1, \ldots, n$ in the vector space $\mathbb{C}^{n^2}$. Show that none of these vectors, say $\mathbf{w}$ can be written as $\mathbf{w} = \mathbf{u} \otimes \mathbf{v}$, where $\mathbf{u}, \mathbf{v} \in \mathbb{C}^n$.

Problem 17. Consider the 2×2 matrices

$$Q := \sigma_3, \quad R := \sigma_2, \quad S := -\frac{1}{\sqrt{2}}(\sigma_2 + \sigma_3), \quad T := -\frac{1}{\sqrt{2}}(-\sigma_2 + \sigma_3)$$

and the entangled state (one of the Bell states)

$$|\psi\rangle = \frac{1}{\sqrt{2}} \begin{pmatrix} 0 \\ 1 \\ -1 \\ 0 \end{pmatrix}.$$

Show that

$$\langle\psi|Q \otimes S|\psi\rangle + \langle\psi|R \otimes S|\psi\rangle + \langle\psi|R \otimes T|\psi\rangle - \langle\psi|Q \otimes T|\psi\rangle = 2\sqrt{2}.$$

Problem 18. Can one find an invertible 8×8 matrix T such that

$$T(I_2 \otimes \sigma_1 \otimes I_2)T^{-1} = I_2 \otimes \sigma_3 \otimes I_2, \quad T(I_2 \otimes \sigma_3 \otimes I_2)T^{-1} = \sigma_3 \otimes \sigma_1 \otimes \sigma_3.$$

Problem 19. Let S_1, S_2, S_3 be the spin matrices for spin $s = 1/2$, $s = 1$, $s = 3/2$, $s = 2$. Find the normalized state

$$\exp(-i\phi S_3) \exp(-i\theta S_2) \exp(-i\psi S_3)|0\rangle$$

where

$$|0\rangle = \begin{pmatrix} 1 \\ 0 \end{pmatrix}, \quad |0\rangle = \begin{pmatrix} 1 \\ 0 \\ 0 \end{pmatrix}, \quad |0\rangle = \begin{pmatrix} 1 \\ 0 \\ 0 \\ 0 \end{pmatrix}, \quad |0\rangle = \begin{pmatrix} 1 \\ 0 \\ 0 \\ 0 \\ 0 \end{pmatrix}$$

for spin $s = 1/2$, $s = 1$, $s = 3/2$, $s = 2$, respectively.

Problem 20. Let $s \in \{1/2, 1, 3/2, 2, \dots\}$ be the spin and S_1, S_2, S_3 be the $(2s+1) \times (2s+1)$ spin matrices with the commutation relations

$$[S_1, S_2] = iS_3, \quad [S_2, S_3] = iS_1, \quad [S_3, S_1] = iS_2$$

and

$$S_1^2 + S_2^2 + S_3^2 = s(s+1)I_{2s+1}$$

where I_{2s+1} is the $(2s+1) \times (2s+1)$ identity matrix. The matrices act on the Hilbert space $\mathbb{C}^{2s+1}$. The Hilbert space $\mathbb{C}^{2s+1}$ is isomorphic to the Hilbert space $\ell_2(\mathbb{Z}_{2s+1})$ where $\mathbb{Z}_{2s+1}$ is the cyclic group of order $2s + 1$. Let $m, n, k \in \{-s, -s+1, \dots, 2s\}$. One defines the linear operators

$$(W_{m,n}\varphi)(k) := \exp\left(-4\frac{i\pi mn}{2s+1} + 4\frac{i\pi nk}{2s+1}\right)\varphi(k - 2m)$$

where the operation $k - 2m$ is mod $(2s+1)$.
(i) Show that the linear operators $W_{m,n}$ are unitary and satisfy

$$W_{m,n}^* = W_{-m,-n}$$

$$W_{m,n}W_{m',n'} = \exp\left(\frac{4i\pi}{2s+1}(m'n - mn')\right)W_{m+m',n+n'}.$$

(ii) Let $s = 1/2$. Find the linear operators $W_{-1/2,-1/2}$, $W_{-1/2,1/2}$, $W_{1/2,-1/2}$, $W_{1/2,1/2}$.

Problem 21. Let σ_1, σ_2, σ_3 be the Pauli spin matrices, I_2 be the 2×2 identity matrix and 0_2 be the 2×2 zero matrix. Let α_1, α_2, α_3, β be the 4×4 matrices

$$\alpha_j = \begin{pmatrix} 0_2 & \sigma_j \\ \sigma_j & 0_2 \end{pmatrix}, \qquad \beta = \begin{pmatrix} I_2 & 0_2 \\ 0_2 & -I_2 \end{pmatrix}$$

where $j = 1, 2, 3$.
(i) Show that these matrices satisfy

$$\alpha_j \alpha_k + \alpha_k \alpha_j = 2\delta_{jk} I_4, \quad j, k = 1, 2, 3$$

$$\alpha_j \beta + \beta \alpha_j = 0_4, \quad j = 1, 2, 3$$

and $\beta^2 = I_4$.
(ii) Let

$$\Psi(t, \mathbf{x}) = \begin{pmatrix} \psi_1(t, \mathbf{x}) \\ \psi_2(t, \mathbf{x}) \\ \psi_3(t, \mathbf{x}) \\ \psi_4(t, \mathbf{x}) \end{pmatrix}.$$

The *Dirac equation* with rest mass m can be written as the 4×4 matrix-valued differential equation

$$i\hbar \frac{\partial}{\partial t} \Psi(t, \mathbf{x}) = \hat{H} \Psi(t, \mathbf{x})$$

where

$$\hat{H} = -i\hbar c \boldsymbol{\alpha} \cdot \nabla + mc^2 \beta \equiv \begin{pmatrix} mc^2 I_2 & -i\hbar c \boldsymbol{\sigma} \cdot \nabla \\ -i\hbar c \boldsymbol{\sigma} \cdot \nabla & -mc^2 I_2 \end{pmatrix}$$

with

$$\boldsymbol{\alpha} \cdot \nabla := \alpha_1 \frac{\partial}{\partial x_1} + \alpha_2 \frac{\partial}{\partial x_2} + \alpha_3 \frac{\partial}{\partial x_3}, \quad \boldsymbol{\sigma} \cdot \nabla := \sigma_1 \frac{\partial}{\partial x_1} + \sigma_2 \frac{\partial}{\partial x_2} + \sigma_3 \frac{\partial}{\partial x_3}.$$

Find the time-evolution of $\Psi^* \Psi$, i.e. calculate

$$\frac{\partial}{\partial t}(\Psi^* \Psi) = \frac{\partial \Psi}{\partial t} \Psi + \Psi^* \frac{\partial \Psi}{\partial t}.$$

Problem 22. Study the spectrum of the Hamilton operator

$$\hat{H} = \hbar\omega(\sigma_1 \otimes \sigma_1 \otimes \sigma_1 + \sigma_2 \otimes \sigma_2 \otimes \sigma_2 + \sigma_3 \otimes \sigma_3 \otimes \sigma_3).$$

First calculate the commutator with

$$K = \sigma_3 \otimes \sigma_3 \otimes I_2 + I_2 \otimes \sigma_3 \otimes \sigma_3 + \sigma_3 \otimes I_2 \otimes \sigma_3.$$

Discuss.

Problem 23. Let $\omega := \exp(2\pi i/4)$. Consider the 4×4 unitary matrices

$$\sigma = \begin{pmatrix} 1 & 0 & 0 & 0 \\ 0 & \omega & 0 & 0 \\ 0 & 0 & \omega^2 & 0 \\ 0 & 0 & 0 & \omega^3 \end{pmatrix}, \qquad \Gamma = \begin{pmatrix} 0 & 0 & 0 & 1 \\ 1 & 0 & 0 & 0 \\ 0 & 1 & 0 & 0 \\ 0 & 0 & 1 & 0 \end{pmatrix}.$$

Let $c > 0$. The four-state *Potts quantum chain* is defined by the Hamilton operator

$$\hat{H} = -\frac{1}{\pi\sqrt{c}} \sum_{j=1}^{N} \left((\sigma_j + \sigma_j^2 + \sigma_j^3) + c(\Gamma_j \Gamma_{j+1}^3 + \Gamma_j^2 \Gamma_{j+1}^2 + \Gamma_j^3 \Gamma_{j+1}) \right)$$

where N is the number of sites and one imposes cyclic boundary conditions $N + 1 \equiv 1$. Let $N = 2$. Find the eigenvalues and eigenvectors of $\hat{H}$. Obviously $\hat{H}$ is a 16×16 matrix.

Problem 24. (i) Consider the 16×16 hermitian matrices (Hamilton operators)

$$\begin{aligned}
H = &\sigma_1 \otimes \sigma_1 \otimes I_2 \otimes I_2 + \sigma_2 \otimes \sigma_2 \otimes I_2 \otimes I_2 + \sigma_3 \otimes \sigma_3 \otimes I_2 \otimes I_2 \\
&+ I_2 \otimes \sigma_1 \otimes \sigma_1 \otimes I_2 + I_2 \otimes \sigma_2 \otimes \sigma_2 \otimes I_2 + I_2 \otimes \sigma_3 \otimes \sigma_3 \otimes I_2 \\
&+ I_2 \otimes I_2 \otimes \sigma_1 \otimes \sigma_1 + I_2 \otimes I_2 \otimes \sigma_2 \otimes \sigma_2 + I_2 \otimes I_2 \otimes \sigma_3 \otimes \sigma_3 \\
&+ \sigma_1 \otimes I_2 \otimes I_2 \otimes \sigma_1 + \sigma_2 \otimes I_2 \otimes I_2 \otimes \sigma_2 + \sigma_3 \otimes I_2 \otimes I_2 \otimes \sigma_3
\end{aligned}$$

and

$$\begin{aligned}
K = &\sigma_1 \otimes I_2 \otimes \sigma_1 \otimes I_2 + \sigma_2 \otimes I_2 \otimes \sigma_2 \otimes I_2 + \sigma_3 \otimes I_2 \otimes \sigma_3 \otimes I_2 \\
&+ I_2 \otimes \sigma_1 \otimes I_2 \otimes \sigma_1 + I_2 \otimes \sigma_2 \otimes I_2 \otimes \sigma_2 + I_2 \otimes \sigma_3 \otimes I_2 \otimes \sigma_3 \\
&+ \sigma_1 \otimes I_2 \otimes \sigma_1 \otimes I_2 + \sigma_2 \otimes I_2 \otimes \sigma_2 \otimes I_2 + \sigma_3 \otimes I_2 \otimes \sigma_3 \otimes I_2.
\end{aligned}$$

Is $[H, K] = 0_{16}$, i.e. do H and K commute? Find the eigenvalues of $\hat{H}$ and $\hat{K}$.
(ii) Find the eigenvalues and eigenvectors of the spin-Hamilton operator (32×32 hermitian matrix)

$$\begin{aligned}
\hat{H} = &J(\sigma_1 \otimes \sigma_1 \otimes I_2 \otimes I_2 \otimes I_2 + \sigma_1 \otimes I_2 \otimes \sigma_1 \otimes I_2 \otimes I_2 \\
&+ \sigma_1 \otimes I_2 \otimes I_2 \otimes \sigma_1 \otimes I_2 + \sigma_1 \otimes I_2 \otimes I_2 \otimes I_2 \otimes \sigma_1) \\
&+ h(\sigma_3 \otimes I_2 \otimes I_2 \otimes I_2 \otimes I_2 + I_2 \otimes \sigma_3 \otimes I_2 \otimes I_2 \otimes I_2 + I_2 \otimes I_2 \otimes \sigma_3 \otimes I_2 \otimes I_2 \\
&+ I_2 \otimes I_2 \otimes I_2 \otimes \sigma_3 \otimes I_2 + I_2 \otimes I_2 \otimes I_2 \otimes I_2 \otimes \sigma_3).
\end{aligned}$$

Problem 25. Find the eigenvalues and eigenvectors of the Hamilton operator

$$\hat{H} = a(\sigma_3 \otimes I_2 \otimes I_2 + I_2 \otimes \sigma_3 \otimes I_2 + I_2 \otimes I_2 \sigma_3) + b(\sigma_1 \otimes \sigma_1 \otimes I_2 + I_2 \otimes \sigma_1 \otimes \sigma_1) + c(\sigma_2 \otimes I_2 \otimes \sigma_2).$$

Problem 26. The spin-1 matrices are given by the hermitian matrices

$$S_1 = \frac{1}{\sqrt{2}} \begin{pmatrix} 0 & 1 & 0 \\ 1 & 0 & 1 \\ 0 & 1 & 0 \end{pmatrix}, \quad S_2 = \frac{1}{\sqrt{2}} \begin{pmatrix} 0 & -i & 0 \\ i & 0 & -i \\ 0 & i & 0 \end{pmatrix}, \quad S_3 = \begin{pmatrix} 1 & 0 & 0 \\ 0 & 0 & 0 \\ 0 & 0 & -1 \end{pmatrix}$$

with the commutation relations $[S_1, S_2] = iS_3$, $[S_2, S_3] = iS_1$, $[S_3, S_1] = iS_2$. The five quadrupole matrices are given by the hermitian matrices

$$U_1 = \begin{pmatrix} 0 & 0 & 1 \\ 0 & 0 & 0 \\ 1 & 0 & 0 \end{pmatrix}, \quad U_2 = \begin{pmatrix} 0 & 0 & -i \\ 0 & 0 & 0 \\ i & 0 & 0 \end{pmatrix},$$

$$V_1 = \frac{1}{\sqrt{2}} \begin{pmatrix} 0 & 1 & 0 \\ 1 & 0 & -1 \\ 0 & -1 & 0 \end{pmatrix}, \quad V_2 = \frac{1}{\sqrt{2}} \begin{pmatrix} 0 & -i & 0 \\ i & 0 & i \\ 0 & -i & 0 \end{pmatrix}, \quad Q_0 = \frac{1}{\sqrt{3}} \begin{pmatrix} 1 & 0 & 0 \\ 0 & -2 & 0 \\ 0 & 0 & 1 \end{pmatrix}.$$

(i) Show that $[U_1, U_2] = 2iS_3$ and $[V_1, V_2] = iS_3$. Show that the eight skew-hermitian matrices

$$iS_1, \; iS_2, \; iS_3, \; iU_1, \; iU_2, \; iV_1, \; iV_2, \; iQ_0$$

form a basis of the semisimple Lie algebra $su(3)$.

(ii) The Hamilton operator for the general quadrupolar interaction of two spin-1 nuclei can be expressed as the hermitian 9×9 matrix

$$\hat{H} = \hbar\omega_0(Q_0 \otimes Q_0) + \hbar\omega_1(V_1 \otimes V_1 + V_2 \otimes V_2) + \hbar\omega_2(U_1 \otimes U_1 + U_2 \otimes U_2).$$

Find the eigenvalues and normalized eigenvectors of $\hat{H}$. Study energy level crossing. Find the 9×9 permutation matrices P such that $P\hat{H}P = \hat{H}$. Study entanglement of the eigenvectors.

Problem 27.   (i) Study the Hamilton operator for the one dimensional spin chain with $N + 1$ lattice points (open end)

$$\hat{H} = -J\sum_{j=0}^{N-1} \sigma_{1,j}\sigma_{1,j+1} - \sum_{j=0}^{N-1} \sqrt{j}\sigma_{3,j}.$$

Show that applying the Jordan-Wigner transformation one obtains the Hamilton operator

$$\hat{H} = -J\sum_{j=0}^{N-1} (c_j^\dagger - c_j)(c_{j+1}^\dagger + c_{j+1}) - \sum_{j=0}^{N-1} \sqrt{j}(2c_j^\dagger c_j - I).$$

(ii) Find the spectrum of the Hamilton operator (open end boundary conditions)

$$\hat{H} = \frac{1}{2}\sum_{j=1}^{N-1} j(\sigma_{1,j}\sigma_{1,j+1} + \sigma_{2,j}\sigma_{2,j+1}).$$

Problem 28.   (i) Let $N \geq 2$ and σ_1, σ_2, σ_3 be the Pauli spin matrices. The XY one-dimensional model with a transversal exterior field and open boundary conditions is given by the hermitian matrix

$$\hat{H} = \sum_{j=1}^{N-1} \mu_j((1+\gamma_j)\sigma_{1,j}\sigma_{1,j+1} + (1-\gamma_j)\sigma_{2,j}\sigma_{2,j+1}) + \sum_{j=1}^{N} \nu_j\sigma_{3,j}.$$

Thus $\hat{H}$ is a hermitian $2^N \times 2^N$ matrix with trace equal to 0. Solve the eigenvalue problem for $N = 2$ and $N = 3$. Extend to higher dimensions.

(ii) Consider the Hamilton operator

$$\hat{H} = J\sum_{k=1}^{N} \sigma_{k,1}\sigma_{k+1,1} + h\sum_{k=1}^{N} \sigma_{k,3}$$

with periodic boundary conditions, i.e. $\sigma_{N+1,3} \equiv \sigma_{1,3}$ and

$$\sigma_{k,1} = I_2 \otimes \cdots \otimes I_2 \otimes \sigma_1 \otimes I_2 \otimes \cdots \otimes I_2$$

where σ_1 is at the k-th position. Thus $\sigma_{k,1}$ is a (hermitian) $2^N \times 2^N$ matrix. Thus the Hamilton operators is also a $2^N \times 2^N$ hermitian matrix. Solve the eigenvalue problem for $N = 3$ and $N = 4$.

(iii) Find the spectrum of the spin Hamilton operator

$$\hat{H} = -\sum_{j=0}^{N-1}(\sigma_{j,3}\sigma_{j+1,3} + \mu_1\sigma_{j,1} + \mu_2\sigma_{j,3})$$

for $N = 2$ and $N = 3$. Study entanglement of the eigenvectors as a function of μ_1 and μ_2.

Problem 29. Let S_1, S_2, S_3 be the spin-1 matrices

$$S_1 = \frac{1}{\sqrt{2}}\begin{pmatrix} 0 & 1 & 0 \\ 1 & 0 & 1 \\ 0 & 1 & 0 \end{pmatrix}, \quad S_2 = \frac{1}{\sqrt{2}}\begin{pmatrix} 0 & -i & 0 \\ i & 0 & -i \\ 0 & i & 0 \end{pmatrix}, \quad S_3 = \begin{pmatrix} 1 & 0 & 0 \\ 0 & 0 & 0 \\ 0 & 0 & -1 \end{pmatrix}$$

with the commutation relations $[S_1, S_2] = iS_3$, $[S_2, S_3] = iS_1$, $[S_3, S_1] = iS_2$. Let $\theta \in [0, \pi/4]$. Consider the Hamilton operator

$$\hat{H}(\theta) = \cos(\theta)\sum_{k=1}^{N}(S_{k,1}S_{k+1} + S_{k,2}S_{k+1,2} + S_{k,3}S_{k+1,3})$$

$$+ \sin(\theta)\sum_{k=1}^{N}(S_{k,1}S_{k+1} + S_{k,2}S_{k+1,2} + S_{k,3}S_{k+1,3})^2$$

with periodic boundary conditions, i.e. $S_{N+1,1} \equiv S_{1,1}$, $S_{N+1,2} \equiv S_{1,2}$, $S_{N+1,3} \equiv S_{1,3}$ and

$$S_{k,1} := I_3 \otimes \cdots I_3 \otimes S_1 \otimes I_3 \otimes \cdots \otimes I_3$$

where the matrix S_1 is at the k-th position. Thus $S_{k,1}$ is a $3^N \times 3^N$ matrix. Let $N = 3$. Find the ground state of the Hamilton operator. Does the Hamilton operator $\hat{H}(\theta)$ commute with

$$\sum_{k=1}^{N}S_{k,3}?$$

Problem 30. (i) A triple spin coupling Hamilton operator is given by

$$\hat{H} = J(\sigma_1 \otimes \sigma_1 \otimes \sigma_1 \otimes I_2 + I_2 \otimes \sigma_1 \otimes \sigma_1 \otimes \sigma_1 + \sigma_1 \otimes I_2 \otimes \sigma_1 \otimes \sigma_1 + \sigma_1 \otimes \sigma_1 \otimes I_2 \otimes \sigma_1).$$

Find the eigenvalues and eigenvectors.
(ii) A Hamilton operator $\hat{H}$ with triple-spin coupling and a transverse field is given by

$$\hat{H} = a(\sigma_1 \otimes \sigma_1 \otimes \sigma_1 \otimes I_2 + I_2 \otimes \sigma_1 \otimes \sigma_1 \otimes \sigma_1)$$
$$+ b(\sigma_3 \otimes I_2 \otimes I_2 \otimes I_2 + I_2 \otimes \sigma_3 \otimes I_2 \otimes I_2 + I_2 \otimes I_2 \otimes \sigma_3 \otimes I_2 + I_2 \otimes I_2 \otimes I_2 \otimes \sigma_3).$$

Find the eigenvalues and eigenvectors of $\hat{H}$.

Problem 31. We identify the states spin and spin down as follows

$$|\uparrow\rangle = \begin{pmatrix} 1 \\ 0 \end{pmatrix}, \quad |\downarrow\rangle = \begin{pmatrix} 0 \\ 1 \end{pmatrix}.$$

We set $|\uparrow\downarrow\uparrow\rangle \equiv |\uparrow\rangle \otimes |\downarrow\rangle \otimes |\uparrow\rangle$ etc. Are the two states

$$|\psi_1\rangle = \frac{1}{\sqrt{6}}(2|\uparrow\downarrow\uparrow\rangle - |\downarrow\uparrow\uparrow\rangle - |\uparrow\uparrow\downarrow\rangle), \quad |\psi_2\rangle = \frac{1}{\sqrt{6}}(-2|\downarrow\uparrow\downarrow\rangle + |\uparrow\downarrow\downarrow\rangle + |\downarrow\downarrow\uparrow\rangle)$$

orthonormal?

Problem 32. Let σ_1, σ_2, σ_3 be the Pauli spin matrices. Let $n \geq 1$ and $j = 1, \ldots, 2n$. The $2n$ matrices of size $2^n \times 2^n$ are defined recursively as

$$\gamma_j^{(n+1)} = \gamma_j^{(n)} \otimes \sigma_3 \quad j = 1, \ldots, 2n$$
$$\gamma_{2n+1}^{(n+1)} = I_{2^n} \otimes \sigma_1, \quad \gamma_{2n+2}^{(n+1)} = I_{2^n} \otimes \sigma_2$$

where $\gamma_1^{(1)} = \sigma_1$ and $\gamma_2^{(1)} = \sigma_2$ and I_{2^n} is the $2^n \times 2^n$ identity matrix. Find $\gamma_1^{(2)}$ and $\gamma_2^{(2)}$.

Problem 33. Let S_1, S_2, S_3 be the spin-$\frac{1}{2}$ matrices. Consider the Hamilton operator

$$\hat{H} = \sum_{j=1}^{2} \hbar\omega_j (S_j \otimes S_j \otimes I_2 \otimes I_2 \otimes I_2 \otimes I_2 + I_2 \otimes S_j \otimes S_j \otimes I_2 \otimes I_2 \otimes I_2$$
$$+ S_j \otimes I_2 \otimes I_2 \otimes S_j \otimes I_2 \otimes I_2 + I_2 \otimes I_2 \otimes S_j \otimes I_2 \otimes S_j \otimes I_2$$
$$+ I_2 \otimes I_2 \otimes I_2 \otimes S_j \otimes S_j \otimes I_2 + I_2 \otimes I_2 \otimes I_2 \otimes S_j \otimes I_2 \otimes S_j + I_2 \otimes I_2 \otimes I_2 \otimes I_2 \otimes S_j \otimes S_j)$$

which describes nearest neighbour interaction for the lattice with six lattice points. Thus $\hat{H}$ is a hermitian 64×64 matrix with trace equal to 0.

Problem 34. Consider the XY-model for the lattice

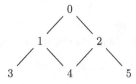

Thus we have six lattice sites and the Hamilton operator is given by the hermitian $2^6 \times 2^6$ matrix

$$\hat{K} = \frac{\hat{H}}{\hbar\omega} = \sum_{j=1}^{2} (\sigma_j \otimes \sigma_j \otimes I_2 \otimes I_2 \otimes I_2 \otimes I_2 + \sigma_j \otimes I_2 \otimes \sigma_j \otimes I_2 \otimes I_2 \otimes I_2$$
$$+ I_2 \otimes \sigma_j \otimes I_2 \otimes \sigma_j \otimes I_2 \otimes I_2 + I_2 \otimes \sigma_j \otimes I_2 \otimes I_2 \otimes \sigma_j \otimes I_2$$
$$+ I_2 \otimes I_2 \otimes \sigma_j \otimes I_2 \otimes \sigma_j \otimes I_2 + I_2 \otimes I_2 \otimes \sigma_j \otimes I_2 \otimes I_2 \otimes \sigma_j).$$

Thus the eigenvalues are real. Actually the matrix is real symmetric. Find the eigenvalues. Note that $\operatorname{tr}(\hat{K}) = 0$, i.e. the sum of the eigenvalues of $\hat{K}$ is equal to 0. The matrix $\hat{K}$ is real symmetric. The *Jacobi method* finds the eigenvalues of a real symmetric matrix by applying a sequence of orthogonal transformations based on the matrix

$$\begin{pmatrix} \cos(\theta) & -\sin(\theta) \\ \sin(\theta) & \cos(\theta) \end{pmatrix}$$

to find the diagonal form of the real symmetric matrix and thus the eigenvalues. Apply the program below `eigenvaluesJacobi.cpp` to find the eigenvalues of $\hat{K}$. The program uses the kron operation (Kronecker product) of SymbolicC++ to find the real symmetric matrix $\hat{K}$ and then the function `rotate` finds the diagonal form. The program also tests whether the eigenvalues add up to 0.

```cpp
// eigenvaluesJacobi.cpp

#include <iostream>
#include <cmath>
#include <cstdio>
#include <cstdlib>
#include "symbolicc++.h"
using namespace std;

void rotate(Symbolic &M,Symbolic &D,int n)
{
   int count = 0;
   double eps = 1e-16;
   while(true)
   {
   double max = 0.0;
   int p = 0; int q = 0;
   for(int i=1;i<n;i++)
     for(int j=0;j<i;j++)
       {
       double h = fabs(double(M(i,j)));
       if(h > max) { max = h; p = i; q = j; }
       }
     if(max < eps) break;
     double theta = (M(q,q)-M(p,p))/(2.0*M(p,q));
     double t = 1.0;
     if(theta > 0.0) t = 1.0/(theta+sqrt(theta*theta+1.0));
     else t = 1.0/(theta-sqrt(theta*theta+1.0));
     double c = 1.0/sqrt(1.0+t*t); double s = t*c;
     M(p,p) = M(p,p)-M(q,p)*t; M(q,q) = M(q,q)+M(q,p)*t; M(p,q) = M(q,p) = 0.0;
     double r = s/(1.0+c);
     for(int j=0;j<n;j++)
     {
     if(j==p || j==q) continue;
     double h = M(p,j)-s*(M(q,j)+r*M(p,j));
     M(q,j) = M(q,j)+s*(M(p,j)-r*M(q,j)); M(p,j) = h;
     }
     for(int i=0;i<n;i++)
     {
     if(i==p || i==q) continue;
     double h = M(i,p)-s*(M(i,q)+r*M(i,p));
     M(i,q) = M(i,q)+s*(M(i,p)-r*M(i,q)); M(i,p) = h;
     }
     count++;
     if(count > 100*n*n*n) { cerr << "Iteration failed"; exit(0); }
   } // end while
   for(int i=0;i<n;i++) D(i) = M(i,i);
} // end method rotate
```

```
int main(void)
{
  using SymbolicConstant::i;
  Symbolic I2("",2,2);
  I2(0,0) = 1.0; I2(0,1) = 0.0; I2(1,0) = 0.0; I2(1,1) = 1.0;
  Symbolic sig1("",2,2);
  sig1(0,0) = 0.0; sig1(0,1) = 1.0; sig1(1,0) = 1.0; sig1(1,1) = 0.0;
  Symbolic sig2("",2,2);
  sig2(0,0) = 0.0; sig2(0,1) = -i; sig2(1,0) = i; sig2(1,1) = 0.0;
  Symbolic K11 = kron(sig1,kron(sig1,kron(I2,kron(I2,kron(I2,I2)))));
  Symbolic K12 = kron(sig2,kron(sig2,kron(I2,kron(I2,kron(I2,I2)))));
  Symbolic K21 = kron(sig1,kron(I2,kron(sig1,kron(I2,kron(I2,I2)))));
  Symbolic K22 = kron(sig2,kron(I2,kron(sig2,kron(I2,kron(I2,I2)))));
  Symbolic K31 = kron(I2,kron(sig1,kron(I2,kron(sig1,kron(I2,I2)))));
  Symbolic K32 = kron(I2,kron(sig2,kron(I2,kron(sig2,kron(I2,I2)))));
  Symbolic K41 = kron(I2,kron(sig1,kron(I2,kron(I2,kron(sig1,I2)))));
  Symbolic K42 = kron(I2,kron(sig2,kron(I2,kron(I2,kron(sig2,I2)))));
  Symbolic K51 = kron(I2,kron(I2,kron(sig1,kron(I2,kron(sig1,I2)))));
  Symbolic K52 = kron(I2,kron(I2,kron(sig2,kron(I2,kron(sig2,I2)))));
  Symbolic K61 = kron(I2,kron(I2,kron(sig1,kron(I2,kron(I2,sig1)))));
  Symbolic K62 = kron(I2,kron(I2,kron(sig2,kron(I2,kron(I2,sig2)))));
  Symbolic K = K11+K12+K21+K22+K31+K32+K41+K42+K51+K52+K61+K62;
  int m = 64;
  Symbolic D("",m);
  rotate(K,D,m);
  cout << "D = " << D << endl; // eigenvalues
  double sumofeig = 0.0;
  for(int j=0;j<m;j++)
  sumofeig = D(j); // check whether sum of eigenvalues is 0
  cout << "sumofeig = " << sumofeig << endl;
  return 0;
}
```

Problem 35. Consider the XY-model for the lattice

with nearest neighbour interaction. There are nine lattice sites and thus the Hamilton operator $\hat{H}$ is a hermitian $2^9 \times 2^9$ matrix with trace equal to 0. Find the eigenvalues.

Problem 36. Let S_1, S_2, S_3 be the spin operators for spin-$\frac{1}{2}$. Consider the three spin Hamilton operators

$$\hat{H}_1 = \sum_{j=1}^{2} \hbar\omega_j (S_j \otimes S_j \otimes I_2 \otimes I_2 + S_j \otimes I_2 \otimes S_j \otimes I_2 + S_j \otimes I_2 \otimes I_2 \otimes S_j)$$

$$\hat{H}_2 = \sum_{j=1}^{2} \hbar\omega_j (S_j \otimes S_j \otimes I_2 \otimes I_2 + I_2 \otimes S_j \otimes S_j \otimes I_2 + I_2 \otimes I_2 \otimes S_j \otimes S_j)$$

$$\hat{H}_3 = \sum_{j=1}^{2} \hbar\omega_j (S_j \otimes S_j \otimes I_2 \otimes I_2 + I_2 \otimes S_j \otimes S_j \otimes I_2 + I_2 \otimes I_2 \otimes S_j \otimes S_j + S_j \otimes I_2 \otimes I_2 \otimes S_j)$$

with nearest neighbour interaction for the following lattices with four vertices

```
     1                                        3 — 2
     |                                        |   |
2 — 0            0 — 1 — 2 — 3                0 — 1
     |
     3
```

Thus the Hamilton operators are 16×16 hermitian matrices with trace equal to 0. Find the spectrum of the Hamilton operators. Compare the ground states for the three different configurations. Study the entanglement of the eigenvectors of the ground states. Discuss.

Problem 37. Let A, H be $n \times n$ hermitian matrices, where H plays the role of the Hamilton operator. The *Heisenberg equation of motion* is given by

$$\frac{dA(t)}{dt} = \frac{i}{\hbar}[H, A(t)].$$

with $A = A(t = 0) = A(0)$ and the solution of the initial value problem

$$A(t) = e^{iHt/\hbar} A e^{-iHt/\hbar}.$$

Let E_j $(j = 1, 2, \ldots, n^2)$ be an orthonormal basis in the *Hilbert space* $\mathcal{H}$ of the $n \times n$ matrices with scalar product

$$\langle X, Y \rangle := \operatorname{tr}(XY^*), \qquad X, Y \in \mathcal{H}.$$

Now $A(t)$ can be expanded using this orthonormal basis as

$$A(t) = \sum_{j=1}^{n^2} c_j(t) E_j$$

and H can be expanded as

$$H = \sum_{j=1}^{n^2} h_j E_j.$$

We find the time evolution for the coefficients $c_j(t)$, i.e. dc_j/dt, where $j = 1, 2, \ldots, n^2$. We have

$$\frac{dA(t)}{dt} = \sum_{j=1}^{n^2} \frac{dc_j}{dt} E_j.$$

Inserting this equation and the expansion for H into the Heisenberg equation of motion we arrive at

$$\sum_{j=1}^{n^2} \frac{dc_j}{dt} E_j = \frac{i}{\hbar} \sum_{k=1}^{n^2} \sum_{j=1}^{n^2} h_k c_j(t) [E_k, E_j].$$

Taking the scalar product of the left and right-hand side of this equation with E_ℓ $(\ell = 1, \ldots, n^2)$ gives

$$\sum_{j=1}^{n^2} \frac{dc_j(t)}{dt} \operatorname{tr}(E_j E_\ell^*) = \frac{i}{\hbar} \sum_{k=1}^{n^2} \sum_{j=1}^{n^2} h_k c_j(t) \operatorname{tr}(([E_k, E_j]) E_\ell^*)$$

where $\ell = 1, 2, \ldots, n^2$. Since $\text{tr}(E_j E_\ell^*) = \delta_{j\ell}$ we obtain

$$\frac{dc_\ell}{dt} = \frac{i}{\hbar} \sum_{k=1}^{n^2} \sum_{j=1}^{n^2} h_k c_j(t) \text{tr}(E_k E_j E_\ell^* - E_j E_k E_\ell^*)$$

where $\ell = 1, 2, \ldots, n^2$. Consider the Hamilton operator

$$\hat{H} = \hbar\omega(\sigma_1 \otimes \sigma_2 \otimes I_2 + I_2 \otimes \sigma_2 \otimes \sigma_3 + \sigma_1 \otimes I_2 \otimes \sigma_3$$

and

$$A = \sigma_1 \otimes \sigma_1 \otimes I_2 + I_2 \otimes \sigma_2 \otimes \sigma_2 + \sigma_3 \otimes I_2 \otimes \sigma_3.$$

(i) Find the time evolution of the $c_j(t)$'s for the standard basis in the Hilbert space of the 8×8 matrices.

(ii) Let

$$X_0 = \frac{1}{\sqrt{2}} \begin{pmatrix} 1 & 0 \\ 0 & 1 \end{pmatrix}, \quad X_1 = \frac{1}{\sqrt{2}} \begin{pmatrix} 0 & 1 \\ 1 & 0 \end{pmatrix}, \quad X_2 = \frac{1}{\sqrt{2}} \begin{pmatrix} 0 & -i \\ i & 0 \end{pmatrix}, \quad X_3 = \frac{1}{\sqrt{2}} \begin{pmatrix} 1 & 0 \\ 0 & -1 \end{pmatrix}$$

be an orthonormal basis in the Hilbert space of the 2×2 matrices. Find the time evolution of the $c_j(t)$'s for the basis given by

$$X_{j_0} \otimes X_{j_1} \otimes X_{j_2}, \qquad j_0, j_1, j_2 \in \{0, 1, 2, 3\}.$$

Apply computer algebra.

Problem 38.   Study the XY-model on the *unit cube* with nearest neighbour interaction. Using the mapping

$$(0,0,0) \to 0, \quad (0,0,1) \to 1, \quad (0,1,0) \to 2, \quad (0,1,1) \to 3$$

$$(1,0,0) \to 4, \quad (1,0,1) \to 5, \quad (1,1,0) \to 6, \quad (1,1,1) \to 7$$

and the 12 interacting pairs

$$(0,1), \quad (0,2), \quad (0,4), \quad (1,3), \quad (1,5), \quad (2,3)$$

$$(2,6), \quad (3,7), \quad (4,5), \quad (4,6), \quad (5,7), \quad (6,7)$$

i.e.

$$((0,0,0),(0,0,1)), \quad ((0,0,0),(0,1,0)), \quad ((0,0,0),(1,0,0)), \quad ((0,0,1),(0,1,1)),$$
$$((0,0,1),(1,0,1)), \quad ((0,1,0),(0,1,1)), \quad ((0,1,0),(1,1,0)), \quad ((0,1,1),(1,1,1)),$$
$$((1,0,0),(1,0,1)), \quad ((1,0,0),(1,1,0)), \quad ((1,0,1),(1,1,1)), \quad ((1,1,0),(1,1,1))$$

the Hamilton operator takes the form

$$\hat{H} = \sum_{j=1}^{2}(\sigma_j \otimes \sigma_j \otimes I_2 \otimes I_2 \otimes I_2 \otimes I_2 \otimes I_2 \otimes I_2 + \sigma_j \otimes I_2 \otimes \sigma_j \otimes I_2 \otimes I_2 \otimes I_2 \otimes I_2 \otimes I_2$$
$$+\sigma_j \otimes I_2 \otimes I_2 \otimes I_2 \otimes \sigma_j \otimes I_2 \otimes I_2 \otimes I_2 + I_2 \otimes \sigma_j \otimes I_2 \otimes \sigma_j \otimes I_2 \otimes I_2 \otimes I_2 \otimes I_2$$
$$+I_2 \otimes \sigma_j \otimes I_2 \otimes I_2 \otimes I_2 \otimes \sigma_j \otimes I_2 \otimes I_2 + I_2 \otimes I_2 \otimes \sigma_j \otimes \sigma_j \otimes I_2 \otimes I_2 \otimes I_2 \otimes I_2$$
$$+I_2 \otimes I_2 \otimes \sigma_j \otimes I_2 \otimes I_2 \otimes I_2 \otimes \sigma_j \otimes I_2 + I_2 \otimes I_2 \otimes I_2 \otimes \sigma_j \otimes I_2 \otimes I_2 \otimes I_2 \otimes \sigma_j$$
$$+I_2 \otimes I_2 \otimes I_2 \otimes I_2 \otimes \sigma_j \otimes \sigma_j \otimes I_2 \otimes I_2 + I_2 \otimes I_2 \otimes I_2 \otimes I_2 \otimes \sigma_j \otimes I_2 \otimes \sigma_j \otimes I_2$$
$$+I_2 \otimes I_2 \otimes I_2 \otimes I_2 \otimes I_2 \otimes \sigma_j \otimes I_2 \otimes \sigma_j + I_2 \otimes I_2 \otimes I_2 \otimes I_2 \otimes I_2 \otimes I_2 \otimes \sigma_j \otimes \sigma_j).$$

Thus $\hat{H}$ is a $2^8 \times 2^8$ hermitian matrix with trace equal to 0.
(i) Does the Hamilton operator $\hat{H}$ commute with

$$\sum_{j=0}^{7} \sigma_{j,3} ?$$

Apply computer algebra.
(ii) Does the Hamilton operator $\hat{H}$ commute with

$$\sigma_3 \otimes \sigma_3 \otimes \sigma_3 \otimes \sigma_3 \otimes \sigma_3 \otimes \sigma_3 \otimes \sigma_3 \otimes \sigma_3 ?$$

Apply computer algebra.

Problem 39. Consider the Hamilton operator

$$\hat{K} = \frac{\hat{H}}{\hbar\omega} = \sigma_1 \cdot \sigma_2 \equiv (\sigma_1 \otimes \sigma_1 + \sigma_2 \otimes \sigma_2 + \sigma_3 \otimes \sigma_3).$$

Show that the Hamilton operator is given by the hermitian 4×4 matrix

$$\hat{K} = \begin{pmatrix} 1 & 0 & 0 & 0 \\ 0 & -1 & 2 & 0 \\ 0 & 2 & -1 & 0 \\ 0 & 0 & 0 & 1 \end{pmatrix}$$

with the eigenvalues -3 (1 times) and 1 (3 times) and the corresponding eigenvectors (for -3)

$$(0 \quad 1 \quad -1 \quad 0)^T$$

and for 1

$$(1 \quad 0 \quad 0 \quad 0)^T, \quad \frac{1}{\sqrt{2}}(0 \quad 1 \quad 1 \quad 0)^T, \quad \frac{1}{\sqrt{2}}(0 \quad 0 \quad 0 \quad 1)^T.$$

Problem 40. (i) Find the eigenvalues of the 4×4 matrices

$$\sigma_1 \otimes \sigma_2 - \sigma_2 \otimes \sigma_1, \quad \sigma_2 \otimes \sigma_3 - \sigma_3 \otimes \sigma_2, \quad \sigma_3 \otimes \sigma_1 - \sigma_1 \otimes \sigma_3.$$

(ii) Find the eigenvalues of the 4×4 matrices

$$\sigma_1 \otimes \sigma_2 + \sigma_2 \otimes \sigma_1, \quad \sigma_2 \otimes \sigma_3 + \sigma_3 \otimes \sigma_2, \quad \sigma_3 \otimes \sigma_1 + \sigma_1 \otimes \sigma_3.$$

Problem 41. Consider the five point XXX model with cyclic boundary conditions

$$\hat{K} = \sum_{j=1}^{3} (\sigma_j \otimes \sigma_j \otimes I_2 \otimes I_2 \otimes I_2 + I_2 \otimes \sigma_j \otimes \sigma_j \otimes I_2 \otimes I_2 + I_2 \otimes I_2 \otimes \sigma_j \otimes \sigma_j \otimes I_2$$
$$+ I_2 \otimes I_2 \otimes I_2 \otimes \sigma_j \otimes \sigma_j + \sigma_j \otimes I_2 \otimes I_2 \otimes I_2 \otimes \sigma_j).$$

(i) Let

$$S_j = \sigma_j \otimes I_2 \otimes I_2 \otimes I_2 \otimes I_2 + I_2 \otimes \sigma_j \otimes I_2 \otimes I_2 \otimes I_2 + I_2 \otimes I_2 \otimes \sigma_j \otimes I_2 \otimes I_2$$
$$+ I_2 \otimes I_2 \otimes I_2 \otimes \sigma_j \otimes I_2 + I_2 \otimes I_2 \otimes I_2 \otimes I_2 \otimes \sigma_j$$

for $j = 1, 2, 3$. Show that K commutes with S_j $(j = 1, 2, 3)$.
(ii) Let

$$\hat{C} = \sum_{j=1}^{3} (\sigma_j \otimes I_2 \otimes \sigma_j \otimes I_2 \otimes I_2 + I_2 \otimes \sigma_j \otimes I_2 \otimes \sigma_j \otimes I_2 + I_2 \otimes I_2 \otimes \sigma_j \otimes I_2 \otimes \sigma_j$$
$$+ \sigma_j \otimes I_2 \otimes I_2 \otimes \sigma_j \otimes I_2 + I_2 \otimes \sigma_j \otimes I_2 \otimes I_2 \otimes \sigma_j).$$

In physics the 32×32 matrix is considered as a *non-local charge*. Show that $[\hat{K}, \hat{C}] = 0_{32}$. Apply computer algebra.

Problem 42. Consider the spin-$\frac{1}{2}$ matrices with $\mathbf{S} = (S_1, S_2, S_3)$ and the Hamilton operator with cyclic boundary condition and four lattice sites

$$\hat{K} = \frac{\hat{H}}{\hbar \omega} = \sum_{j=1}^{4} \mathbf{S}_j \cdot \mathbf{S}_{j+1}.$$

Thus

$$\hat{K} = \sum_{k=1}^{3} (S_k \otimes S_k \otimes I_2 \otimes I_2 + I_2 \otimes S_k \otimes S_k \otimes I_2 + I_2 \otimes I_2 \otimes S_k \otimes S_k + S_k \otimes I_2 \otimes I_2 \otimes S_k).$$

We have a hermitian 16×16 matrix with $\operatorname{tr}(\hat{K}) = 0$. Show that the eigenvalues and normalized eigenvectors are as follows. For the eigenvalue $+1$ we have the five normalized eigenvectors

$$(1, 0, 0, 0, 0, 0, 0, 0, 0, 0, 0, 0, 0, 0, 0, 0)^T$$

$$\frac{1}{2}(0, 1, 1, 0, 1, 0, 0, 0, 1, 0, 0, 0, 0, 0, 0, 0)^T$$

$$\frac{1}{\sqrt{6}}(0, 0, 0, 1, 0, 1, 1, 0, 0, 1, 1, 0, 1, 0, 0, 0)^T$$

$$\frac{1}{2}(0, 0, 0, 0, 0, 0, 0, 1, 0, 0, 0, 1, 0, 1, 1, 0)^T$$

$$(0, 0, 0, 0, 0, 0, 0, 0, 0, 0, 0, 0, 0, 0, 0, 1)^T$$

For the eigenvalue -2 we have one eigenvector

$$\frac{1}{2\sqrt{3}}(0, 0, 0, 1, 0, -2, 1, 0, 0, 1, -2, 0, 1, 0, 0, 0)^T$$

For the eigenvalue -1 we have three eigenvectors

$$\frac{1}{2}(0, 1, -1, 0, 1, 0, 0, 0, -1, 0, 0, 0, 0, 0, 0, 0)^T$$

$$\frac{1}{\sqrt{2}}(0, 0, 0, 0, 0, 1, 0, 0, 0, 0, -1, 0, 0, 0, 0, 0)^T$$

$$\frac{1}{2}(0, 0, 0, 0, 0, 0, 0, 1, 0, 0, 0, -1, 0, 1, -1, 0)^T$$

and for the eigenvalue 0 we have seven eigenvectors

$$\frac{1}{\sqrt{2}}(0, 1, 0, 0, -1, 0, 0, 0, 0, 0, 0, 0, 0, 0, 0, 0)^T$$

$$\frac{1}{\sqrt{2}}(0,0,1,0,0,0,0,0,-1,0,0,0,0,0,0,0)^T$$

$$\frac{1}{\sqrt{2}}(0,0,0,1,0,0,0,0,0,0,0,0,-1,0,0,0)^T$$

$$\frac{1}{\sqrt{2}}(0,0,0,0,0,0,1,0,0,0,0,0,-1,0,0,0)^T$$

$$\frac{1}{\sqrt{2}}(0,0,0,0,0,0,0,1,0,0,0,0,0,-1,0,0)^T$$

$$\frac{1}{\sqrt{2}}(0,0,0,0,0,0,0,0,0,0,1,0,0,-1,0,0,0)^T$$

$$\frac{1}{\sqrt{2}}(0,0,0,0,0,0,0,0,0,0,0,1,0,0,-1,0)^T.$$

Find the eigenvalues and eigenvectors of $\hat{K}$ with the algebraic Bethe ansatz. Discuss.

Problem 43. Consider the *triple spin operator* $K = \sigma_1 \otimes \sigma_2 \otimes \sigma_3$ and

$$T_1 = \sigma_1 \otimes \sigma_1 \otimes \sigma_1, \quad T_2 = \sigma_2 \otimes \sigma_2 \otimes \sigma_2, \quad T_3 = \sigma_3 \otimes \sigma_3 \otimes \sigma_3.$$

Show that $[K, T_1] = [K, T_2] = [K, T_3] = 0_8$. Since $T_1^2 = I_8$ (analogously one can consider $T_2^2 = I_8$ and $T_3^2 = I_8$) we can construct the two projection operators

$$\Pi_1 = \frac{1}{2}(I_8 + T_1), \qquad \Pi_2 = \frac{1}{2}(I_8 - T_1).$$

They project into two four-dimensional subspaces. For Π_1 the basis is given by

$$\frac{1}{\sqrt{2}}(1,0,0,0,0,0,0,1)^T, \quad \frac{1}{\sqrt{2}}(0,1,0,0,0,0,1,0)^T,$$

$$\frac{1}{\sqrt{2}}(0,0,1,0,0,1,0,0)^T, \quad \frac{1}{\sqrt{2}}(0,0,0,1,1,0,0,0)^T.$$

For Π_2 the basis is given by

$$\frac{1}{\sqrt{2}}(1,0,0,0,0,0,0,-1)^T, \quad \frac{1}{\sqrt{2}}(0,1,0,0,0,0,-1,0)^T,$$

$$\frac{1}{\sqrt{2}}(0,0,1,0,0,-1,0,0)^T, \quad \frac{1}{\sqrt{2}}(0,0,0,1,-1,0,0,0)^T.$$

Show that the matrix representation in these two subspaces are given by

$$\begin{pmatrix} 0 & -i & 0 & 0 \\ i & 0 & 0 & 0 \\ 0 & 0 & 0 & i \\ 0 & 0 & -i & 0 \end{pmatrix} = \begin{pmatrix} 0 & -i \\ i & 0 \end{pmatrix} \oplus \begin{pmatrix} 0 & i \\ -i & 0 \end{pmatrix}$$

for Π_1 and

$$\begin{pmatrix} 0 & i & 0 & 0 \\ -i & 0 & 0 & 0 \\ 0 & 0 & 0 & -i \\ 0 & 0 & i & 0 \end{pmatrix} = \begin{pmatrix} 0 & i \\ -i & 0 \end{pmatrix} \oplus \begin{pmatrix} 0 & -i \\ i & 0 \end{pmatrix}$$

for Π_2. Utilize the following SymbolicC++ program for the task.

```cpp
// triplespin2.cpp

#include <iostream>
#include "symbolicc++.h"
using namespace std;

int main(void)
{
using SymbolicConstant::i;
Symbolic sqrt2 = sqrt(Symbolic(2)); // square root of 2
Symbolic I2("I2",2,2);
I2 = I2.identity();   // 2 times 2 identity matrix
Symbolic sig1("sig1",2,2);
sig1(0,0) = 0; sig1(0,1) = 1; sig1(1,0) = 1; sig1(1,1) = 0;
Symbolic sig2("sig2",2,2);
sig2(0,0) = 0; sig2(0,1) = -i; sig2(1,0) = i; sig2(1,1) = 0;
Symbolic sig3("sig3",2,2);
sig3(0,0) = 1; sig3(0,1) = 0; sig3(1,0) = 0; sig3(1,1) = -1;
Symbolic K = kron(sig1,kron(sig2,sig3));
Symbolic T1 = kron(sig1,kron(sig1,sig1));
Symbolic T2 = kron(sig2,kron(sig2,sig2));
Symbolic T3 = kron(sig3,kron(sig3,sig3));
Symbolic C1 = K*T1-T1*K; cout << "C1 = " << C1 << endl;
Symbolic C2 = K*T2-T2*K; cout << "C2 = " << C2 << endl;
Symbolic C3 = K*T3-T3*K; cout << "C3 = " << C3 << endl;
Symbolic v("v",4,8);
v(0,0) = 1/sqrt2; v(0,1) = 0; v(0,2) = 0; v(0,3) = 0;
v(0,4) = 0; v(0,5) = 0; v(0,6) = 0; v(0,7) = 1/sqrt2;
v(1,0) = 0; v(1,1) = 1/sqrt2; v(1,2) = 0; v(1,3) = 0;
v(1,4) = 0; v(1,5) = 0; v(1,6) = 1/sqrt2; v(1,7) = 0;
v(2,0) = 0; v(2,1) = 0; v(2,2) = 1/sqrt2; v(2,3) = 0;
v(2,4) = 0; v(2,5) = 1/sqrt2; v(2,6) = 0; v(2,7) = 0;
v(3,0) = 0; v(3,1) = 0; v(3,2) = 0; v(3,3) = 1/sqrt2;
v(3,4) = 1/sqrt2; v(3,5) = 0; v(3,6) = 0; v(3,7) = 0;
cout << v << endl;
Symbolic vT = v.transpose(); cout << vT << endl;
Symbolic e("e",4,4);
e = v*K*vT; cout << "e = " << e << endl;
Symbolic tr = e.trace(); cout << "tr = " << tr << endl;
Symbolic d = e.determinant(); cout << "d = " << d << endl;
return 0;
}
```

Problem 44. Let $s, n_0 \in \{1/2, 1, 3/2, 2, \ldots\}$ and $s = n_0$ and $|0\rangle$, $|1\rangle$, $\ldots$, $|n_0 + s\rangle$ be the standard basis in $\mathbb{C}^{n_0+s+1}$. Let $\gamma \in \mathbb{R}$. Given s. Is the state

$$|\gamma\rangle = \frac{1}{\sqrt{2s+1}} \sum_{n=n_0-s}^{n_0+s} \exp(in\gamma)|n\rangle$$

normalized?

Problem 45. Let U be a unitary operator on a Hilbert space $\mathcal{H}$. Let Π be the orthogonal projection onto $\{\, v \in \mathcal{H} \,:\, Uv = v \,\}$. Then for any $w \in \mathcal{H}$ one has

$$\lim_{N \to \infty} \frac{1}{N} \sum_{j=0}^{N-1} U^j w = P w$$

where the limit is with respect to the norm implied by the scalar product of the Hilbert space. This is *von Neumann's mean ergodic theorem*. Apply it to the Hilbert space $\mathbb{C}^2$ with $U = \sigma_1$ and

$$\sigma_1 \begin{pmatrix} 1 \\ 1 \end{pmatrix} = \begin{pmatrix} 1 \\ 1 \end{pmatrix}.$$

Apply it to the Hilbert space $\mathbb{C}^4$ with $U = \sigma_1 \otimes \sigma_1$ and

$$(\sigma_1 \otimes \sigma_1)\left(\begin{pmatrix} 1 \\ 1 \end{pmatrix} \otimes \begin{pmatrix} 1 \\ 1 \end{pmatrix}\right) = \begin{pmatrix} 1 \\ 1 \end{pmatrix} \otimes \begin{pmatrix} 1 \\ 1 \end{pmatrix}.$$

Problem 46. Let $N > 2$. Consider a sequence of length N of binary variables (± 1) or Ising spins (with $+1 = \uparrow$ and $-1 = \downarrow$)

$$S = (s_0, s_1, \ldots, s_{N-1}).$$

Thus there are 2^N possible configurations. The autocorrelation function of a given S is defined as

$$C_k(S) := \sum_{j=0}^{N-1} s_j s_{j+k}$$

where all indices are taken modulo N. The *Bernasconi model* is the Hamilton function

$$H(S) = \sum_{k=1}^{N-1} C_k^2(S) \equiv \sum_{i,j=1}^{N-1} \sum_{k=1}^{N-1} s_i s_{i+k} s_j s_{j+k}.$$

So we have a long-range 4-spin interaction. Find the ground state for $N = 3$ and $N = 4$. Write a C++ program that finds the ground state for higher N's by running through all possible (2^N) configurations.

Chapter 3

Fermi Systems

For *Fermi operators* we have to take into account the *Pauli principle*. Let $c_{\mathbf{j}}^{\dagger}$ be Fermi creation operators and let $c_{\mathbf{j}}$ be Fermi annihilation operators, where $\mathbf{j} = (j_1, j_2, \ldots, j_r)$ and j_i $(i = 1, 2, \ldots, r)$ denote the quantum numbers (spin, wave vector, angular momentum, lattice site, etc.). Then we have

$$[c_{\mathbf{j}}^{\dagger}, c_{\mathbf{k}}]_+ \equiv c_{\mathbf{j}}^{\dagger} c_{\mathbf{k}} + c_{\mathbf{k}} c_{\mathbf{j}}^{\dagger} = \delta_{\mathbf{jk}} I$$

$$[c_{\mathbf{j}}, c_{\mathbf{k}}]_+ = [c_{\mathbf{j}}^{\dagger}, c_{\mathbf{k}}^{\dagger}]_+ = 0$$

where $[\,,\,]_+$ denotes the anticommutator and

$$\delta_{\mathbf{jk}} = \delta_{j_1 k_1} \cdots \delta_{j_r k_r}$$

denotes the Kronecker delta and I is the identity operator. A consequence of these equations is

$$c_{\mathbf{j}}^{\dagger} c_{\mathbf{j}}^{\dagger} = 0$$

which describes the Pauli principle, i.e., two particles cannot be in the same state. Here 0 is the zero operator. Thus

$$c_{\mathbf{j}} c_{\mathbf{j}} = 0$$

where 0 is the zero operator. Consider $r = 1$ and $j_1 = 1, 2, \ldots, N$. The states are given by

$$|n_1, n_2, \ldots, n_N\rangle := (c_1^{\dagger})^{n_1} (c_2^{\dagger})^{n_2} \cdots (c_N^{\dagger})^{n_N} |0, \ldots, 0, \ldots, 0\rangle$$

where, because of the Pauli principle, $n_1, n_2, \ldots, n_N \in \{0, 1\}$. We obtain

$$c_j |n_1, \ldots, n_{j-1}, 0, n_{j+1}, \ldots, n_N\rangle = 0 |0, \ldots, 0\rangle$$

and

$$c_j^{\dagger} |n_1, \ldots, n_{j-1}, 1, n_{j+1}, \ldots, n_N\rangle = 0 |0, \ldots, 0\rangle.$$

In the following we write $|\mathbf{0}\rangle \equiv |0, \ldots, 0\rangle$.

3.1   States, Anticommutators, Commutators

Problem 1.   Let I_2 be the 2×2 unit matrix and 0_2 be the 2×2 zero matrix.
(i) Find all 2×2 matrices A and B over $\mathbb{C}$ such that

$$A^2 = 0_2, \quad B^2 = 0_2, \quad [A, B]_+ \equiv AB + BA = I_2, \quad B = A^*.$$

(ii) Show that one can find a 2×2 matrix C over $\mathbb{R}$ such that

$$[C, C^T]_+ \equiv CC^T + C^T C = I_2, \quad C^2 = 0_2$$

where C^T denotes the transpose of C.
(iii) Calculate the vectors

$$C \begin{pmatrix} 1 \\ 0 \end{pmatrix}, \quad C^T \begin{pmatrix} 0 \\ 1 \end{pmatrix}, \quad C \begin{pmatrix} 0 \\ 1 \end{pmatrix}, \quad C^T \begin{pmatrix} 1 \\ 0 \end{pmatrix}.$$

(iv) Find 2×2 matrices X, Y over $\mathbb{C}$ such that

$$X^2 = Y^2 = I_2, \quad [X, Y]_+ \equiv XY + YX = 0_2.$$

Solution 1.   (i) We find the matrices

$$A = \begin{pmatrix} 0 & 0 \\ e^{i\phi} & 0 \end{pmatrix}, \quad B = A^* = \begin{pmatrix} 0 & e^{-i\phi} \\ 0 & 0 \end{pmatrix}$$

where $\phi \in \mathbb{R}$. These matrices are *nonnormal*, i.e. $AA^* \neq A^* A$ and $BB^* \neq B^* B$. However $A + B$ and $A - B$ are normal. Another solution is

$$A = \frac{1}{2} \begin{pmatrix} e^{i\phi_1} & -e^{i(2\phi_1 - \phi_2)} \\ e^{i\phi_2} & -e^{i\phi_1} \end{pmatrix}, \quad B = A^*$$

where $\phi_1, \phi_2 \in \mathbb{R}$. These matrices are also nonnormal.
(ii) One such matrix is the nonnormal matrix

$$C = \begin{pmatrix} 0 & 0 \\ 1 & 0 \end{pmatrix} \Rightarrow C^T = \begin{pmatrix} 0 & 1 \\ 0 & 0 \end{pmatrix}.$$

(iii) The vectors are

$$C \begin{pmatrix} 1 \\ 0 \end{pmatrix} = \begin{pmatrix} 0 \\ 1 \end{pmatrix}, \quad C^T \begin{pmatrix} 0 \\ 1 \end{pmatrix} = \begin{pmatrix} 1 \\ 0 \end{pmatrix}, \quad C \begin{pmatrix} 0 \\ 1 \end{pmatrix} = \begin{pmatrix} 0 \\ 0 \end{pmatrix}, \quad C^T \begin{pmatrix} 1 \\ 0 \end{pmatrix} = \begin{pmatrix} 0 \\ 0 \end{pmatrix}.$$

(iv) We find that $X, Y \in \{ \pm\sigma_1, \pm\sigma_2, \pm\sigma_3 \}$, where σ_1, σ_2, σ_3 are the Pauli spin matrices.

Problem 2.   Find all 2×2 matrices A over $\mathbb{C}$ such that $[A, A^*]_+ = I_2$, $[A, A^*] = \sigma_3$, where $[,]_+$ denotes the anticommutator and $[,]$ denotes the commutator.

Solution 2.   From the two conditions and utilizing addition and subtraction we obtain

$$AA^* = \frac{1}{2}(I_2 + \sigma_3) = \begin{pmatrix} 1 & 0 \\ 0 & 0 \end{pmatrix}, \quad A^* A = \frac{1}{2}(I_2 - \sigma_3) = \begin{pmatrix} 0 & 0 \\ 0 & 1 \end{pmatrix}.$$

Setting

$$A = \begin{pmatrix} r_{11}e^{i\phi_{11}} & r_{12}e^{i\phi_{12}} \\ r_{21}e^{i\phi_{21}} & r_{22}e^{i\phi_{22}} \end{pmatrix}$$

two of the conditions are $r_{21}^2 + r_{22}^2 = 0$, $r_{11}^2 + r_{21}^2 = 0$. This implies $r_{11} = r_{21} = r_{22} = 0$. Furthermore we obtain $r_{12} = 1$. Thus

$$A(\phi) = \begin{pmatrix} 0 & e^{i\phi} \\ 0 & 0 \end{pmatrix} \Rightarrow A^*(\phi) = \begin{pmatrix} 0 & 0 \\ e^{-i\phi} & 0 \end{pmatrix}$$

where we set $\phi = \phi_{12}$.

Problem 3.   (i) Consider the 2×2 matrices A, B

$$A = \begin{pmatrix} 0 & 1 \\ 0 & 0 \end{pmatrix}, \qquad B = \begin{pmatrix} \beta & -\beta^2 \\ 1 & -\beta \end{pmatrix}.$$

Find A^2 and B^2 and the anticommutators $[A, B]_+$, $[A, A]_+$, $[B, B]_+$.
(ii) Consider the 2×2 matrices A, B

$$A = \begin{pmatrix} \alpha & 1 \\ -\alpha^2 & -\alpha \end{pmatrix}, \qquad B = \begin{pmatrix} 0 & 0 \\ 1 & 0 \end{pmatrix}.$$

Find A^2 and B^2 and the anticommutators $[A, B]_+$, $[A, A]_+$, $[B, B]_+$.

Solution 3.   (i) We have $A^2 = 0_2$, $B^2 = 0_2$ and $[A, B]_+ = I_2$, $[A, A]_+ = 0_2$, $[B, B]_+ = 0_2$.
(ii) We have $A^2 = 0_2$, $B^2 = 0_2$ and $[A, B]_+ = I_2$, $[A, A]_+ = 0_2$, $[B, B]_+ = 0_2$.

Problem 4.   Let I_2 be the 2×2 unit matrix. Consider the 4×4 matrices

$$X = \begin{pmatrix} 0 & 0 \\ 1 & 0 \end{pmatrix} \otimes I_2 \ \Rightarrow X^T = \begin{pmatrix} 0 & 1 \\ 0 & 0 \end{pmatrix} \otimes I_2$$

$$Y = \begin{pmatrix} 1 & 0 \\ 0 & -1 \end{pmatrix} \otimes \begin{pmatrix} 0 & 0 \\ 1 & 0 \end{pmatrix} \ \Rightarrow Y^T = \begin{pmatrix} 1 & 0 \\ 0 & -1 \end{pmatrix} \otimes \begin{pmatrix} 0 & 1 \\ 0 & 0 \end{pmatrix}$$

where $\otimes$ denotes the Kronecker product. Find the anticommutators

$$[X, X]_+, \ [X, X^T]_+, \ [Y, Y]_+, \ [Y, Y^T]_+, \ [X, Y]_+, \ [X, Y^T]_+.$$

Solution 4.   Straightforward calculation yields

$$[X, X]_+ = 0_4, \quad [X, X^T]_+ = I_4, \quad [Y, Y]_+ = 0_4,$$

$$[Y, Y^T]_+ = I_4, \quad [X, Y]_+ = 0_4, \quad [X, Y^T]_+ = 0_4$$

where I_4 is the 4×4 identity matrix and 0_4 is the 4×4 zero matrix.

Problem 5.   Let A be a 2×2 matrix with $A^2 = I_2$.
(i) Find the commutator and anticommutator of the 4×4 matrices X and Y

$$X = A \otimes \begin{pmatrix} 0 & 1 \\ 0 & 0 \end{pmatrix}, \qquad Y = A \otimes \begin{pmatrix} 0 & 0 \\ 1 & 0 \end{pmatrix}.$$

(ii) Find the 4×4 matrices X^2 and Y^2.

Solution 5. (i) We have for the commutator

$$[X, Y] = A^2 \otimes \begin{pmatrix} 1 & 0 \\ 0 & 0 \end{pmatrix} - A^2 \otimes \begin{pmatrix} 0 & 0 \\ 0 & 1 \end{pmatrix} = I_2 \otimes \sigma_3$$

and for the anticommutator

$$[X, Y]_+ = A^2 \otimes I_2 = I_2 \otimes I_2 = I_4.$$

(ii) We obtain $X^2 = 0_4$ and $Y^2 = 0_4$.

Problem 6. Consider the four 4×4 matrices

$$A = \begin{pmatrix} 0 & 1 \\ 0 & 0 \end{pmatrix} \otimes I_2 \Rightarrow A^* = \begin{pmatrix} 0 & 0 \\ 1 & 0 \end{pmatrix} \otimes I_2,$$

$$B = \sigma_3 \otimes \begin{pmatrix} 0 & 1 \\ 0 & 0 \end{pmatrix} \Rightarrow B^* = \sigma_3 \otimes \begin{pmatrix} 0 & 0 \\ 1 & 0 \end{pmatrix}.$$

Find the anticommutators $[A, A^*]_+$, $[B, B^*]_+$, $[A, B]_+$, $[A, B^*]_+$, $[A^*, B^*]_+$.

Solution 6. We obtain

$$[A, A^*]_+ = I_2 \otimes I_2, \quad [B, B^*]_+ = I_2 \otimes I_2,$$

$$[A, B]_+ = 0_4, \quad [A, B^*]_+ = 0_4, \quad [A^*, B^*]_+ = 0_4.$$

Problem 7. Let I be the identity operator and $c^\dagger$, c be Fermi creation and annihilation operators with $[c^\dagger, c]_+ = I$, $[c, c]_+ = 0$, $[c^\dagger, c^\dagger]_+ = 0$. A basis is given by

$$\{\, c^\dagger|0\rangle, \quad |0\rangle \,\}$$

and the corresponding dual one by

$$\{\, \langle 0|c, \quad \langle 0| \,\}$$

with $\langle 0|0\rangle = 1$, $c|0\rangle = 0|0\rangle$, $\langle 0|c^\dagger = \langle 0|0$.
(i) Find the matrix representation of $c^\dagger$ and c.
(ii) Find the matrix representation of the *state vectors* $c^\dagger|0\rangle$ and $|0\rangle$.
(iii) The *number operator* $\hat{N}$ is defined by $\hat{N} := c^\dagger c$. Find the matrix representation of $\hat{N}$.
(iv) Apply the operator $c^\dagger + c$ to the normalized states

$$\frac{1}{\sqrt{2}}(c^\dagger|0\rangle + |0\rangle), \quad \frac{1}{\sqrt{2}}(c^\dagger|0\rangle - |0\rangle).$$

Discuss.
(v) Is the hermitian operator $c^\dagger + c$ invertible?
(vi) Consider the linear operators

$$X := \frac{1}{2}(c^\dagger + c), \quad Y := \frac{1}{2}(c^\dagger - c), \quad Z := c^\dagger c - \frac{1}{2}I$$

where I is the identity operator. Calculate the commutators $[X, Y]$, $[Y, Z]$, $[Z, X]$.

(vii) Consider the two-dimensional basis given above. Find the matrix representation of the operators X, Y, Z.
(viii) Find the eigenvalues of these matrices.
(viiii) Find the commutators of these matrices.

Solution 7.   (i) Since $\langle 0|0\rangle = 1$, $\langle 0|cc^\dagger|0\rangle = 1$, $\langle 0|c|0\rangle = 0$, $\langle 0|c^\dagger|0\rangle = 0$, where the vector space under consideration is two-dimensional i.e. $\mathbb{C}^2$. Consequently, $c^\dagger$ and c have the faithful matrix representation

$$c^\dagger = \begin{pmatrix} 0 & 1 \\ 0 & 0 \end{pmatrix} = \frac{1}{2}(\sigma_1 + i\sigma_2) = \frac{1}{2}\sigma_+, \qquad c = \begin{pmatrix} 0 & 0 \\ 1 & 0 \end{pmatrix} = \frac{1}{2}(\sigma_1 - i\sigma_2) = \frac{1}{2}\sigma_-.$$

Here $\sigma_1, \sigma_2, \sigma_3$ are the Pauli matrices and $\sigma_+ = \sigma_1 + i\sigma_2$, $\sigma_- = \sigma_1 - i\sigma_2$.
(ii) The matrix representation for the state vectors are

$$c^\dagger|0\rangle = \begin{pmatrix} 1 \\ 0 \end{pmatrix}, \qquad |0\rangle = \begin{pmatrix} 0 \\ 1 \end{pmatrix}$$

i.e. we have the standard basis.
(iii) The matrix representation of the number operators $\hat{N}$ is

$$\hat{N} := c^\dagger c = \begin{pmatrix} 1 & 0 \\ 0 & 0 \end{pmatrix}$$

with the eigenvalues 0 and 1.
(iv) Since $c^\dagger c^\dagger = 0$, $cc^\dagger = I - c^\dagger c$, $c|0\rangle = 0|0\rangle$ we obtain

$$(c^\dagger + c)\left(\frac{1}{\sqrt{2}}(c^\dagger|0\rangle + |0\rangle)\right) = \frac{1}{\sqrt{2}}c^\dagger|0\rangle + \frac{1}{\sqrt{2}}cc^\dagger|0\rangle = \frac{1}{\sqrt{2}}(c^\dagger|0\rangle + |0\rangle)$$

and

$$(c^\dagger + c)\left(\frac{1}{\sqrt{2}}(c^\dagger|0\rangle - |0\rangle)\right) = -\left(\frac{1}{\sqrt{2}}c^\dagger|0\rangle - \frac{1}{\sqrt{2}}|0\rangle\right).$$

In the first case we have an eigenvalue equation with eigenvalue $+1$ and in the second we have an eigenvalue equation with eigenvalue -1.
(v) Yes. We have

$$(c^\dagger + c)(c^\dagger + c) = c^\dagger c^\dagger + c^\dagger c + cc^\dagger + cc = c^\dagger c + cc^\dagger = I.$$

Thus the inverse of $c^\dagger + c$ is $c^\dagger + c$.
(vi) Using $(c^\dagger)^2 = c^2 = 0$ and $cc^\dagger = I - c^\dagger c$ we find

$$[X, Y] = -Z, \quad [Y, Z] = -X, \quad [Z, X] = Y.$$

(vii) Since $c|0\rangle = 0|0\rangle$, $\langle 0|c^\dagger = \langle 0|0$, $\langle 0|0\rangle = 1$ we obtain the matrix representation

$$X = \frac{1}{2}\begin{pmatrix} 0 & 1 \\ 1 & 0 \end{pmatrix}, \quad Y = \frac{1}{2}\begin{pmatrix} 0 & 1 \\ -1 & 0 \end{pmatrix}, \quad Z = \frac{1}{2}\begin{pmatrix} 1 & 0 \\ 0 & -1 \end{pmatrix}.$$

All three matrices are invertible.
(viii) The eigenvalues of X are $+1/2$, $-1/2$. The eigenvalues of Y are $+i/2$, $-i/2$. The eigenvalues of Z are $+1/2$, $-1/2$.

(viiii) The commutators are

$$[X, Y] = \frac{1}{2} \begin{pmatrix} -1 & 0 \\ 0 & 1 \end{pmatrix} = -Z, \quad [Y, Z] = \frac{1}{2} \begin{pmatrix} 0 & -1 \\ -1 & 0 \end{pmatrix} = -X, \quad [Z, X] = \frac{1}{2} \begin{pmatrix} 0 & 1 \\ -1 & 0 \end{pmatrix} = Y.$$

Problem 8. Let $\hat{N} = c^\dagger c$ be the number operator and $z \in \mathbb{C}$.
(i) Let n be a positive integer with $n \geq 2$. Find

$$(c^\dagger c)^2, \quad (c^\dagger c)^3, \quad \ldots, \quad (c^\dagger c)^n.$$

(ii) Find the commutator $[c^\dagger c, zc^\dagger + \bar{z}c]$. Note that the operators $c^\dagger c$ and $zc^\dagger + \bar{z}c$ are hermitian.
(iii) Find the anticommutator $[c^\dagger c, zc^\dagger + \bar{z}c]_+$.

Solution 8. (i) Since $c^\dagger c^\dagger = 0$ and $cc = 0$ we have

$$(c^\dagger c)^2 = c^\dagger cc^\dagger c = c^\dagger(I - c^\dagger c)c = c^\dagger c.$$

Thus for any positive integer n we have $(c^\dagger c)^n = c^\dagger c$.
(ii) We find for the commutator

$$[c^\dagger c, zc^\dagger + \bar{z}c] = zc^\dagger - \bar{z}c.$$

Note that the operator on the right-hand side is skew-hermitian.
(iii) We find for the anticommutator

$$[c^\dagger c, zc^\dagger + \bar{z}c]_+ = zc^\dagger + \bar{z}c.$$

Problem 9. Let $\hat{N} := c^\dagger c$ be the number operator and I be the identity operator.
(i) Is the operator

$$\hat{\Pi} := (I - \hat{N})$$

a *projection operator?*
(ii) Find $\Pi|0\rangle$ and $\Pi c^\dagger|0\rangle$. Discuss.
(iii) Is $\frac{1}{2}(I + c^\dagger + c)$ a projection operator?

Solution 9. (i) We see that Π is hermitian since $(c^\dagger c)^\dagger = c^\dagger c$. Now with $\hat{N}^2 = \hat{N}$ we find

$$\Pi^2 = (I - \hat{N})(I - \hat{N}) = I - \hat{N} - \hat{N} + \hat{N}^2 = I - \hat{N} = \Pi.$$

Thus Π is a projection operator.
(ii) We obtain

$$\hat{\Pi}|0\rangle = |0\rangle, \qquad \hat{\Pi}c^\dagger|0\rangle = 0.$$

Thus $\hat{\Pi}$ projects into the one-dimensional subspace $\{|0\rangle\}$.
(iii) Yes.

Problem 10. Let $c_\uparrow^\dagger$, $c_\downarrow^\dagger$ be Fermi creation operators with spin up and down, respectively and $c_\uparrow$, $c_\downarrow$ be annihilation operators, respectively. Let

$$\hat{N}_\uparrow := c_\uparrow^\dagger c_\uparrow, \qquad \hat{N}_\downarrow := c_\downarrow^\dagger c_\downarrow$$

be number operators.
(i) Find the operators $(\hat{N}_\uparrow + \hat{N}_\downarrow)^2$, $(\hat{N}_\uparrow - \hat{N}_\downarrow)^2$.
(ii) Consider the state $c_\uparrow^\dagger c_\downarrow^\dagger |0\rangle$. Calculate $\hat{N}_\uparrow c_\uparrow^\dagger c_\downarrow^\dagger |0\rangle$, $\hat{N}_\downarrow c_\uparrow^\dagger c_\downarrow^\dagger |0\rangle$, $\hat{N}_\uparrow \hat{N}_\downarrow c_\uparrow^\dagger c_\downarrow^\dagger |0\rangle$. Discuss.

Solution 10.   (i) Since $\hat{N}_\uparrow \hat{N}_\downarrow = \hat{N}_\downarrow \hat{N}_\uparrow$, $\hat{N}_\uparrow \hat{N}_\uparrow = \hat{N}_\uparrow$ and $\hat{N}_\downarrow \hat{N}_\downarrow = \hat{N}_\downarrow$ we obtain

$$(\hat{N}_\uparrow + \hat{N}_\downarrow)^2 = \hat{N}_\uparrow + \hat{N}_\downarrow + 2\hat{N}_\uparrow \hat{N}_\downarrow, \quad (\hat{N}_\uparrow - \hat{N}_\downarrow)^2 = \hat{N}_\uparrow + \hat{N}_\downarrow - 2\hat{N}_\uparrow \hat{N}_\downarrow.$$

(ii) Using $c_\uparrow |0\rangle = 0|0\rangle$, $c_\downarrow |0\rangle = 0|0\rangle$ we obtain

$$\hat{N}_\uparrow c_\uparrow^\dagger c_\downarrow^\dagger |0\rangle = c_\uparrow^\dagger c_\downarrow^\dagger |0\rangle, \quad \hat{N}_\downarrow c_\uparrow^\dagger c_\downarrow^\dagger |0\rangle = c_\uparrow^\dagger c_\downarrow^\dagger |0\rangle, \quad \hat{N}_\uparrow \hat{N}_\downarrow c_\uparrow^\dagger c_\downarrow^\dagger |0\rangle = c_\uparrow^\dagger c_\downarrow^\dagger |0\rangle.$$

We find the state again. Thus we have three eigenvalue equations for the operators $\hat{N}_\uparrow$, $\hat{N}_\downarrow$, $\hat{N}_\uparrow \hat{N}_\downarrow$ with eigenvalue $+1$.

Problem 11.   Consider the operators

$$\hat{S}_+ := c_\uparrow^\dagger c_\downarrow, \quad \hat{S}_- := c_\downarrow^\dagger c_\uparrow, \quad \hat{S}_3 := \frac{1}{2}(c_\uparrow^\dagger c_\uparrow - c_\downarrow^\dagger c_\downarrow).$$

(i) Find the operator

$$\hat{S}^2 := \hat{S}_3^2 + \frac{1}{2}(\hat{S}_+ \hat{S}_- + \hat{S}_- \hat{S}_+).$$

(ii) Find the commutators $[\hat{S}^2, \hat{S}_+]$, $[\hat{S}^2, \hat{S}_-]$, $[\hat{S}^2, \hat{S}_3]$.
(iii) Let $\hat{N}$ be the number operator

$$\hat{N} = \hat{N}_\uparrow + \hat{N}_\downarrow = c_\uparrow^\dagger c_\uparrow + c_\downarrow^\dagger c_\downarrow.$$

Find the commutator $[\hat{N}, \hat{S}_3]$.

Solution 11.   (i) Since

$$\hat{S}_3^2 = \frac{1}{4}(\hat{N}_\uparrow + \hat{N}_\downarrow - 2\hat{N}_\uparrow \hat{N}_\downarrow), \quad \hat{S}_+ \hat{S}_- + \hat{S}_- \hat{S}_+ = \hat{N}_\uparrow + \hat{N}_\downarrow - 2\hat{N}_\uparrow \hat{N}_\downarrow$$

we end up with

$$\hat{S}^2 = \frac{3}{4}(\hat{N}_\uparrow + \hat{N}_\downarrow) - \frac{3}{2}\hat{N}_\uparrow \hat{N}_\downarrow.$$

(ii) The operator $\hat{S}^2$ commutes with the operators $\hat{S}_+$, $\hat{S}_-$, $\hat{S}_3$. From a Lie algebra point of
view $\hat{S}^2$ is the Casimir operator for the Lie algebra given by $\hat{S}_+$, $\hat{S}_-$, $\hat{S}_3$.
(iii) Since $[\hat{N}_\uparrow, \hat{N}_\downarrow] = 0$ we obtain $[\hat{N}, \hat{S}_3] = 0$.

Problem 12.   (i) Show that $\Pi = I - \hat{N}_\uparrow \hat{N}_\downarrow$ is a *projection operator*.
(ii) Find

$$\Pi c_\uparrow^\dagger c_\downarrow^\dagger |0\rangle, \quad \Pi c_\uparrow^\dagger |0\rangle, \quad \Pi c_\downarrow^\dagger |0\rangle, \quad \Pi |0\rangle.$$

Solution 12.   (i) We have $\Pi = \Pi^\dagger$ and since $(\hat{N}_\uparrow \hat{N}_\downarrow)^2 = \hat{N}_\uparrow \hat{N}_\downarrow$ we arrive at

$$\Pi^2 = (I - \hat{N}_\uparrow \hat{N}_\downarrow)^2 = I - \hat{N}_\uparrow \hat{N}_\downarrow - \hat{N}_\uparrow \hat{N}_\downarrow + \hat{N}_\uparrow \hat{N}_\downarrow = I - \hat{N}_\uparrow \hat{N}_\downarrow = \Pi.$$

Thus Π is a projection operator.

(ii) With $c_\uparrow|0\rangle = 0|0\rangle$ and $c_\downarrow|0\rangle = \langle 0|0$ we find

$$\Pi c_\uparrow^\dagger c_\downarrow^\dagger |0\rangle = 0, \quad \Pi c_\uparrow^\dagger |0\rangle = c_\uparrow^\dagger |0\rangle, \quad \Pi c_\downarrow^\dagger |0\rangle = c_\downarrow^\dagger |0\rangle, \quad \Pi|0\rangle = |0\rangle.$$

Thus Π projects into the three-dimensional subspace $\{c_\uparrow^\dagger|0\rangle, c_\downarrow^\dagger|0\rangle, |0\rangle\}$.

Problem 13. Let $c_\uparrow^\dagger$, $c_\downarrow^\dagger$, $c_\downarrow$, $c_\uparrow$ be Fermi creation and annihilation operators with spin up and down, respectively. Consider the three operators

$$K_+ := c_\uparrow^\dagger c_\downarrow^\dagger, \quad K_- := c_\downarrow c_\uparrow, \quad K_0 := \frac{1}{2}(\hat{N}_\uparrow + \hat{N}_\downarrow - I)$$

where $\hat{N}_\uparrow = c_\uparrow^\dagger c_\uparrow$, $\hat{N}_\downarrow = c_\downarrow^\dagger c_\downarrow$ and I is the identity operator. Note that K_+ and K_- are nonnormal operators.

(i) Calculate the commutators $[K_+, K_-]$, $[K_+, K_0]$, $[K_-, K_0]$.

(ii) Find the commutators $[K_+, c_\uparrow]$, $[K_+, c_\downarrow]$, $[K_-, c_\uparrow^\dagger]$, $[K_-, c_\downarrow^\dagger]$.

Solution 13. (i) Using $(c_\downarrow^\dagger)^2 = c_\downarrow^2 = 0$ and $c_\downarrow c_\downarrow^\dagger = I - c_\downarrow^\dagger c_\downarrow$ we find

$$[K_+, K_-] = -I + \hat{N}_\uparrow + \hat{N}_\downarrow = 2K_0$$

and

$$[K_+, K_0] = -c_\uparrow^\dagger c_\downarrow^\dagger = -K_+, \quad [K_-, K_0] = c_\downarrow c_\uparrow = K_-.$$

(ii) We obtain

$$[K_+, c_\uparrow] = -c_\downarrow^\dagger, \quad [K_+, c_\downarrow] = c_\uparrow^\dagger, \quad [K_-, c_\uparrow^\dagger] = c_\downarrow, \quad [K_-, c_\downarrow^\dagger] = -c_\uparrow.$$

Problem 14. Consider the four dimensional basis

$$c_\uparrow^\dagger c_\downarrow^\dagger |0\rangle, \quad c_\uparrow^\dagger |0\rangle, \quad c_\downarrow^\dagger |0\rangle, \quad |0\rangle$$

with the dual one

$$\langle 0|c_\downarrow c_\uparrow, \quad \langle 0|c_\uparrow, \quad \langle 0|c_\downarrow, \quad \langle 0|.$$

(ii) Find the matrix representation of the states $c_\uparrow^\dagger c_\downarrow^\dagger|0\rangle$, $c_\uparrow^\dagger|0\rangle$, $c_\downarrow^\dagger|0\rangle$, $|0\rangle$.

(ii) Find the matrix representation of the operators $c_\uparrow^\dagger$ and $c_\downarrow^\dagger$.

(iii) Use the result from (ii) to find the matrix representation of $c_\uparrow$ and $c_\downarrow$.

(iv) Use the result from (ii) and (iii) to find the matrix representation of $\hat{N}_\uparrow$, $\hat{N}_\downarrow$, $\hat{N}_\uparrow + \hat{N}_\downarrow$ and $\hat{N}_\uparrow \hat{N}_\downarrow$.

Solution 14. (i) Obviously we obtain the standard basis

$$c_\uparrow^\dagger c_\downarrow^\dagger|0\rangle \mapsto \begin{pmatrix} 1 \\ 0 \\ 0 \\ 0 \end{pmatrix}, \quad c_\uparrow^\dagger|0\rangle \mapsto \begin{pmatrix} 0 \\ 1 \\ 0 \\ 0 \end{pmatrix}, \quad c_\downarrow^\dagger|0\rangle \mapsto \begin{pmatrix} 0 \\ 0 \\ 1 \\ 0 \end{pmatrix}, \quad |0\rangle \mapsto \begin{pmatrix} 0 \\ 0 \\ 0 \\ 1 \end{pmatrix}.$$

(ii) We have to calculate the 16 matrix elements $\langle \psi_j | c_\uparrow^\dagger | \psi_k \rangle$ for the four basis elements $|\psi_j\rangle$ $(j, k = 1, 2, 3, 4)$. For example

$$\langle 0|c_\downarrow c_\uparrow c_\uparrow^\dagger c_\downarrow^\dagger|0\rangle = 1.$$

We obtain the matrix representations

$$c_\uparrow^\dagger \mapsto \begin{pmatrix} 0 & 0 & 1 & 0 \\ 0 & 0 & 0 & 1 \\ 0 & 0 & 0 & 0 \\ 0 & 0 & 0 & 0 \end{pmatrix}, \qquad c_\downarrow^\dagger \mapsto \begin{pmatrix} 0 & -1 & 0 & 0 \\ 0 & 0 & 0 & 0 \\ 0 & 0 & 0 & 1 \\ 0 & 0 & 0 & 0 \end{pmatrix}.$$

(iii) The matrix representations of $c_\uparrow$ and $c_\downarrow$ are given by the transpose and conjugate complex (not necessary here) of the matrix representation of $c_\uparrow^\dagger$ and $c_\downarrow^\dagger$, i.e.

$$c_\uparrow \mapsto \begin{pmatrix} 0 & 0 & 0 & 0 \\ 0 & 0 & 0 & 0 \\ 1 & 0 & 0 & 0 \\ 0 & 1 & 0 & 0 \end{pmatrix}, \qquad c_\downarrow \mapsto \begin{pmatrix} 0 & 0 & 0 & 0 \\ -1 & 0 & 0 & 0 \\ 0 & 0 & 0 & 0 \\ 0 & 0 & 1 & 0 \end{pmatrix}.$$

(iv) Applying matrix multiplication from (i) and (iii) it follows that

$$\hat{N}_\uparrow \mapsto \begin{pmatrix} 1 & 0 & 0 & 0 \\ 0 & 1 & 0 & 0 \\ 0 & 0 & 0 & 0 \\ 0 & 0 & 0 & 0 \end{pmatrix}, \qquad \hat{N}_\uparrow \mapsto \begin{pmatrix} 1 & 0 & 0 & 0 \\ 0 & 0 & 0 & 0 \\ 0 & 0 & 1 & 0 \\ 0 & 0 & 0 & 0 \end{pmatrix},$$

$$\hat{N}_\uparrow + \hat{N}_\downarrow \mapsto \begin{pmatrix} 2 & 0 & 0 & 0 \\ 0 & 1 & 0 & 0 \\ 0 & 0 & 1 & 0 \\ 0 & 0 & 0 & 0 \end{pmatrix}, \qquad \hat{N}_\uparrow \hat{N}_\downarrow \mapsto \begin{pmatrix} 1 & 0 & 0 & 0 \\ 0 & 0 & 0 & 0 \\ 0 & 0 & 0 & 0 \\ 0 & 0 & 0 & 0 \end{pmatrix}.$$

Problem 15. Let $c_\uparrow^\dagger$, $c_\downarrow^\dagger$ be Fermi creation operators with spin up and down, respectively and $c_\uparrow$, $c_\downarrow$ be annihilation operators with spin up and down, respectively. Define the four operators (*hermitian Majorana fermions*)

$$d_1 := c_\uparrow^\dagger + c_\uparrow, \quad d_2 := i(c_\uparrow - c_\uparrow^\dagger), \quad d_3 := c_\downarrow^\dagger + c_\downarrow, \quad d_4 := i(c_\downarrow - c_\downarrow^\dagger).$$

Calculate the anticommutators $[d_j, d_k]_+$.

Solution 15. We obtain $[d_j, d_k]_+ = 2\delta_{jk} I$, where I is the identity operator.

Problem 16. Let $c_1^\dagger$, $c_2^\dagger$ be Fermi creation operators and $\hat{N}_1 = c_1^\dagger c_1$, $\hat{N}_2 = c_2^\dagger c_2$.
(i) Find the commutators $[\hat{N}_1, \hat{N}_1 \hat{N}_2]$, $[\hat{N}_2, \hat{N}_1 \hat{N}_2]$.
(ii) Find the commutators $[c_1^\dagger c_2, c_2^\dagger c_1]$, $[c_1^\dagger c_2, [c_1^\dagger c_2, c_2^\dagger c_1]]$, $[c_2^\dagger c_1, [c_1^\dagger c_2, c_2^\dagger c_1]]$.
(iii) Find the commutators $[c_1^\dagger c_2, c_1^\dagger c_1 c_2^\dagger c_2]$, $[c_2^\dagger c_1, c_1^\dagger c_1 c_2^\dagger c_2]$.
(iv) Consider the operators

$$K_1 = \frac{1}{\sqrt{2}}(c_1^\dagger + c_1), \qquad K_2 = \frac{1}{\sqrt{2}}(c_2^\dagger + c_2)$$

and $H = c_1^\dagger c_2 + c_2^\dagger c_1$. The operators K_1, K_2 and H are hermitian. Find $K_1 K_1^\dagger + K_2 K_2^\dagger$ and $K_1 H K_1^\dagger + K_2 H K_2^\dagger$.

Solution 16. (i) We have $[\hat{N}_1, \hat{N}_1 \hat{N}_2] = [\hat{N}_2, \hat{N}_1 \hat{N}_2] = 0$.

(ii) We obtain

$$[c_1^\dagger c_2, c_2^\dagger c_1] = c_1^\dagger c_1 - c_2^\dagger c_2, \quad [c_1^\dagger c_2, [c_1^\dagger c_2, c_2^\dagger c_1]] = -2c_1^\dagger c_2, \quad [c_2^\dagger c_1, [c_1^\dagger c_2, c_2^\dagger c_1]] = 2c_2^\dagger c_1.$$

(iii) We have $[c_1^\dagger c_2, c_1^\dagger c_1 c_2^\dagger c_2] = 0, \quad [c_2^\dagger c_1, c_1^\dagger c_1 c_2^\dagger c_2] = 0.$
(iv) We obtain
$$K_1 K_1^\dagger + K_2 K_2^\dagger = I, \qquad K_1 H K_1^\dagger + K_2 H K_2^\dagger = 0.$$

Problem 17.   Consider the four-dimensional basis

$$c_2^\dagger c_1^\dagger |0\rangle, \quad c_2^\dagger |0\rangle, \quad c_1^\dagger |0\rangle, \quad |0\rangle.$$

(i) Find the matrix representation of the four nonnormal operators

$$c_2^\dagger, \quad c_1^\dagger, \quad c_2, \quad c_1$$

and show that the 4×4 matrices can be written as Kronecker product of the 2×2 matrices $\sigma_3, I_2, \frac{1}{2}\sigma_+$ and $\frac{1}{2}\sigma_-$.
(ii) Find the matrix representation of the four-dimensional basis.
(iii) Find the matrix representation of the hermitian operator $K = c_2^\dagger c_1^\dagger + c_1 c_2$ and then the eigenvalues of the matrix.

Solution 17.   (i) For the matrix representation of $c_2^\dagger$ we have

$$\langle 0|c_1 c_2 c_2^\dagger c_2^\dagger c_1^\dagger |0\rangle = 0, \quad \langle 0|c_1 c_2 c_2^\dagger c_2^\dagger c_2^\dagger |0\rangle = 0,$$

$$\langle 0|c_1 c_2 c_2^\dagger c_1^\dagger |0\rangle = 1, \quad \langle 0|c_1 c_2 c_2^\dagger |0\rangle = 0,$$

$$\langle 0|c_2 c_2^\dagger c_2^\dagger c_1^\dagger |0\rangle = 0, \quad \langle 0|c_2 c_2^\dagger c_2^\dagger |0\rangle = 0,$$

$$\langle 0|c_2 c_2^\dagger c_1^\dagger |0\rangle = 0, \quad \langle 0|c_2 c_2^\dagger |0\rangle = 1,$$

$$\langle 0|c_1 c_2^\dagger c_2^\dagger c_1^\dagger |0\rangle = 0, \quad \langle 0|c_1 c_2^\dagger c_2^\dagger |0\rangle = 0,$$

$$\langle 0|c_1 c_2^\dagger c_1^\dagger |0\rangle = 0, \quad \langle 0|c_1 c_2^\dagger |0\rangle = 0,$$

$$\langle 0|c_2^\dagger c_2^\dagger c_1^\dagger |0\rangle = 0, \quad \langle 0|c_2^\dagger c_2^\dagger |0\rangle = 0,$$

$$\langle 0|c_2^\dagger c_1^\dagger |0\rangle = 0, \quad \langle 0|c_2^\dagger |0\rangle = 0.$$

Thus the matrix representation of $c_2^\dagger$ is the nonnormal 4×4 matrix

$$c_2^\dagger \mapsto \begin{pmatrix} 0 & 0 & 1 & 0 \\ 0 & 0 & 0 & 1 \\ 0 & 0 & 0 & 0 \\ 0 & 0 & 0 & 0 \end{pmatrix} = \frac{1}{2}\sigma_+ \otimes I_2.$$

Analogously we find for $c_1^\dagger$ the nonnormal 4×4 matrix

$$c_1^\dagger \mapsto \begin{pmatrix} 0 & -1 & 0 & 0 \\ 0 & 0 & 0 & 0 \\ 0 & 0 & 0 & 1 \\ 0 & 0 & 0 & 0 \end{pmatrix} = -\sigma_3 \otimes \frac{1}{2}\sigma_+ = -\begin{pmatrix} 1 & 0 \\ 0 & -1 \end{pmatrix} \otimes \begin{pmatrix} 0 & 1 \\ 0 & 0 \end{pmatrix}.$$

The matrix representation of c_2, c_1 follows from taking the transpose of the 4×4 matrices for $c_2^\dagger$, $c_1^\dagger$, respectively, i.e.

$$c_2 \mapsto \begin{pmatrix} 0 & 0 & 1 & 0 \\ 0 & 0 & 0 & 1 \\ 0 & 0 & 0 & 0 \\ 0 & 0 & 0 & 0 \end{pmatrix} = \frac{1}{2}\sigma_- \otimes I_2 = \begin{pmatrix} 0 & 0 \\ 1 & 0 \end{pmatrix} \otimes \begin{pmatrix} 1 & 0 \\ 0 & 1 \end{pmatrix}$$

$$c_1 \mapsto \begin{pmatrix} 0 & -1 & 0 & 0 \\ 0 & 0 & 0 & 0 \\ 0 & 0 & 0 & 1 \\ 0 & 0 & 0 & 0 \end{pmatrix} = -\sigma_3 \otimes \frac{1}{2}\sigma_- = -\begin{pmatrix} 1 & 0 \\ 0 & -1 \end{pmatrix} \otimes \begin{pmatrix} 0 & 0 \\ 1 & 0 \end{pmatrix}.$$

(ii) Since we have an orthonormal basis the matrix representation is given by the standard basis, i.e.

$$c_2^\dagger c_1^\dagger |0\rangle \mapsto \begin{pmatrix} 1 \\ 0 \\ 0 \\ 0 \end{pmatrix}, \quad c_2^\dagger |0\rangle \mapsto \begin{pmatrix} 0 \\ 1 \\ 0 \\ 0 \end{pmatrix}, \quad c_1^\dagger |0\rangle \mapsto \begin{pmatrix} 0 \\ 0 \\ 1 \\ 0 \end{pmatrix}, \quad |0\rangle \mapsto \begin{pmatrix} 0 \\ 0 \\ 0 \\ 1 \end{pmatrix}.$$

(iii) The matrix representation is given by the hermitian matrix

$$\begin{pmatrix} 0 & 0 & 0 & 1 \\ 0 & 0 & 0 & 0 \\ 0 & 0 & 0 & 0 \\ 1 & 0 & 0 & 0 \end{pmatrix}$$

and the eigenvalues of this matrix are $+1$, -1 and 0 (twice).

Problem 18.   Consider the operator

$$R := (c_1^\dagger + c_1)(c_2^\dagger + c_2) \equiv c_1^\dagger c_2^\dagger + c_1^\dagger c_2 + c_1 c_2^\dagger + c_1 c_2.$$

(i) Show that $R^\dagger = -R$.
(ii) Show that $RR^\dagger = I$, where I is the identity operator.
(iii) Show that

$$Rc_j = -c_j^\dagger R, \qquad Rc_j^\dagger = -c_j R$$

for $j = 1, 2$.

Solution 18.   (i) Since $R = c_1^\dagger c_2^\dagger + c_1^\dagger c_2 + c_1 c_2^\dagger + c_1 c_2$ and $(c_1^\dagger c_2^\dagger)^\dagger = c_2 c_1$, $(c_1^\dagger c_2)^\dagger = c_2^\dagger c_1$ it follows that

$$R^\dagger = c_2 c_1 + c_2^\dagger c_1 + c_2 c_1^\dagger + c_2^\dagger c_1^\dagger = -R.$$

Thus R is skew-hermitian.
(ii) We obtain

$$RR^\dagger = (c_1^\dagger + c_1)(c_2^\dagger + c_2)(c_2 + c_2^\dagger)(c_1 + c_1^\dagger) = (c_1^\dagger + c_1)(c_2^\dagger c_2 + c_2 c_2^\dagger)(c_1 + c_1^\dagger)$$
$$= (c_1^\dagger + c_1)I(c_1 + c_1^\dagger) = c_1^\dagger c_1 + c_1 c_1^\dagger$$
$$= I.$$

(iii) We have

$$Rc_1 = c_1^\dagger c_2^\dagger c_1 + c_1^\dagger c_2 c_1, \qquad Rc_2 = -c_2^\dagger c_1^\dagger c_2 + c_2^\dagger c_2 c_1$$

and

$$c_1^\dagger R = -c_1^\dagger c_2^\dagger c_1 - c_1^\dagger c_2 c_1, \quad c_2^\dagger R = c_2^\dagger c_1^\dagger c_2 - c_2^\dagger c_2 c_1.$$

Analogously we show that $Rc_j^\dagger = -c_j R$.

Problem 19. Consider the hermitian operator

$$K = c_1^\dagger c_2 + c_2^\dagger c_1.$$

(i) Find the commutators $[K, c_1^\dagger c_1]$, $[K, [K, c_1^\dagger c_1]]$.
(ii) Find the commutators $[K, c_2^\dagger c_2]$, $[K, [K, c_2^\dagger c_2]]$.
(iii) Use the results the find the commutator $[K, c_1^\dagger c_1 + c_2^\dagger c_2]$. Discuss.
(iv) Calculate $K^2 = (c_1^\dagger c_2 + c_2^\dagger c_1)^2$.
(v) Find the commutators

$$[c_1, K], \quad [[c_1, K], K], \quad [[[c_1, K], K], K]$$

and

$$[c_2, K], \quad [[c_2, K], K], \quad [[[c_2, K], K], K].$$

Discuss.

Solution 19. (i) We find $[K, c_1^\dagger c_1] = -c_1^\dagger c_2 + c_2^\dagger c_1$. Then

$$[K, [K, c_1^\dagger c_1]] = 2(c_1^\dagger c_1 - c_2^\dagger c_2).$$

(ii) We find

$$[K, c_2^\dagger c_2] = c_1^\dagger c_2 - c_2^\dagger c_1.$$

Then

$$[K, [K, c_2^\dagger c_2]] = 2(-c_1^\dagger c_1 + c_2^\dagger c_2).$$

(iii) Thus $[K, c_1^\dagger c_1] + [K, c_2^\dagger c_2] = [K, c_1^\dagger c_1 + c_2^\dagger c_2] = 0$.
(iv) We have

$$(c_1^\dagger c_2 + c_2^\dagger c_1)^2 = c_1^\dagger c_1 (I - c_2^\dagger c_2) + c_2^\dagger c_2 (I - c_1^\dagger c_1) = \hat{N}_1 + \hat{N}_2 - 2\hat{N}_1 \hat{N}_2.$$

This operator is a projection operator.
(v) We have

$$[c_1, K] = c_2, \quad [[c_1, K], K] = c_1, \quad [[[c_1, K], K], K] = c_2$$

and

$$[c_2, K] = c_1, \quad [[c_2, K], K] = c_2, \quad [[[c_2, K], K], K] = c_1.$$

Problem 20. Let c_j, $c_j^\dagger$ ($j = 1, 2$) be Fermi annihilation and creation operators, respectively. We have $c_j|0\rangle = 0$, where $|0\rangle$ is the *vacuum state*.
(i) Find the state $c_1 c_1^\dagger c_2 c_2^\dagger |0\rangle$. Is the state normalized?
(ii) Find the expectation value $\langle 0|c_1 c_1^\dagger c_2 c_2^\dagger |0\rangle$.

Solution 20. (i) Since $c_j c_j^\dagger = I - c_j^\dagger c_j$, $c_1|0\rangle = 0|0\rangle$ and $c_2|0\rangle = 0|0\rangle$ we have

$$c_1 c_1^\dagger c_2 c_2^\dagger |0\rangle = (I - c_1^\dagger c_1)(I - c_2^\dagger c_2)|0\rangle = |0\rangle.$$

(ii) Using this result we find $\langle 0|c_1 c_1^\dagger c_2 c_2^\dagger|0\rangle = 1$.

Problem 21. Let $\hat{N}_1 := c_1^\dagger c_1$, $\hat{N}_2 := c_2^\dagger c_2$ be number operators.
(i) Is the operator

$$\hat{\Pi} = (I - \hat{N}_1)(I - \hat{N}_2) \equiv I - \hat{N}_1 - \hat{N}_2 + \hat{N}_1 \hat{N}_2$$

a *projection operator?*
(ii) Find

$$\hat{\Pi}|0\rangle, \quad \hat{\Pi}c_1^\dagger|0\rangle, \quad \hat{\Pi}c_2^\dagger|0\rangle, \quad \hat{\Pi}c_2^\dagger c_1^\dagger|0\rangle.$$

Discuss.

Solution 21. (i) We have $\hat{\Pi}^\dagger = \hat{\Pi}$ and

$$\hat{\Pi}\hat{\Pi} = I - \hat{N}_1 - \hat{N}_2 + \hat{N}_1 \hat{N}_2 = \hat{\Pi}.$$

Thus Π is a projection operator.
(ii) We obtain

$$\hat{\Pi}|0\rangle = |0\rangle, \quad \hat{\Pi}c_1^\dagger|0\rangle = 0|0\rangle, \quad \hat{\Pi}c_2^\dagger|0\rangle = 0|0\rangle, \quad \hat{\Pi}c_2^\dagger c_1^\dagger|0\rangle = 0|0\rangle.$$

Thus $\hat{\Pi}$ projects into the one-dimensional subspace $\{\,|0\rangle\,\}$.

Problem 22. Let $j \in \mathbb{Z}$ and

$$X := c_{j+1}^\dagger c_j + c_j^\dagger c_{j+1}, \qquad Y := c_{j-1}^\dagger c_j + c_j^\dagger c_{j-1}.$$

Find the commutator $[X, Y]$.

Solution 22. We have

$$XY = c_{j+1}^\dagger c_{j-1} - c_{j+1}^\dagger c_{j-1} c_j^\dagger c_j - c_{j-1}^\dagger c_{j+1} c_j^\dagger c_j$$

$$YX = c_{j-1}^\dagger c_{j+1} - c_{j-1}^\dagger c_{j+1} c_j^\dagger c_j - c_{j+1}^\dagger c_{j-1} c_j^\dagger c_j.$$

It follows that

$$[X, Y] = c_{j+1}^\dagger c_{j-1} - c_{j-1}^\dagger c_{j+1}.$$

Problem 23. Let $c_1^\dagger$, $c_2^\dagger$, $c_3^\dagger$ be Fermi creation operators.
(i) Find the commutator $[c_1^\dagger c_2^\dagger, c_1 c_2]$.
(ii) Find the commutator $[c_1^\dagger c_2^\dagger c_3^\dagger, c_1 c_2 c_3]$.
(iii) Find the commutators $[c_1^\dagger c_2, c_2^\dagger c_3]$, $[c_2^\dagger c_3, c_3^\dagger c_1]$. Discuss.

Solution 23. (i) We have

$$[c_1^\dagger c_2^\dagger, c_1 c_2] = I - c_1^\dagger c_1 - c_2^\dagger c_2 \equiv I - \hat{N}_1 - \hat{N}_2.$$

(ii) We have

$$[c_1^\dagger c_2^\dagger c_3^\dagger, c_1 c_2 c_3] = I - c_1^\dagger c_1 - c_2^\dagger c_2 - c_3^\dagger c_3 + c_1^\dagger c_1 c_2^\dagger c_2 + c_1^\dagger c_1 c_3^\dagger c_3 + c_2^\dagger c_2 c_3^\dagger c_3 - 2c_1^\dagger c_1 c_2^\dagger c_2 c_3^\dagger c_3.$$

(iii) We obtain

$$[c_1^\dagger c_2, c_2^\dagger c_3] = c_1^\dagger c_3, \qquad [c_2^\dagger c_3, c_3^\dagger c_1] = c_2^\dagger c_1.$$

Problem 24. Let σ_1, σ_2, σ_3 be the Pauli spin matrices. Consider the eight unitary and hermitian 16×16 matrices

$$\gamma_1 = \sigma_2 \otimes \sigma_2 \otimes \sigma_2 \otimes \sigma_2, \quad \gamma_2 = I_2 \otimes \sigma_1 \otimes \sigma_2 \otimes \sigma_2$$
$$\gamma_3 = I_2 \otimes \sigma_3 \otimes \sigma_2 \otimes \sigma_2, \quad \gamma_4 = \sigma_1 \otimes \sigma_2 \otimes I_2 \otimes \sigma_2$$
$$\gamma_5 = \sigma_3 \otimes \sigma_2 \otimes I_2 \otimes \sigma_2, \quad \gamma_6 = \sigma_2 \otimes I_2 \otimes \sigma_1 \otimes \sigma_2$$
$$\gamma_7 = \sigma_2 \otimes I_2 \otimes \sigma_3 \otimes \sigma_2, \quad \gamma_8 = I_2 \otimes I_2 \otimes I_2 \otimes \sigma_1.$$

We define

$$c_j := \frac{1}{2}(\gamma_{2j} + i\gamma_{2j-1}), \qquad c_j^\dagger = \frac{1}{2}(\gamma_{2j} - i\gamma_{2j-1}).$$

Find the anticommutators $(j, k = 1, 2, 3, 4)$ $[c_j, c_k]_+$, $[c_j, c_k^\dagger]_+$.

Solution 24. We obtain $[c_j, c_k]_+ = 0_{16}$ and $[c_j, c_k^\dagger]_+ = \delta_{jk} I_{16}$.

Problem 25. Let $c_j^\dagger$, c_j $(j = 1, \ldots, n)$ be Fermi creation and annihilation operators. Find

$$\langle 0 | c_j c_k^\dagger | 0 \rangle.$$

Solution 25. Since $c_j c_k^\dagger = \delta_{jk} I - c_k^\dagger c_j$ and $c_j | 0 \rangle = 0 | 0 \rangle$ we obtain $\langle 0 | c_j c_k^\dagger | 0 \rangle = \delta_{jk}$, where δ_{jk} is the Kronecker delta.

Problem 26. Let $c_j^\dagger$, c_j $(j = 1, \ldots, n)$ be Fermi creation and annihilation operators, i.e.

$$[c_j, c_k]_+ = 0, \qquad [c_j^\dagger, c_k^\dagger]_+ = 0, \qquad [c_j, c_k^\dagger]_+ = \delta_{jk} I.$$

Let

$$c_k \mapsto c_k' = c_k - i \sum_{j=1}^n (\nu_{kj} c_j + \lambda_{kj} c_j^\dagger), \quad c_k^\dagger \mapsto c_k^{\dagger} - i \sum_{j=1}^n (\nu_{jk} c_j^\dagger + \mu_{jk} c_j).$$

Find the condition on the complex coefficients ν_{kj}, λ_{kj}, ν_{jk} such that c_j' and $c_j'^\dagger$ also satisfy the anticommutation relations given above.

Solution 26. We find $\nu_{jk}^* = \nu_{kj}$, $\lambda_{kj}^* = \mu_{jk}$, $\lambda_{kj} = \lambda_{jk}$.

Problem 27. Consider the four hermitian operators

$$T_{01} = c_0^\dagger c_1 + c_1^\dagger c_0, \quad T_{12} = c_1^\dagger c_2 + c_2^\dagger c_1, \quad T_{23} = c_2^\dagger c_3 + c_3^\dagger c_2, \quad T_{30} = c_3^\dagger c_0 + c_0^\dagger c_3.$$

(i) Find the commutators $[T_{01}, T_{12}]$, $[T_{12}, T_{23}]$, $[T_{23}, T_{30}]$, $[T_{30}, T_{01}]$.
(ii) Do the operators commute with the number operator

$$\hat{N} = c_0^\dagger c_0 + c_1^\dagger c_1 + c_2^\dagger c_2 + c_3^\dagger c_3 ?$$

Solution 27.   (i) We obtain

$$[T_{01}, T_{12}] = c_0^\dagger c_2 - c_2^\dagger c_0, \qquad [T_{12}, T_{23}] = c_1^\dagger c_3 - c_3^\dagger c_1,$$

$$[T_{23}, T_{30}] = c_2^\dagger c_0 - c_0^\dagger c_2, \qquad [T_{30}, T_{01}] = c_3^\dagger c_1 - c_1^\dagger c_3.$$

(ii) Yes.

Problem 28.   Let $c_j^\dagger$ $(j = 0, 1, \ldots, n-1)$ be Fermi creation and annihilation operators. A state can be written as

$$(c_{n-1}^\dagger)^{j_{n-1}} (c_{n-2}^\dagger)^{j_{n-2}} \cdots (c_1^\dagger)^{j_1} (c_0^\dagger)^{j_0} |0\rangle$$

where $j_{n-1}, \ldots, j_0 \in \{0, 1\}$. Let $n = 4$. Find the state $c_0 c_3 c_3^\dagger c_2^\dagger c_1^\dagger c_0^\dagger |0\rangle$.

Solution 28.   Since $c_0 c_0^\dagger = I - c_0^\dagger c_0$ and $c_3 c_3^\dagger = I - c_3^\dagger c_3$ we obtain the normalized state

$$c_0 c_3 c_3^\dagger c_2^\dagger c_1^\dagger c_0^\dagger |0\rangle. = c_3 c_3^\dagger c_2^\dagger c_1^\dagger c_0 c_0^\dagger |0\rangle = c_2^\dagger c_1^\dagger c_3 c_3^\dagger |0\rangle = c_2^\dagger c_1^\dagger |0\rangle.$$

Problem 29.   Consider spinless Fermions. The requirements of the *Pauli principle* are satisfied if the spinless Fermion operators

$$\{ c_k^\dagger, c_j : k, j = 1, 2, \ldots, N \}$$

satisfy the anticommutation relations

$$[c_k^\dagger, c_j]_+ = \delta_{kj} I, \qquad [c_k^\dagger, c_j^\dagger]_+ = [c_k, c_j]_+ = 0$$

for all $k, j = 1, 2, \ldots, N$. The vacuum state $|0\rangle$ is defined by

$$c_k |0\rangle = 0 |0\rangle.$$

The vacuum state is normalized, i.e. $\langle 0|0 \rangle = 1$.
(i) Find a faithful matrix representation of the Fermi operators and of the number operators

$$\hat{N}_j := c_j^\dagger c_j.$$

(ii) Calculate

$$\mathrm{tr} \left(\exp \left(\sum_{k=1}^N \lambda_k \hat{N}_k \right) \right).$$

Solution 29.   (i) Let $N > 1$. Then

$$c_k^\dagger = \overbrace{\sigma_3 \otimes \cdots \otimes \sigma_3}^{N\text{-times}} \otimes \left(\frac{1}{2}\sigma_+ \right) \otimes I_2 \otimes \cdots \otimes I_2$$

$$c_k = \sigma_3 \otimes \cdots \otimes \sigma_3 \otimes \left(\frac{1}{2}\sigma_- \right) \otimes I_2 \otimes \cdots \otimes I_2$$

$$k\text{-th place}$$

where I_2 is the 2×2 unit matrix. One can easily calculate that the anticommutation relations are fulfilled. The *number operator*

$$\hat{N}_k := c_k^\dagger c_k$$

with quantum number k is found to be

$$\hat{N}_k = c_k^\dagger c_k = I_2 \otimes \cdots I_2 \otimes \begin{pmatrix} 1 & 0 \\ 0 & 0 \end{pmatrix} \otimes I_2 \otimes \cdots \otimes I_2.$$

Finally, one has

$$\sum_{k=1}^N \lambda_k \hat{N}_k = \begin{pmatrix} \lambda_1 & 0 \\ 0 & 0 \end{pmatrix} \otimes I_2 \otimes \cdots \otimes I_2 + I_2 \otimes \begin{pmatrix} \lambda_2 & 0 \\ 0 & 0 \end{pmatrix} \otimes I_2 \otimes \cdots \otimes I_2$$
$$+ \cdots + I_2 \otimes \cdots \otimes I_2 \otimes \begin{pmatrix} \lambda_N & 0 \\ 0 & 0 \end{pmatrix}$$

where $\lambda_k \in \mathbb{R}$. The total number operator is defined as

$$\hat{N}_e := \sum_{k=1}^N \hat{N}_k \equiv \sum_{k=1}^N c_k^\dagger c_k.$$

The underlying vector space is given by $\mathbb{C}^{2^N}$.

(ii) The determination of the trace

$$\mathrm{tr}\left(\exp\left(\sum_{k=1}^N \lambda_k \hat{N}_k\right)\right)$$

reduces to the trace calculation in a subspace. In occupation number formalism this subspace is given by the basis

$$\{\, c^\dagger |0\rangle, \quad |0\rangle \,\}$$

and in matrix calculation it is given by the basis

$$\left\{ \begin{pmatrix} 1 \\ 0 \end{pmatrix}, \begin{pmatrix} 0 \\ 1 \end{pmatrix} \right\}.$$

Consequently, we find that

$$\mathrm{tr}\exp\left(\sum_{k=1}^N \lambda_k \hat{N}_k\right) = \prod_{k=1}^N \mathrm{tr}\left(\exp\begin{pmatrix} \lambda_k & 0 \\ 0 & 0 \end{pmatrix}\right) = \prod_{k=1}^N \mathrm{tr}\begin{pmatrix} e^{\lambda_k} & 0 \\ 0 & 1 \end{pmatrix} = \prod_{k=1}^N (e^{\lambda_k} + 1).$$

Problem 30. Spinless Fermi creation and annihilation operators $c_j^\dagger$, c_j $(j = 0, 1, \ldots, N-1)$ obey the anticommutation relations

$$[c_j, c_k]_+ = 0, \qquad [c_j^\dagger, c_k]_+ = \delta_{jk} I$$

where I is the identity operator. A basis for the Hilbert space is given by

$$\prod_{j=0}^{N-1} (c_j^\dagger)^{r_j} |0\rangle, \qquad r_j = 0, 1$$

and the vacuum state $|0\rangle$ is defined by

$$c_j|0\rangle = 0, \qquad j = 0, 1, \ldots, N - 1.$$

Consider the operators

$$\rho_j = \frac{1 - \nu_j}{2} c_j^\dagger c_j + \frac{1 + \nu_j}{2} c_j c_j^\dagger, \qquad \nu_j \in [-1, 1].$$

Show that the ρ_j's are density matrices. Use the matrix representation for $c_j^\dagger$ and c_j

$$c_j^\dagger = \sigma_3 \otimes \cdots \otimes \sigma_3 \otimes \left(\frac{1}{2}\sigma_+\right) \otimes I_2 \otimes \cdots \otimes I_2$$

$$c_j = \sigma_3 \otimes \cdots \otimes \sigma_3 \otimes \left(\frac{1}{2}\sigma_-\right) \otimes I_2 \otimes \cdots \otimes I_2$$

where σ_+ and σ_- are at the j-th position $(j = 0, 1, \ldots, N - 1)$ and

$$\sigma_+ := \sigma_1 + i\sigma_2 = \begin{pmatrix} 0 & 2 \\ 0 & 0 \end{pmatrix}, \qquad \sigma_- := \sigma_1 - i\sigma_2 = \begin{pmatrix} 0 & 0 \\ 2 & 0 \end{pmatrix}.$$

The vacuum state is $|0\rangle = |0\rangle \otimes \cdots \otimes |0\rangle$ with

$$|0\rangle = \begin{pmatrix} 0 \\ 1 \end{pmatrix}.$$

Solution 30.   Since $\sigma_3^2 = I_2$ we have

$$c_j^\dagger c_j = I_2 \otimes \cdots \otimes I_2 \otimes \begin{pmatrix} 1 & 0 \\ 0 & 0 \end{pmatrix} \otimes I_2 \otimes \cdots \otimes I_2$$

and

$$c_j c_j^\dagger = I_2 \otimes \cdots \otimes I_2 \otimes \begin{pmatrix} 0 & 0 \\ 0 & 1 \end{pmatrix} \otimes I_2 \otimes \cdots \otimes I_2.$$

Thus $c_j^\dagger c_j$ and $c_j c_j^\dagger$ are diagonal matrices and the eigenvalues of $c_j^\dagger c_j$ and $c_j c_j^\dagger$ are 1 and 0. Since $\nu_j \in [-1, 1]$ it follows that $(1 - \nu_j)/2$ and $(1 + \nu_j)/2$ cannot be negative. Thus the eigenvalues of ρ_j cannot be negative. Furthermore $\mathrm{tr}(\rho_j) = 1$.

Problem 31.   Let $c_j^\dagger$, c_j $(j = 1, \ldots, n)$ be Fermi creation and annihilation operators. We set

$$c_j = \frac{1}{2}(\gamma_j - i\gamma_{n+j}), \qquad c_j^\dagger = \frac{1}{2}(\gamma_j + i\gamma_{n+j}).$$

Find the anticommutator $[\gamma_j, \gamma_k]_+$, where $j, k = 1, \ldots, 2n$.

Solution 31.   Utilizing $[c_j, c_k^\dagger]_+ = \delta_{jk} I$, where $j, k = 1, \ldots, n$ we obtain $[\gamma_j, \gamma_k]_+ = 2\delta_{jk} I$, where $j, k = 1, \ldots, 2n$.

Problem 32.   Let

$$\sigma_+ = \sigma_1 + i\sigma_2 = \begin{pmatrix} 0 & 2 \\ 0 & 0 \end{pmatrix}, \qquad \sigma_- = \sigma_1 - i\sigma_2 = \begin{pmatrix} 0 & 0 \\ 2 & 0 \end{pmatrix}.$$

For the matrix representation of the spin-less Fermi creation operator $c_j^\dagger$ $(0, 1, \ldots, N - 1)$ (N-lattice sites)

$$c_j^\dagger = \sigma_3 \otimes \cdots \otimes \sigma_3 \otimes \frac{1}{2}\sigma_+ \otimes I_2 \otimes \cdots \otimes I_2$$

where $\frac{1}{2}\sigma_+$ is at the j-position. Thus

$$c_j = \sigma_3 \otimes \cdots \otimes \sigma_3 \otimes \frac{1}{2}\sigma_- \otimes I_2 \otimes \cdots \otimes I_2$$

where $\frac{1}{2}\sigma_-$ is at the j-position. The spin operators $\sigma_{1,j}$, $\sigma_{2,j}$, $\sigma_{3,j}$ are given by

$$\sigma_{1,j} = I_2 \otimes \cdots \otimes I_2 \otimes \sigma_1 \otimes I_2 \otimes \cdots \otimes I_2$$

$$\sigma_{2,j} = I_2 \otimes \cdots \otimes I_2 \otimes \sigma_2 \otimes I_2 \otimes \cdots \otimes I_2$$

$$\sigma_{3,j} = I_2 \otimes \cdots \otimes I_2 \otimes \sigma_3 \otimes I_2 \otimes \cdots \otimes I_2$$

where the Pauli matrix σ_k $(k = 1, 2, 3)$ is at the j-th position.
(i) Find $(j = 0, 1, \ldots, N - 1)$ the matrix representation of the operator

$$c_j^\dagger c_j - c_j c_j^\dagger.$$

(ii) Find $(j = 0, 1, \ldots, N - 2)$ the matrix representation of the operator

$$(c_j^\dagger - c_j)(c_{j+1}^\dagger + c_{j+1}).$$

(iii) Find $(j = 0, 1, \ldots, N - 2)$ the matrix representation of the operator

$$(c_j^\dagger + c_j)(c_{j+1}^\dagger - c_{j+1}).$$

(iv) Find $(j = 0, 1, \ldots, N - 2)$ the matrix representation of the operator

$$(c_j^\dagger - c_j)(c_{j+1}^\dagger - c_{j+1}).$$

(v) Find $(j = 0, 1, \ldots, N - 2)$ the matrix representation of the operator

$$(c_j^\dagger + c_j)(c_{j+1}^\dagger + c_{j+1}).$$

(vi) Find $(j = 0, 1, \ldots, N - 2)$ the matrix representation of the operator

$$c_j^\dagger c_{j+1} + c_{j+1}^\dagger c_j.$$

Solution 32. (i) Using $\sigma_3 \sigma_3 = I_2$ and $\frac{1}{2}\sigma_+ + \frac{1}{2}\sigma_- = \sigma_1$ we find

$$c_j^\dagger c_j - c_j c_j^\dagger = I_2 \otimes \cdots \otimes I_2 \otimes \sigma_3 \otimes I_2 \otimes \cdots \otimes I_2 = \sigma_{3,j}.$$

(ii) We find

$$(c_j^\dagger - c_j)(c_{j+1}^\dagger + c_{j+1}) = I_2 \otimes \cdots \otimes I_2 \otimes (-\sigma_1) \otimes \sigma_1 \otimes I_2 \otimes \cdots \otimes I_2 = -\sigma_{1,j}\sigma_{1,j+1}.$$

(iii) We find

$$(c_j^\dagger + c_j)(c_{j+1}^\dagger - c_{j+1}) = I_2 \otimes \cdots \otimes I_2 \otimes (-i\sigma_2) \otimes (i\sigma_2) \otimes I_2 \otimes \cdots \otimes I_2 = \sigma_{2,j}\sigma_{2,j+1}.$$

(iv) We find

$$(c_j^\dagger - c_j)(c_{j+1}^\dagger - c_{j+1}) = I_2 \otimes \cdots \otimes I_2 \otimes (-\sigma_1) \otimes (-i\sigma_2) \otimes I_2 \otimes \cdots \otimes I_2 = -i\sigma_{1,j}\sigma_{2,j+1}.$$

(v) We find

$$(c_j^\dagger + c_j)(c_{j+1}^\dagger + c_{j+1}) = I_2 \otimes \cdots \otimes I_2 \otimes (-i\sigma_2) \otimes \sigma_1 \otimes I_2 \otimes \cdots \otimes I_2 = -i\sigma_{2,j}\sigma_{1,j+1}.$$

(vi) We find

$$c_j^\dagger c_{j+1} + c_{j+1}^\dagger c_j = I_2 \otimes \cdots \otimes I_2 \otimes \begin{pmatrix} 0 & 0 & 0 & 0 \\ 0 & 0 & -1 & 0 \\ 0 & -1 & 0 & 0 \\ 0 & 0 & 0 & 0 \end{pmatrix} \otimes I_2 \otimes \cdots \otimes I_2.$$

Problem 33. (i) Consider the non-hermitian operators

$$R_+ := -c_{1,\downarrow}^\dagger c_{1,\uparrow}^\dagger + c_{2,\downarrow}^\dagger c_{2,\uparrow}^\dagger, \quad R_- := -c_{1,\uparrow}c_{1,\downarrow} + c_{2,\uparrow}c_{2,\downarrow}$$

and

$$R_3 := \frac{1}{2}(c_{1\uparrow}^\dagger c_{1\uparrow} + c_{2\uparrow}^\dagger c_{2\uparrow} + c_{1\downarrow}^\dagger c_{1\downarrow} + c_{2\downarrow}^\dagger c_{2\downarrow} - I).$$

Note that $(R_+)^\dagger = R_-$. Find the commutators $[R_+, R_-]$, $[R_3, R_+]$, $[R_3, R_-]$. Discuss.

(ii) Consider the operators

$$S_+ := c_{1,\uparrow}^\dagger c_{1,\downarrow} + c_{2,\uparrow}^\dagger c_{2,\downarrow}, \quad S_- := c_{1,\downarrow}^\dagger c_{1,\uparrow} + c_{2,\downarrow}^\dagger c_{2,\uparrow}$$

and

$$S_3 := \frac{1}{2}(c_{1,\uparrow}^\dagger c_{1,\uparrow} + c_{2,\uparrow}^\dagger c_{2,\uparrow} - c_{1,\downarrow}^\dagger c_{1,\downarrow} - c_{2,\downarrow}^\dagger c_{2,\downarrow}).$$

Find the commutators $[S_+, S_-]$, $[S_3, S_+]$, $[S_3, S_+]$. Discuss.

(iii) Consider the non-hermitian operators ($j \neq k$)

$$T_{+,jk} = c_{j\uparrow}^\dagger c_{j\downarrow}^\dagger + c_{j\uparrow}^\dagger c_{k\downarrow}^\dagger + c_{k\uparrow}^\dagger c_{j\downarrow}^\dagger + c_{k\uparrow}^\dagger c_{k\downarrow}^\dagger, \quad T_{-,jk} = c_{j\downarrow}c_{j\uparrow} + c_{k\downarrow}c_{j\uparrow} + c_{j\downarrow}c_{k\uparrow} + c_{k\downarrow}c_{k\uparrow}.$$

Is the commutator $T_{3,jk} := [T_{+,jk}, T_{-,jk}]$ a hermitian operator?

Solution 33. (i) We find

$$[R_+, R_-] = c_{1,\uparrow}^\dagger c_{1,\uparrow}^\dagger + c_{2,\uparrow}^\dagger c_{2,\uparrow}^\dagger + c_{1,\downarrow}^\dagger c_{1,\downarrow}^\dagger + c_{2,\downarrow}^\dagger c_{2,\downarrow}^\dagger - I = 2R_3.$$

Analogously $[R_3, R_+] = R_+$ and $[R_3, R_-] = -R_-$.

(ii) We find

$$[S_+, S_-] = (c_{1,\uparrow}^\dagger c_{1,\uparrow} + c_{2,\uparrow}^\dagger c_{2,\uparrow} - c_{1,\downarrow}^\dagger c_{1,\downarrow} - c_{2,\downarrow}^\dagger c_{2,\downarrow}) = 2S_3.$$

Analogously

$$[S_3, S_+] = S_+, \qquad [S_3, S_-] = -S_-.$$

(iii) We find

$$T_{3,jk} = 2(c_{j\uparrow}^\dagger c_{j\uparrow} + c_{j\downarrow}^\dagger c_{j\downarrow} - I) + 2(c_{k\uparrow}^\dagger c_{k\uparrow} + c_{k\downarrow}^\dagger c_{k\downarrow} - I) + (c_{j\uparrow}^\dagger c_{k\uparrow} + c_{j\downarrow}^\dagger c_{k\downarrow} + j \leftrightarrow k).$$

Thus the operator $T_{3,jk}$ is hermitian.

Problem 34. Let $c_{j,\sigma}^\dagger$, $c_{j,\sigma}$ be Fermi creation and annihilation operators, where $j = 1, \ldots, N$ and $\sigma, \sigma' \in \{\uparrow, \downarrow\}$. Find the expectation value

$$\langle 0 | c_{j,\sigma} c_{k,\sigma'}^\dagger | 0 \rangle.$$

Solution 34. Since $c_{j,\sigma} c_{k,\sigma'}^\dagger = \delta_{jk} \delta_{\sigma,\sigma'} I - c_{k,\sigma'}^\dagger c_j$ and $c_{j,\sigma} | 0 \rangle = 0 | 0 \rangle$ we obtain

$$\langle 0 | c_{j,\sigma} c_{k,\sigma'}^\dagger | 0 \rangle = \delta_{jk} \delta_{\sigma,\sigma'}.$$

Problem 35. Consider the operator

$$S := (c_{1,\uparrow}^\dagger + c_{1,\uparrow})(c_{1,\downarrow}^\dagger + c_{1,\downarrow})(c_{2,\uparrow}^\dagger + c_{2,\uparrow})(c_{2,\downarrow}^\dagger + c_{2,\downarrow}).$$

(i) Find the operator $SS^\dagger$.
(ii) Is

$$Sc_{j,\sigma} = c_{j,\sigma}^\dagger S, \qquad Sc_{j,\sigma}^\dagger = c_{j,\sigma} S$$

where $j = 1, 2$ and $\sigma \in \{\uparrow, \downarrow\}$.

Solution 35. (i) We obtain $SS^\dagger = I$, where I is the identity operator.
(ii) Yes.

Problem 36. Let $c_{j,\sigma}^\dagger$, $c_{j,\sigma}$ be Fermi creation and annihilation operator, where $\sigma \in \{\uparrow, \downarrow\}$. Let $\hat{N}_{j,\sigma} := c_{j,\sigma}^\dagger c_{j,\sigma}$.
(i) Show that $\hat{N}_{j,\sigma}^2 = \hat{N}_{j,\sigma}$.
(ii) Show that

$$\hat{N}_{j,\uparrow} \hat{N}_{j,\downarrow} \equiv 2\alpha(\hat{N}_{j,\uparrow} + \hat{N}_{j,\downarrow}) + \left(\frac{1}{4} - \alpha\right)(\hat{N}_{j,\uparrow} + \hat{N}_{j,\downarrow})^2 - \left(\frac{1}{4} + \alpha\right)(\hat{N}_{j,\uparrow} - \hat{N}_{j,\downarrow})^2$$

for any value of the real parameter α.

Solution 36. (i) Since $[c_{j,\sigma}^\dagger, c_{j,\sigma'}]_+ = \delta_{\sigma,\sigma'} I$ and

$$[c_{j,\sigma}^\dagger, c_{j,\sigma'}^\dagger]_+ = 0, \qquad [c_{j,\sigma}, c_{j,\sigma'}]_+ = 0$$

we find the identity.
(ii) Using the result from (i) we obtain the identity, where we used that $\hat{N}_{j,\uparrow} \hat{N}_{j,\downarrow} = \hat{N}_{j,\downarrow} \hat{N}_{j,\uparrow}$.

Problem 37. Let $c_j^\dagger$, c_j ($j = 1, \ldots, n$) be Fermi creation and annihilation operators. *Majorana fermion operators* are defined as ($k = 1, \ldots, n$)

$$d_{2k-1} = c_k + c_k^\dagger, \qquad d_{2k} = i(c_k - c_k^\dagger).$$

They are hermitian. Find the anticommutator $[d_k, d_\ell]_+$, where $k, \ell = 1, \ldots, 2n$.

Solution 37. We find $[d_k, d_\ell]_+ = 2\delta_{k\ell} I$, where δ_{jk} is the Kronecker delta.

Problem 38. Let $\mathbb{Z}$ be the set of integers and $j, k \in \mathbb{Z}$. The algebra $\mathcal{A}$ of free fermions is generated by c_j, c_j^* satisfying the anticommutation relations

$$[c_j, c_k]_+ = [c_j^*, c_k^*]_+ = 0, \qquad [c_j, c_k^*]_+ = \delta_{jk} I$$

where I is the identity operator. There is a Fock representation $\mathcal{F}$ of this algebra with a vacuum $|0\rangle$ ($\langle 0|0\rangle = 1$) such that

$$c_j|0\rangle = 0 \ (j < 0) \qquad c_j^*|0\rangle = 0 \ (j \geq 0)$$
$$\langle 0|c_j = 0 \ (j \geq 0) \qquad \langle 0|c_j^* = 0 \ (j < 0).$$

Let $p \in \mathbb{Z}$. The states

$$|p\rangle = \begin{cases} c_{p-1} \cdots c_0 |0\rangle & p > 0 \\ |0\rangle & p = 0 \\ c_p^* \cdots c_{-1}^* |0\rangle & p < 0 \end{cases}$$

are $g\ell(\infty)$ highest weight states. The corresponding dual states are given by

$$\langle p| = \begin{cases} \langle 0|c_0^* \cdots c_{p-1}^* & p > 0 \\ \langle 0| & p = 0 \\ \langle 0|c_{-1} \cdots c_p & p < 0 \end{cases}.$$

Let $n \neq 0$. Define the operators

$$\hat{H}_n := \sum_{j \in \mathbb{Z}} c_j c_{j+n}^*.$$

Find the commutators $[\hat{H}_n, \hat{H}_m]$, $[\hat{H}_n, c_j]$, $[\hat{H}_n, c_j^*]$.

Solution 38. We obtain

$$[\hat{H}_n, \hat{H}_m] = n\delta_{n+m,0} I, \qquad [\hat{H}_n, c_j] = c_{j-n}, \qquad [\hat{H}_n, c_j^*] = -c_{j+n}^*.$$

Problem 39. Let $m, n \in \mathbb{Z}$. Let d_n be neutral free fermion satisfying

$$[d_m, d_n]_+ = (-1)^m \delta_{m+n,0}.$$

The vacuum states $|0\rangle$ and $\langle 0|$ are defined by

$$d_n|0\rangle = 0 \quad \text{for } n < 0$$
$$\langle 0|d_n = 0 \quad \text{for } n > 0.$$

The vacuum expectation value is uniquely determined by setting $\langle 0|1|0\rangle = 1$, $\langle 0|d_0|0\rangle = 0$. Find d_0^2.

Solution 39. We obtain $d_0^2 = 1/2$.

Problem 40. Let $p = 1, 2, \ldots$. *Parafermionic operators* c and $c^\dagger$ of order p are defined by

$$(c)^{p+1} = 0 = (c^\dagger)^{p+1}, \quad [[c^\dagger, c], c] = -2c, \quad [[c^\dagger, c], c^\dagger] = 2c^\dagger.$$

(i) We set

$$J_+ = c^\dagger, \quad J_- = c, \quad J_3 = \frac{1}{2}[c^\dagger, c].$$

Find the commutators $[J_+, J_-]$, $[J_+, J_3]$, $J_-, J_3]$.
(ii) Find a matrix representation, where one uses for J_3 the $(p+1) \times (p+1)$ diagonal matrix

$$J_3 = \text{diag}(p/2, p/2 - 1, \ldots, -p/2 + 1, -p/2).$$

Solution 40. (i) We find

$$[J_+, J_-] = 2J_3, \quad [J_3, J_+] = J_+, \quad [J_3, J_-] = -J_-.$$

(ii) $J_+ = c^\dagger$ and $J_- = c$ can be represented by

$$(c)_{j,k} = C_k \delta_{j,k+1}, \qquad (c^\dagger)_{j,k} = C_j \delta_{j+1,k}$$

where $j, k = 1, 2, \ldots, p+1$ and
$$C_k = \sqrt{k(p - k + 1)}.$$

For example for $p = 2$ we have the 3×3 matrices

$$c = \begin{pmatrix} 0 & 0 & 0 \\ \sqrt{2} & 0 & 0 \\ 0 & \sqrt{2} & 0 \end{pmatrix}, \quad c^\dagger = \begin{pmatrix} 0 & \sqrt{2} & 0 \\ 0 & 0 & \sqrt{2} \\ 0 & 0 & 0 \end{pmatrix}.$$

3.2   Fermi Operators and Functions

Problem 41.   Let $c^\dagger$, c be Fermi creation and annihilation operators and I the identity operator. Let $n \in \mathbb{N}$ and $z \in \mathbb{C}$ with $z \neq 0$.
(i) Calculate $(c^\dagger)^n$ and c^n.
(ii) Calculate $(zc^\dagger c)^2$, $(zc^\dagger c)^3$ and $(zc^\dagger c)^n$.

Solution 41.   (i) We have

$$(c^\dagger)^n = 0, \quad c^n = 0 \quad \text{for all } n \geq 2.$$

(ii) Since $(c^\dagger c)^2 = c^\dagger c$ we obtain for all $n \geq 1$

$$(zc^\dagger c)^n = z^n c^\dagger c.$$

Problem 42.   (i) Consider the operator $X := zc^\dagger - \bar{z}c$. Show that X is skew-hermitian. Find the operators X^2, X^3, $\ldots$.
(ii) Let $z_1, z_2 \in \mathbb{C}$. Let $Y := z_1 c^\dagger c + z_2 c^\dagger - \bar{z}_2 c$. Find the operators Y^2 and Y^3.

Solution 42.   (i) We have

$$X^\dagger = (zc^\dagger - \bar{z}c)^\dagger = \bar{z}c - zc^\dagger = -X.$$

Hence X is skew-hermitian. Now

$$\begin{aligned}
X^2 &= -z\bar{z}I \\
X^3 &= -z\bar{z}(zc^\dagger - zc) = -z\bar{z}X \\
X^4 &= (z\bar{z})^2 I \\
X^5 &= (z\bar{z})^2(zc^\dagger - zc) = (z\bar{z})^2 X.
\end{aligned}$$

In general we have

$$X^n = \begin{cases} (-1)^{n/2}(z\bar{z})^{n/2}I & \text{if } n \text{ is even} \\ (-1)^{(n-1)/2}(z\bar{z})^{(n-1)/2}X & \text{if } n \text{ is odd} \end{cases}$$

(ii) Using $c^\dagger c^\dagger = cc = 0$ and $(c^\dagger c)^2 = c^\dagger c$ we obtain for Y^2

$$\begin{aligned}
Y^2 &= (z_1 c^\dagger c + z_2 c^\dagger - \bar{z}_2 c)^2 = z_1^2 c^\dagger c + z_1(z_2 c^\dagger - \bar{z}_2 c) - z_2\bar{z}_2 I \\
&= z_1^2 c^\dagger c + z_1 X - z_2 \bar{z}_2 I.
\end{aligned}$$

For Y^3 we find

$$\begin{aligned}
Y^3 &= z_1^3 c^\dagger c - z_1 z_2 \bar{z}_2 c^\dagger c + z_1^2(z_2 c^\dagger - \bar{z}_2 c) - z_2\bar{z}_2(z_2 c^\dagger - \bar{z}_2 c) - z_1 z_2 \bar{z}_2 I \\
&= (z_1^3 - z_1 z_2 \bar{z}_2)c^\dagger c + (z_1^2 - z_2 \bar{z}_2)X - z_1 z_2 \bar{z}_2 I.
\end{aligned}$$

A general term is of the form $f_1(z_1, z_2, \bar{z}_2)c^\dagger c + f_2(z_1, z_2, \bar{z}_2)X + f_3(z_1, z_2, \bar{z}_2)I$.

Problem 43.   To calculate $\exp(A)$ of a linear operator A in a finite dimensional Hilbert space one can utilize the expansion

$$\exp(A) = \sum_{j=0}^{\infty} \frac{A^j}{j!}.$$

Let $c^\dagger$, c be Fermi creation and annihilation operators, respectively. Let $z \in \mathbb{C}$.
(i) Calculate the operator $\exp(zc^\dagger)$.
(ii) Calculate the operator $\exp(zc)$.
(iii) Calculate $\exp(zc^\dagger)|0\rangle$ and $\exp(zc^\dagger)|1\rangle$.
(iv) Calculate $c\exp(zc^\dagger)|0\rangle$.
(v) Calculate

$$\langle 0|\exp(zc^\dagger)|0\rangle, \quad \langle 0|\exp(zc^\dagger)|1\rangle, \quad \langle 1|\exp(zc^\dagger)|0\rangle, \quad \langle 1|\exp(zc^\dagger)|1\rangle.$$

Thus find the matrix representation of $\exp(zc^\dagger)$.

Solution 43. (i) Since $(c^\dagger)^2 = 0$ we have $\exp(zc^\dagger) = I + zc^\dagger$.
(ii) Since $c^2 = 0$ we have $\exp(zc) = I + zc$.
(iii) From (i) it follows that

$$\exp(zc^\dagger)|0\rangle = (I + zc^\dagger)|0\rangle = |0\rangle + zc^\dagger|0\rangle = |0\rangle + z|1\rangle$$

and from (i) and $c^\dagger|1\rangle = 0$ we find

$$\exp(zc^\dagger)|1\rangle = (I + zc^\dagger)c^\dagger|0\rangle = c^\dagger|0\rangle = |1\rangle.$$

(iv) We obtain

$$c\exp(zc^\dagger)|0\rangle = c(|0\rangle + z|1\rangle) = z|0\rangle.$$

(iv) We find

$$\langle 0|\exp(zc^\dagger)|0\rangle = 1, \quad \langle 0|\exp(zc^\dagger)|1\rangle = 0, \quad \langle 1|\exp(zc^\dagger)|0\rangle = z, \quad \langle 1|\exp(zc^\dagger)|1\rangle = 1.$$

This leads to the matrix representation of $\exp(zc^\dagger)$.

Problem 44. Let $c^\dagger$, c be Fermi creation and annihilation operators, respectively.
(i) Calculate the operator $\exp(zc^\dagger c)$.
(ii) Use the result from (i) to find the operator $\exp(i\pi c^\dagger c)$.
(iii) Find the eigenvalues of the operator $U = \exp(i\pi c^\dagger c)$.
(iv) Find $\exp(i\pi c^\dagger c)$ by setting $f(\epsilon) = \exp(\epsilon c^\dagger c)$. Then find the differential equation for $f(\epsilon)$ and solve it together with the initial conditions.

Solution 44. (i) We obtain

$$e^{zc^\dagger c} = I + c^\dagger c(z + \frac{z^2}{2!} + \frac{z^3}{3!} + \cdots) = I + c^\dagger c(e^z - 1).$$

(ii) With $z = i\pi$ and $e^{i\pi} = -1$ we arrive at

$$e^{i\pi c^\dagger c} = I + c^\dagger c(e^{i\pi} - 1) = I - 2c^\dagger c$$

since $e^{i\pi} = -1$. Obviously the eigenvalues of $I - 2c^\dagger c$ are $+1$ and -1.
(iii) Note that the operator U is unitary and hermitian. The eigenvalues are $+1$ and -1.
(iv) We note that $(c^\dagger c)^2 = c^\dagger c$. Now from $f(\epsilon) = \exp(\epsilon c^\dagger c)$ we obtain

$$\frac{df}{d\epsilon} = e^{\epsilon c^\dagger c}c^\dagger c, \qquad \frac{d^2 f}{d\epsilon^2} = e^{\epsilon c^\dagger c}c^\dagger c = \frac{df}{d\epsilon}.$$

Solving the second order linear differential equation $d^2 f/d\epsilon^2 = df/d\epsilon$ together with the initial conditions $f(0) = I$ and $df(0)/d\epsilon = c^\dagger c$ yields

$$e^{\epsilon c^\dagger c} = (I - c^\dagger c) + e^\epsilon c^\dagger c.$$

Thus since $e^{i\pi} = -1$ we have $e^{i\pi c^\dagger c} = I - 2c^\dagger c$ the result of (ii).

Problem 45.   Let $z \in \mathbb{C}$. Consider the analytic functions $\sin(z)$, $\cos(z)$, $\sinh(z)$, $\cosh(z)$.
(i) Find the operators

$$\sin(zc^\dagger), \quad \sin(zc), \quad \cos(zc^\dagger), \quad \cos(zc).$$

(ii) Find the operators

$$\sinh(zc^\dagger), \quad \sinh(zc), \quad \cosh(zc^\dagger), \quad \cosh(zc).$$

(iii) Find the operators

$$\sinh(zc^\dagger c), \quad \cosh(zc^\dagger c), \quad \sinh(zc^\dagger c) + \sinh(zcc^\dagger).$$

Solution 45.   (i) From the expansion for $\sin(z)$ and $\cos(z)$ and $c^2 = 0$, $(c^\dagger)^2 = 0$ we find

$$\sin(zc^\dagger) = zc^\dagger, \quad \sin(zc) = zc, \quad \cos(zc^\dagger) = I, \quad \cos(zc) = I.$$

(ii) From the expansion for $\sinh(z)$ and $\cosh(z)$ and $c^2 = 0$, $(c^\dagger)^2 = 0$ we find

$$\sinh(zc^\dagger) = zc^\dagger, \quad \sinh(zc) = zc, \quad \cosh(zc^\dagger) = I, \quad \cosh(zc) = I.$$

(iii) Since $(c^\dagger c)^n = c^\dagger c$ and $(cc^\dagger)^n = cc^\dagger$ we obtain

$$\sinh(zc^\dagger c) = zc^\dagger c, \quad \cosh(zc^\dagger c) = I$$

and

$$\sinh(zc^\dagger c) + \sinh(zcc^\dagger) = \sinh(z)c^\dagger c + \sinh(z)cc^\dagger = \sinh(z)I.$$

Problem 46.   Let $z \in \mathbb{C}$. Find the operator-valued function

$$f(z) = e^{zc^\dagger c} c e^{-zc^\dagger c}$$

by calculating the derivative of f with respect to z and solving the resulting linear operator-valued differential equation with the initial condition $f(0) = c$. Then set $z = i\omega t$, where ω is the frequency.

Solution 46.   Since $cc = 0$ and $cc^\dagger = I - c^\dagger c$ we find

$$\frac{df}{dz} = e^{zc^\dagger c} c^\dagger cce^{-zc^\dagger c} - e^{zc^\dagger c}cc^\dagger ce^{-zc^\dagger c} = -e^{zc^\dagger cc^\dagger} ce^{-zc^\dagger c}$$

$$= -e^{zc^\dagger c}ce^{-zc^\dagger c} = -f(z).$$

Together with the initial condition $f(0) = c$ we obtain $f(z) = e^{-z}c$. Therefore $f(i\omega t) = e^{i\omega t}$.

Problem 47. Let $\gamma \in \mathbb{C}$ and $c^\dagger$, c be Fermi creation and annihilation operators. We set

$$X := \gamma c^\dagger - \gamma^* c.$$

(i) Calculate

$$(\gamma c^\dagger - \gamma^* c)^2, \quad (\gamma c^\dagger - \gamma^* c)^3, \quad (\gamma c^\dagger - \gamma^* c)^4.$$

(ii) Find the operator $\exp(\gamma c^\dagger - \gamma^* c)$.
(iii) Calculate the state $\exp(\gamma c^\dagger - \gamma^* c)|0\rangle$.
(iv) Calculate the expectation value $\langle 0| \exp(\gamma c^\dagger - \gamma^* c)|0\rangle$.
(v) Consider a one Fermi system with the matrix representation for the Fermi creation and annihilation operators

$$c^\dagger = \begin{pmatrix} 0 & 1 \\ 0 & 0 \end{pmatrix}, \quad c = \begin{pmatrix} 0 & 0 \\ 1 & 0 \end{pmatrix}, \quad |0\rangle = \begin{pmatrix} 0 \\ 1 \end{pmatrix}, \quad |1\rangle = c^\dagger|0\rangle = \begin{pmatrix} 1 \\ 0 \end{pmatrix}.$$

Find the matrix $\exp(\gamma c^\dagger - \gamma^* c)$ and its eigenvalues and eigenvectors.

Solution 47. (i) Since $c^\dagger c^\dagger = 0$ and $cc = 0$ we have

$$(\gamma c^\dagger - \gamma^* c)^2 = \gamma^2 c^\dagger c^\dagger - \gamma\gamma^* cc^\dagger - \gamma\gamma^* c^\dagger c + \gamma^*\gamma^* cc = -\gamma\gamma^*(c^\dagger c + cc^\dagger) = -\gamma\gamma^* I.$$

From this result we obtain

$$(\gamma c^\dagger - \gamma^* c)^3 = -\gamma\gamma^* X.$$

Using the result for $(\gamma c^\dagger - \gamma^* c)^2$ we find

$$(\gamma c^\dagger - \gamma^* c)^4 = (\gamma\gamma^*)^2 I.$$

(ii) We obtain

$$e^{\gamma c^\dagger - \gamma^* c} = \cos(|\gamma|)I + \frac{\gamma}{|\gamma|}\sin(|\gamma|)c^\dagger - \frac{\gamma^*}{|\gamma|}\sin(|\gamma|)c.$$

(iii) Since $c|0\rangle = 0|0\rangle$ we obtain

$$e^{\gamma c^\dagger - \gamma^* c}|0\rangle = \cos(|\gamma|)|0\rangle + \frac{\gamma}{|\gamma|}c^\dagger|0\rangle.$$

(iv) Since $\langle 0|c^\dagger = \langle 0|0$ we obtain

$$\langle 0|e^{\gamma c^\dagger - \gamma^* c}|0\rangle = \cos(|\gamma|).$$

(v) We set $\gamma = re^{i\phi}$ ($r \geq 0$) and therefore $\gamma^* = re^{-i\phi}$. We obtain

$$\exp(\gamma c^\dagger - \gamma^* c) = \begin{pmatrix} \cos(r) & e^{i\phi}\sin(r) \\ -e^{-i\phi}\sin(r) & \cos(r) \end{pmatrix}.$$

Obviously the matrix is unitary.

Problem 48. Let $z_1, z_2 \in \mathbb{C}$ and consider the operator

$$K = z_1 c^\dagger c + z_2 c^\dagger - \bar{z}_2 c.$$

Then the matrix representation of K with the basis $\{c^\dagger|0\rangle, |0\rangle\}$ is given by

$$K = \begin{pmatrix} z_1 & z_2 \\ -\bar{z}_2 & 0 \end{pmatrix}$$

with the eigenvalues

$$\lambda_\pm = \frac{1}{2}z_1 \pm \sqrt{z_1^2/4 - z_2\bar{z}_2}.$$

Calculate $\exp(K)$ applying the *Cayley-Hamilton theorem*.

Solution 48.   Applying the Cayley-Hamilton theorem we have

$$e^K = a_1 K + a_0 I_2 = \begin{pmatrix} a_1 z_1 + a_0 & a_1 z_2 \\ -a_1 \bar{z}_2 & a_0 \end{pmatrix}$$

and

$$e^{\lambda_+} = a_1\lambda_+ + a_0, \qquad e^{\lambda_-} = a_1\lambda_- + a_0.$$

The last two equations have to be solved for a_0 and a_1 and then to be inserted into the matrix $e^K = a_1 K + a_0 I_2$. Straightforward calculations provide

$$a_1 = \frac{e^{\lambda_+} - e^{\lambda_-}}{\lambda_+ - \lambda_-}, \qquad a_0 = \frac{e^{\lambda_-}\lambda_+ - e^{\lambda_+}\lambda_-}{\lambda_+ - \lambda_-}.$$

It follows that

$$e^K = \frac{e^{\lambda_+} - e^{\lambda_-}}{\lambda_+ - \lambda_-}K + \frac{e^{\lambda_-}\lambda_+ - e^{\lambda_+}\lambda_-}{\lambda_+ - \lambda_-}I_2.$$

Problem 49.   Let $c_1^\dagger$, $c_2^\dagger$ be Fermi creation operators and $\hat{N}_1 = c_1^\dagger c_1$, $\hat{N}_2 = c_2^\dagger c_2$.
(i) Is the hermitian operator

$$\frac{1}{\sqrt{2}}(c_1^\dagger + c_1 + c_2^\dagger + c_2)$$

invertible?
(ii) Calculate $(c_1^\dagger c_1 + c_2^\dagger c_2)^2 \equiv (\hat{N}_1 + \hat{N}_2)^2$.
(iii) Find the operators $(c_1^\dagger c_2)^2$, $(c_2^\dagger c_1)^2$.
(iv) Consider the hermitian operator

$$\hat{K} = c_1^\dagger c_2 + c_2^\dagger c_1.$$

Find $\hat{K}^2$, $\hat{K}^3$, $\hat{K}^4$ and $\hat{K}^n$ with $n \in \mathbb{N}$. Discuss.
(v) Let $z \in \mathbb{C}$. Use the result from (iv) to find

$$\exp(z\hat{K}) = \exp(z(c_1^\dagger c_2 + c_2^\dagger c_1)).$$

Solution 49.   (i) Yes. We have

$$\frac{1}{\sqrt{2}}(c_1^\dagger + c_1 + c_2^\dagger + c_2)\frac{1}{\sqrt{2}}(c_1^\dagger + c_1 + c_2^\dagger + c_2) = I.$$

(ii) Since $(c_1^\dagger c_1)^2 = c_1^\dagger c_1$, $(c_2^\dagger c_2)^2 = c_2^\dagger c_2$ we have

$$(c_1^\dagger c_1 + c_2^\dagger c_2)^2 = c_1^\dagger c_1 + c_2^\dagger c_2 + 2c_1^\dagger c_1 c_2^\dagger c_2 = \hat{N}_1 + \hat{N}_2 + 2\hat{N}_1\hat{N}_2.$$

(iii) Since $c_1^\dagger c_1^\dagger = 0$ and $c_2^\dagger c_2^\dagger = 0$ we obtain

$$(c_1^\dagger c_2)^2 = 0, \qquad (c_2^\dagger c_1)^2 = 0.$$

Note that $c_1^\dagger c_2$ and $c_2^\dagger c_1$ are nonnormal operators.

(iv) Note that
$$(\hat{N}_1 + \hat{N}_2 - 2\hat{N}_1\hat{N}_2)^2 = \hat{N}_1 + \hat{N}_2 - 2\hat{N}_1\hat{N}_2.$$

Thus we obtain
$$\hat{K}^2 = \hat{N}_1 + \hat{N}_2 - 2\hat{N}_1\hat{N}_2, \quad \hat{K}^3 = \hat{K}, \quad \hat{K}^4 = \hat{N}_1 + \hat{N}_2 - 2\hat{N}_1\hat{N}_2.$$

Thus $\hat{K}^n$ is equal to $\hat{K}$ for n odd and equal to $\hat{N}_1 + \hat{N}_2 - 2\hat{N}_1\hat{N}_2$ for n even.

(v) Utilizing the result from (iv) we have
$$e^{z\hat{K}} = I + \left(z + \frac{z^3}{3!} + \frac{z^5}{5!} + \cdots \right) \hat{K} + \left(\frac{z^2}{2!} + \frac{z^4}{4!} + \frac{z^6}{6!} + \cdots \right)(\hat{N}_1 + \hat{N}_2 - 2\hat{N}_1\hat{N}_2).$$

Consequently
$$e^{z\hat{K}} = I + \sinh(z)\hat{K} + (\cosh(z) - 1)(\hat{N}_1 + \hat{N}_2 - 2\hat{N}_1\hat{N}_2).$$

With $z = i\omega t$ we obtain the unitary operator
$$e^{i\omega t\hat{K}} = I + i\sin(\omega t)\hat{K} + (\cos(\omega t) - 1)(\hat{N}_1 + \hat{N}_2 - 2\hat{N}_1\hat{N}_2).$$

Problem 50. Calculate the operator
$$\exp(i\pi(c_1^\dagger c_1 + c_2^\dagger c_2)).$$

Solution 50. Since $[c_1^\dagger c_1, c_2^\dagger c_2] = 0$ we have
$$e^{i\pi(c_1^\dagger c_1 + c_2^\dagger c_2)} = e^{i\pi c_1^\dagger c_1} e^{i\pi c_2^\dagger c_2}.$$

Now
$$e^{i\pi c_1^\dagger c_1} = I - 2c_1^\dagger c_1, \quad e^{i\pi c_2^\dagger c_2} = I - 2c_2^\dagger c_2$$

and therefore
$$e^{i\pi(c_1^\dagger c_1 + c_2^\dagger c_2)} = I - 2c_1^\dagger c_1 - 2c_2^\dagger c_2 + 4c_1^\dagger c_1 c_2^\dagger c_2.$$

Problem 51. Let $c_1^\dagger$, $c_2^\dagger$ be Fermi creation operators and $z \in \mathbb{C}$. Consider the operator
$$\hat{K} = c_1^\dagger c_2 + c_2^\dagger c_1.$$

(i) Calculate $e^{z\hat{K}} c_j e^{-z\hat{K}}$ $(j = 1, 2)$ by considering the operator valued function
$$f_j(z) = e^{z\hat{K}} c_j e^{-z\hat{K}}, \quad j = 1, 2$$

and differentiation with respect to z. Solve the resulting initial value problem for the linear operator valued linear differential equation.

(ii) Apply the same calculations to
$$g_j(z) = e^{z\hat{K}} c_j^\dagger e^{-z\hat{K}}.$$

Finally insert $z = i\omega t$ for the two cases, where ω denotes the frequency.

Solution 51.   (i) The derivative of

$$f_1(z) = e^{z\hat{K}} c_1 e^{-z\hat{K}}$$

with respect to z and applying $c_1 c_1 = 0$ yields

$$\frac{df_1(z)}{dz} = e^{z\hat{K}} \hat{K} c_1 e^{-z\hat{K}} - e^{z\hat{K}} c_1 \hat{K} e^{-z\hat{K}} = -e^{z\hat{K}} c_2 e^{-z\hat{K}}$$

since

$$\hat{K} c_1 - c_1 \hat{K} = -c_2 c_1^\dagger c_1 - c_2 + c_2 c_1^\dagger c_1 = -c_2.$$

Next we calculate the second order derivative of f_1 and find

$$\frac{d^2 f_1(z)}{dz^2} = e^{z\hat{K}} c_1 e^{-z\hat{K}} = e^{z\hat{K}} c_1 e^{-z\hat{K}} = f_1(z).$$

This second order linear operator-valued differential equation with the initial values $f_1(0) = c_1$ and $df_1(0)/dz = -c_2$ yields

$$f_1(z) = \cosh(z) c_1 - \sinh(z) c_2.$$

Analogously we find for $f_2(z) = e^{z\hat{K}} c_2 e^{-z\hat{K}}$ that

$$f_2(z) = \cosh(z) c_2 - \sinh(z) c_1.$$

(ii) For the operator-valued functions $g_1(z) = e^{z\hat{K}} c_1^\dagger e^{-z\hat{K}}$, $g_2(z) = e^{z\hat{K}} c_2^\dagger e^{-z\hat{K}}$ we obtain

$$g_1(z) = \cosh(z) c_1^\dagger + \sinh(z) c_2^\dagger, \quad g_2(z) = \cosh(z) c_2^\dagger + \sinh(z) c_1^\dagger.$$

From the above results we can derive that

$$e^{z\hat{K}} c_1^\dagger c_1 e^{-z\hat{K}} = \cosh^2(z) c_1^\dagger c_1 - c_1^\dagger c_2 \cosh(z) \sinh(z) + c_2^\dagger c_1 \cosh(z) \sinh(z) - c_2^\dagger c_2 \sinh^2(z)$$

$$e^{z\hat{K}} c_2^\dagger c_2 e^{-z\hat{K}} = \cosh^2(z) c_2^\dagger c_2 + c_1^\dagger c_2 \cosh(z) \sinh(z) - c_2^\dagger c_1 \cosh(z) \sinh(z) - c_1^\dagger c_1 \sinh^2(z).$$

Note that $\cosh^2(z) - \sinh^2(z) = 1$. It follows that

$$e^{z\hat{K}} (c_1^\dagger c_1 + c_2^\dagger c_2) e^{-z\hat{K}} = c_1^\dagger c_1 + c_2^\dagger c_2$$

which simply tells us that $\hat{K}$ and $\hat{N} := c_1^\dagger c_1 + c_2^\dagger c_2$ commute.

Problem 52.   Let $z \in \mathbb{C}$. Find the system of differential equation for

$$f_1(z) = \exp(-z c_1^\dagger c_2^\dagger) c_1^\dagger \exp(z c_1^\dagger c_2^\dagger), \quad f_2(z) = \exp(-z c_1^\dagger c_2^\dagger) c_2^\dagger \exp(z c_1^\dagger c_2^\dagger)$$
$$g_1(z) = \exp(-\epsilon c_1^\dagger c_2^\dagger) c_1 \exp(z c_1^\dagger c_2^\dagger), \quad g_2(z) = \exp(-\epsilon c_1^\dagger c_2^\dagger) c_2 \exp(z c_1^\dagger c_2^\dagger)$$

and solve the initial value problem.

Solution 52.   Differentiation of f_1, f_2, g_1, g_2 provides the system of linear differential equations

$$\frac{df_1}{dz} = 0, \quad \frac{df_2}{dz} = 0, \quad \frac{dg_1}{dz} = f_2(z), \quad \frac{dg_2}{dz} = -f_1(z)$$

with the initial values (operators) $f_1(0) = c_1^\dagger$, $f_2(0) = c_2^\dagger$, $g_1(0) = c_1$, $g_2(0) = c_2$. Consequently the solution is

$$f_1(z) = c_1^\dagger, \quad f_2(z) = c_2^\dagger, \quad g_1(z) = zc_2^\dagger + c_1, \quad g_2(z) = -zc_1^\dagger + c_2.$$

Problem 53. Let $c_j^\dagger$, c_j $(j = 1, \ldots, n)$ be Fermi creation and annihilation operators.
(i) Find the eigenvalues of the self-adjoint operator

$$R = \exp\left(i\pi \sum_{j=0}^{N-1} c_j^\dagger c_j \right).$$

Note that $R^2 = I$.
(ii) Find the eigenvalues of the self-adjoint operator

$$Q := \frac{1}{2} \left(I - \exp\left(i\pi \sum_{j=1}^{n} c_j^\dagger c_j \right) \right).$$

Solution 53. (i) The operator R is self-adjoint with $R^2 = I$. The eigenvalues are $+1$ and -1.
(ii) Using the result from (i) we find that the eigenvalues of operator Q are 0 and 1.

Problem 54. Consider the operator

$$\hat{K} = t(c_\uparrow^\dagger c_\uparrow + c_\downarrow^\dagger c_\downarrow) + U c_\uparrow^\dagger c_\downarrow^\dagger c_\uparrow c_\downarrow.$$

One can form the mean-field Bogoliubov-de Gennes Hamilton operator

$$\tilde{K} = t(c_\uparrow^\dagger c_\uparrow + c_\downarrow^\dagger c_\downarrow) + \Delta(e^{i\phi} c_\downarrow c_\uparrow + e^{-i\phi} c_\uparrow^\dagger c_\downarrow^\dagger)$$

where $\Delta e^{i\phi} := U\langle c_\uparrow^\dagger c_\downarrow^\dagger \rangle$. Apply the *Bogoliubov transformation*

$$\tilde{c}_1 = -e^{i\phi} \cos(\theta) c_\uparrow + \sin(\theta) c_\downarrow^\dagger, \quad \tilde{c}_2 = e^{-i\phi} \cos(\theta) c_\downarrow^\dagger + \sin(\theta) c_\uparrow$$

to find the diagonal form of the operator $\tilde{K}$. The operators $\tilde{c}_j^\dagger$, $\tilde{c}_j$ also satisfy the anticommutation relations as one can easily prove.

Solution 54. We obtain the operator

$$\sqrt{t^2 + \Delta^2}(\tilde{c}_1^\dagger \tilde{c}_1 + \tilde{c}_2^\dagger \tilde{c}_2)$$

with $\tan(2\theta) = -\Delta/t$.

Problem 55. Let A, B be linear operators in a finite dimensional Hilbert space with the underlying field $\mathbb{C}$ and $z \in \mathbb{C}$. Then $\exp(zA)B\exp(-zA)$ can be calculated as

$$\exp(zA)B\exp(-zA) = B + \frac{z}{1!}[A, B] + \frac{z^2}{2!}[A, [A, B]] + \frac{z^3}{3!}[A, [A, [A, B]]] + \cdots$$

(i) Apply it to $A = c_1^\dagger c_2 + c_2^\dagger c_1$ and $B = c_1$.

(ii) Apply it to $A = c_1^\dagger c_2 + c_2^\dagger c_1$ and $B = c_2$.

Solution 55.   (i) First we note that

$$[c_1^\dagger c_2 + c_2^\dagger c_1, c_1] = -c_2, \qquad [c_1^\dagger c_2 + c_2^\dagger c_1, c_2] = -c_1.$$

Thus we have

$$[A, B] = [c_1^\dagger c_2 + c_2^\dagger c_1, c_1] = -c_2$$
$$[A, [A, B]] = [c_1^\dagger c_2 + c_2^\dagger c_1, -c_2] = c_1$$
$$[A, [A, [A, B]]] = [c_1^\dagger c_2 + c_2^\dagger c_1, c_2] = -c_2.$$

Thus we see "the pattern" and find

$$e^{zA} B e^{-zA} = (1 + \frac{1}{2!}z^2 + \frac{1}{4!}z^4 + \cdots)c_1 - (\frac{1}{1!}z + \frac{1}{3!}z^3 + \frac{1}{5!}z^5 + \cdots)c_2 = \cosh(z)c_1 - \sinh(z)c_2.$$

(ii) We have

$$[A, B] = [c_1^\dagger c_2 + c_2^\dagger c_1, c_2] = -c_1$$
$$[A, [A, B]] = [c_1^\dagger c_2 + c_2^\dagger c_1, -c_1] = c_2$$
$$[A, [A, [A, B]]] = [c_1^\dagger c_2 + c_2^\dagger c_1, c_2] = -c_1.$$

Thus we see "the pattern" and find

$$e^{zA} B e^{-zA} = \cosh(z)c_1 - \sinh(z)c_2.$$

Problem 56.   Let
$$K = c_1^\dagger c_2 + c_2^\dagger c_1 + c_2^\dagger c_3 + c_3^\dagger c_2.$$
Find $e^{zK} c_1 e^{-zK}$. Note that

$$[K, c_1] = -c_2, \quad [K, c_2] = -c_1 - c_3, \quad [K, c_3] = -c_2.$$

Solution 56.   Using the properties given above provides

$$[K, c_1] = -c_2$$
$$[K, -c_2] = c_1 + c_3$$
$$[K, c_1 + c_3] = -2c_2$$
$$[K, -2c_2] = 2(c_1 + c_3)$$
$$[K, 2(c_1 + c_3)] = -4c_2$$
$$[K, -4c_2] = 4(c_1 + c_3)$$
$$[K, 4(c_1 + c_3)] = -c_2$$

etc. Thus $e^{zK} c_1 e^{-zK}$ is given by

$$c_1 \left(1 + \frac{z^2}{2!} + \frac{2z^4}{4!} + \frac{4z^6}{6!} + \frac{8z^8}{8!} + \cdots \right) - c_2 \left(\frac{z}{1!} + \frac{2z^3}{3!} + \frac{4z^5}{5!} + \frac{8z^7}{7!} + \cdots \right)$$
$$+ c_3 \left(\frac{z^2}{2!} + \frac{2z^4}{4!} + \frac{4z^6}{6!} + \frac{8z^8}{8!} + \cdots \right)$$

Summing up the terms we obtain

$$e^{zK}c_1 e^{-zK} = \frac{1}{2}(\cosh(\sqrt{2}z) + 1)c_1 - \frac{1}{\sqrt{2}}\sinh(\sqrt{2}z)c_2 + \frac{1}{2}(\cosh(\sqrt{2}z) - 1)c_3.$$

Problem 57.   Let σ_1, σ_2, σ_3 be the Pauli spin matrices. Let $N \geq 1$ and $n = 1, \ldots, N$. The *Jordan-Wigner transformation* is given by

$$\sigma_{1,n} = \prod_{m=1}^{n-1}(I - 2c_m^\dagger c_m)(c_n^\dagger + c_n), \quad \sigma_{2,n} = -i\prod_{m=1}^{n-1}(c_n^\dagger - c_n)(I - 2c_m^\dagger c_m)$$

with the convention that for $n = 1$ one has

$$\sigma_{1,1} = c_1^\dagger + c_1, \qquad \sigma_{2,1} = -i(c_1^\dagger - c_1).$$

Let $N = 2$.
(i) Find $\sigma_{1,2}$, $\sigma_{2,2}$ and $\sigma_{1,2}\sigma_{2,2}$.
(ii) Study the matrix representation of $\sigma_{1,2}\sigma_{2,2}$ with the four dimensional basis

$$c_2^\dagger c_1^\dagger|0\rangle, \ c_1^\dagger|0\rangle, \ c_2^\dagger|0\rangle, \ |0\rangle.$$

Compare to $\sigma_{1,2} = I_2 \otimes \sigma_1$, $\sigma_{2,2} = I_2 \otimes \sigma_2$ and $\sigma_{1,2}\sigma_{2,2} = I_2 \otimes \sigma_1\sigma_2$.

Solution 57.   (i) We have

$$\sigma_{1,2} = (I - 2c_1^\dagger c_1)(c_2^\dagger + c_2), \quad \sigma_{2,2} = -i(c_2^\dagger - c_2)(I - 2c_1^\dagger c_1).$$

Since

$$(c_2^\dagger + c_2)(c_2^\dagger - c_2) = I - 2c_2^\dagger c_2$$

we obtain

$$\sigma_{1,2}\sigma_{2,2} = -i(I - 2c_2^\dagger c_2).$$

(ii) With the basis given above we obtain the matrix representation

$$\begin{pmatrix} i & 0 & 0 & 0 \\ 0 & -i & 0 & 0 \\ 0 & 0 & i & 0 \\ 0 & 0 & 0 & -i \end{pmatrix} = (I_2 \otimes \sigma_1)(I_2 \otimes \sigma_2) = I_2 \otimes \sigma_1\sigma_2 = I_2 \otimes (i\sigma_3).$$

Problem 58.   Let $\hat{N}_1 = c_1^\dagger c_1$, $\hat{N}_2 = c_2^\dagger c_2$, $\hat{N}_3 = c_3^\dagger c_3$ and $\hat{N}$ be the number operator

$$\hat{N} = \hat{N}_1 + \hat{N}_2 + \hat{N}_3$$

(i) Let

$$K_1 = c_1^\dagger c_2 + c_2^\dagger c_1 + c_2^\dagger c_3 + c_3^\dagger c_2.$$

Find K_1^2. Apply computer algebra.
(ii) Let

$$K_2 = c_1^\dagger c_2 + c_2^\dagger c_1 + c_2^\dagger c_3 + c_3^\dagger c_2 + c_3^\dagger c_1 + c_1^\dagger c_3.$$

Find K_2^2. Apply computer algebra.

Solution 58.   (i) We obtain for K_1^2

$$K_1^2 = \hat{N}_1 + 2\hat{N}_2 + \hat{N}_3 - \hat{N}_2(\hat{N}_1 + \hat{N}_3) + 2\hat{N}_2(-c_1^\dagger c_3 + c_1 c_3^\dagger) + c_1^\dagger c_3 + c_3^\dagger c_1$$

(ii) We obtain for K_2^2

$$\begin{aligned} K_2^2 = 2\hat{N} &- 2(\hat{N}_1\hat{N}_2 + \hat{N}_2\hat{N}_3 + \hat{N}_1\hat{N}_3) \\ &+ 2\hat{N}_1(-c_2c_3^\dagger - c_2^\dagger c_3) + 2\hat{N}_2(-c_1^\dagger c_3 - c_3^\dagger c_1) + 2\hat{N}_3(-c_1^\dagger c_2 - c_2^\dagger c_1) + K_2. \end{aligned}$$

Problem 59.   Let

$$K = \frac{1}{\sqrt{n}}(c_1^\dagger + c_1 + c_2^\dagger + c_2 + \cdots + c_n^\dagger + c_n).$$

Find K^2, K^3 etc.

Solution 59.   With $c_j^\dagger c_j + c_j c_j^\dagger = I$ and for $j \neq k$ $c_j c_k = -c_k c_j$, $c_j^\dagger c_k^\dagger = -c_k^\dagger c_j^\dagger$, $c_j^\dagger c_k = -c_k c_j^\dagger$ we obtain

$$K^2 = I.$$

Hence $K^3 = K$ etc.

Problem 60.   Let $c_{j\sigma}^\dagger$ and $c_{j\sigma}$ be Fermi creation and annihilation operators, where $\sigma \in \{\uparrow, \downarrow\}$ and $j = 1, 2, \ldots, N$. The Fermi operators obey the anticommutation relations

$$[c_{j\sigma}, c_{k\sigma'}^\dagger]_+ = \delta_{jk}\delta_{\sigma\sigma'}I, \quad [c_{j\sigma}, c_{k\sigma'}]_+ = 0$$

where I is the identity operator and 0 the zero operator. From the second relation it follows that

$$[c_{j\sigma}^\dagger, c_{k\sigma'}^\dagger]_+ = 0.$$

Let

$$\hat{N}_{j\uparrow} := c_{j\uparrow}^\dagger c_{j\uparrow}, \quad \hat{N}_{j\downarrow} := c_{j\downarrow}^\dagger c_{j\downarrow}, \quad \hat{H} = U\sum_{k=1}^{N} \hat{N}_{k\uparrow}\hat{N}_{k\downarrow}$$

where U is a real constant. Calculate

$$e^{\epsilon\hat{H}} c_{j\uparrow}^\dagger e^{-\epsilon\hat{H}}$$

where ϵ is a real positive parameter.

Solution 60.   We set

$$f(\epsilon) = e^{\epsilon\hat{H}} c_{j\uparrow}^\dagger e^{-\epsilon\hat{H}}$$

where f is an operator-valued function. Since $[c_{j\uparrow}, \hat{N}_{k\downarrow}] = 0$ and for $j \neq k$ we have $[c_{j\uparrow}, \hat{N}_{k\uparrow}] = 0$ we obtain

$$f(\epsilon) = e^{\epsilon U \hat{N}_{j\uparrow}\hat{N}_{j\downarrow}} c_{j\uparrow}^\dagger e^{-\epsilon U \hat{N}_{j\uparrow}\hat{N}_{j\downarrow}}.$$

The derivative of f with respect to ϵ yields

$$\frac{df}{d\epsilon} = U e^{\epsilon U \hat{N}_{j\uparrow}\hat{N}_{j\downarrow}} \hat{N}_{j\uparrow}\hat{N}_{j\downarrow} c_{j\uparrow}^\dagger e^{-\epsilon U \hat{N}_{j\uparrow}\hat{N}_{j\downarrow}} - U e^{\epsilon U \hat{N}_{j\uparrow}\hat{N}_{j\downarrow}} c_{j\uparrow}^\dagger \hat{N}_{j\uparrow}\hat{N}_{j\downarrow} e^{-\epsilon U \hat{N}_{j\uparrow}\hat{N}_{j\downarrow}}.$$

Since $\hat{N}_{j\uparrow}\hat{N}_{j\downarrow}c_{j\uparrow}^\dagger = c_{j\uparrow}^\dagger\hat{N}_{j\downarrow}$, $c_{j\uparrow}^\dagger\hat{N}_{j\uparrow}\hat{N}_{j\downarrow} = 0$ we find

$$\frac{df}{d\epsilon} = U e^{\epsilon U \hat{N}_{j\uparrow}\hat{N}_{j\downarrow}} c_{j\uparrow}^\dagger \hat{N}_{j\downarrow} e^{-\epsilon U \hat{N}_{j\uparrow}\hat{N}_{j\downarrow}}.$$

The second derivative yields

$$\frac{d^2 f}{d\epsilon^2} = U^2 e^{\epsilon U \hat{N}_{j\uparrow}\hat{N}_{j\downarrow}} c_{j\uparrow}^\dagger \hat{N}_{j\downarrow} e^{-\epsilon U \hat{N}_{j\uparrow}\hat{N}_{j\downarrow}} = U\frac{df}{d\epsilon}$$

where we used that $\hat{N}_{j\uparrow}\hat{N}_{j\uparrow} = \hat{N}_{j\uparrow}$ and $c_{j\uparrow}^\dagger\hat{N}_{j\uparrow}\hat{N}_{j\downarrow} = 0$. Thus we obtain a linear second order differential equation

$$\frac{d^2 f}{d\epsilon^2} = U\frac{df}{d\epsilon}$$

with the initial conditions

$$f(\epsilon = 0) = c_{j\uparrow}^\dagger, \qquad \frac{df(\epsilon = 0)}{d\epsilon} = U c_{j\uparrow}^\dagger \hat{N}_{j\downarrow}.$$

We solve the linear differential equation with the ansatz $f(\epsilon) = Ae^{k\epsilon}$. We obtain $k^2 - Uk = 0$ with the solutions $k = 0$ and $k = U$. Thus $f(\epsilon) = Ae^{\epsilon U} + B$. Then the initial values provide

$$c_{j\uparrow} = A + B, \qquad U c_{j\uparrow}^\dagger \hat{N}_{j\downarrow} = U A.$$

It follows that

$$A = c_{j\uparrow}^\dagger \hat{N}_{j\downarrow}, \qquad B = c_{j\uparrow}^\dagger - c_{j\uparrow}^\dagger \hat{N}_{j\downarrow}.$$

Therefore

$$f(\epsilon) = (e^{\epsilon U} - 1)c_{j\uparrow}^\dagger \hat{N}_{j\downarrow} + c_{j\uparrow}^\dagger.$$

Thus

$$e^{\epsilon \hat{H}} c_{j\uparrow}^\dagger e^{-\epsilon \hat{H}} = (e^{\epsilon U} - 1)c_{j\uparrow}^\dagger \hat{N}_{j\downarrow} + c_{j\uparrow}^\dagger.$$

3.3   Hamilton Operators

Problem 61.   (i) Consider the Hamilton operator ($\gamma \in \mathbb{R}$ and $\gamma > 0$)

$$\hat{H} = \hbar\omega c^\dagger c + \gamma(c^\dagger + c)$$

and the two-dimensional basis $\{\, |0\rangle, \ c^\dagger|0\rangle \,\}$. Find the matrix representation of $\hat{H}$.
(ii) Calculate the eigenvalues and eigenvectors of this matrix.
(iii) Note that $\{\, c^\dagger|0\rangle, \ |0\rangle \,\}$ is an orthonormal basis in a two-dimensional Hilbert space. Consider the normalized state

$$|\psi\rangle = \alpha c^\dagger|0\rangle + \beta|0\rangle, \quad \alpha, \beta \in \mathbb{C}, \quad \alpha\bar{\alpha} + \beta\bar{\beta} = 1$$

and the hermitian operator

$$\hat{H} = \hbar\omega_1 c^\dagger c + \hbar\omega_2(c^\dagger + c) + \hbar\omega_3(-ic^\dagger + ic)$$

where ω_j ($j = 1, 2, 3$) are frequencies. Calculate the expectation value $\langle\psi|\hat{H}|\psi\rangle$.

Solution 61.   (i) The dual basis is $\{\langle 0|, \ \langle 0|c\}$. Using $c|0\rangle = 0|0\rangle$, $\langle 0|c^\dagger = \langle 0|0$ and the anticommutation relations for Fermi operators we obtain the matrix

$$\begin{pmatrix} \hbar\omega & \gamma \\ \gamma & 0 \end{pmatrix}.$$

(ii) The eigenvalues are

$$E_\pm = \frac{\hbar\omega}{2} \pm \frac{1}{2}\sqrt{\hbar^2\omega^2 + 4\gamma^2}.$$

The non-normalized eigenvectors for E_+ and E_- are

$$\begin{pmatrix} \gamma \\ E_+ - \hbar\omega \end{pmatrix}, \quad \begin{pmatrix} E_- \\ \gamma \end{pmatrix}.$$

(iii) We have

$$\hat{H}|\psi\rangle = (\hbar\omega_1\alpha + \hbar\omega_2\beta - \hbar\omega_3 i\beta)c^\dagger|0\rangle + (\hbar\omega_2\alpha + \hbar\omega_3 i\alpha)|0\rangle.$$

Consequently we obtain the expectation value

$$\langle\psi|\hat{H}|\psi\rangle = \hbar\omega_1\alpha\bar{\alpha} + \hbar\omega_2(\alpha\bar{\beta} + \bar{\alpha}\beta) + \hbar\omega_3(i\alpha\bar{\beta} - i\bar{\alpha}\beta).$$

Problem 62.   Consider Fermi creation and annihilation operators, $c^\dagger$, c. We define the hermitian operator

$$\hat{K} := \frac{1}{2}[c^\dagger, c].$$

(i) Find the commutators $[c, \hat{K}]$ and $[c^\dagger, K]$.
(ii) Apply the operator $\hat{K}$ to the states $|0\rangle$, $c^\dagger|0\rangle$.

Solution 62.   (i) We have

$$\hat{K} = \frac{1}{2}(c^\dagger c - cc^\dagger) = \frac{1}{2}(c^\dagger c - (I - c^\dagger c)) = c^\dagger c - \frac{1}{2}I.$$

Since $c^\dagger c^\dagger = 0$ and $cc = 0$ we obtain $[c, \hat{K}] = c$ and $[c^\dagger, \hat{K}] = -c^\dagger$.
(ii) We find

$$\hat{K}|0\rangle = -\frac{1}{2}|0\rangle, \qquad \hat{K}c^\dagger|0\rangle = \frac{1}{2}c^\dagger|0\rangle.$$

In both cases we have an eigenvalue equation. In the first case the eigenvalue is $-1/2$ and in the second case the eigenvalue is $+1/2$.

Problem 63. Consider the hermitian Hamilton operator

$$\hat{K} = \frac{\hat{H}}{\hbar\omega} = c_1^\dagger c_2 + c_2^\dagger c_1$$

and the *number operator* $\hat{N} := c_1^\dagger c_1 + c_2^\dagger c_2$.
(i) Find the commutators $[\hat{K}, c_1^\dagger c_1]$ and $[\hat{K}, c_2^\dagger c_2]$ and thus show that $[\hat{K}, \hat{N}] = 0$.
(ii) Find

$$\hat{K}c_1^\dagger|0\rangle, \quad (\hat{K})^2 c_1^\dagger|0\rangle, \quad \hat{K}c_2^\dagger|0\rangle, \quad (\hat{K})^2 c_2^\dagger|0\rangle.$$

(iii) Find the matrix representation of the operator $\hat{K}$. Since the operator $\hat{K}$ commutes with the number operator $\hat{N}$ we can consider the four subspaces

$$\{\, c_2^\dagger c_1^\dagger|0\rangle\,\}, \ \{\, c_2^\dagger|0\rangle,\, c_1^\dagger|0\rangle\,\}, \ \{\,|0\rangle\,\}$$

to find the matrix representation. For the dual spaces we have

$$\{\,\langle 0|c_1 c_2\,\}, \ \{\,\langle 0|c_2,\, \langle 0|c_1\,\}, \ \{\,\langle 0|\,\}.$$

Find the matrix representation of each subspace and the eigenvalues.
(iii) Find the commutators

$$[\hat{K}, [\hat{K}, c_1^\dagger c_1]] \quad \text{and} \quad [\hat{K}, [\hat{K}, c_2^\dagger c_2]].$$

(iv) Let

$$\sigma_+ = \begin{pmatrix} 0 & 2 \\ 0 & 0 \end{pmatrix}, \quad \sigma_- = \begin{pmatrix} 0 & 0 \\ 2 & 0 \end{pmatrix}, \quad \sigma_3 = \begin{pmatrix} 1 & 0 \\ 0 & -1 \end{pmatrix}, \quad I_2 = \begin{pmatrix} 1 & 0 \\ 0 & 1 \end{pmatrix}.$$

A matrix representation of $c_1, c_1^\dagger, c_2, c_2^\dagger$ is given by

$$c_1 = \frac{1}{2}\sigma_- \otimes I_2 \Rightarrow c_1^\dagger = \frac{1}{2}\sigma_+ \otimes I_2$$

$$c_2 = \sigma_3 \otimes \frac{1}{2}\sigma_- \Rightarrow c_2^\dagger = \sigma_3 \otimes \frac{1}{2}\sigma_+.$$

Find $\hat{K}$ and $\hat{N}$ for this matrix representation.

Solution 63. (i) We have

$$[c_1^\dagger c_2 + c_2^\dagger c_1, c_1^\dagger c_1] = -c_1^\dagger c_2 + c_2^\dagger c_1, \quad [c_1^\dagger c_2 + c_2^\dagger c_1, c_2^\dagger c_2] = c_1^\dagger c_2 - c_2^\dagger c_1.$$

Thus $[c_1^\dagger c_2 + c_2^\dagger c_1, c_1^\dagger c_1 + c_2^\dagger c_2] = 0$.
(ii) We have

$$\hat{K}c_1^\dagger|0\rangle = c_2^\dagger, \qquad (\hat{K})^2 c_1^\dagger|0\rangle = \hat{K}c_2^\dagger|0\rangle = c_1^\dagger|0\rangle$$

and

$$\hat{K} c_2^\dagger |0\rangle = c_1^\dagger, \qquad (\hat{K})^2 c_2^\dagger |0\rangle = \hat{K} c_1^\dagger |0\rangle = c_2^\dagger |0\rangle.$$

(iii) For the one-dimensional subspace $\{ c_2^\dagger c_1^\dagger |0\rangle \}$ we have $\hat{K} c_2^\dagger c_1^\dagger |0\rangle = 0|0\rangle$ with the eigenvalue 0 and therefore

$$\langle 0 | c_1 c_2 \hat{K} c_2^\dagger c_1^\dagger |0\rangle = 0.$$

For the two-dimensional subspace $\{c_1^\dagger |0\rangle, c_2^\dagger |0\rangle \}$ we have

$$\hat{K} c_1^\dagger |0\rangle = c_2^\dagger |0\rangle, \qquad \hat{K} c_2^\dagger |0\rangle = c_1^\dagger |0\rangle$$

and

$$\langle 0 | c_1 \hat{K} c_1^\dagger |0\rangle = 0, \quad \langle 0 | c_1 \hat{K} c_2^\dagger |0\rangle = 1, \quad \langle 0 | c_2 \hat{K} c_1^\dagger |0\rangle = 1, \quad \langle 0 | c_2 \hat{K} c_2^\dagger |0\rangle = 0$$

with the matrix representation of $\hat{K}$

$$\begin{pmatrix} 0 & 1 \\ 1 & 0 \end{pmatrix} = \sigma_1.$$

The eigenvalues are $+1$ and -1. Finally $\hat{K}|0\rangle = 0|0\rangle$ with the eigenvalue 0.
(iii) We find $[\hat{K}, [\hat{K}, c_1^\dagger c_1]] = \hat{K}$, $[\hat{K}, [\hat{K}, c_2^\dagger c_2]] = \hat{K}$.
(iv) We obtain the symmetric matrix

$$\hat{K} = \begin{pmatrix} 0 & 0 & 0 & 0 \\ 0 & 0 & -1 & 0 \\ 0 & -1 & 0 & 0 \\ 0 & 0 & 0 & 0 \end{pmatrix}$$

with the eigenvalues 0 (twice) and $+1$ and -1. For the number operator we obtain the diagonal matrix

$$\hat{N} = \begin{pmatrix} 2 & 0 & 0 & 0 \\ 0 & 1 & 0 & 0 \\ 0 & 0 & 1 & 0 \\ 0 & 0 & 0 & 0 \end{pmatrix}$$

with the eigenvalues 2, 1 (twice) and 0.

Problem 64.   Consider the Hamilton operator

$$\hat{H} = \hbar\omega(c_1^\dagger c_1 + c_2^\dagger c_2) + t(c_1^\dagger c_2 + c_2^\dagger c_1).$$

The Hamilton operator $\hat{H}$ commutes with the *number operator*

$$\hat{N} := c_1^\dagger c_1 + c_2^\dagger c_2.$$

Owing to $[\hat{H}, \hat{N}] = 0$ we can consider the three subspaces

$$\{c_2^\dagger c_1^\dagger |0\rangle\}, \quad \{c_2^\dagger |0\rangle, c_1^\dagger |0\rangle\}, \quad \{|0\rangle\}$$

to find the eigenvalues of the Hamilton operator. Find the eigenvalues for each subspace.

Solution 64.   For the one-dimensional subspace $\{ c_2^\dagger c_1^\dagger |0\rangle \}$ we have the eigenvalue equation

$$\hat{H} c_2^\dagger c_1^\dagger |0\rangle = 2\hbar\omega c_2^\dagger c_1^\dagger |0\rangle$$

with the eigenvalue $2\hbar\omega$ and the eigenvector $c_2^\dagger c_1^\dagger|0\rangle$. For the two-dimensional subspace $\{c_2^\dagger|0\rangle, c_1^\dagger|0\rangle\}$ we have

$$\hat{H}c_1^\dagger|0\rangle = \hbar\omega c_1^\dagger|0\rangle + tc_2^\dagger|0\rangle, \quad \hat{H}c_2^\dagger|0\rangle = \hbar\omega c_2^\dagger|0\rangle + tc_1^\dagger|0\rangle.$$

Thus

$$\langle 0|c_1\hat{H}c_1^\dagger|0\rangle = \hbar\omega, \quad \langle 0|c_2\hat{H}c_1^\dagger|0\rangle = t, \quad \langle 0|c_1\hat{H}c_2^\dagger|0\rangle = t, \quad \langle 0|c_2\hat{H}c_2^\dagger|0\rangle = \hbar\omega.$$

This provides the symmetric matrix

$$\begin{pmatrix} \hbar\omega & t \\ t & \hbar\omega \end{pmatrix}$$

with the two eigenvalues $\lambda_1 = \hbar\omega + t$, $\lambda_2 = \hbar\omega - t$. Thus the normalized eigenvectors are given by

$$\frac{1}{\sqrt{2}}(c_1^\dagger|0\rangle + c_2^\dagger|0\rangle), \quad \frac{1}{\sqrt{2}}(c_1^\dagger|0\rangle - c_2^\dagger|0\rangle).$$

For the one-dimensional subspace $\{|0\rangle\}$ we have the eigenvalue equation $\hat{H}|0\rangle = 0|0\rangle$ with the eigenvalue 0 and the eigenvector $|0\rangle$.

Problem 65. (i) Consider the hermitian operator

$$\hat{K} = \frac{\hat{H}}{\hbar\omega} = c_2^\dagger c_1^\dagger + c_1 c_2$$

with the four dimensional basis $\{c_2^\dagger c_1^\dagger|0\rangle, c_2^\dagger|0\rangle, c_1^\dagger|0\rangle, |0\rangle\}$. Show that the operator $\hat{K}$ does not commute with the number operator $\hat{N} = c_1^\dagger c_1 + c_2^\dagger c_2$.
(ii) Find

$$\hat{K}c_2^\dagger c_1^\dagger|0\rangle, \quad (\hat{K})^2 c_2^\dagger c_1^\dagger|0\rangle.$$

(iii) Find the matrix representation of $\hat{K}$.
(iv) Let

$$\sigma_+ = \begin{pmatrix} 0 & 2 \\ 0 & 0 \end{pmatrix}, \quad \sigma_- = \begin{pmatrix} 0 & 0 \\ 2 & 0 \end{pmatrix}, \quad \sigma_3 = \begin{pmatrix} 1 & 0 \\ 0 & -1 \end{pmatrix}, \quad I_2 = \begin{pmatrix} 1 & 0 \\ 0 & 1 \end{pmatrix}.$$

A matrix representation of $c_1, c_1^\dagger, c_2, c_2^\dagger$ is given by

$$c_1 = \frac{1}{2}\sigma_- \otimes I_2 \Rightarrow c_1^\dagger = \frac{1}{2}\sigma_+ \otimes I_2$$

$$c_2 = \sigma_3 \otimes \frac{1}{2}\sigma_- \Rightarrow c_2^\dagger = \sigma_3 \otimes \frac{1}{2}\sigma_+.$$

Find $\hat{K}$ and $\hat{N}$ for this matrix representation.
(v) Find the matrix representation of the hermitian Hamilton operator

$$\hat{H} = t(c_1^\dagger c_2 + c_2^\dagger c_1) + \gamma(c_2^\dagger c_1^\dagger + c_1 c_2)$$

with the four dimensional basis $\{c_2^\dagger c_1^\dagger|0\rangle, c_2^\dagger|0\rangle, c_1^\dagger|0\rangle, |0\rangle\}$. Note that $\hat{H}$ does not commute with the number operator $\hat{N} = c_1^\dagger c_1 + c_2^\dagger c_2$.

Solution 65. (i) We have $[\hat{K}, \hat{N}] = 2(-c_2^\dagger c_1^\dagger + c_1 c_2)$. Thus $\hat{N}$ is not a constant of motion.

(ii) We find
$$\hat{K}c_2^\dagger c_1^\dagger |0\rangle = |0\rangle, \qquad (\hat{K})^2 c_2^\dagger c_1^\dagger |0\rangle = \hat{K}|0\rangle = c_2^\dagger c_1^\dagger |0\rangle.$$

(iii) We have
$$\hat{K}c_2^\dagger c_1^\dagger |0\rangle = c_1 c_2 c_2^\dagger c_1^\dagger |0\rangle = |0\rangle, \quad \hat{K}c_2^\dagger|0\rangle = 0 c_2^\dagger|0\rangle, \quad \hat{K}c_1^\dagger|0\rangle = 0 c_1^\dagger|0\rangle, \quad \hat{K}|0\rangle = c_2^\dagger c_1^\dagger |0\rangle.$$

Thus the matrix representation of $\hat{K}$ is given by the hermitian 4×4 matrix
$$\begin{pmatrix} 0 & 0 & 0 & 1 \\ 0 & 0 & 0 & 0 \\ 0 & 0 & 0 & 0 \\ 1 & 0 & 0 & 0 \end{pmatrix}$$

with the eigenvalues 0 (twice), $+1$ and -1.
(iv) We find the matrix representation given in (iii).
(v) We have
$$\hat{H}c_2^\dagger c_1^\dagger |0\rangle = \gamma|0\rangle, \quad \hat{H}c_2^\dagger|0\rangle = tc_1^\dagger|0\rangle, \quad \hat{H}c_1^\dagger|0\rangle = tc_2^\dagger|0\rangle, \quad \hat{H}|0\rangle = \gamma c_2^\dagger c_1^\dagger |0\rangle.$$

Thus the matrix representation of $\hat{H}$ is given by the 4×4 hermitian matrix
$$\begin{pmatrix} 0 & 0 & 0 & \gamma \\ 0 & 0 & t & 0 \\ 0 & t & 0 & 0 \\ \gamma & 0 & 0 & 0 \end{pmatrix}$$

with the eigenvalues $\gamma, -\gamma, t, -t$.

Problem 66. Let c_1, c_2 be Fermi annihilation operators. Consider the hermitian Hamilton operator
$$\hat{H} = \hbar\omega_1 c_1^\dagger c_1 + \hbar\omega_2 c_2^\dagger c_2 + \hbar(Vc_1^\dagger c_2 + V^* c_2^\dagger c_1)$$
where V is a complex coupling constant. Find the Heisenberg equation of motion for the operators c_1 and c_2 given by
$$i\hbar\frac{dc_1}{dt} = [c_1, \hat{H}](t), \quad i\hbar\frac{dc_2}{dt} = [c_2, \hat{H}](t).$$

Solution 66. Inserting the Hamilton operator $\hat{H}$ yields
$$i\frac{dc_1}{dt} = \omega_1 c_1 + Vc_2, \quad i\frac{dc_2}{dt} = \omega_2 c_2 + V^* c_1.$$

This system of linear operator equations can be written in matrix form
$$\begin{pmatrix} dc_1/dt \\ dc_2/dt \end{pmatrix} = \begin{pmatrix} -i\omega_1 & -iV \\ -iV^* & -i\omega_2 \end{pmatrix} \begin{pmatrix} c_1 \\ c_2 \end{pmatrix}.$$

The eigenvalues of the matrix on the right-hand side are given by
$$\lambda_\pm = -\frac{i}{2}(\omega_1 + \omega_2) \pm \frac{i}{2}\sqrt{(\omega_1 - \omega_2)^2 + 4VV^*}.$$

Problem 67. Consider the hermitian Hamilton operator ($\gamma \in \mathbb{R}, \gamma > 0$)

$$\hat{H} = \hbar\omega(c_1^\dagger c_1 + c_2^\dagger c_2) + \gamma(c_1^\dagger + c_1 + c_2^\dagger + c_2)$$

and the basis $c_2^\dagger c_1^\dagger |0\rangle$, $c_2^\dagger |0\rangle$, $c_1^\dagger |0\rangle$, $|0\rangle$. Find the matrix representation of $\hat{H}$. Calculate the eigenvalues of this matrix.

Solution 67. The dual basis is $\langle 0|c_1 c_2$, $\langle 0|c_2$, $\langle 0|c_1$, $\langle 0|$. We obtain

$$\hat{H}c_2^\dagger c_1^\dagger |0\rangle = 2\hbar\omega c_2^\dagger c_1^\dagger |0\rangle - \gamma c_2^\dagger |0\rangle + \gamma c_1^\dagger |0\rangle$$
$$\hat{H}c_2^\dagger |0\rangle = \hbar\omega c_2^\dagger |0\rangle - \gamma c_2^\dagger c_1^\dagger |0\rangle + \gamma |0\rangle$$
$$\hat{H}c_1^\dagger |0\rangle = \hbar\omega c_1^\dagger |0\rangle + \gamma |0\rangle + \gamma c_2^\dagger c_1^\dagger |0\rangle$$
$$\hat{H}|0\rangle = \gamma c_1^\dagger |0\rangle + \gamma c_2^\dagger |0\rangle.$$

Thus the matrix representation of $\hat{H}$ is given by

$$\begin{pmatrix} 2\hbar\omega & -\gamma & \gamma & 0 \\ -\gamma & \hbar\omega & 0 & \gamma \\ \gamma & 0 & \hbar\omega & \gamma \\ 0 & \gamma & \gamma & 0 \end{pmatrix}$$

with the four eigenvalues

$$\frac{1}{2}\left(-\sqrt{\hbar^2\omega^2 + 8\gamma^2} + \hbar\omega\right), \qquad \frac{1}{2}\left(\sqrt{\hbar^2\omega^2 + 8\gamma^2} + \hbar\omega\right),$$

$$\frac{1}{2}\left(-\sqrt{\hbar^2\omega^2 + 8\gamma^2} + 3\hbar\omega\right), \qquad \frac{1}{2}\left(\sqrt{\hbar^2\omega^2 + 8\gamma^2} + 3\hbar\omega\right).$$

Problem 68. Consider the Hamilton operators

$$\hat{K}_1 = \frac{\hat{H}_1}{\hbar\omega} = c_1^\dagger c_2 + c_2^\dagger c_1 + c_2^\dagger c_3 + c_3^\dagger c_2 \quad \text{open end boundary condition}$$

$$\hat{K}_2 = \frac{\hat{H}_2}{\hbar\omega} = c_1^\dagger c_2 + c_2^\dagger c_1 + c_2^\dagger c_3 + c_3^\dagger c_2 + c_3^\dagger c_1 + c_1^\dagger c_3 \quad \text{cyclic boundary conditions.}$$

(i) Find $\hat{K}_1 c_1^\dagger |0\rangle$, $(\hat{K}_1)^2 c_1^\dagger |0\rangle$, $(\hat{K}_1)^3 c_1^\dagger |0\rangle$.
(ii) Find $\hat{K}_2 c_1^\dagger |0\rangle$, $(\hat{K}_2)^2 c_1^\dagger |0\rangle$, $(\hat{K}_2)^3 c_1^\dagger |0\rangle$.
(iii) Given the eight dimensional basis

$$\{c_3^\dagger c_2^\dagger c_1^\dagger |0\rangle, \quad c_3^\dagger c_2^\dagger |0\rangle, \quad c_3^\dagger c_1^\dagger |0\rangle, \quad c_2^\dagger c_1^\dagger |0\rangle\}, \quad c_3^\dagger |0\rangle, \quad c_2^\dagger |0\rangle, \quad c_1^\dagger |0\rangle\}, \quad |0\rangle\}$$

find the matrix representation of $\hat{K}_1$ and $\hat{K}_2$. Find the matrix representation of the number operator $\hat{N}$.
(iv) A matrix representation for the Fermi creation and annihilation operators $c_j^\dagger$, c_j ($j = 1, 2, 3$) is given by

$$c_1 \mapsto \frac{1}{2}\sigma_- \otimes I_2 \otimes I_2 \Rightarrow c_1^\dagger \mapsto \frac{1}{2}\sigma_+ \otimes I_2 \otimes I_2$$

$$c_2 \mapsto \sigma_3 \otimes \frac{1}{2}\sigma_- \otimes I_2 \Rightarrow c_2^\dagger \mapsto \sigma_3 \otimes \frac{1}{2}\sigma_+ \otimes I_2$$

$$c_3 \mapsto \sigma_3 \otimes \sigma_3 \otimes \frac{1}{2}\sigma_- \Rightarrow c_3^\dagger \mapsto \sigma_3 \otimes \sigma_3 \otimes \frac{1}{2}\sigma_+.$$

Find the matrix representation of $\hat{K}_1$ and $\hat{K}_2$ and calculate the eigenvalues of these 8×8 hermitian matrices.

Solution 68.   (i) We have

$$\hat{K}_1 c_1^\dagger|0\rangle = c_2^\dagger|0\rangle, \quad \hat{K}_1 c_2^\dagger|0\rangle = c_1^\dagger|0\rangle + c_3^\dagger|0\rangle, \quad \hat{K}_1(c_1^\dagger|0\rangle + c_3^\dagger|0\rangle) = 2c_2^\dagger|0\rangle.$$

Note that $c_1^\dagger|0\rangle + c_3^\dagger|0\rangle$ is not normalized.
(ii) We have

$$\hat{K}_2 c_1^\dagger|0\rangle = c_2^\dagger|0\rangle + c_3^\dagger|0\rangle, \quad \hat{K}_2(c_2^\dagger|0\rangle + c_3^\dagger) = 2c_1^\dagger|0\rangle + c_2^\dagger|0\rangle + c_3^\dagger|0\rangle$$

and

$$\hat{K}_2(2c_1^\dagger|0\rangle + c_2^\dagger|0\rangle + c_3^\dagger|0\rangle) = 2c_1^\dagger|0\rangle + 3c_2^\dagger|0\rangle + 3c_3^\dagger|0\rangle.$$

(iii) From

$$\hat{K}_1 c_3^\dagger c_2^\dagger c_1^\dagger|0\rangle = 0|0\rangle$$
$$\hat{K}_1 c_3^\dagger c_2^\dagger|0\rangle = c_3^\dagger c_1^\dagger|0\rangle$$
$$\hat{K}_1 c_3^\dagger c_1^\dagger|0\rangle = c_3^\dagger c_2^\dagger|0\rangle + c_2^\dagger c_1^\dagger|0\rangle$$
$$\hat{K}_1 c_2^\dagger c_1^\dagger|0\rangle = c_3^\dagger c_1^\dagger|0\rangle$$
$$\hat{K}_1 c_3^\dagger|0\rangle = c_2^\dagger|0\rangle$$
$$\hat{K}_1 c_2^\dagger|0\rangle = c_1^\dagger|0\rangle + c_3^\dagger|0\rangle$$
$$\hat{K}_1 c_1^\dagger|0\rangle = c_2^\dagger|0\rangle$$
$$\hat{K}_1|0\rangle = 0|0\rangle$$

we obtain the matrix representation for $\hat{K}_1$

$$(0) \oplus \begin{pmatrix} 0 & 1 & 0 \\ 1 & 0 & 1 \\ 0 & 1 & 0 \end{pmatrix} \oplus \begin{pmatrix} 0 & 1 & 0 \\ 1 & 0 & 1 \\ 0 & 1 & 0 \end{pmatrix} \oplus (0)$$

where $\oplus$ denotes the direct sum. The eigenvalues are $\sqrt{2}$ (twice), $-\sqrt{2}$ (twice), 0 (four times).
From

$$\hat{K}_2 c_3^\dagger c_2^\dagger c_1^\dagger|0\rangle = 0|0\rangle$$
$$\hat{K}_2 c_3^\dagger c_2^\dagger|0\rangle = c_3^\dagger c_1^\dagger|0\rangle - c_2^\dagger c_1^\dagger|0\rangle$$
$$\hat{K}_2 c_3^\dagger c_1^\dagger|0\rangle = c_3^\dagger c_2^\dagger|0\rangle + c_2^\dagger c_1^\dagger|0\rangle$$
$$\hat{K}_2 c_2^\dagger c_1^\dagger|0\rangle = c_3^\dagger c_1^\dagger|0\rangle - c_3^\dagger c_2^\dagger|0\rangle$$
$$\hat{K}_2 c_3^\dagger|0\rangle = c_2^\dagger|0\rangle + c_1^\dagger$$
$$\hat{K}_2 c_2^\dagger|0\rangle = c_1^\dagger|0\rangle + c_3^\dagger|0\rangle$$
$$\hat{K}_2 c_1^\dagger|0\rangle = c_2^\dagger|0\rangle + c_3^\dagger$$
$$\hat{K}_2|0\rangle = 0|0\rangle$$

we obtain the matrix representation for $\hat{K}_2$

$$(0) \oplus \begin{pmatrix} 0 & 1 & -1 \\ 1 & 0 & 1 \\ -1 & 1 & 0 \end{pmatrix} \oplus \begin{pmatrix} 0 & 1 & 1 \\ 1 & 0 & 1 \\ 1 & 1 & 0 \end{pmatrix} \oplus (0)$$

where $\oplus$ denotes the direct sum. The eigenvalues are 2, -2, $+1$ (twice), -1 (twice), 0 (twice). The matrix representation for $\hat{N}$ is

$$\text{diag}(3\ 2\ 2\ 2\ 1\ 1\ 1\ 0).$$

(iv) We obtain the matrix representation for $\hat{K}_1$ as

$$(0) \oplus \begin{pmatrix} 0 & -1 & 0 & 0 & 0 & 0 \\ -1 & 0 & 0 & -1 & 0 & 0 \\ 0 & 0 & 0 & 0 & -1 & 0 \\ 0 & -1 & 0 & 0 & 0 & 0 \\ 0 & 0 & -1 & 0 & 0 & -1 \\ 0 & 0 & 0 & 0 & -1 & 0 \end{pmatrix} \oplus (0)$$

with the eigenvalues $\sqrt{2}$ (twice), $-\sqrt{2}$ (twice), 0 (twice). We obtain the matrix representation for $\hat{K}_2$ as

$$(0) \oplus \begin{pmatrix} 0 & -1 & 0 & -1 & 0 & 0 \\ -1 & 0 & 0 & -1 & 0 & 0 \\ 0 & 0 & 0 & 0 & -1 & 1 \\ -1 & -1 & 0 & 0 & 0 & 0 \\ 0 & 0 & -1 & 0 & 0 & -1 \\ 0 & 0 & 1 & 0 & -1 & 0 \end{pmatrix} \oplus (0)$$

with the eigenvalues $+1$ (twice), -1 (twice), $+2$, -2, 0 (twice). For the number operator we obtain the diagonal matrix

$$\text{diag}(3\ 2\ 2\ 1\ 2\ 1\ 1\ 0)$$

as matrix representation. Note that matrix representation differs from the matrix representation given above, but of course the eigenvalues are the same.

Problem 69. (i) Consider the Hamilton operator (open end boundary conditions)

$$\hat{H} = \hbar\omega(c_1^\dagger c_1 + c_2^\dagger c_2 + c_3^\dagger c_3) + t(c_1^\dagger c_2 + c_2^\dagger c_1 + c_2^\dagger c_3 + c_3^\dagger c_2).$$

The Hamilton operator commutes with the number operator

$$\hat{N} = c_1^\dagger c_1 + c_2^\dagger c_2 + c_3^\dagger c_3$$

i.e. $[\hat{H}, \hat{N}] = 0$. Consider the four invariant subspaces spanned by

$$\{c_3^\dagger c_2^\dagger c_1^\dagger |0\rangle\}, \quad \{c_3^\dagger c_2^\dagger |0\rangle, c_3^\dagger c_1^\dagger |0\rangle, c_2^\dagger c_1^\dagger |0\rangle\}, \quad \{c_3^\dagger |0\rangle, c_2^\dagger |0\rangle, c_1^\dagger |0\rangle\}, \quad \{|0\rangle\}.$$

Calculate the eigenvalues and eigenvectors for each subspace. Find the matrix representation of the number operator $\hat{N}$.

(ii) Consider the Hamilton operator (cyclic boundary conditions)

$$\hat{H} = \hbar\omega(c_1^\dagger c_1 + c_2^\dagger c_2 + c_3^\dagger c_3) + t(c_1^\dagger c_2 + c_2^\dagger c_1 + c_2^\dagger c_3 + c_3^\dagger c_2 + c_3^\dagger c_1 + c_1^\dagger c_3).$$

The Hamilton operator $\hat{H}$ commutes with the number operator

$$\hat{N} = c_1^\dagger c_1 + c_2^\dagger c_2 + c_3^\dagger c_3.$$

Consider the four invariant subspaces spanned by

$$\{c_3^\dagger c_2^\dagger c_1^\dagger |0\rangle\}, \quad \{c_3^\dagger c_2^\dagger |0\rangle, c_3^\dagger c_1^\dagger, c_2^\dagger c_1^\dagger |0\rangle\}, \quad \{c_1^\dagger |0\rangle, c_2^\dagger |0\rangle, c_3^\dagger |0\rangle\}, \quad \{|0\rangle\}.$$

Calculate the eigenvalues and eigenvectors for each subspace.

Solution 69.   (i) For the one-dimensional subspace $\{c_3^\dagger c_2^\dagger c_1^\dagger |0\rangle\}$ we have

$$\hat{H} c_3^\dagger c_2^\dagger c_1^\dagger |0\rangle = 3\hbar\omega c_3^\dagger c_2^\dagger c_1^\dagger |0\rangle.$$

Thus the eigenvalues is $3\hbar\omega$ with the eigenvector $c_3^\dagger c_2^\dagger c_1^\dagger |0\rangle$. For the three-dimensional subspace $\{c_3^\dagger c_2^\dagger |0\rangle,\ c_3^\dagger c_1^\dagger |0\rangle,\ c_2^\dagger c_1^\dagger |0\rangle\}$ we obtain

$$\hat{H} c_3^\dagger c_2^\dagger |0\rangle = 2\hbar\omega c_3^\dagger c_2^\dagger |0\rangle + t c_3^\dagger c_1^\dagger |0\rangle$$
$$\hat{H} c_3^\dagger c_1^\dagger |0\rangle = 2\hbar\omega c_3^\dagger c_1^\dagger |0\rangle + t c_3^\dagger c_2^\dagger |0\rangle + t c_2^\dagger c_1^\dagger |0\rangle$$
$$\hat{H} c_2^\dagger c_1^\dagger |0\rangle = 2\hbar\omega c_2^\dagger c_1^\dagger |0\rangle + t c_3^\dagger c_1^\dagger |0\rangle.$$

Thus we obtain the symmetric 3×3 matrix

$$\begin{pmatrix} 2\hbar\omega & t & 0 \\ t & 2\hbar\omega & t \\ 0 & t & 2\hbar\omega \end{pmatrix}$$

with the eigenvalues $2\hbar\omega - \sqrt{2}t$, $2\hbar\omega + \sqrt{2}t$, $2\hbar\omega$. For the subspace with the basis $\{c_3^\dagger |0\rangle,\ c_2^\dagger |0\rangle,\ c_1^\dagger |0\rangle\}$ we have

$$\hat{H} c_3^\dagger |0\rangle = \hbar\omega c_3^\dagger |0\rangle + t c_2^\dagger |0\rangle$$
$$\hat{H} c_2^\dagger |0\rangle = \hbar\omega c_2^\dagger |0\rangle + t c_1^\dagger |0\rangle + t c_3^\dagger |0\rangle$$
$$\hat{H} c_1^\dagger |0\rangle = \hbar\omega c_1^\dagger |0\rangle + t c_2^\dagger |0\rangle$$

with the matrix representation

$$\begin{pmatrix} \hbar\omega & t & 0 \\ t & \hbar\omega & t \\ 0 & t & \hbar\omega \end{pmatrix}$$

and the eigenvalues $\hbar\omega - \sqrt{2}t$, $\hbar\omega + \sqrt{2}t$, $\hbar\omega$. Finally $\hat{H}|0\rangle = 0|0\rangle$ with eigenvalue 0. We have for the number operator $\hat{N}$ the matrix representation

$$\text{diag}(3\ 2\ 2\ 2\ 1\ 1\ 1\ 0).$$

(ii) For the one-dimensional subspace $\{c_3^\dagger c_2^\dagger c_1^\dagger |0\rangle\}$ we have

$$\hat{H}|c_3^\dagger c_2^\dagger c_1^\dagger |0\rangle = 3\hbar\omega c_3^\dagger c_2^\dagger c_1^\dagger |0\rangle.$$

Thus the eigenvalues is $3\hbar\omega$. For the three-dimensional subspace $\{c_3^\dagger c_2^\dagger |0\rangle,\ c_3^\dagger c_1^\dagger |0\rangle,\ c_2^\dagger c_1^\dagger |0\rangle\}$ we obtain

$$\hat{H} c_3^\dagger c_2^\dagger |0\rangle = 2\hbar\omega c_3^\dagger c_2^\dagger |0\rangle + t c_3^\dagger c_1^\dagger |0\rangle - t c_2^\dagger c_1^\dagger |0\rangle$$
$$\hat{H} c_3^\dagger c_1^\dagger |0\rangle = 2\hbar\omega c_3^\dagger c_1^\dagger |0\rangle + t c_3^\dagger c_2^\dagger |0\rangle + t c_2^\dagger c_1^\dagger |0\rangle$$
$$\hat{H} c_2^\dagger c_1^\dagger |0\rangle = 2\hbar\omega c_2^\dagger c_1^\dagger |0\rangle + t c_3^\dagger c_1^\dagger |0\rangle - t c_3^\dagger c_2^\dagger |0\rangle.$$

Thus we obtain the symmetric 3×3 matrix

$$\begin{pmatrix} 2\hbar\omega & t & -t \\ t & 2\hbar\omega & t \\ -t & t & 2\hbar\omega \end{pmatrix}$$

with the eigenvalues $2\hbar\omega - 2t$, $2\hbar\omega + t$ (twice). For the three-dimensional subspace $\{c_3^\dagger|0\rangle, c_2^\dagger|0\rangle, c_1^\dagger|0\rangle\}$ we obtain

$$\hat{H}c_3^\dagger|0\rangle = \hbar\omega c_3^\dagger|0\rangle + tc_2^\dagger|0\rangle + tc_1^\dagger|0\rangle$$
$$\hat{H}c_2^\dagger|0\rangle = \hbar\omega c_2^\dagger|0\rangle + tc_1^\dagger|0\rangle + tc_3^\dagger|0\rangle$$
$$\hat{H}c_1^\dagger|0\rangle = \hbar\omega c_1^\dagger|0\rangle + tc_2^\dagger|0\rangle + tc_3^\dagger|0\rangle.$$

Thus we obtain the symmetric 3×3 matrix

$$\begin{pmatrix} \hbar\omega & t & t \\ t & \hbar\omega & t \\ t & t & 2\hbar\omega \end{pmatrix}$$

with the eigenvalues $2\hbar\omega + 2t$, $\hbar\omega - t$ (twice). Finally $\hat{H}|0\rangle = 0|0\rangle$ with eigenvalue 0 and eigenvector $|0\rangle$.

Problem 70.   Consider the Hamilton operator

$$\hat{H} = \hbar\omega(c_0^\dagger c_1 + c_1^\dagger c_0 + c_2^\dagger c_1 + c_1^\dagger c_2) + V_0\hat{N}_0\hat{N}_1 + V_1\hat{N}_1\hat{N}_2 + V_2\hat{N}_0\hat{N}_1\hat{N}_2$$

where $\hat{N}_j := c_j^\dagger c_j$ ($j = 0, 1, 2$). Find the matrix representation with the eight dimensional basis

$$c_2^\dagger c_1^\dagger c_0^\dagger|0\rangle, \quad c_2^\dagger c_1^\dagger|0\rangle, \quad c_2^\dagger c_0^\dagger|0\rangle, \quad c_1^\dagger c_0^\dagger|0\rangle, \quad c_2^\dagger|0\rangle, \quad c_1^\dagger|0\rangle, \quad c_0^\dagger|0\rangle, \quad |0\rangle.$$

Solution 70.   We obtain the symmetric 8×8 matrix

$$\begin{pmatrix} 0 & 0 & 0 & 0 & 0 & 0 & 0 & 0 \\ 0 & 0 & \hbar\omega & 0 & 0 & 0 & 0 & 0 \\ 0 & \hbar\omega & 0 & \hbar\omega & 0 & 0 & 0 & 0 \\ 0 & 0 & \hbar\omega & 0 & 0 & 0 & 0 & 0 \\ 0 & 0 & 0 & 0 & V_0 & \hbar\omega & 0 & 0 \\ 0 & 0 & 0 & 0 & \hbar\omega & 0 & \hbar\omega & 0 \\ 0 & 0 & 0 & 0 & 0 & \hbar\omega & V_1 & 0 \\ 0 & 0 & 0 & 0 & 0 & 0 & 0 & V_0 + V_1 + V_2 \end{pmatrix}.$$

We see that besides 6 other eigenvalues 0 and $V_0 + V_1 + V_2$ are eigenvalues. The matrix can be written as a direct sum

$$(0) \oplus \begin{pmatrix} 0 & \hbar\omega & 0 \\ \hbar\omega & 0 & \hbar\omega \\ 0 & \hbar\omega & 0 \end{pmatrix} \oplus \begin{pmatrix} V_0 & \hbar\omega & 0 \\ \hbar\omega & 0 & \hbar\omega \\ 0 & \hbar\omega & V_1 \end{pmatrix} \oplus (V_0 + V_1 + V_2)$$

which simplifies the eigenvalue problem to find the eigenvalues and eigenvectors.

Problem 71.   Consider the Hamilton operator

$$\hat{K} = \frac{\hat{H}}{\hbar\omega} = c_0^\dagger c_1 + c_1^\dagger c_0 + c_1^\dagger c_2 + c_2^\dagger c_1 + c_1^\dagger c_3 + c_3^\dagger c_1$$

for the lattice

$$0$$
$$|$$
$$2 \,-\!\!-\, 1 \,-\!\!-\, 3$$

(i) Find

$$\hat{K}c_0^\dagger|0\rangle, \quad \hat{K}c_2^\dagger|0\rangle, \quad \hat{K}c_3^\dagger|0\rangle, \quad \hat{K}c_1^\dagger|0\rangle.$$

(ii) Find the matrix representation of $\hat{K}$ with the six-dimensional basis

$$c_3^\dagger c_2^\dagger|0\rangle, \quad c_3^\dagger c_1^\dagger|0\rangle, \quad c_3^\dagger c_0^\dagger|0\rangle, \quad c_2^\dagger c_1^\dagger|0\rangle, \quad c_2^\dagger c_0^\dagger|0\rangle, \quad c_1^\dagger c_0^\dagger|0\rangle.$$

Solution 71.   (i) We have

$$\hat{K}c_0^\dagger|0\rangle = c_1^\dagger|0\rangle, \quad \hat{K}c_2^\dagger|0\rangle = c_1^\dagger|0\rangle, \quad \hat{K}c_3^\dagger|0\rangle = c_1^\dagger|0\rangle$$

and

$$\hat{K}c_1^\dagger|0\rangle = c_0^\dagger|0\rangle + c_2^\dagger|0\rangle + c_3^\dagger|0\rangle.$$

(ii) Since

$$\hat{K}c_3^\dagger c_2^\dagger|0\rangle = c_3^\dagger c_1^\dagger|0\rangle - c_2^\dagger c_1^\dagger|0\rangle, \quad \hat{K}c_3^\dagger c_1^\dagger|0\rangle = c_3^\dagger c_0^\dagger|0\rangle + c_3^\dagger c_2^\dagger|0\rangle$$
$$\hat{K}c_3^\dagger c_0^\dagger|0\rangle = c_3^\dagger c_1^\dagger|0\rangle + c_1^\dagger c_0^\dagger|0\rangle, \quad \hat{K}c_2^\dagger c_1^\dagger|0\rangle = c_2^\dagger c_0^\dagger|0\rangle - c_3^\dagger c_2^\dagger|0\rangle$$
$$\hat{K}c_2^\dagger c_0^\dagger|0\rangle = c_1^\dagger c_0^\dagger|0\rangle + c_2^\dagger c_1^\dagger|0\rangle, \quad \hat{K}c_1^\dagger c_0^\dagger|0\rangle = c_2^\dagger c_0^\dagger|0\rangle + c_3^\dagger c_0^\dagger|0\rangle$$

we obtain the 6×6 hermitian matrix

$$\begin{pmatrix} 0 & 1 & 0 & 1 & 0 & 0 \\ 1 & 0 & 1 & 0 & 0 & 0 \\ 0 & 1 & 0 & 0 & 0 & 1 \\ 1 & 0 & 0 & 0 & 1 & 0 \\ 0 & 0 & 0 & 1 & 0 & 1 \\ 0 & 0 & 1 & 0 & 1 & 0 \end{pmatrix}$$

with the eigenvalues $+1$ (twice), -1 (twice), $+2$, -2.

Problem 72.   (i) Consider the Hamilton operator

$$\hat{K} = \frac{\hat{H}}{\hbar\omega} = c_0^\dagger c_1 + c_1^\dagger c_0.$$

Apply the *discrete Fourier transform* ($N = 2$)

$$c_j^\dagger = \frac{1}{\sqrt{N}} \sum_{k \in \{0,\pi\}} e^{-ikj} c_k^\dagger \Rightarrow c_j = \frac{1}{\sqrt{N}} \sum_{k \in \{0,\pi\}} e^{ikj} c_k$$

where $j \in \{0,1\}$ and $k \in \{0,\pi\}$. Then the inverse discrete Fourier transform is given by

$$c_k^\dagger = \frac{1}{\sqrt{N}} \sum_{j \in \{0,1\}} e^{ikj} c_j^\dagger \Rightarrow c_k = \frac{1}{\sqrt{N}} \sum_{j \in \{0,1\}} e^{-ikj} c_j.$$

Note that the discrete Fourier transform is a unitary transformation.

(ii) Consider the number operators

$$\hat{N}_0 = c_0^\dagger c_0, \quad \hat{N}_1 = c_1^\dagger c_1, \quad \hat{N} = \hat{N}_0 + \hat{N}_1.$$

Apply the discrete Fourier transform given in (i).

(iii) Apply the discrete Fourier transform to the operator $c_0^\dagger c_0 c_1^\dagger c_1$.

(iv) Consider the Hamilton operator (periodic boundary conditions)

$$\hat{H} = t(c_0^\dagger c_1 + c_1^\dagger c_0 + c_1^\dagger c_2 + c_2^\dagger c_1 + c_2^\dagger c_3 + c_3^\dagger c_2 + c_3^\dagger c_0 + c_0^\dagger c_3).$$

Apply the *discrete Fourier transform* $(N = 4)$

$$c_j^\dagger = \frac{1}{\sqrt{N}} \sum_{k \in \{0, \pi/2, \pi, 3\pi/2\}} e^{-ikj} c_k^\dagger \Rightarrow c_j = \frac{1}{\sqrt{N}} \sum_{k \in \{0, \pi/2, \pi, 3\pi/2\}} e^{ikj} c_k$$

where $j \in \{0, 1, 2, 3\}$ and $k \in \{0, \pi/2, \pi, 3\pi/2\}$. Then the inverse discrete Fourier transform is given by

$$c_k^\dagger = \frac{1}{\sqrt{N}} \sum_{j \in \{0,1\}} e^{ikj} c_j^\dagger \Rightarrow c_k = \frac{1}{\sqrt{N}} \sum_{j \in \{0,1\}} e^{-ikj} c_j.$$

As index set one could also consider $I = \{-\pi/2, 0, \pi/2, \pi\}$ instead of $I = \{0, \pi/2, \pi, 3\pi/2\}$.

Solution 72. (i) Utilizing $e^{i\pi} = e^{-i\pi} = -1$ we obtain

$$c_0^\dagger c_1 = \frac{1}{\sqrt{2}}(c_0^\dagger + c_\pi^\dagger)\frac{1}{\sqrt{2}}(c_0 - c_\pi) = \frac{1}{2}(c_0^\dagger c_0 - c_\pi^\dagger c_\pi - c_0^\dagger c_\pi + c_\pi^\dagger c_0)$$

$$c_1^\dagger c_0 = \frac{1}{\sqrt{2}}(c_0^\dagger - c_\pi^\dagger)\frac{1}{\sqrt{2}}(c_0 + c_\pi) = \frac{1}{2}(c_0^\dagger c_0 - c_\pi^\dagger c_\pi + c_0^\dagger c_\pi - c_\pi^\dagger c_0).$$

Thus since $\cos(\pi) = -1$ we have

$$c_0^\dagger c_1 + c_1^\dagger c_0 = c_0^\dagger c_0 - c_\pi^\dagger c_\pi = \sum_{k \in \{0,\pi\}} \cos(k) c_k^\dagger c_k.$$

Thus the Hamilton operator $\hat{K}$ takes the (diagonal) form after Fourier transform

$$\hat{K} = \sum_{k \in \{0,\pi\}} \cos(k) c_k^\dagger c_k.$$

(ii) We have

$$c_0^\dagger c_0 = \frac{1}{\sqrt{2}}(c_0^\dagger + c_\pi^\dagger)\frac{1}{\sqrt{2}}(c_0 + c_\pi) = \frac{1}{2}(c_0^\dagger c_0 + c_\pi^\dagger c_\pi + c_0^\dagger c_\pi + c_\pi^\dagger c_0)$$

$$c_1^\dagger c_1 = \frac{1}{\sqrt{2}}(c_0^\dagger - c_\pi^\dagger)\frac{1}{\sqrt{2}}(c_0 - c_\pi) = \frac{1}{2}(c_0^\dagger c_0 + c_\pi^\dagger c_\pi - c_0^\dagger c_\pi - c_\pi^\dagger c_0).$$

Thus the number operator $\hat{N}_0 + \hat{N}_1$ takes the form

$$c_0^\dagger c_0 + c_\pi^\dagger c_\pi \equiv \sum_{k \in \{0,\pi\}} c_k^\dagger c_k.$$

(iii) Using the result from (ii) we obtain the operator $c_0^\dagger c_0 c_\pi^\dagger c_\pi$.

(iv) Utilizing

$$e^{i\pi/2} = i, \quad e^{-i\pi/2} = -i, \quad e^{i\pi} = -1, \quad e^{-i\pi} = -1, \quad e^{i3\pi/2} = -i, \quad e^{-i3\pi/2} = i$$

the Hamilton operator $\hat{H}$ takes the form

$$\hat{H} = 2t \sum_{k \in \{0, \pi/2, \pi, 3\pi/2\}} \cos(k) c_k^\dagger c_k.$$

Problem 73. Consider the Hamilton operator given above problem

$$\hat{H} = 2t \sum_{k \in \{0, \pi/2, \pi, 3\pi/2\}} \cos(k) c_k^\dagger c_k \equiv c_0^\dagger c_0 - c_\pi^\dagger c_\pi.$$

The Hamilton operator commutes with the number operator

$$\hat{N} = \sum_{k \in \{0, \pi/2, \pi, 3\pi/2\}} c_k^\dagger c_k.$$

(i) Does the Hamilton operator $\hat{H}$ commute with the total *momentum operator*

$$\hat{P} = \sum_{k \in \{0, \pi/2, \pi, 3\pi/2\}} k c_k^\dagger c_k \equiv \frac{\pi}{2} c_{\pi/2}^\dagger c_{\pi/2} + \pi c_\pi^\dagger c_\pi + \frac{3\pi}{2} c_{3\pi/2}^\dagger c_{3\pi/2}.$$

The eigenvalues of the total momentum operator are 0, $\pi/2$, π, $3\pi/2$.
(ii) Owing to the fact that the Hamilton operator commutes with $\hat{N}$ and $\hat{P}$ we can decompose the Hilbert space into invariant subspaces. Let $N = 2$ (two particles). Then a basis in a six-dimensional Hilbert space is given by

$$c_0^\dagger c_{\pi/2}^\dagger |0\rangle, \quad c_0^\dagger c_\pi^\dagger |0\rangle, \quad c_0^\dagger c_{3\pi/2}^\dagger |0\rangle, \quad c_{\pi/2}^\dagger c_\pi^\dagger |0\rangle, \quad c_{\pi/2}^\dagger c_{3\pi/2}^\dagger |0\rangle, \quad c_\pi^\dagger c_{3\pi/2}^\dagger |0\rangle.$$

Solution 73. (i) Utilizing that

$$\cos(0) = 1, \quad \cos(\pi/2) = 0, \quad \cos(\pi) = -1, \quad \cos(3\pi/2) = 0$$

we find that $[\hat{H}, \hat{P}] = 0$.
(ii) Owing to the modulo 2π addition we have

$$0 + \pi/2 = \pi/2, \quad 0 + \pi = \pi, \quad 0 + 3\pi/2 = 3\pi/2,$$

$$\pi/2 + \pi = 3\pi/2, \quad \pi/2 + 3\pi/2 = 0, \quad \pi + 3\pi/2 = \pi/2.$$

Thus we obtain the four invariant subspaces

$$\{c_{\pi/2}^\dagger c_{3\pi/2}^\dagger |0\rangle\}, \quad \{c_0^\dagger c_{\pi/2}^\dagger |0\rangle, c_\pi^\dagger c_{3\pi/2}^\dagger |0\rangle\}, \quad \{c_0^\dagger c_\pi^\dagger |0\rangle\}, \quad \{c_0^\dagger c_{3\pi/2}^\dagger |0\rangle, c_{\pi/2}^\dagger c_\pi^\dagger |0\rangle\}.$$

Thus for $P = 0$ the subspace is one-dimensional, for $P = \pi/2$ the subspace is two-dimensional, for $P = \pi$ the subspace is one-dimensional and for $P = 3\pi/2$ the subspace is two-dimensional. For $P = 0$ we have

$$\hat{H} c_{\pi/2}^\dagger c_{3\pi/2}^\dagger |0\rangle = 0 |0\rangle$$

with the eigenvalue 0. For $P = \pi/2$ we have

$$\hat{H}c_0^\dagger c_{\pi/2}^\dagger|0\rangle = tc_0^\dagger c_{\pi/2}^\dagger|0\rangle, \quad \hat{H}c_\pi^\dagger c_{3\pi/2}^\dagger|0\rangle = -2tc_\pi^\dagger c_{3\pi/2}|0\rangle$$

with the eigenvalues t and $-t$. For $P = \pi$ we find

$$\hat{H}c_0^\dagger c_\pi^\dagger|0\rangle = c_0^\dagger c_\pi^\dagger|0\rangle - c_0^\dagger c_\pi^\dagger|0\rangle = 0|0\rangle$$

with the eigenvalue 0. For $P = 3\pi/2$ we have

$$\hat{H}c_0^\dagger c_{\pi/2}^\dagger|0\rangle = tc_0^\dagger c_{3\pi/2}^\dagger|0\rangle, \quad \hat{H}c_{\pi/2}^\dagger c_\pi^\dagger|0\rangle = -tc_{\pi/2}^\dagger c_\pi^\dagger|0\rangle$$

with the eigenvalues t and $-t$.

Problem 74. Consider the Hamilton operator

$$\hat{H} = t\sum_{j=0}^{N-1}(c_j^\dagger c_{j+1} + c_{j+1}^\dagger c_j)$$

with cyclic boundary conditions, i.e. $c_N = c_0$, $c_N^\dagger = c_0^\dagger$. Find the Fourier transform

$$c_n^\dagger = \frac{1}{\sqrt{N}}\sum_{k\in I}e^{-ikn}c_k^\dagger, \quad c_n = \frac{1}{\sqrt{N}}\sum_{k\in I}e^{ikn}c_k, \quad n = 0, 1, 2, \ldots, N-1$$

of $\hat{H}$, where the index I is given by

$$I = \{0, 2\pi/N, 4\pi/N, \ldots, 2\pi(N-1)/N\}$$

i.e. $k = 2\pi n/N$ with $n = 0, 1, \ldots, N-1$.

Solution 74. We have

$$c_n^\dagger c_{n+1} = \frac{1}{N}\sum_{k\in I}\sum_{k'\in I}e^{i(k'-k)n}e^{ik'}c_k^\dagger c_k, \quad c_{n+1}^\dagger c_n = \frac{1}{N}\sum_{k\in I}\sum_{k'\in I}e^{i(k'-k)n}e^{-ik}c_k^\dagger c_k'.$$

Since

$$\sum_{n=0}^{N-1}e^{i(k'-k)n} = \delta_{kk'}$$

we have

$$\sum_{n=0}^{N-1}(c_n^\dagger c_{n+1} + c_{n+1}^\dagger c_n) = \sum_{k\in I}\sum_{k'\in I}(e^{ik'}\delta_{kk'}c_k^\dagger c_k + e^{-ik}\delta_{kk'}c_k^\dagger c_k) = \sum_{k\in I}(e^{ik} + e^{-ik})c_k^\dagger c_k$$

$$= 2\sum_{k\in I}\cos(k)c_k^\dagger c_k.$$

Thus the Hamilton operator takes the diagonal form

$$\hat{H} = 2t\sum_{k\in I}\cos(k)c_k^\dagger c_k.$$

Problem 75.   (i) Consider a one-dimensional chain with N lattice points. Let c_j $(c_j^\dagger)$ denotes the Fermi creation (annihilation) operator with $j = 1, \ldots, N$. Consider the Hamilton operator

$$\hat{H} = \sum_{j=1}^{N}(c_j^\dagger c_j - \frac{1}{2}I) - \frac{1}{2}\sum_{j=1}^{N-1}(c_j^\dagger - c_j)(c_{j+1}^\dagger - c_{j+1}) + \frac{1}{2}(1-Q)(c_N^\dagger - c_N)(c_1^\dagger - c_1).$$

Here Q are the eigenvalues of the operator

$$\hat{Q} = \frac{1}{2}(I - \exp(i\pi \sum_{j=1}^{N}c_j^\dagger c_j)).$$

Find the eigenvalues of $\hat{Q}$. Find the commutator $[\hat{H}, \hat{Q}]$.
(ii) Consider the Hamilton operator of a one-dimensional chain

$$\hat{H} = -\sum_{j=1}^{N}\left(c_j^\dagger c_j - \frac{1}{2}I\right) - \frac{1}{2}\sum_{j=1}^{N-1}(c_j^\dagger - c_j)(c_{j+1}^\dagger + c_{j+1}) + \frac{1}{2}P(c_N^\dagger - c_N)(c_1^\dagger + c_1)$$

where N is an even positive integer and $P = \pm 1$. Let

$$\hat{P} = \exp\left(i\pi \sum_{j=1}^{N}c_j^\dagger c_j\right).$$

Find the commutator $[\hat{H}, \hat{P}]$. Find the eigenvalues of $\hat{P}$. Note that $\hat{P}$ is the *parity operator*.

Solution 75.   (i) Since for $j \neq k$ we have $[c_j^\dagger c_j, c_k^\dagger c_k] = 0$ it follows that

$$e^{i\pi(c_j^\dagger c_j + c_k^\dagger c_k)} = e^{i\pi c_j^\dagger c_j}e^{i\pi c_k^\dagger c_k}.$$

Now

$$e^{i\alpha c^\dagger c} = I + c^\dagger c(e^{i\alpha} - 1).$$

Thus for $\alpha = \pi$ we have

$$\exp(i\pi c^\dagger c) = I - 2c^\dagger c.$$

Therefore $\hat{Q}^2 = \hat{Q}$ and the eigenvalues of $\hat{Q}$ are 1 and 0. For the commutator we find $[\hat{H}, \hat{Q}] = 0$.
(ii) We obtain $[\hat{H}, \hat{P}] = 0$. Since $\hat{P}^2 = I$, where I is the identity operator we obtain the eigenvalues ± 1.

Problem 76.   The grand thermodynamic potential Ω divided by the number of lattice sites and the Coulomb repulsion energy U for the Hamilton operator

$$\hat{H} = U\sum_{j=1}^{N}c_{j\uparrow}^\dagger c_{j\uparrow}c_{j\downarrow}^\dagger c_{j\downarrow}$$

is given by $(\beta = 1/kT)$

$$\frac{\Omega}{NU} = -\frac{1}{\beta U}\ln\left(1 + 2e^{\beta\mu} + e^{-\beta U + 2\beta\mu}\right)$$

where μ is the chemical potential. Note that $\beta\mu = (\beta U)(\mu/U)$, where βU and μ/U are dimensionless. Let N_e be the number of electrons. Calculate the chemical potential μ for $N_e/N = 0.5$, $N_e/N = 1$, $N_e/N = 2$, where

$$\frac{1}{N}\frac{\partial\Omega}{\partial\mu} = -\frac{N_e}{N}.$$

Note that $0 \le N_e/N \le 2$.

Solution 76. Taking the derivative of Ω with respect to μ provides

$$\frac{N_e}{N} = \frac{2e^{\beta\mu} + 2e^{-\beta U + 2\beta\mu}}{1 + 2e^{\beta\mu} + e^{-\beta U + 2\beta\mu}}.$$

Setting $\theta = N_e/N$ and $x = e^{\beta\mu}$ we obtain the quadratic equation

$$x^2 + \frac{2(\theta - 1)}{\theta - 2}e^{\beta U}x + \frac{\theta}{\theta - 2}e^{\beta U} = 0.$$

The solution is given

$$x_{1,2} = -\frac{\theta - 1}{\theta - 2}e^{\beta U} \pm \sqrt{\frac{(\theta - 1)^2 e^{2\beta U}}{(\theta - 2)^2} - \frac{\theta e^{\beta U}}{(\theta - 2)}}$$

or

$$x_{1,2} = \frac{e^{\beta U}}{(2 - \theta)}\left((\theta - 1) \mp \sqrt{(\theta - 1)^2 - \theta(\theta - 2)e^{-\beta U}}\right).$$

Since $x = e^{\beta\mu}$ we obtain

$$\beta\mu_{1,2} = \beta U - \ln(2 - \theta) + \ln\left((\theta - 1) \mp \sqrt{(\theta - 1)^2 - \theta(\theta - 2)e^{-\beta U}}\right).$$

Note that $U > 0$, $\beta > 0$, and $0 \le \theta \le 2$.

Problem 77. Let σ_1, σ_2, σ_3 be the Pauli spin matrices. Find the matrix representation of the operator

$$\hat{K} = (\sigma_1 \otimes \sigma_1 + \sigma_2 \otimes \sigma_2 + \sigma_3 \otimes \sigma_3) \otimes c^\dagger + (\sigma_1 \otimes \sigma_1 + \sigma_2 \otimes \sigma_2 + \sigma_3 \otimes \sigma_3) \otimes c$$

with the basis given by

$$\begin{pmatrix}1\\0\\0\\0\end{pmatrix}\otimes|0\rangle, \quad \begin{pmatrix}0\\1\\0\\0\end{pmatrix}\otimes|0\rangle, \quad \begin{pmatrix}0\\0\\1\\0\end{pmatrix}\otimes|0\rangle, \quad \begin{pmatrix}0\\0\\0\\1\end{pmatrix}\otimes|0\rangle,$$

$$\begin{pmatrix}1\\0\\0\\0\end{pmatrix}\otimes c^\dagger|0\rangle, \quad \begin{pmatrix}0\\1\\0\\0\end{pmatrix}\otimes c^\dagger|0\rangle, \quad \begin{pmatrix}0\\0\\1\\0\end{pmatrix}\otimes c^\dagger|0\rangle, \quad \begin{pmatrix}0\\0\\0\\1\end{pmatrix}\otimes c^\dagger|0\rangle.$$

Find the eigenvalues and eigenvectors.

Solution 77.   The 8×8 hermitian matrix is given by

$$\begin{pmatrix}
0 & 1 & 0 & 0 & 0 & 0 & 0 & 0 \\
1 & 0 & 0 & 0 & 0 & 0 & 0 & 0 \\
0 & 0 & 0 & -1 & 0 & 2 & 0 & 0 \\
0 & 0 & -1 & 0 & 2 & 0 & 0 & 0 \\
0 & 0 & 0 & 2 & 0 & -1 & 0 & 0 \\
0 & 0 & 2 & 0 & -1 & 0 & 0 & 0 \\
0 & 0 & 0 & 0 & 0 & 0 & 0 & 1 \\
0 & 0 & 0 & 0 & 0 & 0 & 1 & 0
\end{pmatrix}$$

with the eigenvalues -1 (3 times), $+1$ (3 times), -3, $+3$.

Problem 78.   Let $a, b, c \in \mathbb{R}$ and $b \neq 0$, $c \neq 0$. Consider the tridiagonal $n \times n$ matrix

$$M = \begin{pmatrix}
a & b & & & \\
c & a & b & & \\
& c & \ddots & \ddots & \\
& & \ddots & a & b \\
& & & c & a
\end{pmatrix}$$

and the eigenvalue problem $M\mathbf{v} = \lambda\mathbf{v}$ ($\mathbf{v} \neq \mathbf{0}$). Thus the eigenvalue problem can be written as a linear difference equation

$$cv_{j-1} + (a - \lambda)v_j + bv_{j+1} = 0, \quad j = 1, \dots, n$$

with $v_0 = v_{n+1} = 0$. This linear difference equation can be solved and provides the eigenvalues

$$\lambda_k = a + 2\sqrt{bc}\cos\left(\frac{\pi k}{n+1}\right), \qquad k = 1, \dots, n$$

with the corresponding eigenvectors

$$\mathbf{v}_k = \begin{pmatrix}
(c/b)^{1/2}\sin(\pi k/(n+1)) \\
\vdots \\
(c/b)^{n/2}\sin(\pi n k/(n+1))
\end{pmatrix}, \qquad k = 1, \dots, n.$$

The Hamilton operator

$$\frac{\hat{H}}{\hbar\omega} = c_0^\dagger c_1 + c_1^\dagger c_0 + c_1^\dagger c_2 + c_2^\dagger c_1$$

with the basis $c_0^\dagger|0\rangle$, $c_1^\dagger|0\rangle$, $c_2^\dagger|0\rangle$ provides the matrix representation

$$\begin{pmatrix}
0 & 1 & 0 \\
1 & 0 & 1 \\
0 & 1 & 0
\end{pmatrix}.$$

Find the eigenvalues and normalized eigenvectors of this matrix applying the result given above.

Solution 78.   With $k = 1, 2, 3$ we find the eigenvalues

$$\lambda_1 = \sqrt{2}, \qquad \lambda_2 = 0, \qquad \lambda_3 = -\sqrt{2}.$$

The corresponding normalized eigenvectors are

$$\mathbf{v}_1 = \frac{1}{2}\begin{pmatrix} 1 \\ \sqrt{2} \\ 1 \end{pmatrix}, \qquad \mathbf{v}_2 = \frac{1}{\sqrt{2}}\begin{pmatrix} 1 \\ 0 \\ -1 \end{pmatrix}, \qquad \mathbf{v}_3 = \frac{1}{2}\begin{pmatrix} 1 \\ -\sqrt{2} \\ 1 \end{pmatrix}.$$

Problem 79. Let $c_j^\dagger$, c_j be Fermi creation and annihilation operators with $j = 1, \ldots, N$. Let $\hat{N}_j := c_j^\dagger c_j$. Consider the one-dimensional Hamilton operator with periodic boundary conditions

$$\hat{H} = -J\sum_{j=1}^{N}(c_j^\dagger c_{j+1} + c_{j+1}^\dagger c_j) + V\sum_{j=1}^{N}\hat{N}_j\hat{N}_{j+1} + W\sum_{j=1}^{N}\hat{N}_j.$$

Let $\sigma_1, \sigma_2, \sigma_3$ be the Pauli spin matrices and

$$S_+ := \frac{1}{2}(\sigma_1 + i\sigma_2), \quad S_- := \frac{1}{2}(\sigma_1 - i\sigma_3).$$

Let

$$S_{\pm,j} := I_2 \otimes \cdots \otimes I_2 \otimes S_\pm \otimes I_2 \otimes \cdots \otimes I_2$$

where $S_\pm$ is at the j-th position with $j = 1, \ldots, N$. Apply the *Jordan-Wigner transformation*

$$c_j^\dagger = \left(\exp\left(i\pi\sum_{\ell=1}^{j-1}S_{+,\ell}S_{-,\ell}\right)\right)S_{+,j}, \quad c_j = \left(\exp\left(i\pi\sum_{\ell=1}^{j-1}S_{+,\ell}S_{-,\ell}\right)\right)S_{-,j}.$$

Note that from $c_j^\dagger$ and c_j it follows that

$$\hat{N}_j = \frac{1}{2}(\sigma_{3,j} + I)$$

where I is the $2^N \times 2^N$ identity matrix.

Solution 79. The Hamilton operator $\hat{H}$ takes the form

$$\hat{H} = -\frac{J}{2}\sum_{j=1}^{N}(\sigma_{1,j}\sigma_{1,j+1} + \sigma_{2,j}\sigma_{2,j+1}) - \frac{V}{4}\sum_{j=1}^{N}\sigma_{3,j}\sigma_{3,j} + \frac{1}{2}(W+V)\sum_{j=1}^{N}\sigma_{3,j} + \frac{N}{2}\left(W + \frac{V}{2}\right)I.$$

Problem 80. Consider a square lattice with lattice constant a and periodic boundary conditions. Let $c_{\mathbf{j}}^\dagger$ ($c_{\mathbf{j}}$) denote the creation (annihilation) operators for an electron in the Wannier state at the lattice site $\mathbf{j}$. The Hamilton operator $\hat{H}$ of spinless tight-binding electrons in the presence of a magnetic field can be written as

$$\hat{H} = \sum_{(\mathbf{j}_1\mathbf{j}_2)} t_{\mathbf{j}_1\mathbf{j}_2}c_{\mathbf{j}_1}^\dagger c_{\mathbf{j}_2}$$

with

$$t_{\mathbf{j}_1\mathbf{j}_2} = -t\exp\left(-i\frac{e}{\hbar}\int_{\mathbf{j}_1}^{\mathbf{j}_2}\mathbf{A}\cdot d\mathbf{r}\right)$$

The summation $\langle \mathbf{j}_1, \mathbf{j}_2 \rangle$ runs over the nearest neighbour site on the square lattice. The uniform magnetic field $\mathbf{B}$ is applied in z-direction. Choosing the *Landau gauge*

$$\mathbf{A} = B(0, x, 0)$$

the line integral in $t_{\mathbf{j}_1 \mathbf{j}_2}$ can be written as

$$\frac{e}{\hbar} \int_{\mathbf{j}_1}^{\mathbf{j}_2} \mathbf{A} \cdot d\mathbf{r} = \begin{cases} 0 & \mathbf{j}_1 = (m, n) \quad \mathbf{j}_2 = (m+1, n) \\ -2\pi m \Phi / \Phi_0 & \mathbf{j}_1 = (m, n) \quad \mathbf{j}_2 = (m, n+1) \end{cases}$$

where the integers m and n refer to the $x = 1$, $y = 2$ coordinates of the square lattice sites. $\Phi = Ba^2$ is the magnetic flux through a unit plaquette. Φ_0 stands for the magnetic flux quantum h/e. Find the Hamilton operator in the *Bloch representation*. The Fourier transform is given by

$$c_{\mathbf{k}} = \frac{1}{\sqrt{N}} \sum_{\mathbf{j}} e^{i(k_1 j_1 + k_2 j_2)} c_{\mathbf{j}}$$

with the inverse

$$c_{\mathbf{j}} = \frac{1}{\sqrt{N}} \sum_{\mathbf{k}} e^{-i(k_1 j_1 + k_2 j_2)} c_{\mathbf{k}}$$

where $\mathbf{k}$ runs over the first Brioullin zone.

Solution 80. We obtain the Hamilton operator in Bloch representation

$$\hat{H} = -2t \sum_{\mathbf{k}} \cos(k_1 a) c^{\dagger}(\mathbf{k}) c(\mathbf{k})$$

$$-t \sum_{\mathbf{k}} \left(e^{-ik_2 a} c^{\dagger}(k_1 + 2\frac{\pi \Phi}{a \Phi_0}, k_2) c(k_1, k_2) + e^{ik_2 a} c^{\dagger}(k_1 - 2\frac{\pi \Phi}{a \Phi_0}, k_2) c(k_1, k_2) \right)$$

where $c^{\dagger}(\mathbf{k})$ creates an electron in the Block state with wave vector $\mathbf{k}$ and $c(\mathbf{k})$ annihilates an electron in the Block state with wave vector $\mathbf{k}$. If $\Phi / \Phi_0 = p/q$ is rational, the magnetic Brillouin zone can be reduced to $0 \leq k_1 \leq 2\pi/a$ and $0 \leq k_2 \leq 2\pi/(qa)$. We obtain

$$c \left(k_1 + 2\frac{\pi \Phi}{a \Phi_0} \ell, k_2 \right) = c \left(k_1 + 2\frac{\pi \Phi}{a \Phi_0} (\ell + q), k_2 \right)$$

with ℓ are integers.

3.4 Hubbard Model

Problem 81. Let $c_\uparrow^\dagger, c_\downarrow^\dagger, c_\uparrow, c_\downarrow$ be Fermi creation and annihilation operators with spin up and down, respectively. Let $\hat{N}_\uparrow := c_\uparrow^\dagger c_\uparrow$ and $\hat{N}_\downarrow := c_\downarrow^\dagger c_\downarrow$. The number operator $\hat{N}$ and the spin operator $\hat{S}_3$ are given by

$$\hat{N} = \hat{N}_\uparrow + \hat{N}_\downarrow, \qquad \hat{S}_3 = \frac{1}{2}(\hat{N}_\uparrow - \hat{N}_\downarrow).$$

Consider the hermitian Hamilton operator

$$\hat{H} = \hbar\omega(c_\uparrow^\dagger c_\downarrow + c_\downarrow^\dagger c_\uparrow) + U\hat{N}_\uparrow\hat{N}_\downarrow.$$

(i) Does the Hamilton operator commute with the number operator $\hat{N}$ and the spin operator $\hat{S}_3$.

(ii) Find the matrix representation with the basis given by

$$c_\uparrow^\dagger c_\downarrow^\dagger|0\rangle, \quad c_\uparrow^\dagger|0\rangle, \quad c_\downarrow^\dagger|0\rangle, \quad |0\rangle.$$

(iii) Find the matrix representation of $\hat{N}$ and $\hat{S}_3$.

Solution 81. (i) We have $[\hat{H}, \hat{N}] = 0$ and $[\hat{H}, \hat{S}_3] = 0$.
(ii) We find

$$\hat{H}c_\uparrow^\dagger c_\downarrow^\dagger|0\rangle = (\hbar\omega + U)c_\uparrow^\dagger c_\downarrow^\dagger|0\rangle, \quad \hat{H}c_\uparrow^\dagger|0\rangle = \hbar\omega c_\downarrow^\dagger|0\rangle, \quad \hat{H}c_\downarrow^\dagger|0\rangle = \hbar\omega c_\uparrow^\dagger|0\rangle, \quad \hat{H}|0\rangle = 0|0\rangle$$

with the matrix representation

$$\hat{H} = \begin{pmatrix} \hbar\omega + U & 0 & 0 & 0 \\ 0 & 0 & \hbar\omega & 0 \\ 0 & \hbar\omega & 0 & 0 \\ 0 & 0 & 0 & 0 \end{pmatrix} \equiv (\hbar\omega) \oplus \begin{pmatrix} 0 & \hbar\omega \\ \hbar\omega & 0 \end{pmatrix} \oplus (0).$$

The eigenvalues $\hbar\omega + U$, $\hbar\omega$, $-\hbar\omega$, 0.
(iii) The matrix representation of $\hat{N}$ and $\hat{S}_3$ are

$$\hat{N} \mapsto \begin{pmatrix} 2 & 0 & 0 & 0 \\ 0 & 1 & 0 & 0 \\ 0 & 0 & 1 & 0 \\ 0 & 0 & 0 & 0 \end{pmatrix}, \qquad \hat{S}_3 \mapsto \begin{pmatrix} 0 & 0 & 0 & 0 \\ 0 & 1/2 & 0 & 0 \\ 0 & 0 & -1/2 & 0 \\ 0 & 0 & 0 & 0 \end{pmatrix}$$

with the eigenvalues 2, 1 (twice), 0 for $\hat{N}$ and $1/2$, $-1/2$, 0 (twice) for $\hat{S}_3$.

Problem 82. Let $\hat{N}_\uparrow := c_\uparrow^\dagger c_\uparrow$ and $\hat{N}_\downarrow := c_\downarrow^\dagger c_\downarrow$. The number operator $\hat{N}$ and the spin operator $\hat{S}_3$ are given by

$$\hat{N} = \hat{N}_\uparrow + \hat{N}_\downarrow, \qquad \hat{S}_3 = \frac{1}{2}(\hat{N}_\uparrow - \hat{N}_\downarrow).$$

Consider the Hamilton operator

$$\hat{H} = \hbar\omega(c_\uparrow^\dagger c_\downarrow^\dagger + c_\downarrow c_\uparrow) + U\hat{N}_\uparrow\hat{N}_\downarrow.$$

(i) Does the Hamilton operator commute with the number operator $\hat{N}$ and the spin operator $\hat{S}_3$.

(ii) Find the matrix representation of $\hat{H}$ with the basis given by

$$c_\uparrow^\dagger c_\downarrow^\dagger |0\rangle, \quad c_\uparrow^\dagger |0\rangle, \quad c_\downarrow^\dagger |0\rangle, \quad |0\rangle$$

and the eigenvalues.

Solution 82.   (i) We have $[\hat{H}, \hat{N}] \neq 0$ and $[\hat{H}, \hat{S}_3] \neq 0$.

(ii) We find

$$\hat{H} c_\uparrow^\dagger c_\downarrow^\dagger |0\rangle = \hbar\omega |0\rangle + U c_\uparrow^\dagger c_\downarrow^\dagger |0\rangle, \quad \hat{H} c_\uparrow^\dagger |0\rangle = 0 c_\uparrow^\dagger |0\rangle, \quad \hat{H} c_\downarrow^\dagger |0\rangle = 0 c_\downarrow^\dagger |0\rangle, \quad \hat{H} |0\rangle = \hbar\omega |0\rangle.$$

Thus the second and third equation is an eigenvalue equation both with eigenvalue 0. The matrix representation

$$\hat{H} = \begin{pmatrix} U & 0 & 0 & \hbar\omega \\ 0 & 0 & 0 & 0 \\ 0 & 0 & 0 & 0 \\ \hbar\omega & 0 & 0 & 0 \end{pmatrix}$$

with the eigenvalues 0 (twice) and

$$\frac{1}{2}(-\sqrt{U^2 + 4(\hbar\omega)^2} + U), \qquad \frac{1}{2}(\sqrt{U^2 + 4(\hbar\omega)^2} + U).$$

Problem 83.   The operators $c_{j\uparrow}^\dagger, c_{j\downarrow}^\dagger, c_{j\uparrow}, c_{j\downarrow}$ are Fermi operators. Consider the two point Hubbard Hamilton operator

$$\hat{H} = t(c_{0,\uparrow}^\dagger c_{1,\uparrow} + c_{0,\downarrow}^\dagger c_{1,\downarrow} + c_{1,\uparrow}^\dagger c_{0,\uparrow} + c_{1,\downarrow}^\dagger c_{0,\downarrow}) + U \sum_{j=0}^{1} c_{j,\uparrow}^\dagger c_{j,\uparrow} c_{j,\downarrow}^\dagger c_{j,\downarrow}$$

where $t > 0$ and $U > 0$. Apply the *discrete Fourier transform* $(N = 2)$

$$c_{j,\sigma}^\dagger = \frac{1}{\sqrt{N}} \sum_{k \in \{0,\pi\}} e^{-ikj} c_{k,\sigma}^\dagger \Rightarrow c_{j,\sigma} = \frac{1}{\sqrt{N}} \sum_{k \in \{0,\pi\}} e^{ikj} c_{k,\sigma}$$

where $\sigma \in \{\uparrow, \downarrow\}$, $j \in \{0, 1\}$ and $k \in \{0, \pi\}$. Then the inverse discrete Fourier transform is given by

$$c_{k,\sigma}^\dagger = \frac{1}{\sqrt{N}} \sum_{j \in \{0,1\}} e^{ikj} c_{j,\sigma}^\dagger \Rightarrow c_{k,\sigma} = \frac{1}{\sqrt{N}} \sum_{j \in \{0,1\}} e^{-ikj} c_{j,\sigma}.$$

Solution 83.   For the hopping terms we have

$$c_{0,\uparrow}^\dagger c_{1,\uparrow} = \frac{1}{\sqrt{2}}(c_{0,\uparrow} + c_{\pi,\uparrow})\frac{1}{\sqrt{2}}(c_{0,\uparrow} - c_{\pi,\uparrow}) = \frac{1}{2}(c_{0,\uparrow}^\dagger c_{0,\uparrow} - c_{\pi,\uparrow}^\dagger c_{\pi,\uparrow} - c_{0,\uparrow}^\dagger c_{\pi,\uparrow} + c_{\pi,\uparrow}^\dagger c_{0,\uparrow})$$

$$c_{1,\uparrow}^\dagger c_{0,\uparrow} = \frac{1}{\sqrt{2}}(c_{0,\uparrow} - c_{\pi,\uparrow})\frac{1}{\sqrt{2}}(c_{0,\uparrow} + c_{\pi,\uparrow}) = \frac{1}{2}(c_{0,\uparrow}^\dagger c_{0,\uparrow} - c_{\pi,\uparrow}^\dagger c_{\pi,\uparrow} + c_{0,\uparrow}^\dagger c_{\pi,\uparrow} - c_{\pi,\uparrow}^\dagger c_{0,\uparrow})$$

$$c_{0,\downarrow}^\dagger c_{1,\downarrow} = \frac{1}{\sqrt{2}}(c_{0,\downarrow} + c_{\pi,\downarrow})\frac{1}{\sqrt{2}}(c_{0,\downarrow} - c_{\pi,\downarrow}) = \frac{1}{2}(c_{0,\downarrow}^\dagger c_{0,\downarrow} - c_{\pi,\downarrow}^\dagger c_{\pi,\downarrow} - c_{0,\downarrow}^\dagger c_{\pi,\downarrow} + c_{\pi,\downarrow}^\dagger c_{0,\downarrow})$$

$$c_{1,\downarrow}^\dagger c_{0,\downarrow} = \frac{1}{\sqrt{2}}(c_{0,\downarrow} - c_{\pi,\downarrow})\frac{1}{\sqrt{2}}(c_{0,\downarrow} + c_{\pi,\downarrow}) = \frac{1}{2}(c_{0,\downarrow}^\dagger c_{0,\downarrow} - c_{\pi,\downarrow}^\dagger c_{\pi,\downarrow} + c_{0,\downarrow}^\dagger c_{\pi,\downarrow} - c_{\pi,\downarrow*}^\dagger c_{0,\downarrow}).$$

Summing up the hopping terms provides

$$c_{0,\uparrow}^\dagger c_{0,\uparrow} + c_{0,\downarrow}^\dagger c_{0,\downarrow} - c_{\pi,\uparrow}^\dagger c_{\pi,\uparrow} - c_{\pi,\downarrow}^\dagger c_{\pi,\downarrow}$$

and thus we can write for the hopping term

$$\sum_{k\in\{0,\pi\}} \sum_{\sigma\in\{\uparrow,\downarrow\}} \cos(k) c_{k,\sigma}^\dagger c_{k,\sigma}.$$

Now for the interacting term we have

$$c_{0,\uparrow}^\dagger c_{0,\uparrow} c_{0,\downarrow}^\dagger c_{0,\downarrow} + c_{1,\uparrow}^\dagger c_{1,\uparrow} c_{1,\downarrow}^\dagger c_{1,\downarrow} =$$

$$\frac{1}{2}(c_{0,\uparrow}^\dagger c_{0,\uparrow} c_{0,\downarrow}^\dagger c_{0,\downarrow} + c_{0,\uparrow}^\dagger c_{0,\uparrow} c_{\pi,\downarrow}^\dagger c_{\pi,\downarrow} + c_{0,\uparrow}^\dagger c_{\pi,\uparrow} c_{0,\downarrow}^\dagger c_{\pi,\downarrow} + c_{0,\uparrow}^\dagger c_{\pi,\uparrow} c_{\pi,\downarrow}^\dagger c_{0,\downarrow}$$

$$+c_{\pi,\uparrow}^\dagger c_{0,\uparrow} c_{0,\downarrow}^\dagger c_{\pi,\downarrow} + c_{\pi,\uparrow}^\dagger c_{0,\uparrow} c_{\pi,\downarrow}^\dagger c_{0,\downarrow} + c_{\pi,\uparrow}^\dagger c_{\pi,\uparrow} c_{0,\downarrow}^\dagger c_{0,\downarrow} + c_{\pi,\uparrow}^\dagger c_{\pi,\uparrow} c_{\pi,\downarrow}^\dagger c_{\pi,\downarrow}).$$

Thus we can write for the interacting term

$$\frac{1}{2} \sum_{k_1,k_2,k_3,k_4\in\{0,\pi\}} \delta_{k_1-k_2+k_3-k_4} c_{k_1,\uparrow}^\dagger c_{k_2,\uparrow} c_{k_3,\downarrow}^\dagger c_{k_4,\downarrow}$$

with $\delta(k_1 - k_2 + k_3 - k_4)$ defined as

$$\delta(k_1 - k_2 + k_3 - k_4) := \begin{cases} 1 & \text{if} \quad k_1 - k_2 + k_3 - k_4 = 0 \bmod 2\pi \\ 0 & \text{otherwise} \end{cases}$$

Problem 84. Consider the Hamilton operator (two-point Hubbard model)

$$\hat{H} = t(c_{1\uparrow}^\dagger c_{2\uparrow} + c_{1\downarrow}^\dagger c_{2\downarrow} + c_{2\uparrow}^\dagger c_{1\uparrow} + c_{2\downarrow}^\dagger c_{1\downarrow}) + U(\hat{N}_{1\uparrow}\hat{N}_{1\downarrow} + \hat{N}_{2\uparrow}\hat{N}_{2\downarrow})$$

where $\hat{N}_{j\uparrow} := c_{j\uparrow}^\dagger c_{j\uparrow}$, $\hat{N}_{j\downarrow} := c_{j\downarrow}^\dagger c_{j\downarrow}$.
(i) Show that the Hamilton operator commutes with the total number operator $\hat{N}$ and the total spin operator $\hat{S}_3$ where

$$\hat{N} := \sum_{j=1}^{2}(c_{j\uparrow}^\dagger c_{j\uparrow} + c_{j\downarrow}^\dagger c_{j\downarrow}), \qquad \hat{S}_3 := \frac{1}{2}\sum_{j=1}^{2}(c_{j\uparrow}^\dagger c_{j\uparrow} - c_{j\downarrow}^\dagger c_{j\downarrow}).$$

(ii) We consider the subspace with two particles $N = 2$ and total spin $S_3 = 0$. A basis in this space is given by

$$c_{1\uparrow}^\dagger c_{1\downarrow}^\dagger|0\rangle, \quad c_{1\uparrow}^\dagger c_{2\downarrow}^\dagger|0\rangle, \quad c_{2\uparrow}^\dagger c_{1\downarrow}^\dagger|0\rangle, \quad c_{2\uparrow}^\dagger c_{2\downarrow}^\dagger|0\rangle.$$

Find the matrix representation of $\hat{H}$ for this basis.
(iii) Find the discrete symmetries of $\hat{H}$ and perform a group-theoretical reduction.

Solution 84. (i) Using the Fermi anticommutation relations we obtain $[\hat{H}, \hat{N}] = 0$, $[\hat{H}, \hat{S}_3] = 0$. We also have $[\hat{N}, \hat{S}_3] = 0$.
(ii) Using the Fermi anticommutation relations and $c_{j\uparrow}|0\rangle = 0|0\rangle$, $c_{j\downarrow}|0\rangle = 0|0\rangle$ we obtain the matrix representation of $\hat{H}$ with the given basis

$$\begin{pmatrix} U & t & t & 0 \\ t & 0 & 0 & t \\ t & 0 & 0 & t \\ 0 & t & t & U \end{pmatrix}.$$

(iii) The Hamilton operator admits the discrete symmetry $1 \to 2$, $2 \to 1$, i.e. swapping the sites 1 and 2 leaves the Hamilton operator invariant. Thus we have a finite group and two elements (identity and the swapping of the sites). There are two conjugacy classes and therefore two irreducible representations. We find the two invariant subspaces

$$\left\{ \frac{1}{\sqrt{2}}(c_{1\downarrow}^\dagger c_{1\uparrow}^\dagger |0\rangle + c_{2\downarrow}^\dagger c_{2\uparrow}^\dagger |0\rangle), \quad \frac{1}{\sqrt{2}}(c_{1\downarrow}^\dagger c_{2\uparrow}^\dagger |0\rangle + c_{2\downarrow}^\dagger c_{1\uparrow}^\dagger |0\rangle) \right\}$$

$$\left\{ \frac{1}{\sqrt{2}}(c_{1\downarrow}^\dagger c_{1\uparrow}^\dagger |0\rangle - c_{2\downarrow}^\dagger c_{2\uparrow}^\dagger |0\rangle), \quad \frac{1}{\sqrt{2}}(c_{1\downarrow}^\dagger c_{2\uparrow}^\dagger |0\rangle - c_{2\downarrow}^\dagger c_{1\uparrow}^\dagger |0\rangle) \right\}.$$

Problem 85.   Given the three point Hubbard Hamilton operator with cyclic boundary conditions

$$\hat{H} = t \sum_{\sigma \in \{\uparrow,\downarrow\}} (c_{0\sigma}^\dagger c_{1\sigma} + c_{1\sigma}^\dagger c_{0\sigma} + c_{1\sigma}^\dagger c_{2\sigma} + c_{2\sigma}^\dagger c_{1\sigma} + c_{2\sigma}^\dagger c_{0\sigma} + c_{0\sigma}^\dagger c_{2\sigma}) + U \sum_{j=0}^{2} \hat{N}_{j\uparrow} \hat{N}_{j\downarrow}$$

where $\hat{N}_{j\sigma} := c_{j\sigma}^\dagger c_{j\sigma}$. The Hamilton operator commutes with $[\hat{H}, \hat{N}_e] = 0$ and $[\hat{H}, \hat{S}_3] = 0$. Thus we can consider a 9 dimensional subspace with two particles and total spin equal to 0. A basis is

$$\{ c_{0\uparrow}^\dagger c_{0\downarrow}^\dagger |0\rangle, \quad c_{0\uparrow}^\dagger c_{1\downarrow}^\dagger |0\rangle, \quad c_{0\downarrow}^\dagger c_{1\uparrow}^\dagger |0\rangle, \quad c_{0\uparrow}^\dagger c_{2\downarrow}^\dagger |0\rangle, \quad c_{0\downarrow}^\dagger c_{2\uparrow}^\dagger |0\rangle$$

$$c_{1\uparrow}^\dagger c_{1\downarrow}^\dagger |0\rangle, \quad c_{1\uparrow}^\dagger c_{2\downarrow}^\dagger |0\rangle, \quad c_{1\downarrow}^\dagger c_{2\uparrow}^\dagger |0\rangle, \quad c_{2\uparrow}^\dagger c_{2\downarrow}^\dagger |0\rangle \}.$$

Calculate the matrix representation of $\hat{H}$ using this nine dimensional basis, i.e. $\langle j|\hat{H}|k\rangle$. Find the eigenvalues of this 9×9 matrix.

Solution 85.   The following SymbolicC++ program provides the matrix representation

$$\langle 0|c_{ni\downarrow} c_{mi\uparrow} \hat{H} c_{ib\uparrow}^\dagger c_{jb}^\dagger |0\rangle$$

```
// hubbard.cpp

#include <iostream>
#include "symbolicc++.h"
using namespace std;

const int n=3;
enum { create, annihilate };
enum { up, down };
enum { x, y, z };

Symbolic vacuum = ~Symbolic("|0>");
Symbolic vacuum_dual = ~Symbolic("<0|");
// ~ => non-commutative
Symbolic c[2] = { ~Symbolic("cd",n,2), ~Symbolic("c",n,2) };

using SymbolicConstant::i; // i = sqrt(-1)

Equations fermi_relations(void)
{
  int i, j, k, l, s, t;
```

```
Equations relations;
// <0|0> = 1 and 1/i = -i
relations = (relations,vacuum_dual*vacuum==1,
             1/SymbolicConstant::i==-SymbolicConstant::i,
             ((2*SymbolicConstant::i)^(-1))==-SymbolicConstant::i/2);
// c|0> = 0    <0|cd = 0
for(i=0;i<n;i++)
 for(j=up;j<=down;j++)
  relations = (relations,c[annihilate](i,j)*vacuum==0,
               vacuum_dual*c[create](i,j)==0);
// cd c = I-c cd
for(i=create;i<=annihilate;i++)
  for(j=create;j<=annihilate;j++)
   for(k=up;k<=down;k++)
    for(l=up;l<=down;l++)
     for(s=0;s<n;s++)
      for(t=0;t<n;t++)
      {
      if(i==j && k==l && s==t) relations = (relations,c[i](s,k)*c[i](s,k)==0);
      else
      if(i==annihilate && j==create && k==l && s==t)
       relations = (relations,c[i](s,k)*c[j](s,k)==1-c[j](s,k)*c[i](s,k));
      else
      if((i==annihilate && j==create) || (i==j && t < s) ||
         (i==j && s==t && l==up && k==down))
       relations = (relations,c[i](s,k)*c[j](t,l)==-c[j](t,l)*c[i](s,k));
      }
         return relations;
}

int main(void)
{
 Symbolic t("t"), U("U"), Ni("Ni",n,2);
 Equations relations = fermi_relations();
 for(int j=0;j<n;j++)
 { for(int k=up;k<=down;k++) Ni(j,k) = c[create](j,k)*c[annihilate](j,k); }

 Symbolic H;
 // setup the Hamilton operator used for the commutation relations
 for(int j=0;j<n;j++)
 {
  H += U*Ni(j,up)*Ni(j,down);
  for(int k=up;k<=down;k++)
  {
  H += t*(c[create]((j+1)%n,k)*c[annihilate](j,k)
      +c[create](j,k)*c[annihilate]((j+1)%n,k));
   }
 }
 H = H.subst_all(relations); cout << "H = " << H << endl;
 // find the matrix representation
 cout<<"Matrix representation of H" << endl;
 for(int ib=0;ib<n;ib++)
  for(int jb=0;jb<n;jb++,cout<<endl)
   for(int mi=0;mi<n;mi++)
    for(int ni=0;ni<n;ni++)
```

```
    {
      Symbolic HM =
        vacuum_dual*c[annihilate](ni,down)*c[annihilate](mi,up)
                *H*c[create](ib,up)*c[create](jb,down)*vacuum;
      HM = HM.subst_all(relations); cout << HM << " ";
    }
  cout<<endl;
  return 0;
}
```

Note that

$$(\sum_{j=0}^{2} \hat{N}_{j\uparrow}\hat{N}_{j\downarrow})c_{k\uparrow}^{\dagger}c_{k\downarrow}^{\dagger}|0\rangle = \delta_{jk}c_{k\uparrow}^{\dagger}c_{k\downarrow}^{\dagger}|0\rangle.$$

The program provides the matrix representation

$$\begin{pmatrix} U & t & t & t & 0 & 0 & t & 0 & 0 \\ t & 0 & t & 0 & t & 0 & 0 & t & 0 \\ t & t & 0 & 0 & 0 & t & 0 & 0 & t \\ t & 0 & 0 & 0 & t & t & t & 0 & 0 \\ 0 & t & 0 & t & U & t & 0 & t & 0 \\ 0 & 0 & t & t & t & 0 & 0 & 0 & t \\ t & 0 & 0 & t & 0 & 0 & 0 & t & t \\ 0 & t & 0 & 0 & t & 0 & t & 0 & t \\ 0 & 0 & t & 0 & 0 & t & t & t & U \end{pmatrix}.$$

The nine eigenvalues are given by

$$-\frac{1}{2}(\sqrt{U^2 - 4tU + 36t^2} - U - 2t)$$

$$\frac{1}{2}(\sqrt{U^2 - 4tU + 36t^2} + U + 2t)$$

$$-2t$$

$$-\frac{1}{2}(\sqrt{U^2 + 2tU + 9t^2} - U + t) \text{ (twice)}$$

$$\frac{1}{2}(\sqrt{U^2 + 2tU + 9t^2} + U - t \text{ (twice)}$$

$$t \text{ (twice)}$$

Problem 86. Let $c_{k\sigma}^{\dagger}$ and $c_{k\sigma}$ be Fermi creation and annihilation operators, where $\sigma \in \{\uparrow,\downarrow\}$ and $j = 1, 2, \ldots, N$ and k refers to the momentum. The Fermi operators obey the anticommutation relations

$$[c_{k\sigma}, c_{p\sigma'}^{\dagger}]_+ = \delta_{kp}\delta_{\sigma\sigma'}I, \quad [c_{k\sigma}, c_{k\sigma'}]_+ = 0$$

where I is the identity operator and 0 the zero operator. From the second relation it follows that

$$[c_{k\sigma}^{\dagger}, c_{p\sigma'}^{\dagger}]_+ = 0.$$

Let

$$\hat{N}_{j\uparrow} := c_{k\uparrow}^{\dagger}c_{k\uparrow}, \qquad \hat{N}_{j\downarrow} := c_{k\downarrow}^{\dagger}c_{k\downarrow}.$$

Consider the four point *Hubbard model* in Bloch representation

$$\hat{H} = \sum_{k \in S} \sum_{\sigma \in \{\uparrow\downarrow\}} \epsilon(k) c_{k\sigma}^\dagger c_{k\sigma} + \frac{U}{4} \sum_{k_1,k_2,k_3,k_4 \in S} \delta(k_1 - k_2 + k_3 - k_4) c_{k_1\uparrow}^\dagger c_{k_2\uparrow} c_{k_3\downarrow}^\dagger c_{k_4\downarrow}$$

where $\epsilon(k) = 2t\cos(k)$. Here

$$k, k_1, k_2, k_3 \in S := \{ -\frac{\pi}{2}, \, 0, \, \frac{\pi}{2}, \, \pi \quad \mathrm{mod} \ 2\pi \}$$

and

$$\delta(k_1 - k_2 + k_3 - k_4) = \begin{cases} 1 & \text{if} \quad k_1 - k_2 + k_3 - k_4 = 0 \ \mathrm{mod} \ 2\pi \\ 0 & \text{otherwise} \end{cases}$$

Consider the operators

$$\hat{N}_e = \sum_{k \in S} \sum_\sigma c_{k\sigma}^\dagger c_{j\sigma}, \quad \hat{S}_3 = \frac{1}{2} \sum_{k \in S} (c_{k\uparrow}^\dagger c_{k\uparrow} - c_{k\downarrow}^\dagger c_{k\downarrow}), \quad \hat{P} = \sum_{k \in S} k(c_{k\uparrow}^\dagger c_{k\uparrow} + c_{k\downarrow}^\dagger c_{k\downarrow})$$

where $\hat{N}_e$ is the number operator, $\hat{S}_3$ is the total spin operator in z-direction and $\hat{P}$ the total momentum operator.
(i) Calculate the commutators $[\hat{H}, \hat{N}_e]$, $[\hat{H}, \hat{S}_3]$, $[\hat{H}, \hat{P}]$.
(ii) Find the eigenvalues of the total momentum operator $\hat{P}$.

Solution 86. (i) We find $[\hat{H}, \hat{N}_e] = 0$, $[\hat{H}, \hat{S}_3] = 0$, $[\hat{H}, \hat{P}] = 0$. Thus $\hat{N}_e$, $\hat{S}_3$ and $\hat{P}$ are constants of motion. Thus we can consider subspaces with fixed number of electrons, fixed number of total spin in z-direction and fixed total momentum.
(ii) Obviously we find $-\pi/2$, 0, $\pi/2$, π.

Problem 87. Consider the 4-point Hubbard model in Wannier representation

$$\hat{H} = t \sum_{j=0}^{3} \sum_{\sigma=\uparrow,\downarrow} (c_{j+1\sigma}^\dagger c_{j\sigma} + c_{j\sigma}^\dagger c_{j+1\sigma}) + U \sum_{j=0}^{3} n_{j\uparrow} n_{j\downarrow}$$

with cyclic boundary conditions ($4 \equiv 0$). Given the operator

$$\hat{C} = \sum_{j=0}^{3} ((c_{j\uparrow}^\dagger c_{j-1\uparrow} - c_{j-1\uparrow}^\dagger c_{j\uparrow})(n_{j\downarrow} + n_{j-1\downarrow}) + (c_{j\downarrow}^\dagger c_{j-1\downarrow} - c_{j-1\downarrow}^\dagger c_{j\downarrow})(\hat{N}_{j\uparrow} + \hat{N}_{j-1\uparrow}))$$

$$- \sum_{j=0}^{3} \sum_{\sigma=\uparrow,\downarrow} (c_{j\sigma}^\dagger c_{j-1\sigma} - c_{j-1\sigma}^\dagger c_{j\sigma})$$

where $-1 \equiv 3$ (cyclic boundary conditions). Find the commutator $[\hat{C}, \hat{H}]$.

Solution 87. The calculation is rather lengthy. Thus we apply computer algebra using SymbolicC++.

```
// hubbard1.cpp

#include <iostream>
#include "symbolicc++.h"
using namespace std;
```

```
const int n=4;

enum { create, annihilate };
enum { up, down };
enum { x, y, z };

Symbolic vacuum = ~Symbolic("|0>");
Symbolic vacuum_dual = ~Symbolic("<0|");
// ~ => non-commutative
Symbolic c[2] = { ~Symbolic("cd",n,2),~Symbolic("c",n,2) };

using SymbolicConstant::i; // i = srt(-1)

Equations fermi_relations(void)
{
 int i, j, k, l, s, t;
 Equations relations;
  // <0|0> = 1 and 1/i = -i
 relations = (relations, vacuum_dual * vacuum == 1,
               1/SymbolicConstant::i==-SymbolicConstant::i,
               ((2*SymbolicConstant::i)^(-1))==-SymbolicConstant::i/2);
  // c|0>=0|0>    <0|cd = <0|0
 for(i=0;i<n;i++)
  for(j=up;j<=down;j++)
   relations = (relations,c[annihilate](i,j)*vacuum==0,vacuum_dual*c[create](i,j)==0);
 // cd c = I - c cd
 for(i=create;i<=annihilate;i++)
   for(j=create;j<=annihilate;j++)
    for(k=up;k<=down;k++)
     for(l=up;l<=down;l++)
      for(s=0;s<n;s++)
       for(t=0;t<n;t++)
       {
        if(i==j && k==l && s==t) relations = (relations,c[i](s,k)*c[i](s,k)==0);
        else if(i==annihilate && j==create && k==l && s==t)
         relations = (relations,c[i](s,k)*c[j](s,k)==1-c[j](s,k)*c[i](s,k));
        else
        if((i==annihilate && j==create) || (i==j && t < s) ||
           (i==j && s==t && l==up && k==down))
          relations = (relations,c[i](s,k)*c[j](t,l)==-c[j](t,l)*c[i](s,k));
       }
  return relations;
}

int main(void)
{
 Symbolic t("t"), U("U"), alpha("alpha"),ni("ni",n,2), S("S",n,3);
 Equations relations = fermi_relations();

 // setup n(j,k)
 for(int j=0;j<n;j++)
 { for(int k=up;k<=down;k++) ni(j,k) = c[create](j,k)*c[annihilate](j,k); }

 Symbolic Ne, H, C;
```

```
// setup the operators used for the commutation relations
for(int j=0;j<n;j++)
{
 H += U*ni(j,up)*ni(j,down);
 C += (c[create](j,up)*c[annihilate]((j+n-1)%n,up)
      -c[create]((j+n-1)%n,up)*c[annihilate](j,up))
     *(ni(j,down)+ni((j+n-1)%n,down))
   + (c[create](j,down)*c[annihilate]((j+n-1)%n,down)
      -c[create]((j+n-1)%n,down)*c[annihilate](j,down))
     *(ni(j,up)+ni((j+n-1)%n,up));

 for(int k=up;k<=down;k++)
 {
  Ne += ni(j,k);
  H += t*(c[create]((j+1)%n,k)*c[annihilate](j,k)
         +c[create](j,k)*c[annihilate]((j+1)%n,k));
  C -= c[create](j,k)*c[annihilate]((j+n-1)%n,k)
       -c[create]((j+n-1)%n,k)*c[annihilate](j,k);
 }
}

Ne = Ne.subst_all(relations);
H  = H.subst_all(relations);
C  = C.subst_all(relations);
// determine the commutation relation
Symbolic result = (C*H-H*C).subst_all(relations);
cout << "[C,H]=" << result << endl;
return 0;
}
```

The output is $[\hat{H}, \hat{C}] = 0$.

Problem 88. Let $c_{j\sigma}^{\dagger}$ and $c_{j\sigma}$ be Fermi creation and annihilation operators, where $\sigma \in \{\uparrow, \downarrow\}$ and $j = 1, 2, \ldots, N$. The Fermi operators obey the anticommutation relations

$$[c_{j\sigma}, c_{k\sigma'}^{\dagger}]_+ = \delta_{jk}\delta_{\sigma\sigma'}I, \quad [c_{j\sigma}, c_{k\sigma'}]_+ = 0$$

where I is the identity operator and 0 the zero operator. From the second relation it follows that

$$[c_{j\sigma}^{\dagger}, c_{k\sigma'}^{\dagger}]_+ = 0.$$

Let

$$\hat{N}_{j\uparrow} := c_{j\uparrow}^{\dagger}c_{j\uparrow}, \qquad \hat{N}_{j\downarrow} := c_{j\downarrow}^{\dagger}c_{j\downarrow}.$$

Consider the Hubbard model

$$\hat{H} = t \sum_{i,j} \sum_{\sigma \in \{\uparrow\downarrow\}} c_{i\sigma}^{\dagger}c_{j\sigma} + U \sum_{j=1}^{N} \hat{N}_{j\uparrow}\hat{N}_{j\downarrow}$$

where t and U are real constant. Consider the operators

$$\hat{N}_e = \sum_{j}^{N}\sum_{\sigma} c_{j\sigma}^{\dagger}c_{j\sigma}, \quad \hat{S}_3 = \frac{1}{2}\sum_{j}^{N}(c_{j\uparrow}^{\dagger}c_{j\uparrow} - c_{j\downarrow}^{\dagger}c_{j\downarrow})$$

where $\hat{N}_e$ is the number operator and $\hat{S}_3$ is the total spin operator in z-direction. Calculate the commutators $[\hat{H}, \hat{N}_e]$ and $[\hat{H}, \hat{S}_3]$.

Solution 88. We find $[\hat{H}, \hat{N}_e] = 0$, $[\hat{H}, \hat{S}_3] = 0$. Thus $\hat{N}_e$ and $\hat{S}_3$ are constants of motion. Thus we can consider subspaces with fixed number of electrons and fixed number of total spin in z-direction.

Problem 89. Consider the four point Hubbard model described in Bloch representation

$$\hat{H} = \sum_k \sum_{\sigma \in \{\uparrow\downarrow\}} \epsilon(k) c_{k\sigma}^\dagger c_{k\sigma} + \frac{U}{4} \sum_{k_1,k_2,k_3,k_4} \delta(k_1 - k_2 + k_3 - k_4) c_{k_1\uparrow}^\dagger c_{k_2\uparrow} c_{k_3\downarrow}^\dagger c_{k_4\downarrow}$$

where $\epsilon(k) = 2t\cos(k)$. Here

$$k, k_1, k_2, k_3 \in \{ -\frac{\pi}{2},\ 0,\ \frac{\pi}{2},\ \pi \mod 2\pi \}.$$

Give a basis of the Hilbert space for $N_e = 4$, $S_3 = 0$ and $P = 0$.

Solution 89. Owing to $N_e = 4$, $S_3 = 0$ and the Pauli principle we have the states

$$c_{k_1\uparrow}^\dagger c_{k_2\uparrow}^\dagger c_{k_3\downarrow}^\dagger c_{k_4\downarrow}^\dagger |0\rangle, \quad k_1 < k_2, \ k_3 < k_4, \ k_1 + k_2 + k_3 + k_4 = 0 \mod 2\pi \}.$$

The dimensions of the subspaces with $P = -\pi/2, 0, \pi/2, \pi$ are $8, 10, 8, 10$. For $P = 0$ we obtain the ten states

$$c_{-\pi/2\uparrow}^\dagger c_{0\uparrow}^\dagger c_{-\pi/2\downarrow}^\dagger c_{\pi\downarrow}^\dagger |0\rangle, \quad c_{-\pi/2\uparrow}^\dagger c_{0\uparrow}^\dagger c_{0\downarrow}^\dagger c_{\pi/2\downarrow}^\dagger |0\rangle$$

$$c_{-\pi/2\uparrow}^\dagger c_{\pi\uparrow}^\dagger c_{-\pi/2\downarrow}^\dagger c_{\pi/2\downarrow}^\dagger |0\rangle, \quad c_{-\pi/2\uparrow}^\dagger c_{\pi\uparrow}^\dagger c_{-\pi/2\downarrow}^\dagger c_{0\downarrow}^\dagger |0\rangle$$

$$c_{-\pi/2\uparrow}^\dagger c_{\pi\uparrow}^\dagger c_{\pi/2\downarrow}^\dagger c_{\pi\downarrow}^\dagger |0\rangle, \quad c_{0\uparrow}^\dagger c_{\pi/2\uparrow}^\dagger c_{-\pi/2\downarrow}^\dagger c_{0\downarrow}^\dagger |0\rangle$$

$$c_{0\uparrow}^\dagger c_{\pi/2\uparrow}^\dagger c_{\pi/2\downarrow}^\dagger c_{\pi\downarrow}^\dagger |0\rangle, \quad c_{0\uparrow}^\dagger c_{\pi\uparrow}^\dagger c_{0\downarrow}^\dagger c_{\pi\downarrow}^\dagger |0\rangle$$

$$c_{\pi/2\uparrow}^\dagger c_{\pi\uparrow}^\dagger c_{-\pi/2\downarrow}^\dagger c_{\pi\downarrow}^\dagger |0\rangle, \quad c_{\pi/2\uparrow}^\dagger c_{\pi\uparrow}^\dagger c_{0\downarrow}^\dagger c_{\pi/2\downarrow}^\dagger |0\rangle.$$

Give a basis of the Hilbert space for $N_e = 4$, $S_3 = 0$ and $P = \pi/2$. Give a basis of the Hilbert space for $N_e = 4$, $S_3 = 0$ and $P = \pi$. Give a basis of the Hilbert space for $N_e = 4$, $S_3 = 0$ and $P = 3\pi/2$. For $P = \pi/2$ we obtain the 8 states. For $P = \pi$ we obtain the 10 states. For $P = 3\pi/2$ we obtain the 8 states.

Problem 90. The Hamilton operator of the one-dimensional Hubbard model on a periodic L-site chain is given by

$$\hat{H} = -\sum_{j=1}^{L} \sum_{\sigma=\uparrow,\downarrow} \left(c_{j,\sigma}^\dagger c_{j+1,\sigma} + c_{j+1,\sigma}^\dagger c_{j,\sigma} \right) + U \sum_{j=1}^{L} \left(\hat{N}_{j\uparrow} - \frac{1}{2}I \right) \left(\hat{N}_{j\downarrow} - \frac{1}{2}I \right)$$

$c_{j,\sigma}^\dagger$ and $c_{j,\sigma}$ are creation and annihilation operators of electrons of spin σ at site j (electrons in Wannier states). Let

$$\hat{N}_{j,\sigma} := c_{j,\sigma}^\dagger c_{j,\sigma}$$

be the particle number operator, and $U > 0$ is the coupling constant. Periodic boundary conditions $c_{L+1,\sigma} = c_{1,\sigma}$ are imposed. The Hamilton operator commutes with

$$\hat{N} = \sum_{j=1}^{L} (\hat{N}_{j,\uparrow} + \hat{N}_{j,\downarrow}) \quad \text{and} \quad \hat{S}_3 = \frac{1}{2} \sum_{j=1}^{L} (\hat{N}_{j,\uparrow} - \hat{N}_{j,\downarrow}).$$

Thus the Hamilton operator conserves the number of electrons N and the number of down spins M. The eigenvalue problem can thus be solved for fixed N and M. The Hamilton operator is invariant under a particle hole transformation and under reversal of all spins. Thus we can set $2M \leq N \leq L$. We denote the coordinates and spins of the electrons by x_j and σ_j, respectively, with $x_j = 1, \ldots, L$ and $\sigma_j = \uparrow, \downarrow$. The eigenvector of the Hamilton operator $\hat{H}$ can be written as

$$|N, M\rangle = \frac{1}{\sqrt{N!}} \sum_{x_1, \ldots, x_N = 1} \sum_{\sigma_1, \ldots, \sigma_N = \uparrow, \downarrow} f(x_1, \ldots, x_N; \sigma_1, \ldots, \sigma_N) c_{x_N \sigma_N}^{\dagger} \cdots c_{x_1 \sigma_1}^{\dagger} |0\rangle.$$

Here the function $f(x_1, \ldots, x_N; \sigma_1, \ldots, \sigma_N)$ is the *Bethe ansatz wave function*. It depends on the relative ordering of the coordinates x_j. Any ordering is related to a permutation Q of the numbers $1, \ldots, N$ through the inequality

$$1 \leq x_{Q1} \leq x_{Q2} \leq \ldots \leq x_{QN} \leq L.$$

The set of all permutations of N distinct numbers provides the symmetric group S_N. The inequality divides the configuration space of N electrons into $N!$ sectors. They can be labeled by the permutations Q. The Bethe ansatz wave functions in the sector Q are given as

$$f(x_1, \ldots, x_N; \sigma_1, \ldots, \sigma_N) = \sum_{P \in S_N} \text{sign}(PQ) g_P(\sigma_{Q1}, \ldots, \sigma_{QN}) \exp\left(i \sum_{j=1}^{N} k_{Pj} x_{Qj} \right).$$

The function $\text{sign}(Q)$ is the sign function (or parity) on the symmetric group, which is -1 for odd permutations and $+1$. Find an expression for the spin dependent amplitudes $g_P(\sigma_{Q1}, \ldots, \sigma_{QN})$.

Solution 90. Explicit expressions for the spin dependent amplitudes $g_P(a_{Q1}, \ldots, a_{QN})$ are of the form of the Bethe ansatz wave function

$$f_P(\sigma_{Q1}, \ldots, \sigma_{QN}) = \sum_{\pi \in S_M} A(\lambda_{\pi 1}, \ldots, \lambda_{\pi M}) \prod_{l=1}^{M} F_P(\lambda_{\pi l}; y_l).$$

Here $F_P(\lambda; y)$ is defined as

$$F_P(\lambda; y) := \frac{iU/2}{\lambda - \sin(k_{Py}) + iU/4} \prod_{j=1}^{y-1} \frac{\lambda - \sin(k_{Pj}) - iU/4}{\lambda - \sin(k_{Pj}) + iU/4}$$

and the amplitudes $A(\lambda_1, \ldots, \lambda_M)$ are given by

$$A(\lambda_1, \ldots, \lambda_M) = \prod_{1 \leq m < n \leq M} \frac{\lambda_m - \lambda_n - iU/2}{\lambda_m - \lambda_n}.$$

Here y_i denotes the position of the jth down spin in the sequence $\sigma_{Q1}, \dots \sigma_{QN}$. The y's are coordinates of down spins on electrons. If the number of down spins in the sequence $\sigma_{Q1}, \dots, \sigma_{QN}$ is different from M, the the amplitude $g_P(\sigma_{Q1}, \dots, \sigma_{AN})$ vanishes. The Bethe wave functions depend on two sets of quantum numbers

$$\{k_j : j = 1, \dots, N\} \quad \text{and} \quad \{\lambda_\ell : \ell = 1, \dots, M\}.$$

These quantum numbers may in general be complex. The k_j and λ_l are called charge momenta and spin rapidities, respectively. The charge momenta and spin rapidities satisfy the Lieb-Wu equations

$$e^{ik_j L} = \prod_{l=1}^{M} \frac{\lambda_l - \sin(k_j) - iU/4}{\lambda_l - \sin(k_j) + iU/4}, \quad j = 1, \dots, N$$

$$\prod_{j=1}^{N} \frac{\lambda_l - \sin(k_j) - iU/4}{\lambda_l - \sin(k_j) + iU/4} = \prod_{\substack{m=1 \\ m \neq l}}^{M} \frac{\lambda_l - \lambda_m - iU/2}{\lambda_l - \lambda_m + iU/2}, \quad l = 1, \dots, M.$$

The states $|N, M\rangle$ are joint eigenstates of the Hubbard Hamilton operator and the momentum operator with eigenvalues

$$E = -2 \sum_{j=1}^{N} \cos(k_j) + \frac{U}{4}(L - 2N) , \quad P = \left(\sum_{j=1}^{N} k_j \right) \mod 2\pi.$$

3.5 Supplementary Problems

Problem 1. Let $c_\uparrow^\dagger$, $c_\downarrow^\dagger$ be Fermi creation operators with spin-up and spin-down. Let σ_1, σ_2, σ_3 be the Pauli spin matrices. Consider the four dimensional basis

$$c_\uparrow^\dagger c_\downarrow^\dagger |0\rangle, \quad c_\uparrow^\dagger |0\rangle, \quad c_\downarrow^\dagger |0\rangle, \quad |0\rangle.$$

(i) Find the matrix representation of

$$c_\uparrow^\dagger c_\downarrow, \quad c_\downarrow^\dagger c_\uparrow, \quad c_\uparrow^\dagger c_\uparrow, \quad c_\downarrow^\dagger c_\downarrow.$$

(ii) We define the operators

$$S_j := \frac{1}{2} \begin{pmatrix} c_\uparrow^\dagger & c_\downarrow^\dagger \end{pmatrix} \sigma_j \begin{pmatrix} c_\uparrow \\ c_\downarrow \end{pmatrix}, \quad j = 1, 2, 3.$$

Find the commutators $[S_1, S_2]$, $[S_2, S_3]$, $[S_3, S_1]$. Find the matrix representation with the basis given above.

Problem 2. Consider a two Fermion system with the matrix representation for the Fermi creation and annihilation operators

$$c_1^\dagger = \left(\frac{1}{2}\sigma_+\right) \otimes I_2 = \begin{pmatrix} 0 & 0 & 1 & 0 \\ 0 & 0 & 0 & 1 \\ 0 & 0 & 0 & 0 \\ 0 & 0 & 0 & 0 \end{pmatrix}, \quad c_1 = \left(\frac{1}{2}\sigma_-\right) \otimes I_2 = \begin{pmatrix} 0 & 0 & 0 & 0 \\ 0 & 0 & 0 & 0 \\ 1 & 0 & 0 & 0 \\ 0 & 1 & 0 & 0 \end{pmatrix}$$

$$c_2^\dagger = \sigma_3 \otimes \left(\frac{1}{2}\sigma_+\right) = \begin{pmatrix} 0 & 1 & 0 & 0 \\ 0 & 0 & 0 & 0 \\ 0 & 0 & 0 & -1 \\ 0 & 0 & 0 & 0 \end{pmatrix}, \quad c_2 = \sigma_3 \otimes \left(\frac{1}{2}\sigma_-\right) = \begin{pmatrix} 0 & 0 & 0 & 0 \\ 1 & 0 & 0 & 0 \\ 0 & 0 & 0 & 0 \\ 0 & 0 & -1 & 0 \end{pmatrix}$$

where

$$\frac{1}{2}\sigma_+ := \begin{pmatrix} 0 & 1 \\ 0 & 0 \end{pmatrix}, \quad \frac{1}{2}\sigma_- := \begin{pmatrix} 0 & 0 \\ 1 & 0 \end{pmatrix}.$$

Let $\gamma_1, \gamma_2 \in \mathbb{C}$. Calculate the 4×4 matrices

$$\exp(\gamma_1 c_1^\dagger - \gamma_1^* c_1), \quad \exp(\gamma_2 c_2^\dagger - \gamma_2^* c_2), \quad \exp(\gamma_1 c_1^\dagger - \gamma_1^* c_1 + \gamma_2 c_2^\dagger - \gamma_2^* c_2).$$

Apply the spectral theorem. Note that $c_1^\dagger$, c_1, $c_2^\dagger$, c_2 are nonnormal matrices, but $\gamma_1 c_1^\dagger + \gamma_1^* c_1$ and $\gamma_2 c_2^\dagger - \gamma_2^* c_2$ are normal matrices.

Problem 3. (i) Let $c_j^\dagger$ ($j = 1, 2, 3$) be Fermi creation operators. Consider the hermitian operators

$$T_{12} = c_1^\dagger c_2 + c_2^\dagger c_1, \quad T_{23} = c_2^\dagger c_3 + c_3^\dagger c_2.$$

Find the commutators $[T_{12}, T_{23}]$ and $[[T_{12}, T_{23}], T_{12} + T_{23}]$.
(ii) Consider the operators

$$X_{01} := c_0^\dagger + c_1 + c_1^\dagger + c_0, \quad X_{12} := c_1^\dagger + c_2 + c_2^\dagger + c_1,$$

$$X_{23} := c_2^\dagger + c_3 + c_3^\dagger + c_2, \quad X_{30} := c_3^\dagger + c_0 + c_0^\dagger + c_3.$$

Find the commutators $[X_{01}, X_{12}]$, $[X_{12}, X_{23}]$, $[X_{23}, X_{30}]$, $[X_{30}, X_{01}]$. Do the operators commute with the number operator

$$\hat{N} = c_0^{\dagger}c_0 + c_1^{\dagger}c_1 + c_2^{\dagger}c_2 + c_3^{\dagger}c_3 ?$$

Problem 4.   Let $\hat{N}_{j,\uparrow} = c_{j,\uparrow}^{\dagger}c_{j,\uparrow}$, $\hat{N}_{j,\downarrow} = c_{j,\downarrow}^{\dagger}c_{j,\downarrow}$. Show that

$$\Pi = \prod_{j=1}^{N}(I - \hat{N}_{j,\uparrow}\hat{N}_{j,\downarrow})$$

is a projection operator.

Problem 5.   Consider the Fermi creation and annihilation operators $c_1^{\dagger}$, $c_2^{\dagger}$, c_1, c_2 and the basis

$$\{c_2^{\dagger}c_1^{\dagger}|0\rangle, c_2^{\dagger}|0\rangle, c_1^{\dagger}|0\rangle, |0\rangle\}.$$

(i) Find the matrix representation of the operators $c_1 + c_2 + c_1c_2$, $c_1^{\dagger}c_2^{\dagger} + c_2^{\dagger}c_1^{\dagger}$.
(ii) Find the matrix representation of the hermitian operator $c_1 + c_1^{\dagger} + c_2 + c_2^{\dagger} + c_1c_2 + c_2^{\dagger}c_1^{\dagger}$.
(iii) Find the eigenvalues of the Hamilton operator

$$\hat{H} = \hbar\omega_1 c_1^{\dagger}c_1 + \hbar\omega_2 c_2^{\dagger}c_2 + \hbar\kappa(c_1^{\dagger}c_2 + c_2^{\dagger}c_1)$$

utilizing the basis given above.
(iv) Let $\hat{N}_1 = c_1^{\dagger}c_1$ and $\hat{N}_2 = c_2^{\dagger}c_2$. Study the Hamilton operator

$$\hat{H} = \epsilon(\hat{N}_1 - \hat{N}_2) + t(c_1^{\dagger}c_2 + c_2^{\dagger}c_1) + \frac{1}{2}\kappa(\hat{N}_1 - \hat{N}_2)^2.$$

First show that the total number operator $\hat{N} = \hat{N}_1 + \hat{N}_2$ is a conserved quantity.

Problem 6.   Let $c_{j,\sigma}^{\dagger}$ ($j = 1, 2, 3$) be a Fermi creation operators of an electron with spin σ in the qubit j acting on the vacuum $|0\rangle$. Consider the states

$$|S_{-3/2}\rangle = c_{3,\downarrow}^{\dagger}c_{2,\downarrow}^{\dagger}c_{1,\downarrow}^{\dagger}|0\rangle$$
$$|S_{-1/2}\rangle = (c_{3,\uparrow}^{\dagger}c_{2,\downarrow}^{\dagger}c_{1,\downarrow}^{\dagger} + c_{3,\downarrow}^{\dagger}c_{2,\uparrow}^{\dagger}c_{1,\downarrow}^{\dagger} + c_{3,\downarrow}^{\dagger}c_{2,\downarrow}^{\dagger}c_{1,\uparrow}^{\dagger})|0\rangle$$
$$|S_{1/2}\rangle = (c_{3,\downarrow}^{\dagger}c_{2,\uparrow}^{\dagger}c_{1,\uparrow}^{\dagger} + c_{3,\uparrow}^{\dagger}c_{2,\downarrow}^{\dagger}c_{1,\uparrow}^{\dagger} + c_{3,\uparrow}^{\dagger}c_{2,\uparrow}^{\dagger}c_{1,\downarrow}^{\dagger})|0\rangle$$
$$|S_{3/2}\rangle = c_{3,\uparrow}^{\dagger}c_{2,\uparrow}^{\dagger}c_{1,\uparrow}^{\dagger}|0\rangle.$$

Find the scalar products $\langle S_{-3/2}|S_{-1/2}\rangle$, $\langle S_{-3/2}|S_{1/2}\rangle$, $\langle S_{-3/2}|S_{3/2}\rangle$.

Problem 7.   Consider the operator

$$K = c_1^{\dagger}c_2 + c_2^{\dagger}c_1 + c_2^{\dagger}c_3 + c_3^{\dagger}c_2 + c_3^{\dagger}c_1 + c_1^{\dagger}c_3.$$

Find $\exp(zK)c_j\exp(-zK)$ for $j = 1, 2, 3$. Note that

$$[K, c_1] = -c_2 - c_3, \quad [K, c_2] = -c_1 - c_3, \quad [K, c_3] = -c_1 - c_2.$$

Problem 8. Calculate

$$R(\epsilon_1, \epsilon_2, \epsilon_3) = \exp(\epsilon_1 c_1^\dagger c_2^\dagger + \epsilon_2 c_1 c_2 + \epsilon_3 (c_1^\dagger c_1 + c_2^\dagger c_2))$$

i.e. we want to disentangle the operator R.

Problem 9. Given the Hamilton operator

$$\hat{H} = t \sum_\sigma (c_{1\sigma}^\dagger c_{2\sigma} + c_{2\sigma}^\dagger c_{1\sigma} + c_{2\sigma}^\dagger c_{3\sigma} + c_{3\sigma}^\dagger c_{2\sigma} + c_{3\sigma}^\dagger c_{4\sigma} + c_{4\sigma}^\dagger c_{3\sigma} + c_{4\sigma}^\dagger c_{1\sigma} + c_{1\sigma}^\dagger c_{4\sigma}) + U \sum_{j=1}^{4} n_{j\uparrow} n_{j\downarrow}$$

and the basis

$$\{ |1\rangle = c_{1\uparrow}^\dagger c_{1\downarrow}^\dagger |0\rangle, \quad |2\rangle = c_{2\uparrow}^\dagger c_{2\downarrow}^\dagger |0\rangle, \quad |3\rangle = c_{3\uparrow}^\dagger c_{3\downarrow}^\dagger |0\rangle, \quad |4\rangle = c_{4\uparrow}^\dagger c_{4\downarrow}^\dagger |0\rangle$$

$$|5\rangle = c_{1\uparrow}^\dagger c_{2\downarrow}^\dagger |0\rangle, \quad |6\rangle = c_{1\uparrow}^\dagger c_{3\downarrow}^\dagger |0\rangle, \quad |7\rangle = c_{1\uparrow}^\dagger c_{4\downarrow}^\dagger |0\rangle, \quad |8\rangle = c_{1\downarrow}^\dagger c_{2\uparrow}^\dagger |0\rangle,$$

$$|9\rangle = c_{1\downarrow}^\dagger c_{3\uparrow}^\dagger |0\rangle, \quad |10\rangle = c_{1\downarrow}^\dagger c_{4\uparrow}^\dagger |0\rangle, \quad |11\rangle = c_{2\uparrow}^\dagger c_{3\downarrow}^\dagger |0\rangle, \quad |12\rangle = c_{2\uparrow}^\dagger c_{4\downarrow}^\dagger |0\rangle,$$

$$|13\rangle = c_{2\downarrow}^\dagger c_{3\uparrow}^\dagger |0\rangle, \quad |14\rangle = c_{2\downarrow}^\dagger c_{4\uparrow}^\dagger |0\rangle, \quad |15\rangle = c_{3\uparrow}^\dagger c_{4\downarrow}^\dagger |0\rangle, \quad |16\rangle = c_{3\downarrow}^\dagger c_{4\uparrow}^\dagger |0\rangle \}.$$

(i) Calculate the matrix representation of $\hat{H}$ using this basis.
(ii) Find the eigenvalues of this 16×16 matrix.
(iii) Apply the Hamilton operator to the state

$$|\psi\rangle = \prod_{j=1}^{4} \frac{1}{\sqrt{2}} (c_{j\uparrow}^\dagger + c_{j\downarrow}^\dagger) |0\rangle.$$

Find the expectation value $\langle \psi | \hat{H} | \psi \rangle$.

Problem 10. Let $c_j^\dagger$, c_j ($j = 1, 2, 3$) be Fermi creation and annihilation operators. Let $\hat{N}$ be the number operator

$$\hat{N} = c_1^\dagger c_1 + c_2^\dagger c_2 + c_3^\dagger c_3.$$

(i) Consider the Hamilton operator

$$\hat{H}_1 = t(c_1^\dagger c_2 + c_2^\dagger c_1 + c_2^\dagger c_3 + c_3^\dagger c_2 + c_1^\dagger c_3 + c_3^\dagger c_1) + k_1 c_1^\dagger c_1 + k_2 c_2^\dagger c_2 + k_3 c_3^\dagger c_3.$$

Show that $[\hat{H}, \hat{N}] = 0$. Given a basis with two Fermi particles

$$c_1^\dagger c_2^\dagger |0\rangle, \quad c_1^\dagger c_3^\dagger |0\rangle, \quad c_2^\dagger c_3^\dagger |0\rangle.$$

Find the matrix representation of $\hat{H}$ and $\hat{N}$. Given a basis with one Fermi particle

$$c_1^\dagger |0\rangle, \quad c_2^\dagger |0\rangle, \quad c_3^\dagger |0\rangle.$$

Find the matrix representation of $\hat{H}_1$ and $\hat{N}$.
(ii) Consider the Hamilton operator

$$\hat{H}_2 = t(c_1^\dagger c_2 + c_2^\dagger c_1 + c_2^\dagger c_1 + c_3^\dagger c_2) + k_1 c_1^\dagger c_1 + k_2 c_2^\dagger c_2 + k_3 c_3^\dagger c_3.$$

Compare the spectrum of $\hat{H}_1$ and $\hat{H}_2$.

(iii) Compare the spectrum of the two Hamilton operators

$$K_1 = t(c_1^\dagger c_2 + c_2^\dagger c_1 + c_2^\dagger c_3 + c_3^\dagger c_2 + c_3^\dagger c_1 + c_1^\dagger c_3)$$

and

$$K_2 = t(c_1^\dagger c_2 + c_2^\dagger c_1 + c_2^\dagger c_3 + c_3^\dagger c_2 + c_3^\dagger c_1^\dagger + c_1 c_3)$$

with the basis

$$c_3^\dagger c_2^\dagger c_1^\dagger |0\rangle, \ c_3^\dagger c_2^\dagger |0\rangle, \ c_3^\dagger c_1^\dagger |0\rangle, \ c_2^\dagger c_1^\dagger |0\rangle, \ c_3^\dagger |0\rangle, \ c_2^\dagger |0\rangle, \ c_1^\dagger |0\rangle, \ |0\rangle.$$

(iv) Study the Hamilton operator

$$\hat{H}_3 = \hbar\omega_1 c_1^\dagger c_1 + \hbar\omega_2 c_2^\dagger c_2 + \hbar\omega_3 c_3^\dagger c_3 + \hbar\kappa(c_1^\dagger c_2^\dagger c_3^\dagger + c_3 c_2 c_1).$$

Problem 11.   (i) Study the spectrum of the hermitian operator

$$K = t_1(c_1^\dagger + c_1 + c_2^\dagger + c_2 + c_1^\dagger c_2^\dagger + c_2 c_1) + t_2(c_1^\dagger c_2 + c_2^\dagger c_1)$$

with the basis $\{c_1^\dagger c_2^\dagger |0\rangle, \ c_2^\dagger |0\rangle, \ c_1^\dagger |0\rangle, \ |0\rangle \}$.
(ii) Let $\hat{N}_j := c_j^\dagger c_j$ for $j = 1, 2$ be the number operators. Find the matrix representation of the Hamilton operator

$$\hat{H} = \hbar\omega_1(c_1^\dagger c_1 + c_2^\dagger c_2) + \hbar\omega_2(c_1^\dagger c_2 + c_2^\dagger c_1) + \hbar\omega_3 \hat{N}_1 \hat{N}_2$$

with the basis $\{c_2^\dagger c_1^\dagger |0\rangle, \ c_2^\dagger |0\rangle, \ c_1^\dagger |0\rangle, \ |0\rangle\}$ and then calculate the eigenvalues.

Problem 12.   Let $c_j, c_j^\dagger$ be Fermi annihilation and creation operators ($j = 1, 2, 3$).
(i) Study the eigenvalue problem for the Hamilton operator

$$\hat{H} = \hbar\omega(c_1^\dagger c_1 + c_2^\dagger c_2 + c_3^\dagger c_3) + t_1(c_1^\dagger c_2 + c_2^\dagger c_1) + t_2(c_2^\dagger c_3 + c_3^\dagger c_2) + t_1(c_3^\dagger c_1 + c_1^\dagger c_3)$$

with the basis $\{ c_2^\dagger c_1^\dagger |0\rangle, \ c_3^\dagger c_1^\dagger |0\rangle, \ c_3^\dagger c_2^\dagger |0\rangle \}$ and $t_1, t_2 > 0$ with $t_1 > t_2$.
(ii) Study the operator

$$\hat{K} = c_1^\dagger(c_2 - c_3) + c_2^\dagger(c_3 - c_1) + c_3^\dagger(c_1 - c_2).$$

Does the operator $\hat{K}$ commute with the number operator

$$\hat{N} = c_1^\dagger c_1 + c_2^\dagger c_2 + c_3^\dagger c_3 \ ?$$

Find the eigenvalues and eigenvectors of $\hat{K}$.

Problem 13.   Consider the transfer operators

$$X_{12} := c_1^\dagger c_2 + c_2^\dagger c_1, \quad X_{23} := c_2^\dagger c_3 + c_3^\dagger c_2, \quad X_{34} := c_3^\dagger c_4 + c_4^\dagger c_3, \quad X_{41} := c_4^\dagger c_1 + c_1^\dagger c_4$$

and the Hamilton operator

$$\hat{K} = \frac{\hat{H}}{\hbar\omega} = X_{12} + X_{23} + X_{34}.$$

(i) Let

$$\hat{C} := [X_{12}, X_{23}] + [X_{23}, X_{34}].$$

Find the commutator $[\hat{K}, \hat{C}]$ and anticommutator $[\hat{K}, \hat{C}]_+$.

(ii) Let $\hat{D} := X_{12} + X_{23} + X_{34} + X_{41}$. Find the commutator $[\hat{K}, \hat{D}]$ and anticommutator $[\hat{K}, \hat{D}]_+$.

Problem 14. (i) Consider the Hamilton operator on a one-dimensional lattice

$$\hat{H} = t \sum_{j=0}^{N-1} (c_j^\dagger c_{j+1} + c_{j+1}^\dagger c_j) + h \sum_{j=0}^{N-1} c_j^\dagger c_j$$

with cyclic boundary conditions, i.e. $N \equiv 0$. Solve the eigenvalue problem for two particle states, i.e. the states are given by

$$c_j^\dagger c_k^\dagger |0\rangle, \qquad j, k = 0, 1, \dots, N.$$

(ii) Let $N \geq 2$. Study the Hamilton operator (open ends)

$$\hat{H} = \sum_{j=2}^{N} (c_j^\dagger c_{j-1} + c_{j-1}^\dagger c_j + \frac{1}{2}(2\hat{N}_j - I)(2\hat{N}_{j-1} - I))$$

where $\hat{N}_j := c_j^\dagger c_j$.

Problem 15. Let $c_j^\dagger$, c_j $(j = 0, 1, \dots, N)$ be Fermi creation and annihilation operators and

$$\hat{N} = \sum_{j=0}^{N} c_j^\dagger c_j$$

be the number operator.

(i) Study the one-dimensional Hamilton operator (open ends)

$$\hat{H} = t_1 \sum_{j=0}^{N-1} (c_j^\dagger c_{j+1} + c_{j+1}^\dagger c_j) + t_2 \sum_{j=0}^{N-1} (c_j^\dagger c_{j+1}^\dagger + c_{j+1} c_j).$$

Does the Hamilton operator $\hat{H}$ commute with the number operator?

(ii) Study the Hamilton operator

$$\hat{H} = \hbar\omega \sum_{j=0}^{N} c_j^\dagger c_j + \gamma \sum_{j=0}^{N} (c_j^\dagger + c_j).$$

Problem 16. Consider the hermitian Hamilton operator

$$\hat{H} = t(c_N^\dagger c_{N-1}^\dagger \cdots c_2^\dagger c_1^\dagger + c_1 c_2 \cdots c_{N-1} c_N)$$

and the number operator

$$\hat{N} = \sum_{j=1}^{N} c_j^\dagger c_j.$$

Find the commutator $[\hat{H}, \hat{N}]$. Find the matrix representation of $\hat{H}$ and the eigenvalues and eigenvectors. Utilize

$$\overbrace{\qquad\qquad\qquad\qquad\qquad}^{N\text{-times}}$$

$$c_k^\dagger = \sigma_3 \otimes \sigma_3 \otimes \cdots \otimes \sigma_3 \otimes \left(\frac{1}{2}\sigma_+\right) \otimes I_2 \otimes I_2 \otimes \cdots \otimes I_2$$

$$c_k = \sigma_3 \otimes \sigma_3 \otimes \cdots \otimes \sigma_3 \otimes \left(\frac{1}{2}\sigma_-\right) \otimes I_2 \otimes I_2 \otimes \cdots \otimes I_2.$$

$$k\text{-th place}$$

Consider first the cases $n = 1$ and $n = 2$.

Problem 17.   Let $c_{j\sigma}^\dagger$, $c_{j\sigma}$ be Fermi creation and annihilation operators with spin σ ($\sigma \in \{\uparrow,\downarrow\}$) Consider the Hamilton operator

$$\hat{H} = t \sum_{j=1}^2 (c_{j\uparrow}^\dagger c_{j\uparrow} + c_{j+1\uparrow}^\dagger c_{j\uparrow} + c_{j\uparrow}^\dagger c_{j\uparrow} + c_{j+1\uparrow}^\dagger c_{j\uparrow})$$

with periodic boundary conditions, i.e. $c_{3\sigma}^\dagger = c_{1,\sigma}$. Find the time-evolution of $c_{1,\uparrow}^\dagger$, $c_{2,\uparrow}$.

Problem 18.   Let $c_{j\uparrow}^\dagger$, $c_{j\downarrow}^\dagger$, $c_{j\uparrow}$, $c_{j\downarrow}$ be Fermi creation and annihilation operators with spin up and spin down. Consider the Hamilton operator

$$\hat{K} = \lambda_1 \sum_{j=1}^N \hat{N}_{j\uparrow}\hat{N}_{j\uparrow} + \lambda_2 \sum_{j=1}^N \hat{N}_{j\uparrow} + \lambda_3 \sum_{j=1}^N \hat{N}_{j\downarrow}$$

where $\hat{N}_{j\sigma} := c_{j\sigma}^\dagger c_{j\sigma}$. Consider the density matrix

$$W = \frac{e^{-\beta\hat{K}}}{\mathrm{tr}\, e^{-\beta\hat{K}}}.$$

Calculate the trace

$$\mathrm{tr}(a_1 a_2 \ldots a_n W)$$

where

$$a_i \in \{\, c_{j\uparrow}^+, c_{j\uparrow}, c_{j\downarrow}^+, c_{j\downarrow} \; : \; j = 1,2,\ldots,N\}.$$

Note that $\hat{K}$ commutes with the operators

$$\sum_{i=1}^N \hat{N}_{i\uparrow} \quad \text{and} \quad \sum_{i=1}^N \hat{N}_{i\downarrow}$$

the trace vanishes unless $(a_1 a_2 \ldots a_n)$ contains an equal number of creation and annihilation operators with spin up and also an equal number of creation and annihilation operators with spin down. Terms like

$$\mathrm{tr}(c_{j\uparrow}^+ c_{j\downarrow} W)$$

therefore vanish. In the expression

$$X := \mathrm{tr}(a_1 a_2 \ldots a_n W)$$

one now anticommutes a_1 successively to the right, each time extracting the anticommutator (either 0 or 1) from the trace. In particular show the prove the following three special cases

$$\text{tr}\left(\hat{N}_{i\uparrow}W\right) = \frac{e^{\beta(\lambda_1+\lambda_3)}+1}{Z_0}\text{tr}W$$

$$\text{tr}\left(\hat{N}_{i\downarrow}W\right) = \frac{e^{\beta(\lambda_1+\lambda_2)}+1}{Z_0}\text{tr}W$$

$$\text{tr}\left(\hat{N}_{i\uparrow}\hat{N}_{i\downarrow}W\right) = \frac{1}{Z_0}\text{tr}(W).$$

Problem 19. Consider the Hamilton operator

$$\hat{H} = t(c_{1\uparrow}^\dagger c_{2\uparrow} + c_{1\downarrow}^\dagger c_{2\downarrow} + c_{2\uparrow}^\dagger c_{1\uparrow} + c_{2\downarrow}^\dagger c_{1\downarrow}) + U_1 n_{1\uparrow} n_{1\downarrow} + U_2 n_{2\uparrow} n_{2\downarrow}.$$

Consider the six dimensional basis

$$c_{1\uparrow}^\dagger c_{1\downarrow}^\dagger|0\rangle, \quad c_{2\uparrow}^\dagger c_{2\downarrow}^\dagger|0\rangle, \quad c_{1\uparrow}^\dagger c_{2\uparrow}^\dagger|0\rangle, \quad c_{1\downarrow}^\dagger c_{2\downarrow}^\dagger|0\rangle, \quad c_{1\uparrow}^\dagger c_{2\downarrow}^\dagger|0\rangle, \quad c_{1\downarrow}^\dagger c_{2\uparrow}^\dagger|0\rangle.$$

Find the matrix representation of $\hat{H}$ and the eigenvalues of the matrix.

Problem 20. Study the Fermi Hamilton operator with three lattice sites and open ends

$$\hat{H} = \sum_{j=0}^{2}\sum_{\sigma\in\{\uparrow,\downarrow\}}(c_{j,\sigma}^\dagger c_{j+1,\sigma} + c_{j+1,\sigma}^\dagger c_{j,\sigma}) + V\sum_{j=0}^{2}(c_{j,\uparrow}^\dagger c_{j,\downarrow}^\dagger c_{j+1,\downarrow} c_{j+1,\uparrow} + c_{j+1,\uparrow}^\dagger c_{j+1,\downarrow}^\dagger c_{j,\downarrow} c_{j,\uparrow}).$$

Is the Hamilton operator hermitian? Does the Hamilton operator $\hat{H}$ commute with the number operator

$$\hat{N} = \sum_{j=0}^{2}\sum_{\sigma\in\{\uparrow,\downarrow\}}c_{j,\sigma}^\dagger c_{j,\sigma}\ ?$$

Problem 21. Consider the Fermi creation and annihilation operators

$$c_{j,\sigma}^\dagger \quad c_{j,\sigma}, \quad j=1,2 \quad \sigma\in\{\uparrow,\downarrow\}$$

and the basis with the 6 elements

$$c_{1,\uparrow}^\dagger c_{1,\downarrow}^\dagger|0\rangle, \quad c_{2,\uparrow}^\dagger c_{2,\downarrow}^\dagger|0\rangle, \quad c_{1,\uparrow}^\dagger c_{2,\uparrow}^\dagger|0\rangle, \quad c_{1,\downarrow}^\dagger c_{2,\downarrow}^\dagger|0\rangle, \quad c_{1,\uparrow}^\dagger c_{2,\downarrow}^\dagger|0\rangle, \quad c_{1,\downarrow}^\dagger c_{2,\uparrow}^\dagger|0\rangle.$$

(i) Study the Hamilton operator with spin flip terms

$$\hat{H} = \hbar\omega(c_{1,\uparrow}^\dagger c_{2,\downarrow} + c_{1,\downarrow}^\dagger c_{2,\uparrow} + c_{2,\uparrow}^\dagger c_{1,\uparrow} + c_{2,\uparrow}^\dagger c_{1,\downarrow})(\hat{N}_{1,\uparrow} + \hat{N}_{1,\downarrow} + \hat{N}_{2,\uparrow} + \hat{N}_{2,\downarrow} - 2\hat{N}_{1,\uparrow}\hat{N}_{2,\downarrow}).$$

(ii) Study the Hamilton operator with spin flip terms

$$\hat{H}_1 = t_1(c_{1,\uparrow}^\dagger c_{2,\uparrow} + c_{2,\uparrow}^\dagger c_{1,\uparrow} + c_{1,\downarrow} c_{2,\downarrow} + c_{2,\downarrow} c_{1,\downarrow}) + t_2(c_{1,\uparrow}^\dagger c_{2,\downarrow} + c_{2,\downarrow}^\dagger c_{1,\uparrow} c_{1,\downarrow}^\dagger c_{2,\uparrow} + c_{2,\uparrow}^\dagger c_{1,\downarrow})$$

and

$$\hat{H}_2 = t_1(c_{1,\uparrow}^\dagger c_{2,\uparrow} + c_{2,\uparrow}^\dagger c_{1,\uparrow}) + c_{1,\downarrow}^\dagger c_{2,\downarrow} + c_{2,\downarrow}^\dagger c_{1,\downarrow} + t_2(c_{1,\uparrow}^\dagger c_{1,\downarrow}^\dagger + c_{2,\uparrow}^\dagger c_{2,\downarrow}).$$

(iii) Study the Hamilton operator

$$\hat{H} = t(c^\dagger_{1,\uparrow}c_{2,\uparrow} + c^\dagger_{2,\uparrow}c_{1,\uparrow} + c^\dagger_{1,\downarrow}c_{2,\downarrow} + c^\dagger_{2,\downarrow}c_{1,\downarrow}).$$

Problem 22. Let $\epsilon \in [0,1]$. Consider the one-dimensional Hamilton operator (open end boundary condition and L even)

$$\hat{H} = -\sum_{j=1}^{L-1} \left(\frac{1}{2}(1+\epsilon)\sigma_{1,j}\sigma_{1,j+1} + \frac{1}{2}(1-\epsilon)\sigma_{2,j}\sigma_{2,j+1} \right).$$

Show that under the *Jordan-Wigner transformation* one finds the Hamilton operator

$$\hat{H} = -\sum_{j=1}^{L-1} \left(\left(c^\dagger_j c_{j+1} + c^\dagger_{j+1}c_j \right) + \epsilon \left(c^\dagger_j c^\dagger_{j+1} + c_{j+1}c_j \right) \right)$$

where $c^\dagger_j$, c_j are Fermi creation and annihilation operators, respectively at lattice site j.

Problem 23. Let $\mathbb{Z}$ be the set of integers. Consider the Hamilton operator

$$\hat{H} = \sum_{j\in\mathbb{Z}} \epsilon_j c^\dagger_j c_j + \sum_{j\in\mathbb{Z}} V_{j+1,j}(c^\dagger_{j+1}c_j + c^\dagger_j c_{j+1})$$

where $c^\dagger_j$, c_j are Fermi creation and annihilation operators. and $V_{j,j+1}$ (real and positive) are the tunneling matrix connecting lattice site j to the lattice site $j + 1$. Show that the corresponding eigenvalue equation is

$$\epsilon_j C_j + V_{j+1,j}C_{j+1} + V_{j,j-1}C_{j-1} = EC_j, \quad j \in \mathbb{Z}.$$

So we have a linear bounded operator in the Hilbert space $\ell_2(\mathbb{Z})$.

Problem 24. Let $\mathbb{Z}$ be the set of integers. Consider the Hamilton operator

$$\hat{H} = \sum_{n\in\mathbb{Z}} (\epsilon_n c^\dagger_n c_n + V_n(c^\dagger_{n+1}c_n + c^\dagger_n c_{n+1}))$$

and the wave function

$$|\psi\rangle = \sum_{j\in\mathbb{Z}} f_j c^\dagger_j |0\rangle.$$

Find $\hat{H}|\psi\rangle$.

Problem 25. Let $j \in \mathbb{Z}$. Let $c^\dagger_j$, c_j be Fermi creation and annihilation operators at a lattice site j of a one-dimensional infinite chain. Let $W, t > 0$. The one-dimensional *Aubry Hamilton operator* is given by

$$\hat{H} = \sum_{j=-\infty}^{\infty} (W\cos(kn)c^\dagger_j c_j + t(c^\dagger_{j+1}c_j + c^\dagger_j c_{j+1})).$$

The operator commutes with the number operator $\hat{N}$.

(i) Consider the self-adjoint operators

$$X_{j,j+1} := c_{j+1}^\dagger c_j + c_j^\dagger c_{j+1}, \quad X_{j-1,j} := c_{j-1}^\dagger c_j + c_j^\dagger c_{j-1}.$$

Find the commutator $[X_{j,j+1}, X_{j-1,j}]$.

(ii) Consider the basis $\{c_j^\dagger|0\rangle\}$ ($j \in \mathbb{Z}$). Show that the matrix representation of $\hat{H}$ is

$$H_{jj} = W\cos(kn), \quad H_{j,j+1} = H_{j+1,j} = t, \quad H_{jk} = 0\,, \text{otherwise}$$

where $j = -\infty, \ldots, -1, 0, 1, \ldots, +\infty$.

Problem 26. Consider the hermitian Hamilton operator

$$\hat{H} = \hbar\omega(c_0^\dagger c_1 + c_1^\dagger c_0 + c_0^\dagger c_2 + c_2^\dagger c_0 + c_0^\dagger c_3 + c_3^\dagger c_0 + c_1^\dagger c_2 + c_2^\dagger c_1 + c_1^\dagger c_3 + c_3^\dagger c_1 + c_2^\dagger c_3 + c_3^\dagger c_2)$$

for the *tetrahedron*

The Hamilton operator $\hat{H}$ commutes with the number operator

$$\hat{N} = \sum_{j=0}^{3} c_j^\dagger c_j.$$

Consider the six-dimensional basis for two Fermi particles

$$c_1^\dagger c_0^\dagger|0\rangle, \quad c_2^\dagger c_0^\dagger|0\rangle, \quad c_3^\dagger c_0^\dagger|0\rangle, \quad c_2^\dagger c_1^\dagger|0\rangle, \quad c_3^\dagger c_1^\dagger|0\rangle, \quad c_3^\dagger c_2^\dagger|0\rangle.$$

Find the matrix representation for the Hamilton operator $\hat{H}$. Find the eigenvalues and eigenvectors of the 6×6 matrix. Discuss the symmetry of the tetrahedron and these solutions.

Problem 27. Study the Hamilton operator

$$\begin{aligned}
\hat{H} = \hbar\omega(&c_{-1,1}^\dagger c_{0,1} + c_{0,1}^\dagger c_{-1,1} + c_{0,1}^\dagger c_{1,1} + c_{1,1}^\dagger c_{0,1} \\
+&c_{-1,0}^\dagger c_{0,0} + c_{0,0}^\dagger c_{-1,0} + c_{0,0}^\dagger c_{1,0} + c_{1,0}^\dagger c_{0,0} \\
+&c_{-1,-1}^\dagger c_{-1,0} + c_{-1,0}^\dagger c_{-1,-1} + c_{-1,0}^\dagger c_{-1,1} + c_{-1,1}^\dagger c_{-1,0} \\
+&c_{-1,1}^\dagger c_{-1,0} + c_{-1,0}^\dagger c_{-1,1} + c_{-1,0}^\dagger c_{-1,-1} + c_{-1,-1}^\dagger c_{-1,0} \\
+&c_{0,1}^\dagger c_{0,0} + c_{0,0}^\dagger c_{0,1} + c_{0,0}^\dagger c_{-1,0} + c_{-1,0}^\dagger c_{0,0} \\
+&c_{1,1}^\dagger c_{1,0} + c_{1,0}^\dagger c_{1,1} + c_{1,0}^\dagger c_{-1,1} + c_{-1,1}^\dagger c_{1,0})
\end{aligned}$$

for the lattice

$$
\begin{array}{ccc}
(-1,1) \rule[0.5ex]{1em}{0.4pt} (0,1) \rule[0.5ex]{1em}{0.4pt} (1,1) \\
| \quad\quad | \quad\quad | \\
(-1,0) \rule[0.5ex]{1em}{0.4pt} (0,0) \rule[0.5ex]{1em}{0.4pt} (1,0) \\
| \quad\quad | \quad\quad | \\
(-1,-1) \rule[0.5ex]{1em}{0.4pt} (0,-1) \rule[0.5ex]{1em}{0.4pt} (1,-1)
\end{array}
$$

Problem 28.   Let $n \geq 2$. Consider the *tridiagonal matrix*

$$A_n = \begin{pmatrix} a_1 & b_1 & 0 & & & \\ c_1 & a_2 & b_2 & & & \\ 0 & c_2 & \ddots & & \ddots & \\ & & & \ddots & \ddots & b_{n-1} \\ & & & & c_{n-1} & a_n \end{pmatrix}.$$

We set $D_0 = 1$ and $D_1 = a_1$. For $n \geq 2$ we set $D_n = \det(A_n)$, i.e. the determinant of A_n. Then the determinant D_n $(n \geq 2)$ satisfies the recurrence relation

$$D_n = a_n D_{n-1} - c_{n-1} b_{n-1} D_{n-2}, \quad n = 2, 3, \dots$$

If we set $a_1 = a_2 = \cdots = a_n = -\lambda$ (λ will be the eigenvalue) then we obtain the characteristic polynomial for the matrix

$$M_n = \begin{pmatrix} 0 & b_1 & 0 & & & \\ c_1 & 0 & b_2 & & & \\ 0 & c_2 & \ddots & & \ddots & \\ & & & \ddots & \ddots & b_{n-1} \\ & & & & c_{n-1} & 0 \end{pmatrix}.$$

The matrix $(b_j = c_j = 1)$

$$K = \begin{pmatrix} 0 & 1 & 0 & & & \\ 1 & 0 & 1 & & & \\ 0 & 1 & \ddots & & \ddots & \\ & & & \ddots & \ddots & 1 \\ & & & & 1 & 0 \end{pmatrix}$$

appears for the matrix representation of Fermi systems of the form

$$\sum_{j=1}^{N-1} (c_j^\dagger c_{j+1} + c_{j+1}^\dagger c_j).$$

Solve the characteristic equation and thus find the eigenvalues of K.

Problem 29.   Let $n \geq 3$. Study the eigenvalue problem for the Hamilton operator of the one-dimensional chain (open ends)

$$\hat{H} = t \sum_{j=1}^{n-1} c_j^\dagger c_{j+1} + U c_1^\dagger c_1 c_2^\dagger c_2 \cdots c_n^\dagger c_n.$$

First show that the Hamilton operator $\hat{H}$ commutes with the number operator

$$\hat{N} = \sum_{j=1}^{n} c_j^\dagger c_j$$

and therefore we can consider the subspaces with a fixed number of Fermi operators.

Problem 30. Given the Hubbard Hamilton operator

$$\hat{H} = t \sum_{\sigma \in \{\uparrow,\downarrow\}} (c_{1\sigma}^\dagger c_{2\sigma} + c_{2\sigma}^\dagger c_{1\sigma} + c_{2\sigma}^\dagger c_{3\sigma} + c_{3\sigma}^\dagger c_{2\sigma} + c_{3\sigma}^\dagger c_{4\sigma} + c_{4\sigma}^\dagger c_{3\sigma} + c_{4\sigma}^\dagger c_{1\sigma} + c_{1\sigma}^\dagger c_{4\sigma})$$

$$+ U \sum_{j=1}^{4} \hat{N}_{j\uparrow} \hat{N}_{j\downarrow}$$

where $\hat{N}_{j\sigma} := c_{j\sigma}^\dagger c_{j\sigma}$ with $\sigma \in \{\uparrow,\downarrow\}$.
(i) Show that $[\hat{H}, \hat{N}_e] = 0$ and $[\hat{H}, \hat{S}_3] = 0$.
(ii) Given the basis

$$\{ |1\rangle = c_{1\uparrow}^\dagger c_{2\uparrow}^\dagger |0\rangle, \quad |2\rangle = c_{1\uparrow}^\dagger c_{3\uparrow}^\dagger |0\rangle, \quad |3\rangle = c_{1\uparrow}^\dagger c_{4\uparrow}^\dagger |0\rangle,$$

$$|4\rangle = c_{2\uparrow}^\dagger c_{3\uparrow}^\dagger |0\rangle, \quad |5\rangle = c_{2\uparrow}^\dagger c_{4\uparrow}^\dagger |0\rangle, \quad |6\rangle = c_{3\uparrow}^\dagger c_{4\uparrow}^\dagger |0\rangle \}.$$

Calculate the matrix representation of $\hat{H}$ using this basis, i.e. $\langle j|\hat{H}|k\rangle$. Find the eigenvalues of this 6×6 matrix.

Problem 31. Let L be the number of lattice sites counting from $j = 0$ to $j = L - 1$. Consider the Hamilton operator (one-dimensional Hubbard model)

$$\hat{H} = t \sum_{j=0}^{L-1} \sum_{\sigma \in \{\uparrow,\downarrow\}} (c_{j,\sigma}^\dagger c_{j+1,\sigma} + c_{j+1,\sigma}^\dagger c_{j,\sigma}) + U \sum_{j=0}^{L-1} \hat{N}_{j,\uparrow} \hat{N}_{j,\downarrow}$$

with cyclic boundary conditions, i.e. $c_L = c_0$ and $c_L^\dagger = c_0^\dagger$.
(i) Consider the operators

$$\hat{R}_- := \sum_{j=0}^{L-1} (-1)^{j+1} c_{j\uparrow} c_{j\downarrow}, \quad \hat{R}_+ := \sum_{j=0}^{L-1} (-1)^{j+1} c_{j\downarrow}^\dagger c_{j\uparrow}^\dagger$$

and

$$\hat{R}_3 := \frac{1}{2} \sum_{j=0}^{L-1} (\hat{N}_{j\uparrow} + \hat{N}_{j\downarrow} - I).$$

Find the commutators $[R_+, \hat{H}]$, $[R_-, \hat{H}]$, $[R_3, \hat{H}]$. Discuss.
(ii) Consider the operators

$$\hat{S}_- := \sum_{j=0}^{L-1} c_{j\uparrow}^\dagger c_{j\downarrow}, \quad \hat{S}_+ := \sum_{j=0}^{L-1} c_{j\downarrow}^\dagger c_{j\uparrow}, \quad \hat{S}_3 = \frac{1}{2} \sum_{j=0}^{L-1} (n_{j\uparrow} + n_{j\downarrow} - I).$$

Find the commutators $[S_+, \hat{H}]$, $[S_-, \hat{H}]$, $[S_3, \hat{H}]$. Discuss.
(iii) The *particle hole operation* is given by

$$C \begin{pmatrix} c_{j\uparrow} \\ c_{j\downarrow} \end{pmatrix} C^{-1} = (-1)^j \begin{pmatrix} c_{j\uparrow}^\dagger \\ c_{j\downarrow}^\dagger \end{pmatrix}.$$

Apply it to the one-dimensional Hubbard model.

Chapter 4

Lie Algebras

4.1 Lie Algebras and Bose Operators

Problem 1. Let $b^\dagger$, b Bose creation and annihilation operators with $[b, b^\dagger] = I$, where I is the identity operator. The *harmonic oscillator Lie algebra ho(1)* is spanned by

$$\{\, I, \ b, \ b^\dagger, \ b^\dagger b \,\}.$$

The Lie subalgebra spanned by

$$\{\, I, \ b, \ b^\dagger \,\}$$

is known as the *Heisenberg-Weyl Lie algebra h(1)*.
(i) Find the commutators for the harmonic oscillator algebra.
(ii) Is the harmonic oscillator Lie algebra semisimple?
(iii) Is the Heisenberg-Weyl Lie algebra semisimple?

Solution 1. (i) We have $[b^\dagger b, b] = -b$, $[b^\dagger b, b^\dagger] = b^\dagger$ and

$$[b, b^\dagger] = I, \quad [b, I] = [b^\dagger, I] = [b^\dagger b, I] = 0.$$

(ii) Owing to the element I the harmonic oscillator Lie algebra is not semisimple.
(iii) Owing to the element I the Heisenberg-Weyl Lie algebra is not semisimple.

Problem 2. Let $b^\dagger$, b Bose creation and annihilation operators with $[b, b^\dagger] = I$, where I is the identity operator.
(i) Consider the operators

$$b^\dagger b, \quad b^\dagger, \quad b, \quad I.$$

Find the commutators.
(ii) Show that there is a non-hermitian faithful representation by 3×3 matrices

$$b^\dagger b \to M_{22} = \begin{pmatrix} 0 & 0 & 0 \\ 0 & 1 & 0 \\ 0 & 0 & 0 \end{pmatrix}, \quad I \to M_{13} = \begin{pmatrix} 0 & 0 & 1 \\ 0 & 0 & 0 \\ 0 & 0 & 0 \end{pmatrix},$$

$$b^\dagger \to M_{23} = \begin{pmatrix} 0 & 0 & 0 \\ 0 & 0 & 1 \\ 0 & 0 & 0 \end{pmatrix}, \quad b \to M_{12} = \begin{pmatrix} 0 & 1 & 0 \\ 0 & 0 & 0 \\ 0 & 0 & 0 \end{pmatrix}.$$

Note that the identity operator I is not mapped into the 3×3 identity matrix. Which of the four matrices are normal, i.e. $MM^* = M^*M$?

Solution 2.   (i) For the commutators we find

$$[b^\dagger b, b^\dagger] = b^\dagger, \quad [b^\dagger b, b] = -b, \quad [b^\dagger, b] = -I.$$

All the other commutators are 0. Thus we have a basis of a Lie algebra.
(ii) For the commutators of the matrices M_{22}, M_{23}, M_{12} and M_{13} we find

$$[M_{22}, M_{23}] = M_{23}, \quad [M_{22}, M_{12}] = -M_{12}, \quad [M_{22}, M_{13}] = 0$$

$$[M_{23}, M_{12}] = -M_{13}, \quad [M_{23}, M_{13}] = 0, \quad [M_{12}, M_{13}] = 0.$$

Thus we have a faithful representation of the Lie algebra given by the operators $b^\dagger b$, $b^\dagger$, b, I. Only the matrix M_{22} is normal.

Problem 3.   (i) Let $b^\dagger$, b be Bose creation and annihilation operators with $[b, b^\dagger] = I$, where I is the identity operator. Show that the operators

$$\hat{N} = b^\dagger b, \quad b^\dagger + b, \quad b^\dagger - b, \quad I$$

form a basis of a Lie algebra. Classify the Lie algebra.
(ii) Let $|n\rangle$ $(n = 0, 1, \dots)$ be number states and $|\beta\rangle$ be coherent states $(\beta \in \mathbb{C})$. Find

$$\langle n|b^\dagger b|n\rangle, \quad \langle \beta|b^\dagger b|\beta\rangle,$$

$$\langle n|(b^\dagger + b)|n\rangle, \quad \langle \beta|(b^\dagger + b)|\beta\rangle,$$

$$\langle n|(b^\dagger - b)|n\rangle, \quad \langle \beta|(b^\dagger - b)|\beta\rangle.$$

Solution 3.   (i) We obtain the commutators

$$[b^\dagger b, b^\dagger + b] = b^\dagger - b, \quad [b^\dagger b, b^\dagger - b] = b^\dagger + b, \quad [b^\dagger + b, b^\dagger - b] = 2I.$$

Obviously the identity operator I commutes with all other operators. So the Lie algebra is not semi-simple.
(ii) We have

$$\langle n|b^\dagger b|n\rangle = n, \quad \langle n|(b^\dagger + b)|n\rangle = 0, \quad \langle n|(b^\dagger - b)|n\rangle = 0$$

$$\langle \beta|b^\dagger b|\beta\rangle = \beta\beta^*, \quad \langle \beta|(b^\dagger + b)|\beta\rangle = 2\Re(\beta), \quad \langle \beta|(b^\dagger - b)|\beta\rangle = -2\Im(\beta).$$

Problem 4.   Let $b^\dagger$, b be Bose creation and annihilation operators, respectively.
(i) Calculate the commutators

$$[b^\dagger b, b^\dagger + b], \quad [b^\dagger b, b^\dagger - b], \quad [b^\dagger + b, b^\dagger - b].$$

(ii) Consider the matrices

$$N_2 = \begin{pmatrix} 0 & 0 \\ 0 & 1 \end{pmatrix}, \quad B_2 = \begin{pmatrix} 0 & 1 \\ 1 & 0 \end{pmatrix}, \quad C_2 = \begin{pmatrix} 0 & -1 \\ 1 & 0 \end{pmatrix}.$$

These matrices appear when we truncate the infinite dimensional unbounded matrices $b^\dagger b$, $b^\dagger + b$ and $b^\dagger - b$ to 2×2 matrices. Find the commutator $[N_2, B_2]$, $[N_2, C_2]$, $[B_2, C_2]$. Discuss.
(ii) Consider the 3×3 matrices

$$
N_3 = \begin{pmatrix} 0 & 0 & 0 \\ 0 & 1 & 0 \\ 0 & 0 & 2 \end{pmatrix}, \quad
B_3 = \begin{pmatrix} 0 & 1 & 0 \\ 1 & 0 & \sqrt{2} \\ 0 & \sqrt{2} & 0 \end{pmatrix}, \quad
C_3 = \begin{pmatrix} 0 & -1 & 0 \\ 1 & 0 & -\sqrt{2} \\ 0 & \sqrt{2} & 0 \end{pmatrix}.
$$

These matrices appear when we truncate the infinite dimensional unbounded matrices $b^\dagger b$, $b^\dagger + b$, $b^\dagger - b$ to 3×3 matrices. Find the commutator $[N_3, B_3]$, $[N_3, C_3]$, $[B_3, C_3]$. Discuss.
(iii) Consider the 4×4 matrices

$$
N_4 = \begin{pmatrix} 0 & 0 & 0 & 0 \\ 0 & 1 & 0 & 0 \\ 0 & 0 & 2 & 0 \\ 0 & 0 & 0 & 3 \end{pmatrix}, \quad
B_4 = \begin{pmatrix} 0 & 1 & 0 & 0 \\ 1 & 0 & \sqrt{2} & 0 \\ 0 & \sqrt{2} & 0 & \sqrt{3} \\ 0 & 0 & \sqrt{3} & 0 \end{pmatrix},
$$

$$
C_4 = \begin{pmatrix} 0 & -1 & 0 & 0 \\ 1 & 0 & -\sqrt{2} & 0 \\ 0 & \sqrt{2} & 0 & -\sqrt{3} \\ 0 & 0 & \sqrt{3} & 0 \end{pmatrix}.
$$

These matrices appear when we truncate the infinite dimensional unbounded matrices $b^\dagger b$ and $b^\dagger + b$ to 4×4 matrices. Find the commutators $[N_4, B_4]$, $[N_4, C_4]$, $[B_4, C_4]$. Discuss.

Solution 4. (i) Using $bb^\dagger = I + b^\dagger b$ we obtain

$$
[b^\dagger b, b^\dagger b] = b^\dagger - b, \quad [b^\dagger b, b^\dagger - b] = b^\dagger + b, \quad [b^\dagger + b, b^\dagger - b] = 2I.
$$

(ii) We obtain

$$
[N_2, B_2] = \begin{pmatrix} 0 & -1 \\ 1 & 0 \end{pmatrix} = C_2 = -i\sigma_2
$$

$$
[N_2, C_2] = \begin{pmatrix} 0 & 1 \\ 1 & 0 \end{pmatrix} = B_2 = \sigma_1
$$

$$
[B_2, C_2] = 2 \begin{pmatrix} 1 & 0 \\ 0 & -1 \end{pmatrix} = 2I_1 \oplus (-1) = 2\sigma_3
$$

where I_1 is the 1×1 identity matrix and $\oplus$ denotes the direct sum.
(iii) We obtain

$$
[N_3, B_3] = \begin{pmatrix} 0 & -1 & 0 \\ 1 & 0 & -\sqrt{2} \\ 0 & \sqrt{2} & 0 \end{pmatrix} = C_3
$$

$$
[N_3, C_3] = \begin{pmatrix} 0 & 1 & 0 \\ 1 & 0 & \sqrt{2} \\ 0 & \sqrt{2} & 0 \end{pmatrix} = B_3
$$

$$
[B_3, C_3] = 2 \begin{pmatrix} 1 & 0 & 0 \\ 0 & 1 & 0 \\ 0 & 0 & -2 \end{pmatrix} = 2I_2 \oplus 2(-2)
$$

where I_2 is the 2×2 identity matrix.

(iv) We obtain

$$[N_4, B_4] = \begin{pmatrix} 0 & -1 & 0 & 0 \\ 1 & 0 & -\sqrt{2} & 0 \\ 0 & \sqrt{2} & 0 & -\sqrt{3} \\ 0 & 0 & \sqrt{3} & 0 \end{pmatrix} = C_4$$

$$[N_4, C_4] = \begin{pmatrix} 0 & 1 & 0 & 0 \\ 1 & 0 & \sqrt{2} & 0 \\ 0 & \sqrt{2} & 0 & \sqrt{3} \\ 0 & 0 & \sqrt{3} & 0 \end{pmatrix} = B_4$$

$$[B_4, C_4] = 2 \begin{pmatrix} 1 & 0 & 0 & 0 \\ 0 & 1 & 0 & 0 \\ 0 & 0 & 1 & 0 \\ 0 & 0 & 0 & -3 \end{pmatrix} = 2I_3 \oplus 2(-3)$$

where I_3 is the 3×3 identity matrix.

Problem 5. Let b, $b^\dagger$ be Bose annihilation and creation operators. Let $z \in \mathbb{C}$. Consider the operators

$$J_+(z) := (b^\dagger)^2 b - z b^\dagger, \quad J_0(z) := b^\dagger b - \frac{z}{2} I, \quad J_-(z) := b.$$

(i) Show that the operators span the $s\ell(2)$ Lie algebra.
(ii) Find the *Casimir operator*. The Casimir operator commutes with the operators $J_+(z)$, $J_-(z)$ and $J_0(z)$.
(iii) If $z \in \mathbb{Z}^+$, then the operators $J_+(z)$, $J_-(z)$, $J_0(z)$ possess a finite-dimensional, irreducible representation in Fock space leaving invariant the linear space of polynomials in $b^\dagger$ acting on the vacuum state $|0\rangle$

$$\mathcal{P}_z(b^\dagger) = \langle 1, b^\dagger, (b^\dagger)^2, \dots, (b^\dagger)^n |0\rangle$$

of dimension $\dim(\mathcal{P}_z) = (z + 1)$. Find the matrix representation for the case $z = 1$.

Solution 5. (i) We obtain the commutation relations

$$[J_0(z), J_\pm(z)] = \pm J_\pm, \quad [J_+(z), J_-(z)] = -2J_0(z).$$

Thus $J_0(z)$, $J_+(z)$, $J_-(0)$ span the Lie algebra $s\ell(2)$.
(ii) The Casimir operator is given by

$$C_2 = \frac{1}{2}[J_+(z), J_-(z)]_+ - J_0(z)J_0(z) = -\frac{z}{2}\left(\frac{z}{2} + 1\right) I.$$

(iii) With $|1\rangle = b^\dagger|0\rangle$, $b|0\rangle = 0$, $\langle 0|b^\dagger = 0$ we have

$$\langle 0|J_+(1)|0\rangle = 0, \quad \langle 0|J_+(1)|1\rangle = 0, \quad \langle 1|J_+(1)|0\rangle = 1, \quad \langle 1|J_+(1)|1\rangle = 0.$$

Thus

$$J_+ \to \begin{pmatrix} 0 & 0 \\ 1 & 0 \end{pmatrix}.$$

Analogously we have

$$J_- \to \begin{pmatrix} 0 & 1 \\ 0 & 0 \end{pmatrix}, \quad J_0 \to \begin{pmatrix} -1/2 & 0 \\ 0 & 1/2 \end{pmatrix}.$$

Problem 6.   Show that the three operators

$$J_3 = \frac{1}{4}(b^\dagger b + bb^\dagger), \quad B_1 = \frac{1}{4}(b^\dagger b^\dagger + bb), \quad B_2 = \frac{i}{4}(b^\dagger b^\dagger - bb)$$

form a basis of a Lie algebra under the commutator. Note that $bb^\dagger = I + b^\dagger b$.

Solution 6.   We obtain for the commutators $[B_1, B_2] = -iJ_3$, $[J_3, B_1] = -iB_2$, $[J_3, B_2] = iB_1$. Thus we have basis of a simple Lie algebra.

Problem 7.   What Lie algebra is generated by the operators $b^\dagger b^\dagger$ and bb?

Solution 7.   For the commutators we have

$$[b^\dagger b^\dagger, bb] = -2I - 4b^\dagger b, \quad [b^\dagger b^\dagger, b^\dagger b] = -2b^\dagger b^\dagger, \quad [bb, b^\dagger b] = 2bb$$

where I is the identity operator. Thus we have a three dimensional Lie algebra with the basis

$$b^\dagger b^\dagger, \qquad bb, \qquad I + 2b^\dagger b.$$

However, we could also consider a four dimensional Lie algebra with the basis

$$b^\dagger b^\dagger, \quad bb, \quad b^\dagger b, \quad I.$$

Problem 8.   Let k be a nonnegative integer and I the identity operator. Consider the three operators

$$K_- = (2kI + b^\dagger b)^{1/2}b, \quad K_+ = b^\dagger(2kI + b^\dagger b)^{1/2}, \quad K_0 = kI + b^\dagger b.$$

Note that $b^\dagger b$ can be represented by an infinite dimensional unbounded diagonal matrix with diagonal elements $(0, 1, 2, \ldots)$. Thus the square root is well-defined.
(i) Find the commutators. Discuss.
(ii) Find the Casimir operator.

Solution 8.   Since $[b, b^\dagger] = I$ and $b^\dagger b$ and $(2kI + b^\dagger b)^{1/2}$ commute we obtain

$$\begin{aligned}
[K_+, K_-] &= b^\dagger(2kI + b^\dagger b)b - (2kI + b^\dagger b)^{1/2}(I + b^\dagger b)(2kI + b^\dagger b)^{1/2} \\
&= 2kb^\dagger b + b^\dagger b^\dagger bb - 2kI - b^\dagger b - b^\dagger b(2kI + b^\dagger b) \\
&= b^\dagger b^\dagger bb - 2kI - b^\dagger b - b^\dagger(I + b^\dagger b)b = -2kI - 2b^\dagger b \\
&= -2K_0.
\end{aligned}$$

Analogously we find $[K_0, K_+] = K_+$ and $[K_0, K_-] = -K_-$. Thus we have a realization of the simple Lie algebra $su(1, 1)$.
(ii) The Casimir operator C is

$$C = K_0^2 - \frac{1}{2}(K_+K_- + K_-K_+).$$

Problem 9.   Let $b_1^\dagger$, b_1, $b_2^\dagger$, b_2 be Bose creation and annihilation operators. What Lie algebra is generated by the operators $b_1^\dagger b_2$ and $b_2^\dagger b_1$?

Solution 9. First we note that $[b_1^\dagger b_2, b_2^\dagger b_1] = b_1^\dagger b_1 - b_2^\dagger b_2$. Now

$$[b_1^\dagger b_2, b_1^\dagger b_1] = -b_1^\dagger b_2, \quad [b_1^\dagger b_2, b_2^\dagger b_2] = b_1^\dagger b_2, \quad [b_2^\dagger b_1, b_1^\dagger b_1] = b_2^\dagger b_1, \quad [b_2^\dagger b_1, b_2^\dagger b_2] = -b_2^\dagger b_1$$

and therefore

$$[b_1^\dagger b_2, b_1^\dagger b_1 - b_2^\dagger b_2] = -2b_1^\dagger b_2, \quad [b_2^\dagger b_1, b_1^\dagger b_1 - b_2^\dagger b_2] = 2b_2^\dagger b_1.$$

Thus we could consider a three-dimensional Lie algebra with basis $\{\, b_1^\dagger b_2, \, b_2^\dagger b_1, \, b_1^\dagger b_1 - b_2^\dagger b_2 \,\}$ or a four dimensional Lie algebra with basis $\{\, b_1^\dagger b_1, \, b_2^\dagger b_2, b_1^\dagger b_2, \, b_2^\dagger b_1 \,\}$.

Problem 10. Let b_1 and b_2 denotes the Boson annihilation operators for the pump and signal mode, respectively. Consider the operators

$$J_1 = \frac{1}{2}\left(b_1^\dagger b_2 + b_2^\dagger b_1\right), \quad J_2 = -\frac{i}{2}\left(b_1^\dagger b_2 - b_2^\dagger b_1\right), \quad J_3 = \frac{1}{2}\left(b_1^\dagger b_1 - b_2^\dagger b_2\right)$$

and the total *number operator* ($\hat{N}_1 = b_1^\dagger b_1, \hat{N}_2 = b_2^\dagger b_2$)

$$\hat{N} = \hat{N}_1 + \hat{N}_2 = b_1^\dagger b_1 + b_2^\dagger b_2.$$

(i) Show that J_1, J_2, J_3 satisfy the usual angular-momentum commutation relations.
(ii) Find the *Casimir operator* $J^2 = J_1^2 + J_2^2 + J_3^2$.
(iii) Does the number operator $\hat{N}$ commute with J_1, J_2, J_3?
(iv) Find the states

$$J_3(|n\rangle \otimes |0\rangle), \quad J^2(|n\rangle \otimes |0\rangle)$$

where $|n\rangle$ are the number states.

Solution 10. (i) We have $[J_1, J_2] = \frac{i}{2}(b_1^\dagger b_1 - b_2^\dagger b_2) = iJ_3$. Analogously $[J_2, J_3] = iJ_1$ and $[J_3, J_1] = iJ_2$.
(ii) Since

$$J_1^2 + J_2^2 = \frac{1}{2}(b_1^\dagger b_1 + b_2^\dagger b_2) + b_1^\dagger b_1 b_2^\dagger b_2, \quad J_3^2 = \frac{1}{4}(b_1^\dagger b_1 b_1^\dagger b_1 + b_2^\dagger b_2 b_2^\dagger b_2 - 2b_1^\dagger b_1 b_2^\dagger b_2)$$

we obtain

$$J^2 = J_1^2 + J_2^2 + J_3^2 = \frac{1}{2}\hat{N}\left(\frac{1}{2}\hat{N} + I\right).$$

(iii) Yes.
(iv) With $b_1 = b \otimes I$ and $b_2 = I \otimes b$ we obtain

$$J_3(|n\rangle \otimes |0\rangle) = \frac{n}{2}|n\rangle \otimes |0\rangle, \quad J^2(|n\rangle \otimes |0\rangle) = \frac{n}{2}\left(\frac{n}{2} + 1\right)|n\rangle \otimes |0\rangle.$$

Problem 11. Consider the operators

$$S_0 = b_1^\dagger b_1 + b_2^\dagger b_2, \quad S_1 = b_1^\dagger b_1 - b_2^\dagger b_2,$$

$$S_2 = b_1^\dagger b_1 e^{i\phi} + b_2^\dagger b_2 e^{-i\phi}, \quad S_3 = ib_1^\dagger b_1 e^{-i\phi} - ib_2^\dagger b_2 e^{i\phi}.$$

Find the commutators. Note that S_0 is the number operator.

Solution 11. We have $[S_1, S_2] = 2iS_3$, $[S_2, S_3] = 2iS_1$, $[S_3, S_1] = 2iS_2$ and $[S_j, S_0] = 0$ for $j = 1, 2, 3$.

Problem 12.   The nine operators $b_j^\dagger b_k$ $(j,k = -1,0,1)$ form a basis of a Lie algebra. It contains an $so(2)$ Lie algebra

$$L = b_1^\dagger b_1 - b_{-1}^\dagger b_{-1}.$$

Find the commutator $[L, b_j^\dagger]$. Discuss.

Solution 12.   We find

$$[L, b_j^\dagger] = \text{ad}(L)(b_j^\dagger) = jb_j^\dagger.$$

Thus we have an eigenvalue equation with eigenvalues -1, 0, 1. For $j = 0$ we have $[L, b_0^\dagger] = 0$.

Problem 13.   Let $b_1^\dagger$, $b_2^\dagger$ be Bose creation operator and let I be the identity operator. The semi-simple Lie algebra $su(1,1)$ is generated by

$$K_+ := b_1^\dagger b_2^\dagger, \quad K_- := b_1 b_2, \quad K_0 := \frac{1}{2}(b_1^\dagger b_1 + b_2^\dagger b_2 + I)$$

with the commutation relations

$$[K_0, K_+] = K_+, \quad [K_0, K_-] = -K_-, \quad [K_-, K_+] = 2K_0.$$

We use the ordering K_+, K_-, K_0 for the basis.
(i) Find the *adjoint representation* for the Lie algebra.
(ii) Find the *Killing form* and metric tensor (g_{ij}) for this Lie algebra.
(iii) Find the *Casimir invariant* C using

$$C = \sum_{i=1}^{3} \sum_{j=1}^{3} g^{ij} X_i X_j$$

where (g^{ij}) is the inverse of the matrix (g_{ij}). We set $X_1 = K_+$, $X_2 = K_-$, $X_3 = K_0$.

Solution 13.   (i) Using the ordering given above for the basis we have

$$(\text{ad}K_+)K_+ = [K_+, K_+] = 0$$
$$(\text{ad}K_+)K_- = [K_+, K_-] = -2K_0$$
$$(\text{ad}K_+)K_0 = [K_+, K_0] = -K_+.$$

Thus we write in matrix notation

$$(K_+ \quad K_- \quad K_0) \begin{pmatrix} 0 & 0 & -1 \\ 0 & 0 & 0 \\ 0 & -2 & 0 \end{pmatrix} = (0 \quad -2K_0 \quad -K_+).$$

Since K_+, K_-, K_0 is a basis of the Lie algebra it follows that the representation is

$$\text{ad}K_+ = \begin{pmatrix} 0 & 0 & -1 \\ 0 & 0 & 0 \\ 0 & -2 & 0 \end{pmatrix}.$$

Analogously we find

$$\text{ad}K_- = \begin{pmatrix} 0 & 0 & 0 \\ 0 & 0 & 1 \\ 2 & 0 & 0 \end{pmatrix}, \quad \text{ad}K_0 = \begin{pmatrix} 1 & 0 & 0 \\ 0 & -1 & 0 \\ 0 & 0 & 0 \end{pmatrix}.$$

With $X_1 = K_+$, $X_2 = K_-$, $X_3 = K_0$ we have

$$g_{ij} = K(X_i, X_j) = \text{tr}(\text{ad}X_i \text{ad}X_j), \quad i,j = 1,2,3$$

where tr denotes the trace. We obtain

$$\text{tr}(\text{ad}K_+ \text{ad}K_+) = 0, \quad \text{tr}(\text{ad}K_+ \text{ad}K_-) = -4, \quad \text{tr}(\text{ad}K_+ \text{ad}K_0) = 0$$

$$\text{tr}(\text{ad}K_- \text{ad}K_+) = 0, \quad \text{tr}(\text{ad}K_- \text{ad}K_-) = 0, \quad \text{tr}(\text{ad}K_- \text{ad}K_0) = 0$$

$$\text{tr}(\text{ad}K_0 \text{ad}K_+) = 0, \quad \text{tr}(\text{ad}K_0 \text{ad}K_-) = 0, \quad \text{tr}(\text{ad}K_0 \text{ad}K_0) = 2.$$

Thus we obtain the metric tensor

$$g = \begin{pmatrix} 0 & -4 & 0 \\ -4 & 0 & 0 \\ 0 & 0 & 2 \end{pmatrix}$$

with the inverse matrix

$$g^{-1} = \begin{pmatrix} 0 & -1/4 & 0 \\ -1/4 & 0 & 0 \\ 0 & 0 & 1/2 \end{pmatrix}.$$

Thus the Casimir operator is

$$C = \sum_{i=1}^{3}\sum_{j=1}^{3} g^{ij} X_i X_j = \frac{1}{8}((b_1^\dagger b_1 - b_2^\dagger b_2)^2 - I) = \frac{1}{2}\left(K_0^2 - \frac{1}{2}(K_+ K_- + K_- K_+)\right)$$

where I is the identity operator.

Problem 14. Let b_j, $b_k^\dagger$ be Bose annihilation and creation operators with $j, k = 1, \ldots, N-1$. We define the operators

$$E_j := b_j^\dagger b_{j+1}, \quad F_j := b_j b_{j+1}^\dagger, \quad H_j := b_j^\dagger b_j - b_{j+1}^\dagger b_{j+1}.$$

Find the commutators $[E_i, F_j]$, $[H_i, E_j]$, $[H_i, F_j]$, $[E_i, E_j]$, $[H_i, H_j]$. Discuss.

Solution 14. We obtain for the commutators

$$[E_i, F_j] = \delta_{ij} H_j, \quad [H_i, E_j] = a_{ij} E_j, \quad [H_i, F_j] = -a_{ij} F_j$$

and

$$[E_i, E_j] = 0 \quad \text{for} \quad |i - j| > 1.$$

Here a_{ij} is the *Cartan matrix* of the Lie algebra $su(N)$, i.e. $a_{ii} = 2$, $a_{i,i\pm1} = -1$ and $a_{ij} = 0$ for $|i-j| > 1$. We also have $[H_i, H_j] = 0$. The Cartan subalgebra is generated by the elements of H_j.

Problem 15. Consider the Hamilton operator

$$\hat{H} = \epsilon(b_1^\dagger b_1 + b_2^\dagger b_2) + \gamma_1(b_1^\dagger b_1^\dagger b_1 b_1 + b_2^\dagger b_2^\dagger b_2 b_2) + \gamma_2 b_1^\dagger b_2^\dagger b_1 b_2.$$

The Hamilton operator describes a quantum system of two nonlinear interacting oscillators. The two-mode representation of the Lie algebra $su(1,1)$ is given by

$$J_- = b_1 b_2, \quad J_+ = b_1^\dagger b_2^\dagger, \quad J_0 = \frac{1}{2}(b_1^\dagger b_1 + b_2^\dagger b_2 + I).$$

Express the Hamilton operator $\hat{H}$ using J_-, J_+, J_0.

Solution 15.   We obtain

$$\hat{H} = (2\gamma_1 - \epsilon)I + (2\epsilon - 6\gamma_1)J_0 + 4\gamma_1 J_0^2 + (\gamma_2 - 2\gamma_1)J_+ J_-.$$

Problem 16.   Let $i, j \in \{1, 2, \ldots, n\}$. Consider the Lie algebra spanned by the operators $E_{ij} = (E_{ij})^\dagger$, $B_i^\dagger$, $B_i = (B_i^\dagger)^\dagger$, I (identity operator) and the non-zero commutators

$$[E_{ij}, E_{k\ell}] = \delta_{jk} E_{i\ell} - \delta_{i\ell} E_{kj}, \quad [B_i, B_j^\dagger] = \delta_{ij} I,$$

$$[E_{ij}, B_k^\dagger] = \delta_{jk} B_i^\dagger, \quad [E_{ij}, B_k] = -\delta_{ik} B_j.$$

We define the operators

$$\widetilde{E}_{ij} := E_{ij} - B_i^\dagger B_j.$$

Find the commutators $[\widetilde{E}_{ij}, B_k^\dagger]$, $[\widetilde{E}_{ij}, B_k]$, $[\widetilde{E}_{ij}, \widetilde{E}_{k\ell}]$.

Solution 16.   We obtain $[\widetilde{E}_{ij}, B_k^\dagger] = 0$, $[\widetilde{E}_{ij}, B_k] = 0$, $[\widetilde{E}_{ij}, \widetilde{E}_{k\ell}] = \delta_{jk} \widetilde{E}_{i\ell} - \delta_{i\ell} \widetilde{E}_{kj}$.

Problem 17.   Let b_1, b_2, b_3 be Bose annihilation operators. Show that

$$H_1 = b_1^\dagger b_1 - b_2^\dagger b_2, \qquad H_2 = b_2^\dagger b_2 - b_3^\dagger b_3$$

$$E_{12} = b_1^\dagger b_2, \quad E_{23} = b_2^\dagger b_3, \quad E_{13} = b_1^\dagger b_3$$

$$E_{21} = b_2^\dagger b_1, \quad E_{32} = b_3^\dagger b_2, \quad E_{31} = b_3^\dagger b_1$$

are a representation of the Lie algebra $su(3)$.

Solution 17.   The commutators are $[H_1, H_2] = 0$ and

$$[H_1, E_{12}] = 2E_{12}, \quad [H_1, E_{23}] = -E_{23}, \quad [H_1, E_{13}] = E_{13}$$

$$[H_2, E_{12}] = -E_{12}, \quad [H_2, E_{23}] = 2E_{23}, \quad [H_2, E_{13}] = E_{13}$$

$$[E_{12}, E_{21}] = H_1, \quad [E_{23}, E_{32}] = H_2, \quad [E_{12}, E_{23}] = E_{13}.$$

Problem 18.   Given a semisimple Lie algebra with the generators A_1, A_2, A_3, B_1, B_2, B_3, C_1, C_2 and the commutation relations

$$[C_1, C_2] = 0$$
$$[C_1, A_1] = A_1, \quad [C_1, A_2] = A_2, \quad [C_1, A_3] = 2A_3$$
$$[C_2, A_1] = 3A_1, \quad [C_2, A_2] = -3A_2, \quad [C_2, A_3] = 0$$
$$[A_1, A_2] = 2A_3, \quad [A_1, A_3] = 0, \quad [A_2, A_3] = 0$$
$$[A_1, B_1] = 2(C_1 + C_2), \quad [A_1, B_2] = 0, \quad [A_1, B_3] = -2B_2$$
$$[A_2, B_2] = 2(C_1 - C_2), \quad [A_2, B_3] = 2B_1, \quad [A_3, B_3] = 4C_1.$$

Let b_1, b_2, b_3 be Bose annihilation operators and I the identity operator in the product Hilbert space. Show that

$$A_1 = b_1 - b_2^\dagger b_3, \quad A_2 = b_2 + b_1^\dagger b_3, \quad A_3 = b_3$$

$$B_1 = 4\lambda b_1^\dagger - 4b_1^\dagger b_1^\dagger b_1 - 2(b_3^\dagger - b_1^\dagger b_2^\dagger)b_2 - 2(b_3^\dagger + b_1^\dagger b_2^\dagger)b_1^\dagger b_3$$

$$B_2 = 4\mu b_2^\dagger - 4b_2^\dagger b_2^\dagger b_2 + 2(b_3^\dagger + b_1^\dagger b_2^\dagger)b_1 - 2(b_3^\dagger - b_1^\dagger b_3^\dagger)b_2^\dagger b_3$$

$$B_3 = 4(\lambda + \mu)b_3^\dagger + 4(\lambda - \mu)b_1^\dagger b_2^\dagger$$
$$\qquad - 4b_3^\dagger(b_1^\dagger b_1 + b_2^\dagger b_2) - 4b_1^\dagger b_2^\dagger(b_1^\dagger b_1 - b_2^\dagger b_2) - 4(b_3^\dagger b_3^\dagger + b_1^\dagger b_1^\dagger b_2^\dagger b_2^\dagger)b_3$$

$$C_1 = (\lambda + \mu)I - b_1^\dagger b_1 - b_2^\dagger b_2 - 2b_3^\dagger b_3$$

$$C_2 = (\lambda - \mu)I - 3b_1^\dagger b_1 + 3b_2^\dagger b_2$$

where $\lambda, \mu \in \mathbb{C}$.

Solution 18. Since $[b_j, b_k^\dagger] = \delta_{jk}I$ we find for the commutator $[A_1, A_2]$

$$[A_1, A_2] = (b_1 - b_2^\dagger b_3)(b_2 + b_1^\dagger b_3) - (b_2 + b_1^\dagger b_3)(b_1 - b_2^\dagger b_3)$$
$$= (b_1 b_1^\dagger - b_1^\dagger b_1)b_3 + (b_2 b_2^\dagger - b_2^\dagger b_2)b_3 = 2b_3$$
$$= 2A_3.$$

Analogously we calculate the other commutators.

Problem 19. (i) Let $b_j^\dagger$, b_j be Bose creation and annihilation operators, where $j = 1, 2, \ldots, n$. Consider the vector space of the operators $b_i^\dagger b_j$, where $i, j = 1, 2, \ldots, n$. Calculate the commutator

$$[b_i^\dagger b_j, b_k^\dagger b_\ell].$$

Do we have a basis of a Lie algebra?
(ii) Consider the vector space of the vector fields

$$x_i \frac{\partial}{\partial x_j}, \quad i, j = 1, 2, \ldots, n.$$

Calculate the commutator

$$\left[x_i \frac{\partial}{\partial x_j}, x_k \frac{\partial}{\partial x_\ell} \right].$$

Do we have a basis of a Lie algebra?
(iii) In both case we have a Lie algebra. Are the Lie algebras isomorphic?

Solution 19. (i) We obtain the commutator

$$[b_i^\dagger b_j, b_k^\dagger b_\ell] = -\delta_{\ell i} b_k^\dagger b_j + \delta_{jk} b_i^\dagger b_\ell.$$

Owing to the right-hand side we have a basis of a Lie algebra. The Lie algebra contains the number operator

$$\hat{N} = \sum_{j=1}^{n} b_j^\dagger b_j$$

which commutes with all the operators $b_k^\dagger b_\ell$.

(ii) We obtain

$$\left[x_i\frac{\partial}{\partial x_j}, x_k\frac{\partial}{\partial x_\ell}\right] = -\delta_{\ell i}x_k\frac{\partial}{\partial x_j} + \delta_{jk}x_i\frac{\partial}{\partial x_\ell}.$$

Owing to the right-hand side we have a basis of a Lie algebra. The Lie algebra contains the element

$$\sum_{j=1}^{n} x_j\frac{\partial}{\partial x_j}$$

which commutes with all the elements of the Lie algebra.

(iii) Yes the two Lie algebras are isomorphic. We have

$$x_i\frac{\partial}{\partial x_j} \leftrightarrow b_i^\dagger b_j.$$

Problem 20. Consider the Lie algebra $u(1,1)$ with the generators

$$H, \quad A_+, \quad A_-, \quad \Xi$$

and the commutation relations

$$[A_+, A_-] = \Xi + 2H, \quad [H, A_\pm] = \pm A_\pm, \quad [\Xi, H] = [\Xi, A_+] = [\Xi, A_-] = 0.$$

(i) Show that a representation of the Lie algebra $u(1,1)$ is given by

$$\hat{H} = \sum_{j=0}^{\infty} j b_j^\dagger b_j, \qquad \hat{\Xi} = 2s\sum_{j=0}^{\infty} b_j^\dagger b_j$$

$$\hat{A}_+ = \sum_{j=0}^{\infty}\sqrt{(j+1)(2s+j)}b_j^\dagger b_{j+1}, \qquad \hat{A}_- = \sum_{j=0}^{\infty}\sqrt{j(2s+j-1)}b_j^\dagger b_{j-1}$$

where $s = 1/2, 1, 3/2, \ldots$. Give an interpretation of the operators.

(ii) Let $|0\rangle$ be the vacuum state. Find $\hat{H}|0\rangle$, $\hat{\Xi}|0\rangle$, $\hat{A}_+|0\rangle$, $\hat{A}_-|0\rangle$.

Solution 20.   (i) Using the commutator

$$[b_j^\dagger b_k, b_\ell^\dagger b_m] = -\delta_{mj}b_\ell^\dagger b_k + \delta_{\ell k}b_j^\dagger b_m$$

we find that the operators satisfy the commutation relation given above. Here $\hat{E} = \hbar\omega\hat{H}$ can be considered as the total energy operator and $\hat{N} = \frac{1}{2s}\hat{\Xi}$ may be considered as the total number operator.

(ii) Since $b_j|0\rangle = 0|0\rangle$ we obtain $\hat{H}|0\rangle = 0|0\rangle$, $\hat{\Xi}|0\rangle = 0|0\rangle$, $\hat{A}_+|0\rangle = 0|0\rangle$, $\hat{A}_-|0\rangle = 0|0\rangle$.

Problem 21. Show that the operators

$$b_j^\dagger b_k, \quad b_j^\dagger, \quad b_j, \quad I, \quad j,k = 1,2,\ldots,n$$

form a basis of a Lie algebra under the commutator. Is the Lie algebra semisimple?

Solution 21. For the nonzero commutators we find

$$[b_\ell, b_j^\dagger b_k] = \delta_{\ell j}b_k, \qquad [b_\ell^\dagger, b_j^\dagger b_k] = -\delta_{k\ell}b_j^\dagger,$$

$$[b_j^\dagger b_k, b_m^\dagger b_n] = b_j^\dagger b_n \delta_{km} - b_m^\dagger b_k \delta_{nj}, \quad [b_j, b_k^\dagger] = \delta_{jk} I, \quad [b_j^\dagger b_k, I] = [b_j, I] = [b_j^\dagger, I] = 0.$$

Problem 22. Consider the six linear operators I, P, Q, P^2, QP, Q^2, with

$$[P, Q] = cI$$

where I is the identity operator and $c \neq 0$. Show that they form a basis of a Lie algebra.

Solution 22. We have to calculate all commutators. We find the nonzero commutators

$$[P, Q] = cI, \quad [P^2, Q^2] = 2c^2 I + 4cQP, \quad [P^2, Q] = 2cP, \quad [Q^2, P] = -2cQ$$

$$[P^2, QP] = 2cP^2, \quad [Q^2, QP] = -2cQ^2, \quad [Q, QP] = -cQ, \quad [P, QP] = cP.$$

The right-hand sides can be expressed again as linear combinations of the six operators. Thus we have a basis on a Lie algebra.

Problem 23. Consider the Lie algebra $so(3,1)$ with the basis $L_1, L_2, L_3, B_1, B_2, B_3$ and the commutation relations ($\epsilon_{123} = +1$)

$$[L_j, L_k] = i\hbar\epsilon_{jk\ell}L_\ell, \quad [L_j, B_k] = i\hbar\epsilon_{jk\ell}B_\ell, \quad [B_j, B_k] = -i\hbar\epsilon_{jk\ell}L_\ell.$$

Obviously the elements L_1, L_2, L_3 form a simple Lie subalgebra of $so(3,1)$. Find a bilinear realization using Bose creation and annihilation operators.

Solution 23. Since $[b_j^\dagger b_k, b_\ell^\dagger b_m] = -\delta_{mj} b_\ell^\dagger b_k + \delta_{k\ell} b_j^\dagger b_m$ we have for the elements L_1, L_2, L_3 the representation

$$L_j = \frac{\hbar}{2}\left(\begin{pmatrix} b_1^\dagger & b_2^\dagger \end{pmatrix} \sigma_j \begin{pmatrix} b_1 \\ b_2 \end{pmatrix} + \begin{pmatrix} b_3^\dagger & b_4^\dagger \end{pmatrix} \sigma_j \begin{pmatrix} b_3 \\ b_4 \end{pmatrix} \right)$$

where $\sigma_1, \sigma_2, \sigma_3$ are the Pauli spin matrices. Thus we have

$$L_1 = \frac{\hbar}{2}(b_1^\dagger b_2 + b_2^\dagger b_1 + b_3^\dagger b_4 + b_4^\dagger b_3)$$

$$L_2 = \frac{\hbar}{2}(-ib_1^\dagger b_2 + ib_2^\dagger b_1 - ib_3^\dagger b_4 + ib_4^\dagger b_3)$$

$$L_3 = \frac{\hbar}{2}(b_1^\dagger b_1 - b_2^\dagger b_2 + b_3^\dagger b_3 - b_4^\dagger b_4).$$

Since

$$[b_j^\dagger b_k^\dagger, b_\ell b_m] = -\delta_{mj}\delta_{k\ell} I - \delta_{mk}\delta_{j\ell} I - \delta_{mj} b_k^\dagger b_\ell - \delta_{mk} b_j^\dagger b_\ell - \delta_{\ell j} b_k^\dagger b_m - \delta_{\ell k} b_j^\dagger b_m$$

we have for B_j

$$B_j = \frac{\hbar}{2}\left(\begin{pmatrix} b_1^\dagger & b_2^\dagger \end{pmatrix} \sigma_j \begin{pmatrix} 0 & -1 \\ 1 & 0 \end{pmatrix} \begin{pmatrix} b_3^\dagger \\ b_4^\dagger \end{pmatrix} - \begin{pmatrix} b_1 & b_2 \end{pmatrix} \begin{pmatrix} 0 & -1 \\ 1 & 0 \end{pmatrix} \sigma_j \begin{pmatrix} b_3 \\ b_4 \end{pmatrix} \right).$$

Thus

$$B_1 = \frac{\hbar}{2}(b_1^\dagger b_3^\dagger - b_2^\dagger b_4^\dagger + b_1 b_3 - b_2 b_4)$$

$$B_2 = \frac{\hbar}{2}(-ib_1^\dagger b_3^\dagger - ib_2^\dagger b_4^\dagger + ib_1 b_3 + ib_2 b_4)$$

$$B_3 = \frac{\hbar}{2}(-b_1^\dagger b_4^\dagger - b_2^\dagger b_3^\dagger - b_1 b_4 - b_2 b_3).$$

Problem 24. The Lie algebra $sp(2n, \mathbb{R})$ is spanned by the generators

$$E_{ij} = (E_{ji})^\dagger, \qquad D_{ij}^\dagger = D_{ji}^\dagger, \qquad D_{ij} = D_{ji} = (D_{ij}^\dagger)^\dagger$$

where $i, j = 1, \ldots, n$. The non-vanishing commutation relations are

$$[E_{ij}, E_{k\ell}] = \delta_{jk} E_{i\ell} - \delta_{i\ell} E_{kj}$$
$$[E_{ij}, D_{k\ell}^\dagger] = \delta_{jk} D_{i\ell}^\dagger + \delta_{j\ell} D_{ik}^\dagger$$
$$[E_{ij}, D_{k\ell}] = -\delta_{ik} D_{j\ell} - \delta_{j\ell} D_{jk}$$
$$[D_{ij}, D_{k\ell}^\dagger] = \delta_{ik} E_{\ell j} + \delta_{i\ell} E_{kj} + \delta_{jk} E_{\ell i} + \delta_{j\ell} E_{ki}.$$

The generators E_{ij} form the sub Lie algebra $u(n)$. Find a faithful representation with Bose creation and annihilation operators.

Solution 24. We find

$$E_{ij} = \sum_{s=1}^{m} b_{is}^\dagger b_{js} + \frac{1}{2} m \delta_{ij} I, \quad D_{ij} = \sum_{s=1}^{m} b_{is} b_{js}, \quad D_{ij}^\dagger = b_{is}^\dagger b_{js}^\dagger.$$

Problem 25. Let $b_1^\dagger$, $b_2^\dagger$, $b_3^\dagger$, b_1, b_2, b_3 be Bose creation and annihilation operators. Show that a basis for the Lie algebra $su(3)$ is given by

$$H_1 = b_1^\dagger b_1 - b_2^\dagger b_2, \qquad H_2 = b_2^\dagger b_2 - b_3^\dagger b_3$$

$$E_{12} = b_1^\dagger b_2, \quad E_{23} = b_2^\dagger b_3, \quad E_{13} = b_1^\dagger b_3$$

$$E_{21} = b_2^\dagger b_1, \quad E_{32} = b_3^\dagger b_2, \quad E_{31} = b_3^\dagger b_1.$$

Solution 25. The commutators are

$$[H_1, E_{12}] = 2E_{12}, \quad [H_1, E_{23}] = -E_{23}, \quad [H_1, E_{13}] = E_{13}$$

$$[H_2, E_{12}] = -E_{12}, \quad [H_2, E_{23}] = 2E_{23}, \quad [H_2, E_{13}] = E_{13}$$

$$[E_{12}, E_{21}] = H_1, \quad [E_{23}, E_{32}] = H_2, \quad [E_{12}, E_{23}] = E_{13}$$

and $[H_1, H_2] = 0$.

Problem 26. Consider the Lie algebra $so(4)$ with a basis given by bilinear combinations of Bose operators

$$J_+ = b_1^\dagger b_2, \quad J_- = b_2^\dagger b_1, \quad J_z = \frac{1}{2}(b_1^\dagger b_1 - b_2^\dagger b_2)$$

$$K_+ = b_3^\dagger b_4, \quad K_- = b_4^\dagger b_3, \quad K_z = \frac{1}{2}(b_3^\dagger b_3 - b_4^\dagger b_4).$$

Find the commutators. Discuss.

Solution 26. Obviously we have two Lie subalgebras $so(3)$ given by

$$\{ J_+, \ J_-, \ J_z \}, \qquad \{ K_+, \ K_-, \ K_z \}.$$

Problem 27. The three generators K_0, K_+, K_- of the noncompact Lie group $SU(1,1)$ obey the commutation relation

$$[K_0, K_\pm] = \pm K_\pm, \qquad [K_+, K_-] = -2K_0.$$

Find the disentanglement coefficients β, γ, δ in the relation

$$\exp(\alpha K_+ - \alpha^* K_-) = \exp(\beta K_+)\exp(\gamma K_0)\exp(\delta K_-).$$

Solution 27. We set

$$V = \exp(\beta K_+)\exp(\gamma K_0)\exp(\delta K_-).$$

Let $|\alpha|$ and θ be the modulus and argument of α. We obtain

$$e^{\alpha K_+ - \alpha^* K_-} K_0 e^{-\alpha K_+ + \alpha^* K_-} = \cosh(2|\alpha|)K_0 - \frac{1}{2}\sinh(2|\alpha|)(e^{i\theta}K_+ + e^{-i\theta}K_-)$$
$$e^{\alpha K_+ - \alpha^* K_-} K_+ e^{-\alpha K_+ + \alpha^* K_-} = -e^{-i\theta}\sinh(2|\alpha|)K_0 + \cosh^2(|\alpha|)K_+ + e^{-i2\theta}\sinh^2(|\alpha|)K_-$$
$$e^{\alpha K_+ - \alpha^* K_-} K_- e^{-\alpha K_+ + \alpha^* K_-} = -e^{i\theta}\sinh(2|\alpha|)K_0 + \cosh^2(|\alpha|)K_- + e^{i2\theta}\sinh^2(|\alpha)K_+$$

and

$$V K_- V^{-1} = -2\beta e^{-\gamma}K_0 + e^{-\gamma}K_- + e^{-\gamma}\beta^2 K_+$$
$$V K_0 V^{-1} = (1 - 2e^{-\gamma}\beta\delta)K_0 + \delta e^{-\gamma}K_- - \beta(1 - e^{-\gamma}\beta\delta)K_+.$$

Comparing these two sets of equations with respect to the basis elements (generators) K_0, K_+, K_- we obtain the disentanglement coefficients

$$\gamma = -2\ln(\cosh(|\alpha|)), \qquad \beta = e^{i\theta}\tanh(|\alpha), \qquad \delta = -e^{-i\theta}\tanh(|\alpha|).$$

Problem 28. Consider the Lie algebra $su(1,1)$ of the noncompact Lie group $SU(1,1)$ with a basis given by K_+, K_-, K_3 satisfying the commutation relations

$$[K_3, K_+] = K_+, \quad [K_3, K_-] = -K_-, \quad [K_+, K_-] = -2K_3.$$

A faithful matrix representation is given by

$$K_+ = \begin{pmatrix} 0 & 1 \\ 0 & 0 \end{pmatrix}, \quad K_- = \begin{pmatrix} 0 & 0 \\ -1 & 0 \end{pmatrix}, \quad K_3 = \frac{1}{2}\begin{pmatrix} 1 & 0 \\ 0 & -1 \end{pmatrix}.$$

(i) Let $x \in \mathbb{R}$. Calculate $\exp(x(K_- - K_+))$.
(ii) Let α, β, γ. Calculate $\exp(\alpha K_+)\exp(\beta K_3)\exp(\gamma K_-)$.
(iii) Solve

$$\exp(x(K_- - K_+)) = \exp(\alpha K_+)\exp(\beta K_3)\exp(\gamma K_-)$$

for α, β, γ.
(iv) The result derived in (iii) is valid for all faithful representations. Let b, $b^\dagger$ be Bose annihilation and creation operators. Consider the faithful representation

$$K_+ = \frac{1}{2}b^\dagger b^\dagger, \quad K_- = \frac{1}{2}bb, \quad K_3 = \frac{1}{4}(b^\dagger b + bb^\dagger).$$

Find

$$\exp(\tanh(x)K_+)b\exp(-\tanh(x)K_+), \quad \exp(\tanh(x)K_-)b^\dagger\exp(-\tanh(x)K_-)$$

and
$$\exp(-2\ln(\cosh(x))K_3)(b^\dagger + \tanh(x)b)\exp(2\ln(\cosh(x))K_3).$$

(v) Let $|0\rangle$ be the vacuum state, i.e. $b|0\rangle = 0|0\rangle$. Calculate the state
$$\exp(-2\ln(\cosh(x))K_3)|0\rangle.$$

Solution 28. (i) We obtain
$$\exp(x(K_+ - K_-)) = \begin{pmatrix} \cosh(x) & -\sinh(x) \\ -\sinh(x) & \cosh(x) \end{pmatrix}.$$

(ii) We obtain
$$\exp(\alpha K_+)\exp(\beta K_3)\exp(\gamma K_-) = \begin{pmatrix} \exp(\beta/2) - \alpha\gamma\exp(-\beta/2) & \alpha\exp(-\beta/2) \\ -\gamma\exp(-\beta/2) & \exp(-\beta/2) \end{pmatrix}.$$

(iii) Solving the equation
$$\begin{pmatrix} \cosh(x) & -\sinh(x) \\ -\sinh(x) & \cosh(x) \end{pmatrix} = \begin{pmatrix} \exp(\beta/2) - \alpha\gamma\exp(-\beta/2) & \alpha\exp(-\beta/2) \\ -\gamma\exp(-\beta/2) & \exp(-\beta/2) \end{pmatrix}$$

for α, β, γ yields
$$\alpha = -\tanh(x), \quad \beta = -2\ln(\cosh(x)), \quad \gamma = \tanh(x).$$

(iv) We obtain
$$\exp(\tanh(x)K_+)b\exp(-\tanh(x)K_+) = b - \tanh(x)b^\dagger$$
$$\exp(\tanh(x)K_-)b^\dagger\exp(-\tanh(x)K_-) = b^\dagger + \tanh(x)b$$

and
$$\exp(-2\ln(\cosh(x))K_3)(b^\dagger + \tanh(x)b)\exp(2\ln(\cosh(x))K_3) = \operatorname{sech}(x)b^\dagger + \sinh(x)b.$$

(v) Since $b|0\rangle = 0|0\rangle$ we obtain
$$\exp(-2\ln(\cosh(x))K_3)|0\rangle = (\operatorname{sech}(x))^{1/2}|0\rangle.$$

Problem 29. Consider the Lie algebra $su(1,1)$ with the basis J_+, J_-, J_0 and the commutation relation
$$[J_+, J_-] = -2J_0, \qquad [J_0, J_\pm] = \pm J_\pm.$$

A two-mode representation of $su(1,1)$ is given by
$$J_- = b_1 b_2, \quad J_+ = b_1^\dagger b_2^\dagger, \quad J_0 = \frac{1}{2}(b_1^\dagger b_1 + b_2^\dagger b_2 + I)$$

with the Casimir operator
$$C = -\frac{1}{4}I + \frac{1}{4}(b_1^\dagger b_1 - b_2^\dagger b_2)^2.$$

A single bosonic realization is
$$J_- = (2kI + b^\dagger b)^{1/2}b, \quad J_+ = b^\dagger(2kI + b^\dagger b)^{1/2}, \quad J_0 = kI + b^\dagger b$$

where $k = 0, 1, 2, \ldots$ and I is the identity operator. The Casimir operator is $C = k(k-1)I$. Consider the Hamilton operator

$$\hat{H}(t) = a_1(t)J_+ + a_2(t)J_0 + a_3(t)J_-$$

where a_1, a_2, a_3 are smooth functions of t. The Schrödinger equation is

$$\hat{H}(t)|\psi(t)\rangle = i\hbar \frac{\partial}{\partial t}|\psi(t)\rangle.$$

We define the evolution operator $U(t, 0)$ such that

$$|\phi(t)\rangle = U(t)|\phi(0)\rangle$$

where $|\phi(0)\rangle$ is the state at time $t = 0$. Thus

$$\hat{H}(t)U(t) = i\hbar \frac{\partial U(t)}{\partial t}, \qquad U(0) = I.$$

Since J_+, J_-, J_0 form a basis of the finite dimensional Lie algebra $su(1, 1)$, the evolution operator can be expressed in the form

$$U(t) = \exp(c_1(t)J_+) \exp(c_2(t)J_0) \exp(c_3(t)J_-).$$

Find the system of ordinary differential equations for c_1, c_2, c_3 depending on $a_1(t)$, $a_2(t)$, $a_3(t)$.

Solution 29. By differentiation of $U(t)$ with respect to t we obtain

$$\frac{\partial U(t)}{\partial t} = \left(\left(\frac{dc_1}{dt} - c_1 \frac{dc_2}{dt} + c_1^2 \exp(-c_2) \frac{dc_3}{dt} \right) J_+ \right.$$
$$\left. + \left(\frac{dc_2}{dt} - 2c_1 \exp(-c_2) \frac{dc_3}{dt} \right) J_0 + \exp(-c_2) \frac{dc_3}{dt} J_- \right) U(t).$$

Inserting the Hamilton operator $\hat{H}$, the ansatz for $U(t)$, and $\partial U(t)/\partial t$ into the evolution equation for $U(t)$ and comparing the coefficients we obtain the system of first ordinary differential equations

$$\frac{dc_1}{dt} - c_1 \frac{dc_2}{dt} + c_1^2 \exp(-c_2) \frac{dc_3}{dt} = \frac{1}{i\hbar} a_1(t)$$

$$\frac{dc_2}{dt} - 2c_1 \exp(-c_2) \frac{dc_3}{dt} = \frac{1}{i\hbar} a_2(t)$$

$$\exp(-c_2) \frac{dc_3}{dt} = \frac{1}{i\hbar} a_3(t).$$

It follows that

$$\frac{dc_1}{dt} = \frac{1}{i\hbar}(a_1(t) + a_2(t)c_1 + a_3(t)c_1^2)$$

$$\frac{dc_2}{dt} = \frac{1}{i\hbar}(a_2(t) + 2a_3(t)c_1)$$

$$\frac{dc_3}{dt} = \frac{1}{i\hbar} a_3(t) \exp(c_2)$$

with the initial conditions $c_1(0) = c_2(0) = c_3(0) = 0$. The first equation is a Ricatti equation for c_1. After solving this equation we obtain c_2 and c_3 as

$$c_2(t) = \frac{1}{i\hbar} \int_0^t (a_2(s) + 2a_3(s)c_1(s))ds, \quad c_3(t) = \frac{1}{i\hbar} \int_0^t a_3(s) \exp(c_2(s))ds.$$

Problem 30.   Consider the quadratic Hamilton operator

$$\hat{H}(t) = f_3(t)\frac{\hat{p}^2}{2m} + \frac{\omega_0}{2} f_2(t)(\hat{q}\hat{p} + \hat{p}\hat{q}) + f_1(t)\frac{m\omega^2}{2}\hat{q}^2$$

where f_j are smooth real functions with $f_1(0) = f_3(0) = 1$ and $f_2(0) = 0$. Express the Hamilton operator with the Bose creation and annihilation operators

$$b = \frac{1}{\sqrt{2m\omega_0\hbar}}(m\omega_0\hat{q} + i\hat{p}), \qquad b^\dagger = \frac{1}{\sqrt{2m\omega_0\hbar}}(m\omega_0\hat{q} - i\hat{p}).$$

Utilize the operators

$$\hat{K}_0 = \frac{1}{2}\left(b^\dagger b + \frac{1}{2}I\right), \quad \hat{K}_+ = \frac{1}{2}b^\dagger b^\dagger, \quad \hat{K}_- = \frac{1}{2}bb.$$

Find the commutators $[\hat{K}_0, \hat{K}_+]$, $\hat{K}_0, \hat{K}_-]$, $[\hat{K}_+, \hat{K}_-]$. Consider the time-dependent Bose annihilation and creation operators

$$b(t) = \mu(t)b + \nu(t)b^\dagger, \qquad b^\dagger(t) = \nu^*(t)b + \mu^*(t)b^\dagger.$$

Find the conditions on $\mu(t)$ and $\nu(t)$ such that $[b(t), b^\dagger(t)] = I$.

Solution 30.   We find

$$\hat{H}(t) = \hbar\omega_0\left(f_1 + f_3)\hat{K}_0 + \left(\frac{f_1 - f_3}{2} + if_2\right)\hat{K}_+ + \left(\frac{f_1 - f_3}{2} - if_2\right)\hat{K}_-\right).$$

We obtain the commutators $[\hat{K}_0, \hat{K}_+] = K_+$, $[\hat{K}_0, \hat{K}_-] = -\hat{K}_-$, $[\hat{K}_+, \hat{K}_-] = -2K_0$. We find the condition

$$|\mu(t)|^2 - |\nu(t)|^2 = 1.$$

4.2 Lie Algebras and Spin Operators

Problem 31. Let σ_1, σ_2, σ_3 be the Pauli spin matrices. Note that

$$\det(\sigma_1) = \det(\sigma_2) = \det(\sigma_3) = -1.$$

Consider the three matrices

$$A_1 = -i\sigma_1 = \begin{pmatrix} 0 & -i \\ -i & 0 \end{pmatrix}, \quad A_2 = -i\sigma_2 = \begin{pmatrix} 0 & -1 \\ 1 & 0 \end{pmatrix}, \quad A_3 = -i\sigma_3 = \begin{pmatrix} -i & 0 \\ 0 & i \end{pmatrix}.$$

(i) Are the three matrices elements of the compact Lie group $SU(2)$? We have to check that the matrices are unitary and the determinant is equal to 1.

(ii) Are the three matrices elements of the Lie algebra $su(2)$? We have to check whether the trace is equal to 0 and that the matrices are skew-hermitian.

(iii) Find the scalar products

$$\langle A_j, A_k \rangle := \mathrm{tr}(A_j A_k^*), \quad j, k = 1, 2, 3 \ j \neq k$$

(iv) Find the commutator $[A_1, A_2]$, $[A_2, A_3]$, $[A_3, A_1]$.

Solution 31. (i) Yes the three matrices are unitary, i.e. $A_j^* = A_j^{-1}$ ($j = 1, 2, 3$) and $\det(A_j) = 1$ for $j = 1, 2, 3$.

(ii) Yes the three matrices have trace equal to 0 and are skew-hermitian. The eigenvalues for all three matrices are $+i$ and $-i$.

(iii) We have $\mathrm{tr}(A_j A_k^*) = 0$ for $j \neq k$. The three matrices A_1, A_2, A_3 form a basis of the simple Lie algebra $su(2)$.

(iv) The commutators are $[A_1, A_2] = 2A_3$, $[A_2, A_3] = 2A_1$, $[A_3, A_1] = 2A_2$. Find the adjoint representation.

Problem 32. Let σ_1, σ_2, σ_3 be the Pauli spin matrices. A basis for the simple Lie algebra $su(2)$ is given by the skew-hermitian and unitary matrices

$$A_1 = -i\sigma_1 = \begin{pmatrix} 0 & -i \\ -i & 0 \end{pmatrix}, \quad A_2 = -i\sigma_2 = \begin{pmatrix} 0 & -1 \\ 1 & 0 \end{pmatrix}, \quad A_3 = -i\sigma_3 = \begin{pmatrix} -i & 0 \\ 0 & i \end{pmatrix}$$

with the commutators $[A_1, A_2] = 2A_3$, $[A_2, A_3] = 2A_1$, $[A_3, A_1] = 2A_2$. Find the 2×2 matrices K_1, K_2, K_3 such that $A_1 = e^{K_1}$, $A_2 = e^{K_2}$, $A_3 = e^{K_3}$. Find the commutators $[K_1, K_2]$, $[K_2, K_3]$, $[K_3, K_1]$.

Solution 32. We apply the *spectral representation* of A_1, A_2, A_3 to find K_1, K_2, K_3. The eigenvalues of A_1 are $\lambda_1 = -i$, $\lambda_2 = +i$ with the corresponding normalized eigenvectors

$$\mathbf{v}_1 = \frac{1}{\sqrt{2}} \begin{pmatrix} 1 \\ 1 \end{pmatrix}, \quad \mathbf{v}_2 = \frac{1}{\sqrt{2}} \begin{pmatrix} 1 \\ -1 \end{pmatrix}.$$

Then

$$A_1 = \lambda_1 \mathbf{v}_1 \mathbf{v}_1^* + \lambda_2 \mathbf{v}_2 \mathbf{v}_2^* = -i\frac{1}{\sqrt{2}} \begin{pmatrix} 1 \\ 1 \end{pmatrix} \frac{1}{\sqrt{2}} (1 \ \ 1) + i\frac{1}{\sqrt{2}} \begin{pmatrix} 1 \\ -1 \end{pmatrix} \frac{1}{\sqrt{2}} (1 \ -1).$$

Since $-i = e^{-i\pi/2}$, $i = e^{i\pi/2}$ we have $\ln(-i) = -i\pi/2$, $\ln(i) = i\pi/2$. Consequently the spectral representation for K_1 is

$$K_1 = -\frac{i\pi}{2} \frac{1}{\sqrt{2}} \begin{pmatrix} 1 \\ 1 \end{pmatrix} \frac{1}{\sqrt{2}} (1 \ \ 1) + \frac{i\pi}{2} \frac{1}{\sqrt{2}} \begin{pmatrix} 1 \\ -1 \end{pmatrix} \frac{1}{\sqrt{2}} (1 \ -1)$$

$$= \frac{\pi}{2} \begin{pmatrix} 0 & -i \\ -i & 0 \end{pmatrix} = \frac{\pi}{2} A_1.$$

Analogously we do the calculation for K_2 and K_3 and obtain

$$K_2 = \frac{\pi}{2} A_2, \qquad K_3 = \frac{\pi}{2} A_3.$$

Thus the commutators are $[K_1, K_2] = \pi K_3$, $[K_2, K_3] = \pi K_1$, $[K_3, K_1] = \pi K_2$.

Problem 33. Let σ_1, σ_2, σ_3 be the Pauli spin matrices. A basis for the simple Lie algebra $su(2)$ is given by the skew-hermitian and unitary matrices

$$A_1 = -i\sigma_1 = \begin{pmatrix} 0 & -i \\ -i & 0 \end{pmatrix}, \quad A_2 = -i\sigma_2 = \begin{pmatrix} 0 & -1 \\ 1 & 0 \end{pmatrix}, \quad A_3 = -i\sigma_3 = \begin{pmatrix} -i & 0 \\ 0 & i \end{pmatrix}$$

with the commutators $[A_1, A_2] = 2A_3$, $[A_2, A_3] = 2A_1$, $[A_3, A_1] = 2A_2$. Find the 2×2 matrices Q_1, Q_2, Q_3 such that $Q_1 = \sqrt{A_1}$, $Q_2 = \sqrt{A_2}$, $Q_3 = \sqrt{A_3}$. Find the commutators $[Q_1, Q_2]$, $[Q_2, Q_3]$, $[Q_3, Q_1]$.

Solution 33. We apply the *spectral representation* to A_1, A_2, A_3 in order to find Q_1, Q_2, Q_3. The eigenvalues of A_1 are $\lambda_1 = -i$, $\lambda_2 = +i$ with the corresponding normalized eigenvectors

$$\mathbf{v}_1 = \frac{1}{\sqrt{2}} \begin{pmatrix} 1 \\ 1 \end{pmatrix}, \qquad \mathbf{v}_2 = \frac{1}{\sqrt{2}} \begin{pmatrix} 1 \\ -1 \end{pmatrix}.$$

Then

$$A_1 = \lambda_1 \mathbf{v}_1 \mathbf{v}_1^* + \lambda_2 \mathbf{v}_2 \mathbf{v}_2^* = -i\frac{1}{\sqrt{2}} \begin{pmatrix} 1 \\ 1 \end{pmatrix} \frac{1}{\sqrt{2}} (1 \ \ 1) + i\frac{1}{\sqrt{2}} \begin{pmatrix} 1 \\ -1 \end{pmatrix} \frac{1}{\sqrt{2}} (1 \ \ -1).$$

Since

$$\sqrt{-i} = e^{-i\pi/4} \equiv \frac{1}{\sqrt{2}}(1 - i), \qquad \sqrt{i} = e^{i\pi/4} \equiv \frac{1}{\sqrt{2}}(1 + i)$$

we find the spectral representation for Q_1 as

$$Q_1 = -\frac{1}{\sqrt{2}}(1 - i)\frac{1}{\sqrt{2}} \begin{pmatrix} 1 \\ 1 \end{pmatrix} \frac{1}{\sqrt{2}} (1 \ \ 1) + \frac{1}{\sqrt{2}}(1 + i)\frac{1}{\sqrt{2}} \begin{pmatrix} 1 \\ -1 \end{pmatrix} \frac{1}{\sqrt{2}} (1 \ \ -1)$$

$$= \frac{1}{\sqrt{2}} \begin{pmatrix} 1 & -i \\ -i & 1 \end{pmatrix} = \frac{1}{\sqrt{2}}(I_2 + A_1).$$

Analogously we do the calculation for Q_2 and Q_3 and obtain

$$Q_2 = \frac{1}{\sqrt{2}} \begin{pmatrix} 1 & -1 \\ 1 & 1 \end{pmatrix} = \frac{1}{\sqrt{2}}(I_2 + A_2), \qquad Q_3 = \frac{1}{\sqrt{2}} \begin{pmatrix} 1 - i & 0 \\ 0 & 1 + i \end{pmatrix} = \frac{1}{\sqrt{2}}(I_2 + A_3).$$

Thus the commutators are $[Q_1, Q_2] = A_3$, $[Q_2, Q_3] = A_1$, $[Q_3, Q_1] = A_2$.

Problem 34. (i) Consider the spin-1 matrices S_1, S_2, S_3

$$S_1 = \frac{1}{\sqrt{2}} \begin{pmatrix} 0 & 1 & 0 \\ 1 & 0 & 1 \\ 0 & 1 & 0 \end{pmatrix}, \quad S_2 = \frac{1}{\sqrt{2}} \begin{pmatrix} 0 & -i & 0 \\ i & 0 & -i \\ 0 & i & 0 \end{pmatrix}, \quad S_3 = \begin{pmatrix} 1 & 0 & 0 \\ 0 & 0 & 0 \\ 0 & 0 & -1 \end{pmatrix}$$

with $S_1^2 + S_2^2 + S_3^2 = 2I_3$. The matrices are hermitian, but non-invertible.
(i) Do the spin-1 matrices form a basis of a Lie algebra?

(ii) Are the matrices

$$A_1 = -iS_1, \qquad A_2 = -iS_2, \qquad A_3 = -iS_3$$

elements of the Lie algebra $su(3)$? Find the commutators.
(iii) Are the matrices $\exp(\alpha A_j)$ $(j = 1, 2, 3)$ with $\alpha \in \mathbb{R}$ unitary? Are the matrices elements of the Lie group $SU(3)$?

Solution 34. (i) We have the commutators $[S_1, S_2] = iS_3$, $[S_2, S_3] = iS_1$, $[S_3, S_1] = iS_2$. Thus we have a basis of a Lie algebra. The Lie algebra is semisimple.
(ii) The three matrices are skew-hermitian and the trace is equal to 0. Thus the matrices are elements of the Lie algebra $su(3)$. The commutators are

$$[A_1, A_2] = A_3, \quad [A_2, A_3] = A_1, \quad [A_3, A_1] = A_2.$$

(iii) Yes the matrices are unitary. For any $n \times n$ matrix M the identity

$$\det(\exp(M)) \equiv \exp(\operatorname{tr}(M))$$

is valid. Thus if $\operatorname{tr}(M) = 0$ one has $\det(\exp(M)) = 1$ and therefore the matrices A_1, A_2, A_3 are elements of the compact Lie group $SU(3)$.

Problem 35. Let σ_1, σ_2, σ_3 be the Pauli spin matrices. Consider the nonnormal matrices

$$A = \begin{pmatrix} 1 & 1 \\ 0 & 1 \end{pmatrix}, \quad B = \frac{1}{2}\begin{pmatrix} 3 & -1 \\ 1 & 1 \end{pmatrix}, \quad C = \frac{1}{2}\begin{pmatrix} 3 & i \\ i & 1 \end{pmatrix}$$

where $\det(A) = \det(B) = \det(C) = 1$, i.e. A, B, C are elements of the Lie group $SL(2, \mathbb{C})$.
(i) Find the commutators $[A, A^*]$, $[B, B^*]$ and $[C, C^*]$.
(ii) Consider the unitary matrices

$$U = \frac{1}{\sqrt{2}}\begin{pmatrix} 1 & 1 \\ 1 & -1 \end{pmatrix}, \quad V = \begin{pmatrix} 1 & 0 \\ 0 & i \end{pmatrix}.$$

Find UAU^* and VBV^*.
(iii) Consider the nonnormal and noninvertible matrices

$$X = \begin{pmatrix} 0 & 1 \\ 0 & 0 \end{pmatrix}, \quad Y = \frac{1}{2}\begin{pmatrix} 1 & -1 \\ 1 & -1 \end{pmatrix}, \quad Z = \frac{1}{2}\begin{pmatrix} 1 & i \\ i & -1 \end{pmatrix}.$$

All have trace zero and thus are elements of the Lie algebra $s\ell(2, \mathbb{C})$. Find

$$[X, X^*], \quad [Y, Y^*], \quad [Z, Z^*].$$

Hint. Obviously we have $X = A - I_2$, $Y = B - I_2$, $Z = C - I_2$.
(iv) Show that $Y = UXU^*$, $Z = VYV^*$.
(v) Study the commutators

$$[X \otimes X, X^* \otimes X^*], \qquad [X \otimes X^*, X^* \otimes X],$$

$$[Y \otimes Y, Y^* \otimes Y^*], \qquad [Y \otimes Y^*, Y^* \otimes Y],$$

$$[Z \otimes Z, Z^* \otimes Z^*], \qquad [Z \otimes Z^*, Z^* \otimes Z]$$

where $\otimes$ denotes the Kronecker product.

Solution 35.   (i) We obtain the Pauli spin matrices

$$[A, A^*] = \sigma_3, \quad [B, B^*] = \sigma_1, \quad [C, C^*] = \sigma_2.$$

(ii) We obtain $B = UAU^*$ and $C = VBV^*$.
(iii) We obtain $[X, X^*] = \sigma_3$, $[Y, Y^*] = \sigma_1$, $[Z, Z^*] = \sigma_2$.
(iv) We obtain $Y = UXU^*$, $Z = VYV^*$.
(v) Since

$$X^* = \begin{pmatrix} 0 & 0 \\ 1 & 0 \end{pmatrix}, \quad Y^* = \frac{1}{2}\begin{pmatrix} 1 & 1 \\ -1 & -1 \end{pmatrix}, \quad Z^* = \frac{1}{2}\begin{pmatrix} 1 & -i \\ -i & -1 \end{pmatrix}$$

we arrive at

$$[X \otimes X, X^* \otimes X^*] = \begin{pmatrix} 1 & 0 & 0 & 0 \\ 0 & 0 & 0 & 0 \\ 0 & 0 & 0 & 0 \\ 0 & 0 & 0 & -1 \end{pmatrix}, \quad [X \otimes X^*, X^* \otimes X] = \begin{pmatrix} 0 & 0 & 0 & 0 \\ 0 & 1 & 0 & 0 \\ 0 & 0 & -1 & 0 \\ 0 & 0 & 0 & 0 \end{pmatrix},$$

$$[Y \otimes Y, Y^* \otimes Y^*] = \frac{1}{2}\begin{pmatrix} 0 & 1 & 1 & 0 \\ 1 & 0 & 0 & 1 \\ 1 & 0 & 0 & 1 \\ 0 & 1 & 1 & 0 \end{pmatrix}, \quad [Y \otimes Y^*, Y^* \otimes Y] = \frac{1}{2}\begin{pmatrix} 0 & -1 & 1 & 0 \\ -1 & 0 & 0 & 1 \\ 1 & 0 & 0 & -1 \\ 0 & 1 & -1 & 0 \end{pmatrix},$$

$$[Z \otimes Z, Z^* \otimes Z^*] = \frac{i}{2}\begin{pmatrix} 0 & -1 & -1 & 0 \\ 1 & 0 & 0 & -1 \\ 1 & 0 & 0 & -1 \\ 0 & 1 & 1 & 0 \end{pmatrix}, \quad [Z \otimes Z^*, Z^* \otimes Z] = \frac{i}{2}\begin{pmatrix} 0 & 1 & -1 & 0 \\ -1 & 0 & 0 & -1 \\ 1 & 0 & 0 & 1 \\ 0 & 1 & -1 & 0 \end{pmatrix}.$$

Problem 36.   Let

$$J = \begin{pmatrix} 1 & 0 \\ 0 & -1 \end{pmatrix} = \sigma_3.$$

The Lie algebra $su(2)$ is defined as the 2×2 matrices X over $\mathbb{C}$

$$su(2) := \{\, X^* = -X \, : \, \mathrm{tr}(X) = 0 \,\}.$$

Let σ_1, σ_2, σ_3 be the Pauli spin matrices with the commutation relation

$$[\sigma_1, \sigma_2] = 2i\sigma_3, \quad [\sigma_2, \sigma_3] = 2i\sigma_1, \quad [\sigma_3, \sigma_1] = 2i\sigma_2.$$

Thus we have

$$su(2) = \mathrm{span}\{\, i\sigma_1, i\sigma_2, i\sigma_3 \,\}.$$

Find the commutators

$$[\sigma_1 \otimes I_2 + I_2 \otimes \sigma_2, \sigma_2 \otimes I_2 + I_2 \otimes \sigma_3]$$
$$[\sigma_2 \otimes I_2 + I_2 \otimes \sigma_3, \sigma_3 \otimes I_2 + I_2 \otimes \sigma_1]$$
$$[\sigma_3 \otimes I_2 + I_2 \otimes \sigma_1, \sigma_2 \otimes I_2 + I_2 \otimes \sigma_3].$$

Discuss.

Solution 36.   Since $\sigma_1\sigma_2 = i\sigma_3$, $\sigma_2\sigma_3 = i\sigma_1$, $\sigma_3\sigma_1 = i\sigma_2$, $\sigma_2\sigma_1 = -i\sigma_3$, $\sigma_3\sigma_2 = -i\sigma_1$, $\sigma_1\sigma_3 = -i\sigma_2$ we obtain

$$[\sigma_1 \otimes I_2 + I_2 \otimes \sigma_2, \sigma_2 \otimes I_2 + I_2 \otimes \sigma_3] = 2i(\sigma_3 \otimes I_2 + I_2 \otimes \sigma_1)$$

$$[\sigma_2 \otimes I_2 + I_2 \otimes \sigma_3, \sigma_3 \otimes I_2 + I_2 \otimes \sigma_1] = 2i(\sigma_1 \otimes I_2 + I_2 \otimes \sigma_2)$$
$$[\sigma_3 \otimes I_2 + I_2 \otimes \sigma_1, \sigma_2 \otimes I_2 + I_2 \otimes \sigma_3] = -2i(\sigma_1 \otimes I_2 + I_2 \otimes \sigma_2).$$

Problem 37. Let $\sigma_1, \sigma_2, \sigma_3$ be the Pauli spin matrices. The matrices

$$A_1 = -i\sigma_1, \quad A_2 = -i\sigma_2, \quad A_3 = -i\sigma_3$$

are elements of the Lie algebra $su(2)$ since $A_j^* = -A_j$ $(j = 1, 2, 3)$ (skew-hermitian) and $\det(\sigma_j) = 1$ for $j = 1, 2, 3$. Are the nine 4×4 matrices

$$A_j \otimes A_k, \quad j, k = 1, 2, 3$$

elements of the Lie algebra $su(4)$?

Solution 37. No. The nine matrices are not skew-hermitian. The nine matrices are unitary and hermitian.

Problem 38. Consider the matrices

$$T_{12} = \sigma_1 \otimes \sigma_1 \otimes I_2, \quad T_{23} = I_2 \otimes \sigma_1 \otimes \sigma_1, \quad T_{31} = \sigma_1 \otimes I_2 \otimes \sigma_1.$$

Find the commutators $[T_{12}, T_{23}]$, $[T_{23}, T_{31}]$, $[T_{31}, T_{12}]$. Discuss.

Solution 38. Note that $\sigma_1^2 = I_2$. For all three cases the commutator vanishes

$$[T_{12}, T_{23}] = 0_8, \quad [T_{23}, T_{31}] = 0_8, \quad [T_{31}, T_{12}] = 0_8.$$

Thus we have a basis of a three-dimensional commutative Lie algebra.

Problem 39. Let $\sigma_1, \sigma_2, \sigma_3$ be the Pauli spin matrices and $\sigma_1 \otimes \sigma_2$, $\sigma_2 \otimes \sigma_3$, $\sigma_3 \otimes \sigma_1$. Find the commutators

$$[\sigma_1 \otimes \sigma_2, \sigma_2 \otimes \sigma_3], \quad [\sigma_2 \otimes \sigma_3, \sigma_3 \otimes \sigma_1], \quad [\sigma_3 \otimes \sigma_1, \sigma_1 \otimes \sigma_2].$$

Discuss.

Solution 39. The three commutators vanish. Thus we have a three dimensional basis of a commutative Lie algebra.

Problem 40. Consider the spin operators S_+, S_-, S_3 with the commutation relations

$$[S_3, S_+] = S_+, \quad [S_3, S_-] = -S_-, \quad [S_+, S_-] = 2S_3.$$

Let $b^\dagger$, b be Bose creation and annihilation operators, respectively. Consider the *Dyson-Maleev transformation*

$$S_+ \mapsto b^\dagger(I - b^\dagger b), \quad S_- \mapsto b, \quad S_3 \mapsto b^\dagger b - \frac{1}{2}I$$

where I is the identity operator. Show that the commutation relations are preserved.

Solution 40. We have

$$[S_+, S_-] \mapsto [b^\dagger(I - b^\dagger b), b] = [b^\dagger, b] - [b^\dagger b^\dagger b, b] = -I + 2b^\dagger b \mapsto 2S_3.$$

Analogously we prove the other commutation relations.

Problem 41.   Consider the spin matrices

$$S_+ = \begin{pmatrix} 0 & 1 \\ 0 & 0 \end{pmatrix}, \quad S_- = \begin{pmatrix} 0 & 0 \\ 1 & 0 \end{pmatrix}, \quad S_3 = \frac{1}{2}\begin{pmatrix} 1 & 0 \\ 0 & -1 \end{pmatrix}$$

with $S_+ = S_1 + iS_2$, $S_- = S_1 - iS_2$ and the commutation relations

$$[S_3, S_\pm] = \pm S_\pm, \quad [S_+, S_-] = 2S_3.$$

Let $b^\dagger$, b be Bose creation and annihilation operators. Consider the map (*Dyson-Maleev transformation*)

$$S_+ \to b^\dagger(I - b^\dagger b), \quad S_- \to b, \quad S_3 \to b^\dagger b - \frac{1}{2}I$$

where I is the identity operator. Show that the unbounded operators

$$b^\dagger(I - b^\dagger b), \quad b, \quad b^\dagger b - \frac{1}{2}I$$

satisfy the same commutation relations as S_+, S_-, S_3.

Solution 41.   We have

$$\begin{aligned}
[b^\dagger(I - b^\dagger b), b] &= [b^\dagger, b] - [b^\dagger b^\dagger b, b] \\
&= -I - b^\dagger b^\dagger + b b^\dagger b^\dagger b = -I - b^\dagger b^\dagger b b + (I + b^\dagger b) b^\dagger b \\
&= -I - b^\dagger b^\dagger b b + b^\dagger b + b^\dagger (I + b^\dagger b) b = -I + 2 b^\dagger b \\
&= 2(b^\dagger b - \frac{1}{2}I).
\end{aligned}$$

Analogously we find $[b^\dagger b - I/2, b] = -b$, $[b^\dagger b - I/2, b(I - b^\dagger b)] = b^\dagger(I - b^\dagger b)$.

Problem 42.   Find all nonzero 2×2 matrices J_+, J_-, J_3 such that

$$[J_3, J_+] = J_+, \quad [J_3, J_-] = -J_-, \quad [J_+, J_-] = 2J_3$$

where $(J_+)^* = J_-$.

Solution 42.   Since $\mathrm{tr}([A, B]) = 0$ for any two $n \times n$ matrices A, B we find that

$$\mathrm{tr}(J_+) = \mathrm{tr}(J_-) = \mathrm{tr}(J_3) = 0.$$

The matrices are given by

$$J_+ = \begin{pmatrix} 0 & 1 \\ 0 & 0 \end{pmatrix}, \quad J_- = \begin{pmatrix} 0 & 0 \\ 1 & 0 \end{pmatrix}, \quad J_3 = \frac{1}{2}\begin{pmatrix} 1 & 0 \\ 0 & -1 \end{pmatrix}.$$

Problem 43.   Consider the Pauli spin matrices σ_0, σ_1, σ_2, σ_3, where $\sigma_0 = I_2$. One has $\sigma_1\sigma_2 = i\sigma_3$, $\sigma_2\sigma_3 = i\sigma_1$, $\sigma_3\sigma_1 = i\sigma_2$. Let $\mu, \nu = 0, 1, 2, 3$. Consider the sixteen 4×4 matrices

$$M_{\mu\nu} = \sigma_\mu \otimes \sigma_\nu$$

where $\otimes$ is the Kronecker product. Are the matrices elements of the Lie group $SU(4)$?

Solution 43. Since the Pauli spin matrices σ_μ are unitary the matrices of the Kronecker product are also unitary and we have $\det(M_{\mu\nu}) = 1$. The 16 matrices are elements of the compact Lie group $SU(4)$.

Problem 44. Consider the Lie group $SL(2, \mathbb{C})$ and the 2×2 matrix

$$J := i\sigma_2 \equiv \begin{pmatrix} 0 & 1 \\ -1 & 0 \end{pmatrix}.$$

Then any 2×2 matrix $A \in SL(2, \mathbb{C})$ satisfies $A^T J A = J$, where A^T is the transpose of A.
(i) Let A be a 2×2 matrix with $A \in SL(2, \mathbb{C})$. Find $(A \otimes A)^T (J \otimes J)(A \otimes A)$.
(ii) Find $(A \oplus A)^T (J \oplus J)(A \oplus A)$.

Solution 44. (i) We obtain $(A \otimes A)^T (J \otimes J)(A \otimes A) = J \otimes J$.
(ii) We obtain $(A \oplus A)^T (J \oplus J)(A \oplus A) = J \oplus J$.

Problem 45. Let σ_1, σ_2, σ_3 be the Pauli spin matrices. Let $\hat{p}$ be the differential operator $\hat{p} = -i\hbar d/dx$. We define the linear operators

$$\hat{Q}_1 := \frac{1}{2}(\sigma_1\hat{p} + \sigma_2 W(x)), \qquad \hat{Q}_2 := \frac{1}{2}(\sigma_2\hat{p} - \sigma_1 W(x))$$

where W is a smooth function, $W : \mathbb{R} \to \mathbb{R}$. Let

$$\hat{H} = \frac{1}{2}\left(\hat{p}^2 + (W(x))^2 + \hbar\sigma_3\frac{dW}{dx}\right)$$

which is the Hamilton operator of the system. Find the anticommutators

$$[\hat{Q}_1, \hat{Q}_1]_+, \quad [\hat{Q}_2, \hat{Q}_2]_+, \quad [\hat{Q}_1, \hat{Q}_2]_+$$

and the commutators $[\hat{H}, \hat{Q}_1]$, $[\hat{H}, \hat{Q}_2]$.

Solution 45. We obtain

$$[\hat{Q}_1, \hat{Q}_1]_+ = [\hat{Q}_2, \hat{Q}_2]_+ = \hat{H}, \quad [\hat{Q}_1, \hat{Q}_2]_+ = 0$$

and

$$[\hat{H}, \hat{Q}_1] = [\hat{H}, \hat{Q}_2] = 0.$$

Problem 46. Let σ_1, σ_2, σ_3 be the Pauli spin matrices. What Lie algebra is generated by the three 4×4 matrices

$$\sigma_1 \otimes \sigma_1, \qquad \sigma_2 \otimes \sigma_2, \qquad \sigma_3 \otimes \sigma_3?$$

Solution 46. The commutators vanish, i.e.

$$[\sigma_1 \otimes \sigma_1, \sigma_2 \otimes \sigma_2] = 0_4, \quad [\sigma_2 \otimes \sigma_2, \sigma_3 \otimes \sigma_3] = 0_4, \quad [\sigma_3 \otimes \sigma_3, \sigma_1 \otimes \sigma_1] = 0_4.$$

Thus we have a commutative Lie algebra.

Problem 47. Let σ_1, σ_2, σ_3 be the Pauli spin matrices with the commutation relation

$$[\sigma_1, \sigma_2] = 2i\sigma_3, \quad [\sigma_2, \sigma_3] = 2i\sigma_1, \quad [\sigma_3, \sigma_1] = 2i\sigma_2.$$

Let b_1, b_2 be Bose annihilation operators. Consider the operators

$$\hat{S}_j := \frac{1}{2} \begin{pmatrix} b_1^\dagger & b_2^\dagger \end{pmatrix} \sigma_j \begin{pmatrix} b_1 \\ b_2 \end{pmatrix}$$

where $j = 1, 2, 3$. Find the commutators $[\hat{S}_1, \hat{S}_2]$, $[\hat{S}_2, \hat{S}_3]$, $[\hat{S}_3, \hat{S}_1]$. Discuss.

Solution 47.   We obtain

$$\hat{S}_1 = \frac{1}{2}(b_1^\dagger b_2 + b_2^\dagger b_1), \quad \hat{S}_2 = -\frac{i}{2}(b_1^\dagger b_2 + b_2^\dagger b_1), \quad \hat{S}_3 = \frac{1}{2}(b_1^\dagger b_1 - b_2^\dagger b_2).$$

Thus $[\hat{S}_1, \hat{S}_2] = i\hat{S}_3$, $[\hat{S}_2, \hat{S}_3] = i\hat{S}_1$, $[\hat{S}_3, \hat{S}_1] = i\hat{S}_2$.

Problem 48.   Let $b^\dagger$, b be Bose creation and annihilation operators with $[b, b^\dagger] = I_B$ and $\sigma_0 = I_2$, σ_1, σ_2, σ_3 be the Pauli spin matrices. The *Rabi model* is given by the Hamilton operator

$$\hat{H} = \hbar\omega b^\dagger b \otimes I_2 + \hbar\Omega I_B \otimes \sigma_3 + g(b^\dagger + b) \otimes \sigma_1.$$

Find the Lie algebra generated by the operators

$$b^\dagger b \otimes I_2, \quad I_B \otimes \sigma_3, \quad (b^\dagger + b) \otimes \sigma_1.$$

Solution 48.   Let $K_1 = b^\dagger b \otimes I_2$, $K_2 = I_b \otimes \sigma_3$ and $K_3 = (b^\dagger + b) \otimes \sigma_1$. We find

$$\begin{aligned}
[K_1, K_2] &= 0 \\
[K_1, K_3] &= (b^\dagger - b) \otimes \sigma_1 := K_4 \\
[K_2, K_3] &= 2i(b^\dagger + b) \otimes \sigma_2 := 2iK_5 \\
[K_1, K_4] &= K_3 \\
[K_1, K_5] &= K_4 \\
[K_2, K_4] &= 2i(b^\dagger - b) \otimes \sigma_2 := 2iK_6 \\
[K_2, K_5] &= -2iK_3 \\
[K_3, K_4] &= 2I_b \otimes \sigma_1 := 2K_7 \\
[K_3, K_5] &= 2i(b^\dagger + b) \otimes \sigma_3 := 2iK_8 \\
[K_4, K_5] &= -2i(b^{\dagger 2} - b^2) \otimes \sigma_1 := -2iK_9.
\end{aligned}$$

Problem 49.   Let X, Y be 2×2 matrices over $\mathbb{C}$. We define a 4×4 matrix via (star product)

$$X \star Y := \begin{pmatrix} x_{11} & 0 & 0 & x_{12} \\ 0 & y_{11} & y_{12} & 0 \\ 0 & y_{21} & y_{22} & 0 \\ x_{21} & 0 & 0 & x_{22} \end{pmatrix}.$$

Consider the basis

$$A_1 = -i\sigma_1 = \begin{pmatrix} 0 & -i \\ -i & 0 \end{pmatrix}, \quad A_2 = -i\sigma_2 = \begin{pmatrix} 0 & -1 \\ 1 & 0 \end{pmatrix}, \quad A_3 = -i\sigma_3 = \begin{pmatrix} -i & 0 \\ 0 & i \end{pmatrix}$$

of the Lie algebra $su(2)$. Do the matrices $A_1 \star A_1$, $A_2 \star A_2$, $A_3 \star A_3$ form a basis of a Lie algebra under the commutator? Discuss.

Solution 49. We obtain the skew-hermitian matrices

$$
A_1 \star A_1 = \begin{pmatrix} 0 & 0 & 0 & -i \\ 0 & 0 & -i & 0 \\ 0 & -i & 0 & 0 \\ -i & 0 & 0 & 0 \end{pmatrix}, \quad
A_2 \star A_2 = \begin{pmatrix} 0 & 0 & 0 & -1 \\ 0 & 0 & -1 & 0 \\ 0 & 1 & 0 & 0 \\ 1 & 0 & 0 & 0 \end{pmatrix},
$$

$$
A_3 \star A_3 = \begin{pmatrix} -i & 0 & 0 & 0 \\ 0 & -i & 0 & 0 \\ 0 & 0 & i & 0 \\ 0 & 0 & 0 & i \end{pmatrix}
$$

with the commutators

$$
[A_1 \star A_1, A_2 \star A_2] = 2A_3 \star A_3, \quad [A_2 \star A_2, A_3 \star A_3] = 2A_1 \star A_1,
$$

$$
[A_3 \star A_3, A_1 \star A_1] = 2A_2 \star A_2.
$$

Consequently we have a representation of the Lie algebra $su(2)$.

Problem 50. Let σ_1, σ_2, σ_3 be the Pauli spin matrices. Let $N \geq 2$ and $j = 1, \dots, N - 1$. Let $\sigma_{1,j}$, $\sigma_{2,j}$, $\sigma_{3,j}$ be the Pauli matrices acting on site j, i.e.

$$
\sigma_{1,j} = I_2 \otimes \cdots \otimes I_2 \otimes \sigma_1 \otimes I_2 \otimes \cdots \otimes I_2
$$

where σ_1 is at j-th position and we have N terms. Thus $\sigma_{1,j}$ is a unitary and hermitian $2^N \times 2^N$ matrix. Analogously we define $\sigma_{2,j}$ and $\sigma_{3,j}$. Let $N \geq 2$ and $q > 0$. A *Temperley-Lieb algebra* $T_N(q)$ is defined by the following relations on the generators E_j, $(j = 1, 2, \dots, N - 1)$

$$
E_j E_j = (q + q^{-1}) E_j
$$
$$
E_j E_{j \pm 1} E_j = E_j
$$
$$
E_j E_k = E_k E_j \quad (k \neq j \pm 1).
$$

The $2^N \times 2^N$ matrices

$$
E_\ell =
$$
$$
-\frac{1}{2} \left(\sigma_{1,\ell} \sigma_{1,\ell+1} + \sigma_{2,\ell} \sigma_{2,\ell+1} + \frac{q + q^{-1}}{2} (\sigma_{3,\ell} \sigma_{3,\ell+1} - I) + \frac{q - q^{-1}}{2} (\sigma_{3,\ell} - \sigma_{3,\ell+1}) \right)
$$

are representations of the Temperley-Lieb algebra, where $\ell = 1, \dots, N - 1$ and I is the $2^N \times 2^N$ identity matrix. Let $N = 3$. Write down the 8×8 matrices for E_1 and E_2 and test the relations.

Solution 50. Since

$$
\sigma_{1,1} \sigma_{1,2} = (\sigma_1 \otimes I_2 \otimes I_2)(I_2 \otimes \sigma_1 \otimes I_2) = \sigma_1 \otimes \sigma_1 \otimes I_2
$$
$$
\sigma_{2,1} \sigma_{2,2} = (\sigma_2 \otimes I_2 \otimes I_2)(I_2 \otimes \sigma_2 \otimes I_2) = \sigma_2 \otimes \sigma_2 \otimes I_2
$$
$$
\sigma_{3,1} \sigma_{3,2} = (\sigma_3 \otimes I_2 \otimes I_2)(I_2 \otimes \sigma_3 \otimes I_2) = \sigma_3 \otimes \sigma_3 \otimes I_2
$$
$$
\sigma_{1,2} \sigma_{1,3} = (I_2 \otimes \sigma_1 \otimes I_2)(I_2 \otimes I_2 \otimes \sigma_1) = I_2 \otimes \sigma_1 \otimes \sigma_1
$$
$$
\sigma_{2,2} \sigma_{2,3} = (I_2 \otimes \sigma_2 \otimes I_2)(I_2 \otimes I_2 \otimes \sigma_2) = I_2 \otimes \sigma_2 \otimes \sigma_2
$$
$$
\sigma_{3,2} \sigma_{3,3} = (I_2 \otimes \sigma_3 \otimes I_2)(I_2 \otimes I_2 \otimes \sigma_3) = I_2 \otimes \sigma_3 \otimes \sigma_3
$$
$$
\sigma_{3,1} = \sigma_3 \otimes I_2 \otimes I_2
$$
$$
\sigma_{3,2} = I_2 \otimes \sigma_3 \otimes I_2
$$
$$
\sigma_{3,3} = I_2 \otimes I_2 \otimes \sigma_3
$$

and

$$E_1 = -\frac{1}{2}(\sigma_{1,1}\sigma_{1,2} + \sigma_{2,1}\sigma_{2,2} + \frac{q+q^{-1}}{2}(\sigma_{3,1}\sigma_{3,2} - I_8) + \frac{q-q^{-1}}{2}(\sigma_{3,1} - \sigma_{3,2}))$$

$$E_2 = -\frac{1}{2}(\sigma_{1,2}\sigma_{1,3} + \sigma_{2,2}\sigma_{2,3} + \frac{q+q^{-1}}{2}(\sigma_{3,2}\sigma_{3,3} - I_8) + \frac{q-q^{-1}}{2}(\sigma_{3,2} - \sigma_{3,3}))$$

we obtain the 8×8 matrices

$$E_1 = \begin{pmatrix} 0 & 0 & 0 & 0 & 0 & 0 & 0 & 0 \\ 0 & 0 & 0 & 0 & 0 & 0 & 0 & 0 \\ 0 & 0 & 1/q & 0 & -1 & 0 & 0 & 0 \\ 0 & 0 & 0 & 1/q & 0 & -1 & 0 & 0 \\ 0 & 0 & -1 & 0 & q & 0 & 0 & 0 \\ 0 & 0 & 0 & -1 & 0 & q & 0 & 0 \\ 0 & 0 & 0 & 0 & 0 & 0 & 0 & 0 \\ 0 & 0 & 0 & 0 & 0 & 0 & 0 & 0 \end{pmatrix}$$

and

$$E_2 = \begin{pmatrix} 0 & 0 & 0 & 0 & 0 & 0 & 0 & 0 \\ 0 & 1/q & -1 & 0 & 0 & 0 & 0 & 0 \\ 0 & -1 & q & 0 & 0 & 0 & 0 & 0 \\ 0 & 0 & 0 & 0 & 0 & 0 & 0 & 0 \\ 0 & 0 & 0 & 0 & 0 & 0 & 0 & 0 \\ 0 & 0 & 0 & 0 & 0 & 1/q & -1 & 0 \\ 0 & 0 & 0 & 0 & 0 & -1 & q & 0 \\ 0 & 0 & 0 & 0 & 0 & 0 & 0 & 0 \end{pmatrix}.$$

To test that $E_1 E_2 E_1 = E_1$ and $E_2 E_1 E_2 = E_2$ we apply the Maxima program

```
/* Temperley.mac */

I2: matrix([1,0],[0,1]);
sig1: matrix([0,1],[1,0]);
sig2: matrix([0,-%i],[%i,0]);
sig3: matrix([1,0],[0,-1]);
T110: kronecker_product(sig1,kronecker_product(sig1,I2));
T220: kronecker_product(sig2,kronecker_product(sig2,I2));
T330: kronecker_product(sig3,kronecker_product(sig3,I2));
T011: kronecker_product(I2,kronecker_product(sig1,sig1));
T022: kronecker_product(I2,kronecker_product(sig2,sig2));
T033: kronecker_product(I2,kronecker_product(sig3,sig3));
T300: kronecker_product(sig3,kronecker_product(I2,I2));
T030: kronecker_product(I2,kronecker_product(sig3,I2));
T003: kronecker_product(I2,kronecker_product(I2,sig3));
I8: kronecker_product(I2,kronecker_product(I2,I2));
E1: -(T110 + T220 + (q+1/q)/2*(T330-I8) + (q-1/q)/2*(T300-T030))/2;
E2: -(T011 + T022 + (q+1/q)/2*(T033-I8) + (q-1/q)/2*(T030-T003))/2;
R1: E1 . E2 . E1 - E1;
R1: expand(R1);
R2: E2 . E1 . E2 - E2;
R2: expand(R2);
```

4.3   Lie Algebras and Fermi Operators

Problem 51.   Let $c^\dagger$, c be Fermi creation and annihilation operators.
(i) Calculate the commutator

$$[c^\dagger c, c^\dagger + c].$$

(ii) Do the operators $c^\dagger c$, $(c^\dagger + c)$ form a basis of a Lie algebra?
(iii) Can the set of operators be extended so that we have a basis of a Lie algebra?
(iv) Give the matrix representation of the operators $c^\dagger c$ and $c^\dagger + c$ with respect to the basis $c^\dagger|0\rangle$, $|0\rangle$. Then calculate the commutator of these matrices.

Solution 51.   (i) Since $c^\dagger c + cc^\dagger = I$, where I is the identity operator we find

$$[c^\dagger c, c^\dagger + c] = c^\dagger - c.$$

(ii) We see that we do not have a basis of a Lie algebra.
(iii) We add the operator $(c^\dagger - c)$ to the set and calculate all the commutators. We obtain

$$[c^\dagger c, c^\dagger - c] = c^\dagger + c, \qquad [c^\dagger + c, c^\dagger - c] = 2(I - 2c^\dagger c).$$

From this result we see that

$$\{ c^\dagger c,\ c^\dagger + c,\ c^\dagger - c,\ I \}$$

is a basis of a four dimensional Lie algebra. Thus we could also use the basis

$$\{ c^\dagger c,\ c^\dagger,\ c,\ I \}$$

by forming linear combinations. Is the Lie algebra semisimple?
(iv) We have

$$c^\dagger c \mapsto \begin{pmatrix} 1 & 0 \\ 0 & 0 \end{pmatrix}, \qquad c^\dagger + c \mapsto \begin{pmatrix} 0 & 1 \\ 1 & 0 \end{pmatrix} = \sigma_3$$

and

$$\begin{pmatrix} 1 & 0 \\ 0 & 0 \end{pmatrix}\begin{pmatrix} 0 & 1 \\ 1 & 0 \end{pmatrix} - \begin{pmatrix} 0 & 1 \\ 1 & 0 \end{pmatrix}\begin{pmatrix} 1 & 0 \\ 0 & 0 \end{pmatrix} = \begin{pmatrix} 0 & 1 \\ -1 & 0 \end{pmatrix} \mapsto c^\dagger - c.$$

Problem 52.   Let $c^\dagger$, c be Fermi creation and annihilation operators, respectively.
(i) Show that the operators

$$\left\{ c^\dagger,\ c,\ c^\dagger c - \frac{1}{2}I \right\}$$

form a basis of a Lie algebra under the commutator.
(ii) Find the adjoint representation.

Solution 52.   (i) We obtain for the commutators

$$[c^\dagger, c] = 2\left(c^\dagger c - \frac{1}{2}I\right), \qquad [c^\dagger c - \frac{1}{2}I, c] = -c, \qquad [c^\dagger c - \frac{1}{2}I, c^\dagger] = c^\dagger.$$

The Lie algebra is semi-simple.
(ii) We set $x_1 = c^\dagger$, $x_2 = c$ and $x_3 = c^\dagger c - I/2$ and $\mathrm{ad}x_j(x_k) := [x_j, x_k]$. Then with $[x_1, x_2] = 2x_3$, $[x_2, x_3] = x_2$, $[x_3, x_1] = x_1$ and

$$\begin{matrix} (\,x_1 & x_2 & x_3\,)\,\mathrm{ad}x_1 = (\,0 & 2x_3 & -x_1\,) \\ (\,x_1 & x_2 & x_3\,)\,\mathrm{ad}x_2 = (-2x_3 & 0 & x_2\,) \\ (\,x_1 & x_2 & x_3\,)\,\mathrm{ad}x_3 = (\,x_1 & -x_2 & 0\,) \end{matrix}$$

we obtain the adjoint representation

$$\text{ad}x_1 = \begin{pmatrix} 0 & 0 & -1 \\ 0 & 0 & 0 \\ 0 & 2 & 0 \end{pmatrix}, \quad \text{ad}x_2 = \begin{pmatrix} 0 & 0 & 0 \\ 0 & 0 & 1 \\ -2 & 0 & 0 \end{pmatrix}, \quad \text{ad}x_3 = \begin{pmatrix} 1 & 0 & 0 \\ 0 & -1 & 0 \\ 0 & 0 & 0 \end{pmatrix}.$$

Problem 53. The matrix representation for the Fermi operators are

$$c^\dagger \mapsto A = \begin{pmatrix} 0 & 1 \\ 0 & 0 \end{pmatrix}, \quad c \mapsto B = \begin{pmatrix} 0 & 0 \\ 1 & 0 \end{pmatrix}.$$

Find the Lie algebra generated by the matrices A and B.

Solution 53. The commutator of A and B yields

$$[A, B] = \begin{pmatrix} 1 & 0 \\ 0 & -1 \end{pmatrix} = C = \sigma_3.$$

Then $[C, A] = 2A$ and $[C, B] = 2B$. The matrices A, B, C are a basis of the Lie algebra $s\ell(2, \mathbb{R})$.

Problem 54. Let $c^\dagger$, c be Fermi creation and annihilation operators with $[c, c^\dagger]_+ = I$, $[c, c]_+ = 0$, $[c^\dagger, c^\dagger]_+ = 0$, where I is the identity operator. Show that the operators

$$\hat{N} = c^\dagger c, \quad c^\dagger + c, \quad c^\dagger - c, \quad I$$

form a basis of a Lie algebra. Classify the Lie algebra.

Solution 54. We obtain the commutators

$$[c^\dagger c, c^\dagger + c] = c^\dagger - c, \quad [c^\dagger c, c^\dagger - c] = c^\dagger + c, \quad [c^\dagger + c, c^\dagger - c] = 2I - 4c^\dagger c.$$

Obviously the identity operator I commutes with all other operators. So the Lie algebra is not semi-simple.

Problem 55. Find the Lie algebra generated by the Fermi operators with spin

$$c_\uparrow^\dagger, \quad c_\downarrow^\dagger, \quad c_\uparrow, \quad c_\downarrow.$$

Solution 55. We find the zero commutators

$$[c_\uparrow^\dagger, c_\uparrow^\dagger] = 0, \quad [c_\downarrow^\dagger, c_\downarrow^\dagger] = 0, \quad [c_\uparrow, c_\uparrow] = 0, \quad [c_\downarrow, c_\downarrow] = 0$$

and the nonzero commutators

$$[c_\downarrow^\dagger, c_\downarrow] = 2c_\downarrow^\dagger - I, \quad [c_\uparrow^\dagger, c_\uparrow] = 2c_\uparrow^\dagger - I, \quad [c_\uparrow^\dagger, c_\downarrow] = 2c_\uparrow^\dagger c_\downarrow, \quad [c_\downarrow, c_\uparrow] = 2c_\downarrow^\dagger c_\uparrow,$$

$$[c_\uparrow^\dagger, c_\downarrow^\dagger] = 2c_\uparrow^\dagger c_\downarrow^\dagger, \quad [c_\uparrow, c_\downarrow] = 2c_\uparrow c_\downarrow$$

$$[c_\uparrow^\dagger c_\downarrow^\dagger, c_\uparrow c_\downarrow] = I - c_\uparrow^\dagger c_\uparrow - c_\downarrow^\dagger c_\downarrow, \quad [c_\uparrow^\dagger c_\downarrow, c_\downarrow^\dagger c_\uparrow] = c_\uparrow^\dagger c_\uparrow - c_\downarrow^\dagger c_\downarrow.$$

Thus we have to add the elements

$$c_\uparrow c_\downarrow, \quad c_\uparrow^\dagger c_\downarrow^\dagger, \quad c_\uparrow^\dagger c_\downarrow, \quad c_\downarrow^\dagger c_\uparrow, \quad c_\uparrow^\dagger c_\uparrow, \quad c_\downarrow^\dagger c_\downarrow, \quad I$$

for a basis of a Lie algebra.

Problem 56. Do the operators $S_+ = c_\uparrow^\dagger c_\downarrow$, $S_- = c_\downarrow^\dagger c_\uparrow$, $S_3 = \frac{1}{2}(c_\uparrow^\dagger c_\uparrow - c_\downarrow^\dagger c_\downarrow)$ form a Lie algebra?

Solution 56. We find for the commutators $[S_+, S_-] = 2S_3$, $[S_3, S_+] = S_+$, $[S_3, S_-] = S_-$. Thus we have a basis of the Lie algebra. The Lie algebra is semi-simple.

Problem 57. Find the Lie algebra generated by the two operators $c_\uparrow^\dagger c_\downarrow^\dagger$, $c_\uparrow c_\downarrow$.

Solution 57. We have

$$[c_\uparrow^\dagger c_\downarrow^\dagger, c_\uparrow c_\downarrow] = I - c_\uparrow^\dagger c_\uparrow - c_\downarrow^\dagger c_\downarrow, \quad [c_\uparrow^\dagger c_\downarrow^\dagger, c_\uparrow^\dagger c_\uparrow] = -c_\uparrow^\dagger c_\downarrow^\dagger, \quad [c_\uparrow^\dagger c_\downarrow^\dagger, c_\downarrow^\dagger c_\downarrow] = -c_\uparrow^\dagger c_\downarrow^\dagger.$$

Thus we have a three dimensional Lie algebra with the basis

$$c_\uparrow^\dagger c_\downarrow^\dagger, \qquad c_\uparrow c_\downarrow, \qquad I - c_\uparrow^\dagger c_\uparrow - c_\downarrow^\dagger c_\downarrow.$$

The Lie algebra is semisimple. Since I is the identity operator we can also consider the four dimensional Lie algebra with the basis

$$c_\uparrow^\dagger c_\downarrow^\dagger, \qquad c_\uparrow c_\downarrow, \quad I, \quad c_\uparrow^\dagger c_\uparrow + c_\downarrow^\dagger c_\downarrow.$$

Problem 58. Consider the trace-less 2×2 matrices

$$X_+ = \begin{pmatrix} 0 & 1 \\ 0 & 0 \end{pmatrix}, \quad X_- = \begin{pmatrix} 0 & 0 \\ 1 & 0 \end{pmatrix}, \quad H = \begin{pmatrix} 1 & 0 \\ 0 & -1 \end{pmatrix}.$$

(i) Calculate the commutators and thus show that we have a basis of a Lie algebra.
(ii) Let $c_\uparrow^\dagger$ and $c_\downarrow^\dagger$ Fermi operators with spin up and down down, respectively. Find the operators

$$\hat{A}_+ = \begin{pmatrix} c_\uparrow^\dagger & c_\downarrow^\dagger \end{pmatrix} X_+ \begin{pmatrix} c_\uparrow \\ c_\downarrow \end{pmatrix}, \quad \hat{A}_- = \begin{pmatrix} c_\uparrow^\dagger & c_\downarrow^\dagger \end{pmatrix} X_- \begin{pmatrix} c_\uparrow \\ c_\downarrow \end{pmatrix}, \quad \hat{H} = \begin{pmatrix} c_\uparrow^\dagger & c_\downarrow^\dagger \end{pmatrix} H \begin{pmatrix} c_\uparrow \\ c_\downarrow \end{pmatrix}.$$

Find the commutators of these operators and discuss.

Solution 58. (i) We find

$$[H, X_+] = 2X_+, \quad [H, X_-] = -2X_-, \quad [X_+, X_-] = H.$$

(ii) We obtain

$$\hat{A}_+ = c_\uparrow^\dagger c_\downarrow, \quad \hat{A}_- = c_\downarrow^\dagger c_\uparrow, \quad \hat{H} = c_\uparrow^\dagger c_\uparrow - c_\downarrow^\dagger c_\downarrow$$

and the commutators are given by

$$[A_+, A_-] = c_\uparrow^\dagger c_\uparrow - c_\downarrow^\dagger c_\downarrow, = \hat{H}, \quad [\hat{H}, A_+] = 2c_\uparrow^\dagger c_\downarrow = 2A_+, \quad [\hat{H}, A_-] = -2c_\downarrow^\dagger c_\uparrow = -2A_-.$$

Problem 59.   Consider the trace-less 2×2 matrices

$$X_+ = \begin{pmatrix} 0 & 1 \\ 0 & 0 \end{pmatrix}, \quad X_- = \begin{pmatrix} 0 & 0 \\ 1 & 0 \end{pmatrix}, \quad H = \begin{pmatrix} 1 & 0 \\ 0 & -1 \end{pmatrix}.$$

Find the operators

$$\hat{B}_+ = \begin{pmatrix} c_\uparrow^\dagger & c_\downarrow \end{pmatrix} X_+ \begin{pmatrix} c_\uparrow \\ c_\downarrow^\dagger \end{pmatrix}, \quad \hat{B}_- = \begin{pmatrix} c_\uparrow^\dagger & c_\downarrow \end{pmatrix} X_- \begin{pmatrix} c_\uparrow \\ c_\downarrow^\dagger \end{pmatrix}, \quad \hat{K} = \begin{pmatrix} c_\uparrow^\dagger & c_\downarrow \end{pmatrix} H \begin{pmatrix} c_\uparrow \\ c_\downarrow^\dagger \end{pmatrix}.$$

Find the commutators of these operators and discuss.

Solution 59.   We obtain $\hat{B}_+ = c_\uparrow^\dagger c_\downarrow^\dagger$, $\hat{B}_- = c_\downarrow c_\uparrow$, $\hat{K} = c_\uparrow^\dagger c_\uparrow + c_\downarrow^\dagger c_\downarrow - I$. The commutators are given by

$$[B_+, B_-] = c_\uparrow^\dagger c_\uparrow + c_\downarrow^\dagger c_\downarrow = \hat{K} + I, \quad [\hat{K}, B_+] = 2c_\uparrow^\dagger c_\downarrow^\dagger = 2B_+, \quad [\hat{K}, B_-] = 2c_\uparrow c_\downarrow = -2B_-.$$

Problem 60.   (i) Do the operators

$$\hat{N} = c_\uparrow^\dagger c_\uparrow + c_\downarrow^\dagger c_\downarrow, \quad \hat{K} = c_\uparrow^\dagger c_\downarrow + c_\downarrow^\dagger c_\uparrow$$

form a basis of a Lie algebra?
(ii) Do the operators

$$\hat{N} = c_{1\uparrow}^\dagger c_{1\uparrow} + c_{2\uparrow}^\dagger c_{2\uparrow} + c_{1\downarrow}^\dagger c_{1\downarrow} + c_{2\downarrow}^\dagger c_{2\downarrow}, \quad \hat{S}_3 = \frac{1}{2}(c_{1\uparrow}^\dagger c_{1\uparrow} + c_{2\uparrow}^\dagger c_{2\uparrow} - c_{1\downarrow}^\dagger c_{1\downarrow} - c_{2\downarrow}^\dagger c_{2\downarrow})$$

and

$$\hat{K} = c_{1\uparrow}^\dagger c_{2\uparrow} + c_{2\uparrow}^\dagger c_{1\uparrow} + c_{1\downarrow}^\dagger c_{2\downarrow} + c_{2\downarrow}^\dagger c_{1\downarrow}$$

form a basis of a Lie algebra?

Solution 60.   (i) We have $[\hat{N}, \hat{K}] = 0$. So we have a basis of a two-dimensional abelian (commutative) Lie algebra.
(ii) We have $[\hat{N}, \hat{K}] = 0$, $[\hat{N}, \hat{S}_3] = 0$, $[\hat{K}, \hat{S}_3] = 0$. So we have a basis of a three-dimensional abelian (commutative) Lie algebra.

Problem 61.   Find the commutator of the operators

$$\hat{X} := c_1^\dagger c_2 + c_2^\dagger c_1, \quad \hat{Y} := c_1^\dagger c_1 c_2^\dagger c_2.$$

Then study which Lie algebra is generated by the two operators A and B.

Solution 61.   Note that both $\hat{X}$ and $\hat{Y}$ commute with the number operator $\hat{N} = c_1^\dagger c_1 + c_2^\dagger c_2$. We have $[\hat{X}, \hat{Y}] = 0$. Thus we have a basis of a two-dimensional abelian (commutative) Lie algebra.

Problem 62.   Let $n \geq 2$ and $j = 0, 1, \ldots, n-1$. Let $c_j^\dagger$, c_j be Fermi creation and annihilation operators, respectively. Consider the set of operators

$$\{ c_0^\dagger c_1, \ c_1^\dagger c_2, \ \ldots, c_{n-2}^\dagger c_{n-1}, \ c_{n-1}^\dagger c_0 \}.$$

Does this set form a Lie algebra under the commutator? If not can it be extended to form a Lie algebra.

Solution 62. Using $c_j^\dagger c_k = \delta_{jk} I - c_k c_j^\dagger$ we have $[c_0^\dagger c_1, c_1^\dagger c_2] = c_0^\dagger c_2$. Thus the set does not form a Lie algebra under the commutator. The extension to

$$\{\, c_j^\dagger c_k \,:\, j, k = 0, 1, \ldots, n-1 \,\}$$

provides a Lie algebra.

Problem 63. (i) Consider the Fermi creation and annihilation operators $c_j^\dagger$, $c_\ell^\dagger$, c_k, c_m, where $j, \ell, k, m = 1, 2, \ldots, N$.
(i) Find the commutator

$$[c_j^\dagger c_k, c_\ell^\dagger c_m].$$

Does the set $\{\, c_j^\dagger c_k \,\}$ $(j, k = 1, 2, \ldots, N$ form a Lie algebra under the commutator?
(ii) A basis of a N^2 dimensional representation of the Lie algebra $g\ell(n, \mathbb{C})$ is given by

$$\{\, c_j^\dagger c_k \,:\, j, k = 1, \ldots, N \,\}$$

with the commutation relation $[c_j^\dagger c_k, c_\ell^\dagger c_m] = \delta_{\ell k} c_j^\dagger c_\ell - \delta_{jm} c_\ell^\dagger c_k$. Is the Lie algebra semi-simple?

Solution 63. (i) We have

$$[c_j^\dagger c_k, c_\ell^\dagger c_m] = -\delta_{mj} c_\ell^\dagger c_k + \delta_{\ell k} c_j^\dagger c_m.$$

Thus the set $\{\, c_j^\dagger c_k \,\}$ $(j, k = 1, 2, \ldots, N)$ forms a Lie algebra under the commutator.
(ii) The Lie algebra is not semi-simple but it is reductive, i.e.

$$g\ell(n, \mathbb{C}) = s\ell(n, \mathbb{C}) \oplus \mathbb{C} I_1$$

where $s\ell(n, \mathbb{C})$ is a semi-simple Lie algebra and

$$I_1 = \sum_{j=1}^{n} c_j^\dagger c_j.$$

This means I_1 (the number operator) commutes with all elements of the Lie algebra.

Problem 64. Let I be the identity operator. Do the operators

$$c_1^\dagger c_2^\dagger, \quad c_1 c_2, \quad c_1^\dagger c_1, \quad c_2^\dagger c_2, \quad I$$

form a basis of a Lie algebra under the commutator?

Solution 64. Yes. We have

$$[c_1^\dagger c_2^\dagger, c_1 c_2] = I - c_1^\dagger c_2^\dagger, \quad [c_1^\dagger c_2^\dagger, c_1^\dagger c_1] = 0, \quad [c_1^\dagger c_2^\dagger, c_2^\dagger c_2] = 0, \quad [c_1^\dagger c_2^\dagger, I] = 0$$

$$[c_1 c_2, c_1^\dagger c_1] = c_1 c_2, \quad [c_1 c_2, c_2^\dagger c_2] = c_1 c_2, \quad [c_1 c_2, I] = 0$$

$$[c_1^\dagger c_1, c_2^\dagger c_2] = 0, \quad [c_1^\dagger c_1, I] = 0, \quad [c_2^\dagger c_2, I] = 0.$$

Problem 65. Consider the operators

$$\hat{S}_+ = \sum_{k \in I} c^\dagger_{k\uparrow} c_{k\downarrow}, \quad \hat{S}_- = \sum_{k \in I} c^\dagger_{k\downarrow} c_{k\uparrow}, \quad \hat{S}_3 = \frac{1}{2} \sum_{k \in I} \left(c^\dagger_{k\uparrow} c_{k\uparrow} - c^\dagger_{k\downarrow} c_{k\downarrow} \right)$$

where $c_{k\uparrow}$, $c_{k\downarrow}$ are Fermi annihilation operators with spin-up and spin-down and wave vector k, respectively. The index set I is given by (one-dimensional chain of length N with periodic boundary conditions counting from 0)

$$I = \{ k = 2n\pi/(Na) : n = 0, 1, \ldots, N - 1 \}.$$

Show that these operators form a basis of a Lie algebra under the commutator.

Solution 65. We have $[S_+, S_-] = 2S_3$, $[S_+, S_3] = -S_+$, $[S_-, S_3] = S_-$. The Lie algebra is semi-simple.

Problem 66. Let $c^\dagger_1$, $c^\dagger_2$, $c^\dagger_3$, c_1, c_2, c_3 be Fermi creation and annihilation operators. Do the six operators

$$c^\dagger_1 c_2, \quad c^\dagger_2 c_1, \quad c^\dagger_2 c_3, \quad c^\dagger_3 c_2, \quad c^\dagger_3 c_1, \quad c^\dagger_1 c_3$$

form a basis of a Lie algebra under the commutator?

Solution 66. No. We have to add the operators

$$c^\dagger_1 c_1 - c^\dagger_2 c_2, \quad c^\dagger_2 c_2 - c^\dagger_3 c_3, \quad c^\dagger_3 c_3 - c^\dagger_1 c_1.$$

All the elements of these Lie algebra commute with the number operator

$$\hat{N} = c^\dagger_1 c_1 + c^\dagger_2 c_2 + c^\dagger_3 c_3.$$

Problem 67. Let σ_1, σ_2, σ_3 be the Pauli spin matrices. Let c_1, c_2 be Fermi annihilation operators. Consider the operators

$$\hat{S}_j := \begin{pmatrix} c^\dagger_1 & c^\dagger_2 \end{pmatrix} \sigma_j \begin{pmatrix} c_1 \\ c_2 \end{pmatrix}$$

where $j = 1, 2, 3$.
(i) Find the commutators $[\hat{S}_1, \hat{S}_2]$, $[\hat{S}_2, \hat{S}_3]$, $[\hat{S}_3, \hat{S}_1]$. Discuss.
(ii) Find the anticommutators $[\hat{S}_1, \hat{S}_2]_+$, $[\hat{S}_2, \hat{S}_3]_+$, $[\hat{S}_3, \hat{S}_1]_+$. Discuss.

Solution 67. (i) For the commutators we find $[S_1, S_2] = iS_3$, $[S_2, S_3] = iS_1$, $[S_3, S_1] = iS_2$.
(ii) The anticommutators vanish, i.e. $[S_1, S_2]_+ = 0$, $[S_2, S_3] = 0$, $[S_3, S_1] = 0$.

Problem 68. Let $c^\dagger_1$, $c^\dagger_2$, $c^\dagger_3$, c_1, c_2, c_3 be Fermi creation and annihilation operators. Find the Lie algebra generated by the operators

$$\{ c^\dagger_3 c^\dagger_2 c^\dagger_1, \ c_1 c_2 c_3 \}$$

under the commutator.

Solution 68. We have

$$[c^\dagger_3 c^\dagger_2 c^\dagger_1, c_1 c_2 c_3] = 2c^\dagger_3 c^\dagger_2 c^\dagger_1 c_1 c_2 c_3 - c^\dagger_2 c^\dagger_1 - c^\dagger_3 c^\dagger_2 c_2 c_3 - c^\dagger_3 c^\dagger_1 c_1 c_3 + c^\dagger_1 c_1 + c^\dagger_2 c_2 + c^\dagger_3 c_3 - I.$$

Next we have

$$[c_1 c_2 c_3, [c_3^\dagger c_2^\dagger c_1^\dagger, c_1 c_2 c_3]] = 2c_1 c_2 c_3, \qquad [c_3^\dagger c_2^\dagger c_1^\dagger, [c_3^\dagger c_2^\dagger c_1^\dagger]] = -2c_3^\dagger c_2^\dagger c_1^\dagger.$$

Thus the three operators

$$c_1 c_2 c_3, \quad c_3^\dagger c_2^\dagger c_1^\dagger, \quad [c_3^\dagger c_2^\dagger c_1^\dagger, c_1 c_2 c_3]$$

form a basis of a Lie algebra that is isomorphic to the simple Lie algebra $s\ell(2, \mathbb{R})$.

Problem 69. Let $1 \leq j < k \leq n$. Show that the operators $L_{jk} = i(c_j^\dagger c_k - c_k^\dagger c_j)$ form a basis of the Lie algebra $so(n)$.

Solution 69. We have

$$[L_{j_1 k_1}, L_{j_2 k_2}] =$$
$$(c_{k_2}^\dagger c_{j_1} - c_{j_1}^\dagger c_{k_2})\delta_{k_1 j_2} + (c_{j_2}^\dagger c_{k_1} - c_{k_1}^\dagger c_{j_2})\delta_{j_1 k_2} + (c_{j_1}^\dagger c_{j_2} - c_{j_2}^\dagger c_{j_1})\delta_{k_1 k_2} + (c_{k_1}^\dagger c_{k_2} - c_{k_2}^\dagger c_{k_1})\delta_{j_1 j_2}.$$

Problem 70. *Parafermion* c, $c^\dagger$ are defined by the trilinear relations

$$c^3 = 0, \quad cc^\dagger c = 2c, \quad c^2 c^\dagger + c^\dagger c^2 = 2c.$$

Is a matrix representation given by

$$c^\dagger \mapsto \sqrt{2} \begin{pmatrix} 0 & 1 & 0 \\ 0 & 0 & 1 \\ 0 & 0 & 0 \end{pmatrix} \Rightarrow c \mapsto \sqrt{2} \begin{pmatrix} 0 & 0 & 0 \\ 1 & 0 & 0 \\ 0 & 1 & 0 \end{pmatrix} ?$$

Solution 70. Yes a matrix representation is given by

$$c^\dagger \mapsto \sqrt{2} \begin{pmatrix} 0 & 1 & 0 \\ 0 & 0 & 1 \\ 0 & 0 & 0 \end{pmatrix} \Rightarrow c \mapsto \sqrt{2} \begin{pmatrix} 0 & 0 & 0 \\ 1 & 0 & 0 \\ 0 & 1 & 0 \end{pmatrix}.$$

First we note that

$$\begin{pmatrix} 0 & 0 & 0 \\ 1 & 0 & 0 \\ 0 & 1 & 0 \end{pmatrix}^3 = \begin{pmatrix} 0 & 0 & 0 \\ 0 & 0 & 0 \\ 0 & 0 & 0 \end{pmatrix}.$$

For the commutator $[c^\dagger, c]$ and anticommutator $[c^\dagger, c]_+$ we have

$$[c^\dagger, c] \mapsto 2 \begin{pmatrix} 1 & 0 & 0 \\ 0 & 0 & 0 \\ 0 & 0 & -1 \end{pmatrix}, \quad [c^\dagger, c]_+ \mapsto 2 \begin{pmatrix} 1 & 0 & 0 \\ 0 & 2 & 0 \\ 0 & 0 & 1 \end{pmatrix}.$$

4.4   Lie Superalgebra

Problem 71.   The Lie superalgebra $osp(1|2)$ has five generators K_0, K_+, K_-, F_+, F_-. The commutation relations are given by

$$[K_0, K_\pm] = \pm K_\pm, \qquad [K_+, K_-] = -2K_0.$$

$$[K_0, F_\pm] = \pm \frac{1}{2} F_\pm, \quad [K_\pm, F_\pm] = 0, \quad [K_\pm, F_\mp] = \mp F_\pm.$$

The anticommutation relations are given by

$$[F_\pm, F_\pm]_+ = K_\pm, \quad [F_+, F_-]_+ = K_0.$$

Thus the Lie superalgebra contains the sub Lie algebra $su(1,1)$ spanned by K_0, K_+, K_-.
(i) Find the Casimir operator of the Lie superalgebra.
(ii) Find a faithful representation with Bose operators $b^\dagger$, b and the identity operator I.
(iii) Introduce the number states (Fock states) $|n\rangle$ $(n = 0, 1, \ldots,)$ and apply the operators from (ii) to the number states.
(iv) Let $\beta, \gamma \in \mathbb{C}$ and $\gamma = se^{i\theta}$. Introduce the operators

$$D(\beta) := \exp(\beta F_+ - \beta^* F_-), \qquad S(\gamma) := \exp(\gamma K_+ - \gamma^* K_-).$$

Calculate

$$D^\dagger(\beta) \left(\frac{F_-}{F_+} \right) D(\beta), \qquad S^\dagger(\gamma) \left(\frac{F_-}{F_+} \right) S(\gamma),$$

$$D^\dagger(\beta) \left(\frac{K_-}{K_+} \right) D(\beta), \qquad S^\dagger(\gamma) \left(\frac{K_-}{K_+} \right) S(\gamma).$$

(v) Consider the state

$$|\beta\gamma\rangle = S(\gamma)D(\beta)|0\rangle.$$

Calculate $|\beta\gamma\rangle$ using the number states $|n\rangle$ and the identity

$$\sum_{n=0}^{\infty} \frac{(t/2)^n}{n!} H_n(x) H_n(y) \equiv \frac{1}{\sqrt{1-t^2}} \exp((1-t)^{-1}(2xyt - (x^2 + y^2)t))$$

where H_n $(n = 0, 1, \ldots)$ are the Hermite polynomials.

Solution 71.   (i) We obtain

$$C_2 = K_0^2 - \frac{1}{2}(K_+ K_- + K_- K_+) + \frac{1}{2}(F_+ F_- - F_- F_+).$$

(ii) Using the $[b, b^\dagger] = I$ we have

$$K_+ = \frac{1}{2}(b^\dagger)^2, \qquad K_- = \frac{1}{2}b^2, \qquad K_0 = \frac{1}{2}\left(b^\dagger b + \frac{1}{2}I\right), \qquad F_+ = \frac{1}{2}b^\dagger, \qquad F_- = \frac{1}{2}b.$$

(iii) Applying the operators to the number state yields

$$K_+|n\rangle = \frac{1}{2}\sqrt{(n+1)(n+2)}|n+2\rangle,$$

$$K_-|n\rangle = \frac{1}{2}\sqrt{n(n-1)}|n-2\rangle,$$

$$K_0|n\rangle = \frac{1}{2}\left(n + \frac{1}{2}\right)|n\rangle,$$

$$F_+|n\rangle = \frac{1}{2}\sqrt{n+1}|n+1\rangle,$$

$$F_-|n\rangle = \frac{1}{2}\sqrt{n}|n-1\rangle.$$

(iv) Using the Baker-Campbell-Hausdorff formula yields

$$D^\dagger(\beta)\begin{pmatrix} F_- \\ F_+ \end{pmatrix} D(\beta) = \begin{pmatrix} F_- + \frac{1}{4}\beta I \\ F_+ + \frac{1}{4}\beta^* I \end{pmatrix}$$

$$S^\dagger(\gamma)\begin{pmatrix} F_- \\ F_+ \end{pmatrix} S(\gamma) = \begin{pmatrix} \cosh(r) & e^{i\theta}\sinh(r) \\ e^{-i\theta}\sinh(r) & \cosh(r) \end{pmatrix}\begin{pmatrix} F_- \\ F_+ \end{pmatrix}$$

$$D^\dagger(\beta)\begin{pmatrix} K_- \\ K_+ \end{pmatrix} D(\beta) = \begin{pmatrix} K_- + \beta F_- + \frac{1}{8}\beta^2 I \\ K_+ + \beta^* F_+ + \frac{1}{8}(\beta^*)^2 I \end{pmatrix}$$

$$S^\dagger(\gamma)\begin{pmatrix} K_- \\ K_+ \end{pmatrix} S(\gamma) = \begin{pmatrix} \cosh^2(s) & e^{2i\theta}\sinh^2(s) \\ e^{-2i\theta}\sinh^2(s) & \cosh^2(s) \end{pmatrix}\begin{pmatrix} K_- \\ K_+ \end{pmatrix} + \cosh(s)\sinh(s)\begin{pmatrix} e^{i\theta}I \\ e^{-i\theta}I \end{pmatrix}.$$

(v) We obtain

$$|\beta\gamma\rangle = \sum_{n=0}^{\infty}(n!\cosh(s))^{-1/2}\left(\frac{1}{2}e^{i\theta}\tanh(s)\right)^{n/2}$$

$$\times \exp\left(-\frac{1}{8}(|\beta|^2 - \beta^2 e^{i\theta}\tanh(s))\right)H_n\left(\frac{\beta}{2}(e^{i\theta}\sinh(s)^{-1/2})\right)|n\rangle.$$

Problem 72. Let $b^\dagger$, b be Bose creation and annihilation operator and $c^\dagger$ c be Fermi creation and annihilation operators. Let I_B be the identity operator for the Bose system and I_F be the identity operator for the Fermi system. Consider the operators

$$\hat{N} := b^\dagger b \otimes I_F + I_B \otimes c^\dagger c, \qquad \hat{M} := \frac{1}{2}(b^\dagger b \otimes I_F - I_B \otimes c^\dagger c + I_B \otimes I_F),$$

and $\hat{Q} := b \otimes c^\dagger$. Thus $\hat{Q}^\dagger = b^\dagger \otimes c$.
(i) Find the commutators $[\hat{M}, \hat{Q}]$, $[\hat{M}, \hat{Q}]$, $[\hat{M}, \hat{N}]$, $[\hat{Q}, \hat{N}]$, $[\hat{Q}^\dagger, \hat{N}]$.
(ii) Find the anticommutators $[\hat{Q}^\dagger, \hat{Q}]_+$, $[\hat{Q}^\dagger, \hat{Q}^\dagger]_+$, $[\hat{Q}, \hat{Q}]_+$.

Solution 72. (i) We obtain

$$[\hat{M}, \hat{Q}] = -\hat{Q}, \quad [\hat{M}, \hat{Q}^\dagger] = \hat{Q}^\dagger, \quad [\hat{M}, \hat{N}] = [\hat{Q}, \hat{N}] = [\hat{Q}^\dagger, \hat{N}] = 0_B \otimes 0_F.$$

(ii) We obtain for the anticommutators $[\hat{Q}^\dagger, \hat{Q}]_+ = \hat{N}$, $[\hat{Q}^\dagger, \hat{Q}^\dagger]_+ = [\hat{Q}, \hat{Q}]_+ = 0_B \otimes 0_F$.

Problem 73. In supersymmetric quantum mechanics the realization of the algebra

$$\hat{Q}^2 = (\hat{Q}^*)^2 = 0, \quad [\hat{Q}, \hat{Q}^*]_+ = \hat{H}, \quad [\hat{H}, \hat{Q}] = [\hat{H}, \hat{Q}^*] = 0$$

can be given in terms of linear differential operators. Let σ_1, σ_2 be the Pauli spin matrices, $\otimes$ the tensor product, $\hat{P} := -i d/dx$ and $V : \mathbb{R} \to \mathbb{R}$ an analytic function (potential). Let

$$\hat{Q} := \frac{1}{2^{3/2}}(\hat{P} - iV(x)) \otimes (\sigma_1 + i\sigma_2).$$

Find $\hat{H}$ (i.e. the Hamilton operator).

Solution 73. We obtain

$$\hat{H} = [\hat{Q}, \hat{Q}^*]_+ = \frac{1}{2}(\hat{P}^2 + V^2(x)) \otimes I_2 + \frac{1}{2}\left(\frac{dV}{dx} \otimes \sigma_3\right)$$

where $\hat{P}^2 = -d^2/dx^2$.

Problem 74. Let σ_1, σ_2, σ_3 be the Pauli spin matrices. Let $\oplus$ denote the direct sum. Consider a representation of the superalgebra $su(2|1)$. It generators are given by the 3×3 matrices

$$L_1 = \sigma_1 \oplus (0) \equiv \begin{pmatrix} 0 & 1 & 0 \\ 1 & 0 & 0 \\ 0 & 0 & 0 \end{pmatrix}, \quad L_2 = \sigma_2 \oplus (0) \equiv \begin{pmatrix} 0 & -i & 0 \\ i & 0 & 0 \\ 0 & 0 & 0 \end{pmatrix},$$

$$L_3 = \sigma_3 \oplus (0) = \begin{pmatrix} 1 & 0 & 0 \\ 0 & -1 & 0 \\ 0 & 0 & 0 \end{pmatrix}, \quad L_4 = \frac{1}{2}\begin{pmatrix} 1 & 0 & 0 \\ 0 & 1 & 0 \\ 0 & 0 & 2 \end{pmatrix}$$

$$V_1 = \frac{1}{2}\begin{pmatrix} 0 & 0 & 1 \\ 0 & 0 & 0 \\ 1 & 0 & 0 \end{pmatrix}, \quad V_2 = \frac{1}{2}\begin{pmatrix} 0 & 0 & -i \\ 0 & 0 & 0 \\ i & 0 & 0 \end{pmatrix},$$

$$W_1 = \frac{1}{2}\begin{pmatrix} 0 & 0 & 0 \\ 0 & 0 & 1 \\ 0 & 1 & 0 \end{pmatrix}, \quad W_2 = \frac{1}{2}\begin{pmatrix} 0 & 0 & 0 \\ 0 & 0 & -i \\ 0 & i & 0 \end{pmatrix}.$$

Find the commutators and thus show that L_1, L_2, L_3, L_4 are the generators for the Lie subalgebra $su(2) \otimes u(1)$ and V_1, V_2, W_1, W_2 are the supergenerators.
(ii) Let

$$L_\pm = \frac{1}{2}(L_1 \pm iL_2), \quad L_0 = L_3 + L_4, \quad V_\pm = V_1 \pm iV_2, \quad W_\pm = W_1 \pm W_2.$$

Here L_1, L_2, L_3, L_4 are the generators forming the Lie subalgebra $su(2) \otimes u(1)$ of $su(2|1)$ and V_1, V_2, W_1, W_2 are the supergenerators. Find the commutators.
(iii) Let $c^\dagger$, c be Fermi creation and annihilation operators. Let $b_1^\dagger$, $b_2^\dagger$ be Bose creation operators. Let I_B be the identity operator for the Bose operators and I_F be the identity operator for the Fermi operators. A Bose-Fermion realization is given by

$$L_+ = b_1^\dagger b_2 \otimes I_F, \quad L_- = b_2^\dagger b_1 \otimes I_F$$

$$L_3 = \frac{1}{2}(b_1^\dagger b_1 - b_2^\dagger b_2) \otimes I_F, \quad L_0 = b_1^\dagger b_1 \otimes I_F + I_B \otimes c^\dagger c$$

$$V_+ = b_1^\dagger \otimes c, \quad V_- = b_1 \otimes c^\dagger, \quad W_+ = b_2^\dagger \otimes c, \quad W_- = b_2 \otimes c^\dagger.$$

Show that the representations are isomorphic.

Solution 74. (i) We have

$$[L_1, L_2] = 2iL_3, \quad [L_2, L_3] = 2iL_1, \quad [L_3, L_1] = 2iL_2$$

and $[L_4, L_1] = [L_4, L_2] = [L_4, L_3] = 0_3$. Furthermore

$$[V_1, V_2] = \frac{1}{2}\begin{pmatrix} i & 0 & 0 \\ 0 & 0 & 0 \\ 0 & 0 & -i \end{pmatrix}, \quad [W_1, W_2] = \frac{1}{2}\begin{pmatrix} 0 & 0 & 0 \\ 0 & i & 0 \\ 0 & 0 & -i \end{pmatrix}, \quad [V_1, W_1] = \frac{1}{4}\begin{pmatrix} 0 & 1 & 0 \\ -1 & 0 & 0 \\ 0 & 0 & 0 \end{pmatrix},$$

$$[V_1, W_2] = \frac{1}{4}\begin{pmatrix} 0 & i & 0 \\ i & 0 & 0 \\ 0 & 0 & 0 \end{pmatrix}, \quad [V_2, W_1] = \frac{1}{4}\begin{pmatrix} 0 & -i & 0 \\ -i & 0 & 0 \\ 0 & 0 & 0 \end{pmatrix}, \quad [V_2, W_2] = \frac{1}{4}\begin{pmatrix} 0 & 1 & 0 \\ -1 & 0 & 0 \\ 0 & 0 & 0 \end{pmatrix}.$$

(ii) We have

$$L_+ = \frac{1}{2}(L_1 + iL_2) = \begin{pmatrix} 0 & 1 & 0 \\ 0 & 0 & 0 \\ 0 & 0 & 0 \end{pmatrix}, \quad L_- = \frac{1}{2}(L_1 - iL_2) = \begin{pmatrix} 0 & 0 & 0 \\ 1 & 0 & 0 \\ 0 & 0 & 0 \end{pmatrix},$$

$$V_+ = V_1 + iV_2 = \begin{pmatrix} 0 & 0 & 1 \\ 0 & 0 & 0 \\ 0 & 0 & 0 \end{pmatrix}, \quad V_- = V_1 - iV_2 = \begin{pmatrix} 0 & 0 & 0 \\ 0 & 0 & 0 \\ 1 & 0 & 0 \end{pmatrix}.$$

$$W_+ = \begin{pmatrix} 0 & 0 & 0 \\ 0 & 0 & 1 \\ 0 & 0 & 0 \end{pmatrix}, \quad W_- = \begin{pmatrix} 0 & 0 & 0 \\ 0 & 0 & 0 \\ 0 & 1 & 0 \end{pmatrix}.$$

Consequently

$$[L_+, L_-] = L_3, \quad [L_+, L_3] = -2L_+, \quad [L_-, L_3] = 2L_-$$

$$[V_+, V_-] = \begin{pmatrix} 1 & 0 & 0 \\ 0 & 0 & 0 \\ 0 & 0 & -1 \end{pmatrix}, \quad [W_+, W_-] = \begin{pmatrix} 0 & 0 & 0 \\ 0 & 1 & 0 \\ 0 & 0 & -1 \end{pmatrix}.$$

Thus for the anticommutator $[V_+, V_-]_+$ we find

$$[V_+, V_-]_+ = \begin{pmatrix} 1 & 0 & 0 \\ 0 & 0 & 0 \\ 0 & 0 & 1 \end{pmatrix} = L_3 + L_4 = L_0.$$

(iii) For the Bose-Fermi realization we have

$$[L_+, L_-] = [b_1^\dagger b_2 \otimes I_F, b_2^\dagger b_1 \otimes I_F] = (b_1^\dagger b_1 - b_2^\dagger b_2) \otimes I_F = 2L_3$$
$$[L_+, L_3] = [b_1^\dagger b_2 \otimes I_F, \frac{1}{2}(b_1^\dagger b_1 - b_2^\dagger b_2 \otimes I_F] = -b_1^\dagger b_2 = -L_+$$
$$[L_-, L_3] = [b_2^\dagger b_1 \otimes I_F, \frac{1}{2}(b_1^\dagger b_1 - b_2^\dagger b_2 \otimes I_F] = b_2^\dagger b_1 = L_-$$

$$[b_1^\dagger \otimes c, b_1 \otimes c^\dagger]_+ = b_1^\dagger b_1 \otimes I_F + I_B \otimes c^\dagger c = L_0.$$

Do the operators commute with the total number operator

$$\hat{N} = I_B \otimes c^\dagger c + (b_1^\dagger b_1 + b_2^\dagger b_2) \otimes I_F ?$$

Problem 75. Let $b^\dagger$, b be Bose creation and annihilation operators and $c^\dagger$, c be Fermi creation and annihilation operators.

(i) Consider the operators

$$\hat{Q} = b \otimes c^\dagger c \quad \Rightarrow \quad \hat{Q}^\dagger = b^\dagger \otimes c^\dagger c.$$

Find the commutator $[\hat{Q}, \hat{Q}^\dagger]$. Discuss.
(ii) Consider the operators

$$\hat{Q} = b^\dagger b \otimes c \quad \Rightarrow \quad \hat{Q}^\dagger = b^\dagger b \otimes c^\dagger.$$

Find the anticommutator $[\hat{Q}, \hat{Q}^\dagger]_+$.

Solution 75.   Since $c^\dagger c c^\dagger c = c^\dagger c$ and $[b, b^\dagger] = I_B$ the commutator is given by

$$[\hat{Q}, \hat{Q}^\dagger] = b b^\dagger \otimes c^\dagger c - b^\dagger b \otimes c^\dagger c = I_B \otimes c^\dagger c.$$

Thus $\hat{Q}$, $\hat{Q}^\dagger$, $I_B \otimes c^\dagger c$ form a basis of a three-dimensional Lie algebra.
(ii) With $c^\dagger c + c c^\dagger = I_F$ we obtain for the anticommutator $[\hat{Q}, \hat{Q}^\dagger]_+ = b^\dagger b b^\dagger b \otimes I_F$.

Problem 76.   Consider the operators

$$S = b \otimes c^\dagger \quad \Rightarrow \quad S^\dagger = b^\dagger \otimes c.$$

(i) Find the commutator $[S, S^\dagger]$.
(ii) Find the anticommutator $[S, S^\dagger]_+$.
(iii) Find the operator $[[S, S]_+, S^\dagger] + [[S, S^\dagger]_+, S] + [[S^\dagger, S]_+, S]$.

Solution 76.   (i) Since $b b^\dagger = I_B + b^\dagger b$ and $c c^\dagger = I_F - c^\dagger c$ we obtain

$$[S, S^\dagger] = [b \otimes c^\dagger, b^\dagger \otimes c] = I_B \otimes c^\dagger c - b^\dagger b \otimes I_F + 2 b^\dagger b \otimes c^\dagger c.$$

(ii) Since $b b^\dagger = I_B + b^\dagger b$ and $c c^\dagger = I_F - c^\dagger c$ we obtain

$$[S, S^\dagger]_+ = [b \otimes c^\dagger, b^\dagger \otimes c]_+ = I_B \otimes c^\dagger c + b^\dagger b \otimes I_F.$$

(iii) Using the properties from (i) and (ii) we obtain

$$[[S, S]_+, S^\dagger] + [[S, S^\dagger]_+, S] + [[S^\dagger, S]_+, S] = 0_B \otimes 0_F.$$

Problem 77.   Let $b^\dagger$, b be Bose creation and annihilation operators. Let X be a 2×2 matrix with $X^2 = I_2$ and $X = X^*$, for example the Pauli spin matrices. Consider the operators

$$J_+ = -\frac{1}{2} b^\dagger b^\dagger \otimes I_2, \quad J_- = \frac{1}{2} b b \otimes I_2, \quad J_0 = \frac{1}{2}\left(b^\dagger b + \frac{1}{2} I_B\right) \otimes I_2$$

and

$$V_+ = \frac{1}{\sqrt{2}} b^\dagger \otimes X, \quad V_- = \frac{1}{\sqrt{2}} b \otimes X.$$

(i) Find the anticommutators $[V_+, V_-]_+$, $[V_+, V_+]_+$, $[V_-, V_-]_+$.
(ii) Find the commutators $[J_+, J_-]$, $[J_0, J_+]$, $[J_0, J_+]$.

Solution 77.   (i) Utilizing $b b^\dagger = I_B + b^\dagger b$ and $X^2 = I_2$ we obtain for the anticommutators

$$[V_+, V_-]_+ = 2 J_0, \quad [V_+, V_+]_+ = -2 J_+, \quad [V_-, V_-]_+ = 2 J_-.$$

(ii) Utilizing $b b^\dagger = I_B + b^\dagger b$ we find for the commutators

$$[J_+, J_-] = 2 J_0, \quad [J_0, J_+] = J_+, \quad [J_0, J_-] = -J_0.$$

4.5 · Supplementary Problems

Problem 1. Let $b^\dagger$, b be Bose creation and annihilation operators and $\hat{N} = b^\dagger b$ be the number operator.
(i) Show that the generators of the Lie algebra $so(2,1)$ expressed in Bose creation and annihilation operators are given by

$$K_+ = -\frac{i}{2} b^\dagger b^\dagger, \quad K_- = \frac{i}{2} bb, \quad K_0 = \frac{1}{2}(b^\dagger b + \frac{1}{2} I).$$

(ii) The Lie algebra $su(1,1)$ consists of the three basis elements $\{K_0, K_+, K_-\}$ which satisfy the commutation relations

$$[K_0, K_\pm] = \pm K_\pm, \qquad [K_-, K_+] = 2K_0.$$

The *Casimir invariant* is

$$C = K_0^2 - \frac{1}{2}(K_+ K_- + K_- K_+).$$

Let $b^\dagger$, b be Bose creation and annihilation operators. Let $k = 1, 2, \ldots$. Show that

$$K_0 = b^\dagger b + kI, \quad K_+ = b^\dagger (2kI + b^\dagger b)^{1/2}, \quad K_- = (2kI + b^\dagger b)^{1/2} b$$

is a representation. Show that the eigenvalues of C are $k(k-1)$.
(iii) Let $\hat{N} = b^\dagger b$ be the number operator. Do the matrices

$$K_+ = \hat{N}^{1/2} b^\dagger, \quad K_- = b\hat{N}^{1/2}, \quad K_0 = \hat{N} + \frac{1}{2} I$$

form a basis of a Lie algebras? This means calculate the commutators

$$[K_+, K_-], \quad [K_+, K_0], \quad [K_-, K_0].$$

Show that $(\beta \in \mathbb{C})$

$$\exp(\beta K_+ - \bar{\beta} K_-)|0\rangle = (1 - |z|^2)^{1/2} \sum_{n=0}^{\infty} z^n |n\rangle$$

where

$$z := \frac{\beta}{|\beta|} \tanh(|\beta|), \quad |z| < 1.$$

Problem 2. (i) Let $b_1, b_2, \ldots, b_n$ be the Bose annihilation operators. Let

$$\hat{N}_j := b_j^\dagger b_j, \quad j = 1, 2, \ldots, n.$$

Consider the operators

$$L_1 := \frac{1}{2}(b_1^\dagger b_2^\dagger \cdots b_{n-1}^\dagger b_n + b_1 b_2 \cdots b_{n-1} b_n^\dagger)$$

$$L_2 := \frac{1}{2i}(b_1^\dagger b_2^\dagger \cdots b_{n-1}^\dagger b_n - b_1 b_2 \cdots b_{n-1} b_n^\dagger)$$

$$L_3 := \frac{1}{2}((\hat{N}_1 + I) \prod_{j=1}^{n-1} \hat{N}_j - \hat{N}_n \prod_{j=1}^{n-1} (\hat{N}_j + I)).$$

Find the commutation relations $[L_1, L_2]$, $[L_2, L_3]$, $[L_3, L_1]$.

(ii) Let $b_1, b_2, \ldots, b_n$ be the Bose annihilation operators. Let

$$\hat{N}_j := b_j^\dagger b_j, \quad j = 1, 2, \ldots, n.$$

Consider the operators

$$H_1 := \frac{1}{2}(b_1^\dagger b_2^\dagger \cdots b_n^\dagger + b_1 b_2 \cdots b_n)$$

$$H_2 := \frac{1}{2i}(b_1^\dagger b_2^\dagger \cdots b_n^\dagger - b_1 b_2 \cdots b_n)$$

$$H_3 := \frac{1}{2}\left(\prod_{j=1}^n \hat{N}_j - \prod_{j=1}^n (\hat{N}_j + I)\right).$$

Find the commutation relations $[H_1, H_2]$, $[H_2, H_3]$, $[H_3, H_1]$. Discuss.

Problem 3. Show that the ten operators

$$\hat{O}_1 = b_1^\dagger b_1^\dagger + b_1^\dagger b_1 + b_1 b_1^\dagger + b_1 b_1$$

$$\hat{O}_2 = b_2^\dagger b_2^\dagger + b_2^\dagger b_2 + b_2 b_2^\dagger + b_2 b_2$$

$$\hat{O}_3 = b_1^\dagger b_1^\dagger - b_1^\dagger b_1 - b_1 b_1^\dagger + b_1 b_1$$

$$\hat{O}_4 = b_2^\dagger b_2^\dagger - b_2^\dagger b_2 - b_2 b_2^\dagger + b_2 b_2$$

$$\hat{O}_5 = b_1^\dagger b_2^\dagger + b_1^\dagger b_2 + b_1 b_2^\dagger + b_1 b_2$$

$$\hat{O}_6 = b_1^\dagger b_2^\dagger - b_1^\dagger b_2 - b_1 b_2^\dagger + b_1 b_2$$

$$\hat{O}_7 = i(b_1^\dagger b_1^\dagger - b_1 b_1)$$

$$\hat{O}_8 = i(b_2^\dagger b_2^\dagger - b_2^\dagger b_2)$$

$$\hat{O}_9 = i(b_1^\dagger b_2^\dagger - b_1^\dagger b_2 + b_1 b_2^\dagger - b_1 b_2)$$

$$\hat{O}_{10} = i(b_1^\dagger b_2^\dagger + b_1^\dagger b_2 - b_1 b_2^\dagger - b_1 b_2)$$

form a basis of the Lie algebra $o(3, 2)$.

Problem 4.   Let $j, k \in \{1, 2\}$ and $s \in \{1, 2, \ldots, n\}$. Let $b_{js}^\dagger$, b_{js} be Bose creation and annihilation operators and I the identity operator. Consider the operators

$$C_{jk} := \sum_{s=1}^n b_{js}^\dagger b_{ks} + \frac{1}{2} n \delta_{jk} I, \qquad B_{jk} := \sum_{s=1}^n b_{js} b_{ks}$$

and thus

$$B_{jk}^\dagger = \sum_{s=1}^n b_{js}^\dagger b_{ks}^\dagger.$$

Show that the operators C_{jk}, B_{jk} and $B_{jk}^\dagger$ form a basis of the non-compact symplectic Lie algebra $sp(4, \mathbb{R})$. Note that $B_{jk} = B_{kj}$ and $B_{jk}^\dagger = B_{kj}^\dagger$ with $(j \neq k)$. Thus the dimension of the Lie algebra is $n(2n + 1) = 10$ where $n = 2$.

Problem 5.   Consider the operators

$$\hat{K}_- = b_1 b_2, \quad \hat{K}_+ = b_1^\dagger b_2^\dagger, \quad \hat{K}_0 = \frac{1}{2}(b_1^\dagger b_1 + b_2^\dagger b_2 + I)$$

and the Hamilton operator

$$\hat{H} = \hbar\omega_1(b_1^\dagger b_1 + b_2^\dagger b_2) + \hbar\omega_2 b_1^\dagger b_1 b_2^\dagger b_2 + \hbar\omega_3(b_1^\dagger b_1^\dagger b_1 b_1 + b_2^\dagger b_2^\dagger b_2 b_2).$$

Show that the Hamilton operator can be expressed as

$$\hat{H} = (2\hbar\omega_1 - \hbar\omega_3)I + (2\hbar\omega_1 - 6\hbar\omega_3)\hat{K}_0 + 4\hbar\omega_3\hat{K}_0^2 + (\hbar\omega_2 - 2\hbar\omega_3)\hat{K}_+\hat{K}_-.$$

Problem 6. Let b_j, $b_j^\dagger$ be Bose annihilation and creation operators.
(i) Show that a basis of the Lie algebra $sp(2n, \mathbb{R})$ is given by the $n(2n+1)$ bilinear operators

$$H_j = \frac{1}{2}(b_j^\dagger b_j + b_j b_j^\dagger) \quad j = 1, \ldots, n$$

$$C_{jk} = b_j^\dagger b_k \quad j \neq k, \quad j, k = 1, \ldots, n$$

$$b_j^\dagger b_k^\dagger, \quad b_j b_k \quad j \leq k, j = 1, \ldots, n.$$

(ii) Show that the n^2 operators H_j and C_{jk} form a basis of the Lie algebra $u(n)$ which is a sub Lie algebra of $sp(2n, \mathbb{R})$.

Problem 7. Let $b_{jn}^\dagger$, b_{km} $(j, k = 1, 2, 3; m, n = 1, \ldots, N)$ be Bose creation and annihilation operators, respectively with the commutation relation $[b_{km}, b_{jn}] = \delta_{jk}\delta_{mn}I$, where I is the identity operator. Show that a realization of the non-compact symplectic Lie algebra $sp(2n, \mathbb{R})$ $(n = 3)$ is given by

$$A_{jk} = \sum_{n=1}^N b_{jn}^\dagger b_{kn}^\dagger, \quad B_{jk} = \sum_{n=1}^N b_{jn} b_{kn}, \quad C_{jk} = \frac{1}{2}\sum_{n=1}^N (b_{jn}^\dagger b_{kn} + b_{kn} b_{jn}^\dagger).$$

Problem 8. Let $b_k^\dagger$, b_k be Bose creation and annihilation operators. Consider the operators

$$P_+ := -\frac{1}{2}\sum_{k=1}^n b_k^\dagger b_k^\dagger, \quad P_- := \frac{1}{2}\sum_{k=1}^n b_k b_k, \quad P_3 := \frac{1}{4}\sum_{k=1}^n (b_k^\dagger b_k + b_k b_k^\dagger).$$

Find the commutators. Discuss.

Problem 9. Show that the 10 operators

$$J_0 = \frac{1}{2}(b_1^\dagger b_1 + b_2^\dagger b_2 + I)$$

$$J_1 = \frac{1}{2}(b_1^\dagger b_2 + b_2^\dagger b_1)$$

$$J_2 = \frac{i}{2}(b_2^\dagger b_1 - b_1^\dagger b_2)$$

$$J_3 = \frac{1}{2}(b_1^\dagger b_1 - b_2^\dagger b_2)$$

$$K_1 = \frac{1}{4}((b_1^\dagger)^2 - b_1^2 + (b_2^\dagger)^2 - (b_2)^2)$$

$$K_2 = -\frac{i}{4}((b_1^\dagger)^2 - b_1^2 + (b_2^\dagger)^2 - b_2^2)$$

$$K_3 = -\frac{1}{2}(b_1^\dagger b_2^\dagger + b_1 b_2)$$

$$L_1 = \frac{i}{4}((b_1^\dagger)^2 - b_1^2 - (b_2^\dagger)^2 + b_2^2)$$

$$L_2 = \frac{1}{4}((b_1^\dagger)^2 + b_1^2 + (b_2^\dagger)^2 + b_2^2)$$

$$L_3 = -\frac{i}{2}(b_1^\dagger b_2^\dagger - b_1 b_2)$$

form a basis of a Lie algebra. Find all Lie sub-algebras.

Problem 10.   Consider the eigenvalue problem $\hat{H}u(x) = Eu(x)$ with the Hamilton operator

$$\hat{H} = -\frac{\hbar^2}{2m}\frac{d^2}{dx^2} + \frac{1}{2}m\omega^2 x^2 + \frac{g^2}{x^2}$$

(i) Show that introducing the dimensionless variables

$$\tilde{u}(\tilde{x}(x)) = u(x), \quad \tilde{x}(x) = \left(\frac{m\omega}{\hbar}\right)^{1/2}x, \quad k^2 = \frac{mg^2}{\hbar^2}$$

we obtain the second order linear differential equation

$$\left(-\hbar\omega\frac{d^2}{d\tilde{x}^2} + \frac{1}{2}\hbar\omega\tilde{x}^2 + \hbar\omega\frac{k^2}{\tilde{x}^2}\right)\tilde{u}(\tilde{x}) = E\tilde{u}(\tilde{x})$$

or

$$\left(-\frac{d^2}{d\tilde{x}^2} + \frac{1}{2}\tilde{x}^2 + \frac{k^2}{\tilde{x}^2}\right)\tilde{u}(\tilde{x}) = \tilde{E}\tilde{u}(\tilde{x})$$

where $\tilde{E} = E/(\hbar\omega)$.
(ii) Define the differential operators

$$b = \frac{1}{\sqrt{2}}\left(\tilde{x} + \frac{d}{d\tilde{x}}\right), \quad b^\dagger = \frac{1}{\sqrt{2}}\left(\tilde{x} - \frac{d}{d\tilde{x}}\right)$$

with $[b, b^\dagger] = I$ and

$$K_- = \frac{1}{2}\left(b^2 - \frac{k^2}{\tilde{x}^2}\right), \quad K_+ = \frac{1}{2}\left((b^\dagger)^2 - \frac{k^2}{\tilde{x}^2}\right), \quad K_0 = \frac{1}{2}\left(b^\dagger b + \frac{1}{2}I + \frac{k^2}{\tilde{x}^2}\right).$$

Show that $[K_0, K_\pm] = \pm K_\pm$ and $[K_-, K_+] = 2K_0$.

Problem 11.   Let $b_1^\dagger$, $b_2^\dagger$, $b_3^\dagger$ be Bose creation operators and $\hat{N}_1 := b_1^\dagger b_1$, $\hat{N}_2 := b_2^\dagger b_2$, $\hat{N}_3 := b_3^\dagger b_3$.
(i) Consider the operators

$$L_x := \frac{1}{2}(b_1^\dagger b_2^\dagger b_3 + b_1 b_2 b_3^\dagger), \quad L_y := \frac{1}{2i}(b_1^\dagger b_2^\dagger b_3 - b_1 b_2 b_3^\dagger)$$

and

$$L_z := \frac{1}{2}(\hat{N}_1\hat{N}_2(\hat{N}_3 + I) - (\hat{N}_1 + I)(\hat{N}_2 + I)\hat{N}_3).$$

Find the commutators $[L_x, L_y]$, $[L_z, L_x]$, $[L_y, L_z]$.

(ii) Consider the operators

$$H_x = \frac{1}{2}(b_1^\dagger b_2^\dagger b_3^\dagger + b_1 b_2 b_3), \quad H_y = \frac{1}{2i}(b_1^\dagger b_2^\dagger b_3^\dagger - b_1 b_2 b_3)$$

and

$$H_z = \frac{1}{2}(\hat{N}_1 \hat{N}_2 \hat{N}_3 - (\hat{N}_1 + I)(\hat{N}_2 + I)(\hat{N}_3 + I)).$$

Find the commutators $[H_x, H_y]$, $[H_z, H_x]$, $[H_y, H_z]$.

Problem 12.   (i) Consider the operators

$$K_+ = \frac{1}{2}(b_1^\dagger b_1^\dagger + b_2^\dagger b_2^\dagger + b_3^\dagger b_3^\dagger), \quad K_- = \frac{1}{2}(b_1 b_1 + b_2 b_2 + b_3 b_3)$$

and

$$K_3 = \frac{1}{2}\left(b_1^\dagger b_1 + b_2^\dagger b_2 + b_3^\dagger b_3 + \frac{3}{2}I\right).$$

Show that $[K_3, K_\pm] = \pm K_\pm$, $[K_+, K_-] = -2K_3$ and hence we have a basis of the Lie algebra $su(1,1)$.

(ii) What Lie algebra is generated by the operators

$$\hat{K} = (b_1^\dagger)^2 b_2, \quad \hat{K}^\dagger = b_2^\dagger b_1^2 ?$$

Problem 13.   Consider the Lie algebra $s\ell(2,\mathbb{R})$. Consider the operators $\hat{P}$, $\hat{Q}$ and the Lie algebra generated by the elements $\hat{P}^2$, $\hat{Q}^2$ and $\hat{P}\hat{Q}+\hat{Q}\hat{P}$. Show that the isomorphism between $s\ell(2,\mathbb{R})$ and the Lie algebra generated by the elements $\hat{P}^2$, $\hat{Q}^2$ and $\hat{P}\hat{Q}+\hat{Q}\hat{P}$ has the form

$$\begin{pmatrix} a & b \\ c & -a \end{pmatrix} \leftrightarrow \frac{1}{2}(a(\hat{P}\hat{Q}+\hat{Q}\hat{P}) + b\hat{Q}^2 - c\hat{P}^2).$$

Problem 14.   Let σ_1, σ_2, σ_3 be the Pauli spin matrices. Show that the irreducible matrix representation of the *Clifford algebra* $C(5,0)$, modulo overall sign, is unique up to the unitary transformation and can be given by the 4×4 matrices

$$\alpha_j = I_2 \otimes \sigma_j, \quad j = 1, 3$$

$$\alpha_2 = \sigma_2 \otimes \sigma_2, \quad \alpha_4 = \sigma_1 \otimes \sigma_2, \quad \alpha_5 = \sigma_3 \otimes \sigma_2.$$

Find the anticommutator $[\alpha_4, \alpha_5]_+$.

Problem 15.   The real Lie algebra $s\ell(2,\mathbb{C})$ is a six-dimensional Lie algebra with the basis

$$J_\pm, \quad J_0, \quad K_\pm, \quad K_0.$$

The non-vanishing commutators are

$$[J_0, J_+] = J_+, \quad [J_0, K_+] = K_+, \quad [K_0, K_+] = -J_+, \quad [K_0, J_+] = K_+,$$

$$[J_0, J_-] = -J_-, \quad [J_0, K_-] = -K_-, \quad [K_0, K_-] = J_-, \quad [K_0, J_-] = -K_-,$$

$$[J_+, J_-] = 2J_0, \quad [J_+, K_-] = 2K_0, \quad [K_+, K_-] = -2J_0, \quad [J_-, K_+] = 2K_0.$$

Show that the two quadratic Casimir operators are given by

$$C_1 = J_0^2 + \frac{1}{2}(J_+J_- + J_-J_+) - K_0^2 - \frac{1}{2}(K_+K_- + K_-K_+)$$

$$C_2 = J_0K_0 + \frac{1}{2}(J_+K_- + J_-K_+) + K_0J_0 + \frac{1}{2}(K_+J_- + K_-J_+).$$

Let σ_1, σ_2, σ_3 be the Pauli spin matrices, and $\sigma_\pm := \frac{1}{2}(\sigma_1 \pm i\sigma_2)$ and

$$\mathbf{z} = (z_1 \quad z_2), \quad \bar{\mathbf{z}} = \begin{pmatrix} \bar{z}_1 \\ \bar{z}_2 \end{pmatrix}, \quad \frac{\partial}{\partial \mathbf{z}} = \begin{pmatrix} \partial/\partial z_1 \\ \partial/\partial z_2 \end{pmatrix}, \quad \frac{\partial}{\partial \bar{\mathbf{z}}} = (\partial/\partial\bar{z}_1 \quad \partial/\partial\bar{z}_2).$$

A realization for J_+, J_-, K_+, K_- is given by

$$J_+ = \mathbf{z}\sigma_+ \frac{\partial}{\partial \mathbf{z}} - \frac{\partial}{\partial \bar{\mathbf{z}}}\sigma_+ \bar{\mathbf{z}}, \qquad J_- = \mathbf{z}\sigma_- \frac{\partial}{\partial \mathbf{z}} - \frac{\partial}{\partial \bar{\mathbf{z}}}\sigma_- \bar{\mathbf{z}}$$

$$K_+ = i(\mathbf{z}\sigma_+ \frac{\partial}{\partial \mathbf{z}} + \frac{\partial}{\partial \bar{\mathbf{z}}}\sigma_+ \bar{\mathbf{z}}), \qquad K_- = i(\mathbf{z}\sigma_- \frac{\partial}{\partial \mathbf{z}} + \frac{\partial}{\partial \bar{\mathbf{z}}}\sigma_- \bar{\mathbf{z}}).$$

Calculate the other operators from the commutation relations.

Problem 16. Let $c_1^\dagger$, $c_2^\dagger$, c_1, c_2 be Fermi creation and annihilation operators. Let $\hat{N}_1 = c_1^\dagger c_1$, $\hat{N}_2 = c_2^\dagger c_2$.
(i) Construct the Lie algebra generated by the operators

$$c_1^\dagger c_1, \quad c_2^\dagger c_2, \quad c_1^\dagger c_2, \quad c_2^\dagger c_1.$$

(ii) Construct the Lie algebra generated by the operators

$$c_1^\dagger c_1, \quad c_2^\dagger c_2, \quad c_1^\dagger c_2, \quad c_2^\dagger c_1, \quad \hat{N}_1\hat{N}_2.$$

(iii) Construct the Lie algebra generated by the operators

$$c_{1,\uparrow}^\dagger c_{1,\uparrow}, \quad c_{2,\uparrow}^\dagger c_{2,\uparrow}, \quad c_{1,\downarrow}^\dagger c_{1,\downarrow}, \quad c_{2,\downarrow}^\dagger c_{2,\downarrow}, \quad c_{1,\uparrow}^\dagger c_{2,\uparrow}, \quad c_{2,\uparrow}^\dagger c_{1,\uparrow}, \quad c_{1,\downarrow}^\dagger c_{2,\downarrow}, \quad c_{2,\downarrow}^\dagger c_{1,\downarrow}.$$

Problem 17. Let σ_1, σ_2, σ_3 be the Pauli spin matrices. Let $N > 2$ and $j = 1, \ldots, N-1$. Let $\sigma_{1,j}$, $\sigma_{2,j}$, $\sigma_{3,j}$ be the Pauli matrices acting on site j, i.e.

$$\sigma_{1,j} = I_2 \otimes \cdots \otimes I_2 \otimes \sigma_1 \otimes I_2 \otimes \cdots \otimes I_2$$

where σ_1 is at j-th position and we have N terms. Thus $\sigma_{1,j}$ is a $2^N \times 2^N$ matrix. Analogously we define $\sigma_{2,j}$ and $\sigma_{3,j}$. Let

$$E_j := -\frac{1}{2}(\sigma_{1,j}\sigma_{1,j+1} + \sigma_{2,j}\sigma_{2,j+1} + \frac{1}{2}(q + q^{-1})(\sigma_{3,j}\sigma_{3,j+1} - I) + \frac{1}{2}(q - q^{-1})(\sigma_{3,j} - \sigma_{3,j+1})).$$

Show that

$$E_j E_j = (q + q^{-1})E_j$$
$$E_j E_{j\pm 1} E_j = E_j$$
$$E_j E_k = E_k E_j \quad (k \neq j \pm 1).$$

The $2^N \times 2^N$ E_j $(j = 1, \ldots, N-1)$ matrices are generators of a *Temperley-Lieb algebra*.

Problem 18. Let $c_{j\sigma}^\dagger$, $c_{j\sigma}$ be Fermi creation and annihilation operators with spin σ.
(i) Consider operators

$$S = c_{1\uparrow}^\dagger c_{1\downarrow} + c_{2\uparrow}^\dagger c_{2\downarrow}, \qquad S^\dagger = c_{1\downarrow}^\dagger c_{1\uparrow} + c_{2\downarrow}^\dagger c_{2\uparrow}$$

and

$$S_3 = \frac{1}{2}(c_{1\downarrow}^\dagger c_{1\downarrow} + c_{2\downarrow}^\dagger c_{2\downarrow} - c_{1\uparrow}^\dagger c_{1\uparrow} - c_{2\uparrow}^\dagger c_{2\uparrow}).$$

Find the commutators $[S, S^\dagger]$, $[S, S_3]$, $[S^\dagger, S_3]$. Discuss.
(iii) Consider the Hamilton operator

$$\hat{H} = -t \sum_{\sigma \in \{\uparrow, \downarrow\}} (c_{1\sigma}^\dagger c_{2\sigma} + c_{2\sigma}^\dagger c_{1\sigma}).$$

Find the commutators $[\hat{H}, S]$, $[\hat{H}, S^\dagger]$, $[\hat{H}, S_3]$. Discuss.
(iv) Let $c_{j\sigma}^\dagger$, $c_{j\sigma}$ be Fermi creation and annihilation operators with spin σ. Consider operators

$$R_+ = -c_{1\uparrow} c_{1\downarrow} + c_{2\uparrow} c_{2\downarrow}, \qquad R_- = -c_{1\downarrow}^\dagger c_{1\uparrow}^\dagger + c_{2\downarrow}^\dagger c_{2\uparrow}^\dagger$$

and

$$R_3 = \frac{1}{2}(c_{1\uparrow}^\dagger c_{1\uparrow} + c_{2\uparrow}^\dagger c_{2\uparrow} + c_{1\downarrow}^\dagger c_{1\downarrow} + c_{2\downarrow}^\dagger c_{2\downarrow} - I).$$

Find the commutators $[R_+, R_-]$, $[R_+, R_3]$, $[R_-, R_3]$. Discuss.
(v) Consider the Hamilton operator

$$\hat{H} = -t \sum_{\sigma \in \{\uparrow, \downarrow\}} (c_{1\sigma}^\dagger c_{2\sigma} + c_{2\sigma}^\dagger c_{1\sigma}).$$

Find the commutators $[\hat{H}, R_+]$, $[\hat{H}, R_-]$, $[\hat{H}, R_3]$. Discuss.

Problem 19. Let $c_1^\dagger$, $c_2^\dagger$, $c_3^\dagger$ be Fermi creation operators.
(i) Consider the two operators

$$L_1 = \frac{1}{2}(c_1^\dagger c_2^\dagger c_3 + c_1 c_2 c_3^\dagger), \qquad L_2 = \frac{1}{2i}(c_1^\dagger c_2^\dagger c_3 - c_1 c_2 c_3^\dagger).$$

Find the Lie algebra generated by L_1 and L_2. Set $[L_1, L_2] = iL_3$.
(ii) Consider the two operators

$$L_1 = \frac{1}{2}(c_1^\dagger c_2^\dagger c_3^\dagger + c_1 c_2 c_3), \qquad L_2 = \frac{1}{2i}(c_1^\dagger c_2^\dagger c_3^\dagger - c_1 c_2 c_3).$$

Find the Lie algebra generated by L_1 and L_2. Set $[L_1, L_2] = iL_3$.

Problem 20. Let $c_1^\dagger$, $c_2^\dagger$, c_1, c_2 be Fermi creation and annihilation operators. Let I be the unit operator.
(i) Find the Lie algebra generated by the operators

$$c_1^\dagger c_1, \quad c_2^\dagger c_2, \quad c_1^\dagger + c_1, \quad c_2^\dagger + c_2.$$

(ii) Do the operators

$$\{c_1^\dagger c_2^\dagger, c_1^\dagger c_2, c_1 c_2^\dagger, c_1^\dagger c_1, c_2^\dagger c_2, c_1 c_2, I\}$$

form a basis of a Lie algebra under the commutator? For example we have

$$[c_1^\dagger c_2^\dagger, c_1 c_2] = I - c_1^\dagger c_1 - c_2^\dagger c_2.$$

(iii) Consider the operators

$$S_0 = c_1^\dagger c_1 + c_2^\dagger c_2, \quad S_1 = c_1^\dagger c_1 - c_2^\dagger c_2,$$

$$S_2 = c_1^\dagger c_1 e^{i\phi} + c_2^\dagger c_2 e^{-i\phi}, \quad S_3 = i c_1^\dagger c_1 e^{-i\phi} - i c_2^\dagger c_2 e^{i\phi}.$$

Find the commutators. Discuss.
(iv) Find the Lie algebra generated by the operators

$$c_1^\dagger c_1, \quad c_2^\dagger c_2, \quad c_1^\dagger + c_1, \quad c_2^\dagger + c_2, \quad c_1^\dagger + c_2, \quad c_1 + c_2^\dagger.$$

Problem 21. Consider the Pauli spin matrices $\sigma_0, \sigma_1, \sigma_2, \sigma_3$, where $\sigma_0 = I_2$. One has $\sigma_1 \sigma_2 = i\sigma_3$, $\sigma_2 \sigma_3 = i\sigma_1$, $\sigma_3 \sigma_1 = i\sigma_2$. Let $\mu, \nu = 0, 1, 2, 3$. Consider the sixteen 4×4 matrices

$$M_{\mu\nu} = \sigma_\mu \otimes \sigma_\nu$$

where $\otimes$ is the Kronecker product.
(i) Find the commutators $[M_{\mu\nu}, M_{\alpha\beta}]$. Discuss.
(ii) Find the anticommutators $[M_{\mu\nu}, M_{\alpha\beta}]_+$. Discuss.

Problem 22. The $osp(1|2)$ superalgebra has the five generators

$$J_0, \quad J_+, \quad J_-, \quad V_+, \quad V_-$$

with the commutation relations

$$[J_0, J_\pm] = \pm J_\pm, \quad [J_+, J_-] = 2J_0, \quad [J_0, V_\pm] = \pm\frac{1}{2}V_\pm, \quad [J_\pm, V_\pm] = 0, \quad [J_\pm, V_\mp] = V_\pm$$

and anticommutation relations

$$[V_\pm, V_\pm]_+ = \pm\frac{1}{2}J_\pm, \quad [V_+, V_-]_+ = -\frac{1}{2}J_0.$$

(i) Let $b^\dagger$, b be Bose creation and annihilation operators and I the identity operator. Show that

$$J_+ = -\frac{1}{2}(b^\dagger)^2, \quad J_- = \frac{1}{2}b^2, \quad J_0 = \frac{1}{2}b^\dagger b + \frac{1}{4}I$$

$$V_+ = \frac{i}{2\sqrt{2}}b^\dagger, \quad V_- = \frac{i}{2\sqrt{2}}b.$$

is a realization of the superalgebra.
(ii) Let $E_{(jk)}$ be a 3×3 matrix having 1 at the position of the jth row and k column and 0 otherwise. Show that a representation using these matrices and linear combinations of it is given by

$$H = \frac{1}{2}(E_{(11)} - E_{(33)}), \quad J_+ = E_{(13)}, \quad J_- = E_{(31)},$$

$$V_+ = \frac{1}{2}(E_{(12)} + E_{(23)}), \quad V_- = \frac{1}{2}(-E_{(21)} + E_{(32)}).$$

Problem 23. The superalgebra $u(1|1)$ is defined as follows: The bosonic and fermionic bilinear combinations $b^\dagger b$ and $c^\dagger c$ generate the Lie algebras of $u_B(1)$ and $u_F(1)$, respectively. The Bose-Fermi bilinears

$$b \otimes c^\dagger, \qquad b^\dagger \otimes c$$

close with the set $\{b^\dagger b, c^\dagger c\}$ under the anticommutations

$$[b \otimes c^\dagger, b^\dagger \otimes c]_+ = b^\dagger b \otimes I_F + I_B \otimes c^\dagger c, \quad [b \otimes c^\dagger, b \otimes c^\dagger]_+ = [b \otimes c, b \otimes c]_+ = 0_B \otimes 0_F.$$

Note that

$$[b^\dagger b \otimes I_F, I_B \otimes c^\dagger c] = 0_B \otimes 0_F.$$

Hence, the bilinear combinations $b \otimes c^\dagger$ and $b^\dagger \otimes c$ are the odd generators and $b^\dagger b \otimes I_F$, $I_B \otimes c^\dagger c$ are the even generators of the Lie superalgebra $u(1|1)$. Find the states

$$(b \otimes c^\dagger)(|n\rangle \otimes |0\rangle_F), \quad (b \otimes c^\dagger)(|n\rangle \otimes c^\dagger |0\rangle_F), \quad (b^\dagger \otimes c)(|n\rangle \otimes |0\rangle_F), \quad (b^\dagger \otimes c)(|n\rangle \otimes c^\dagger |0\rangle_F).$$

Chapter 5

Bose-Spin Systems

5.1 Solved Problems

Problem 1. We consider a product Hilbert space

$$\mathcal{H} = \mathcal{H}_1 \otimes \mathcal{H}_2$$

where $\mathcal{H}_2 = \mathbb{C}^2$. The Pauli spin matrices

$$\sigma_1 = \begin{pmatrix} 0 & 1 \\ 1 & 0 \end{pmatrix}, \quad \sigma_2 = \begin{pmatrix} 0 & -i \\ i & 0 \end{pmatrix}, \quad \sigma_3 = \begin{pmatrix} 1 & 0 \\ 0 & -1 \end{pmatrix}$$

act in the Hilbert space $\mathcal{H}_2 = \mathbb{C}^2$ and Bose creation and annihilation operators $b^\dagger$, b and the identity operator I_B act in the Hilbert space $\mathcal{H}_1$. We denote by 0_B the zero operator in the Hilbert space $\mathcal{H}_1$. The zero operator 0_2 in the Hilbert space $\mathcal{H}_2 = \mathbb{C}^2$ is the 2×2 zero matrix. Thus $0_B \otimes 0_2$ is the zero operator in the product Hilbert space. We have

$$\sigma_1^2 = \sigma_2^2 = \sigma_3^2 = I_2, \quad \sigma_1\sigma_2 = i\sigma_3, \quad \sigma_2\sigma_3 = i\sigma_1, \quad \sigma_3\sigma_1 = i\sigma_2,$$

$$[\sigma_1, \sigma_2]_+ = [\sigma_2, \sigma_3]_+ = [\sigma_3, \sigma_1]_+ = 0_2$$

$$[\sigma_1, \sigma_2] = 2i\sigma_3, \quad [\sigma_2, \sigma_3] = 2i\sigma_1, \quad [\sigma_3, \sigma_1] = 2i\sigma_2$$

$$[b, b^\dagger] = I_B$$

and

$$\hat{N} = b^\dagger b$$

is the number operator for the Bose system. Consider the six operators

$$b \otimes \sigma_1, \quad b^\dagger \otimes \sigma_1, \quad b \otimes \sigma_2, \quad b^\dagger \otimes \sigma_2, \quad b \otimes \sigma_3, \quad b^\dagger \otimes \sigma_3$$

in the product Hilbert space.
(i) Find the commutators for these operators. Discuss.
(ii) Find the anticommutators for these operators. Discuss.

Solution 1. (i) For the commutators we obtain

$$[b \otimes \sigma_1, b^\dagger \otimes \sigma_1] = I_B \otimes I_2$$
$$[b \otimes \sigma_1, b \otimes \sigma_2] = 2ib^2 \otimes \sigma_3$$
$$[b \otimes \sigma_1, b^\dagger \otimes \sigma_2] = i(I_B + 2b^\dagger b) \otimes \sigma_3$$
$$[b \otimes \sigma_1, b \otimes \sigma_3] = -2ib^2 \otimes \sigma_2$$
$$[b \otimes \sigma_1, b^\dagger \otimes \sigma_3] = -i(I_B + 2b^\dagger b) \otimes \sigma_2$$
$$[b^\dagger \otimes \sigma_1, b \otimes \sigma_2] = i(I_B + 2b^\dagger b) \otimes \sigma_3$$
$$[b^\dagger \otimes \sigma_1, b^\dagger \otimes \sigma_2] = 2i(b^\dagger)^2 \otimes \sigma_3$$
$$[b^\dagger \otimes \sigma_1, b \otimes \sigma_3] = -i(I_B + 2b^\dagger b) \otimes \sigma_2$$
$$[b^\dagger \otimes \sigma_1, b^\dagger \otimes \sigma_3] = -2i(b^\dagger)^2 \otimes \sigma_2$$
$$[b \otimes \sigma_2, b^\dagger \otimes \sigma_2] = I_B \otimes I_2$$
$$[b \otimes \sigma_2, b \otimes \sigma_3] = 2ib^2 \otimes \sigma_1$$
$$[b \otimes \sigma_2, b^\dagger \otimes \sigma_3] = i(I_B + 2b^\dagger b) \otimes \sigma_1$$
$$[b^\dagger \otimes \sigma_2, b \otimes \sigma_3] = i(I_B + 2b^\dagger b) \otimes \sigma_1$$
$$[b^\dagger \otimes \sigma_2, b^\dagger \otimes \sigma_3] = 2i(b^\dagger)^2 \otimes \sigma_1$$
$$[b \otimes \sigma_3, b^\dagger \otimes \sigma_3] = I_B \otimes I_2.$$

Thus for the anticommutators we obtain

$$[b \otimes \sigma_1, b^\dagger \otimes \sigma_1]_+ = (I_B + 2b^\dagger b) \otimes I_2.$$
$$[b \otimes \sigma_1, b \otimes \sigma_2]_+ = 0_B \otimes 0_2$$
$$[b \otimes \sigma_1, b^\dagger \otimes \sigma_2]_+ = i(I_B \otimes \sigma_3)$$
$$[b \otimes \sigma_1, b \otimes \sigma_3]_+ = 0_B \otimes 0_2$$
$$[b \otimes \sigma_1, b^\dagger \otimes \sigma_3]_+ = -i(I_B \otimes \sigma_2)$$
$$[b^\dagger \otimes \sigma_1, b \otimes \sigma_2]_+ = i(I_B \otimes \sigma_3)$$
$$[b^\dagger \otimes \sigma_1, b^\dagger \otimes \sigma_2]_+ = 0_B \otimes 0_2$$
$$[b^\dagger \otimes \sigma_1, b \otimes \sigma_3]_+ = i(I_B \otimes \sigma_2)$$
$$[b^\dagger \otimes \sigma_1, b^\dagger \otimes \sigma_3]_+ = 0_B \otimes 0_2$$
$$[b \otimes \sigma_2, b^\dagger \otimes \sigma_2]_+ = (I_B + 2b^\dagger b) \otimes I_2$$
$$[b \otimes \sigma_2, b \otimes \sigma_3]_+ = 0_B \otimes 0_2$$
$$[b \otimes \sigma_2, b^\dagger \otimes \sigma_3]_+ = i(I_B \otimes \sigma_1)$$
$$[b^\dagger \otimes \sigma_2, b \otimes \sigma_3]_+ = -i(I_B \otimes \sigma_1)$$
$$[b^\dagger \otimes \sigma_2, b^\dagger \otimes \sigma_3]_+ = 0_B \otimes 0_2$$
$$[b \otimes \sigma_3, b^\dagger \otimes \sigma_3]_+ = (I_B + 2b^\dagger b) \otimes I_2.$$

Problem 2. Given the spin matrices for spin-$\frac{1}{2}$

$$S_+ = \frac{1}{2}(\sigma_1 + i\sigma_2) = \begin{pmatrix} 0 & 1 \\ 0 & 0 \end{pmatrix}, \qquad S_- = \frac{1}{2}(\sigma_1 - i\sigma_2) = \begin{pmatrix} 0 & 0 \\ 1 & 0 \end{pmatrix},$$

$$S_3 = \frac{1}{2}\sigma_3 = \frac{1}{2}\begin{pmatrix} 1 & 0 \\ 0 & -1 \end{pmatrix}$$

and the operators

$$b^\dagger \otimes S_+, \quad b^\dagger \otimes S_-, \quad b \otimes S_+, \quad b \otimes S_-, \quad b^\dagger \otimes S_3, \quad b \otimes S_3$$

in the product Hilbert space.
(i) Find the commutators of these operators.
(ii) Find the anticommutators of these operators.

Solution 2.   (i) Note that $S_+S_- - S_-S_+ = \sigma_3 = 2S_3$, $S_-S_3 - S_3S_- = S_-$, $S_+S_3 - S_3S_+ = -S_+$. For the commutators we have

$$[b^\dagger \otimes S_+, b^\dagger \otimes S_-] = 2(b^\dagger)^2 \otimes S_3$$
$$[b^\dagger \otimes S_+, b \otimes S_+] = 0_B \otimes 0_2$$
$$[b^\dagger \otimes S_+, b \otimes S_-] = -I_B \otimes \begin{pmatrix} 0 & 0 \\ 0 & 1 \end{pmatrix} + 2b^\dagger b \otimes S_3$$
$$[b^\dagger \otimes S_+, b^\dagger \otimes S_3] = -(b^\dagger)^2 \otimes S_+$$
$$[b^\dagger \otimes S_+, b \otimes S_3] = -\frac{1}{2}I_B \otimes S_+ - b^\dagger b \otimes S_+$$
$$[b^\dagger \otimes S_-, b \otimes S_+] = -I_B \otimes \begin{pmatrix} 1 & 0 \\ 0 & 0 \end{pmatrix} + 2b^\dagger b \otimes S_3$$
$$[b^\dagger \otimes S_-, b \otimes S_-] = 0_B \otimes 0_2$$
$$[b^\dagger \otimes S_-, b^\dagger \otimes S_3] = (b^\dagger)^2 \otimes S_-$$
$$[b^\dagger \otimes S_-, b \otimes S_3] = \frac{1}{2}I_B \otimes S_- + 2b^\dagger b \otimes S_-$$
$$[b \otimes S_+, b \otimes S_-] = 2b^2 \otimes S_3$$
$$[b \otimes S_+, b^\dagger \otimes S_3] = -\frac{1}{2}I_B \otimes S_+ - b^\dagger b \otimes S_+$$
$$[b \otimes S_+, b \otimes S_3] = -b^2 \otimes S_+$$
$$[b \otimes S_-, b^\dagger \otimes S_3] = \frac{1}{2}I_B \otimes S_+ + b^\dagger B \otimes S_-$$
$$[b \otimes S_-, b \otimes S_3] = b^2 \otimes S_-$$
$$[b^\dagger \otimes S_3, b \otimes S_3] = -\frac{1}{4}I_B \otimes I_2.$$

(ii) Note that $S_+S_3 + S_3S_+ = 0_2$, $S_-S_3 + S_3S_- = 0_2$, $S_+S_- + S_-S_+ = I_2$. Thus for the anticommutators we obtain

$$[b^\dagger \otimes S_+, b^\dagger \otimes S_-]_+ = (b^\dagger)^2 \otimes I_2$$
$$[b^\dagger \otimes S_+, b \otimes S_+]_+ = 0_B \otimes 0_2$$
$$[b^\dagger \otimes S_+, b \otimes S_-]_+ = b^\dagger b \otimes I_2 + I_B \otimes \begin{pmatrix} 0 & 0 \\ 0 & 1 \end{pmatrix}$$
$$[b^\dagger \otimes S_+, b^\dagger \otimes S_3]_+ = 0_B \otimes 0_2$$
$$[b^\dagger \otimes S_+, b \otimes S_3]_+ = \frac{1}{2}I_B \otimes S_+$$
$$[b^\dagger \otimes S_-, b \otimes S_+]_+ = b^\dagger b \otimes I_2 + I_B \otimes \begin{pmatrix} 1 & 0 \\ 0 & 0 \end{pmatrix}$$
$$[b^\dagger \otimes S_-, b \otimes S_-]_+ = 0_B \otimes 0_2$$
$$[b^\dagger \otimes S_-, b^\dagger \otimes S_3]_+ = 0_B \otimes 0_2$$
$$[b^\dagger \otimes S_-, b \otimes S_3]_+ = -\frac{1}{2}I_B \otimes S_-$$
$$[b \otimes S_+, b \otimes S_-]_+ = b^2 \otimes I_2$$
$$[b \otimes S_+, b^\dagger \otimes S_3]_+ = -\frac{1}{2}I_B \otimes S_+$$
$$[b \otimes S_+, b \otimes S_3]_+ = 0_B \otimes 0_2$$
$$[b \otimes S_-, b^\dagger \otimes S_3]_+ = \frac{1}{2}I_B \otimes S_-$$
$$[b \otimes S_-, b \otimes S_3]_+ = 0_B \otimes 0_2$$
$$[b^\dagger \otimes S_3, b \otimes S_3]_+ = \frac{1}{2}b^\dagger b \otimes I_2 + \frac{1}{4}I_B \otimes I_2.$$

Problem 3. Consider the operator

$$\hat{N} = b^\dagger b \otimes I_2 + I_B \otimes S_3$$

where

$$b^\dagger b = \text{diag}(0\,1\,2\,3\,\ldots), \qquad S_3 = \frac{1}{2}\begin{pmatrix} 1 & 0 \\ 0 & -1 \end{pmatrix}, \qquad I_2 = \begin{pmatrix} 1 & 0 \\ 0 & 1 \end{pmatrix}$$

and I_B is the infinite dimensional identity matrix.
(i) Find the eigenvalues of $\hat{N}$.
(ii) Does the operator $\hat{N}$ commute with the operator

$$\hat{K} = b^\dagger \otimes S_- + b \otimes S_+ ?$$

Solution 3. (i) We obtain the diagonal matrix for $\hat{N}$

$$\hat{N} = \begin{pmatrix} 1/2 & 0 \\ 0 & -1/2 \end{pmatrix} \oplus \begin{pmatrix} 3/2 & 0 \\ 0 & 1/2 \end{pmatrix} \oplus \begin{pmatrix} 5/2 & 0 \\ 0 & 3/2 \end{pmatrix} \oplus \begin{pmatrix} 7/2 & 0 \\ 0 & 5/2 \end{pmatrix} \oplus \cdots$$

where $\oplus$ denotes the direct sum. Therefore the eigenvalues are $-1/2$, $1/2$ (twice), $3/2$ (twice), $5/2$ (twice), $7/2$ (twice) etc. Hence the lowest eigenvalue is not degenerate.
(ii) Since $[S_-, S_3] = S_-$, $[S_+, S_3] = -S_+$ and $bb^\dagger = I + b^\dagger b$ we end up with

$$[\hat{K}, \hat{N}] = 0_B \otimes 0_2$$

where 0_2 is the 2×2 zero matrix, i.e. $\hat{K}$ and $\hat{N}$ commute.

Problem 4. Let $z \in \mathbb{C}$ and σ_1 the first Pauli spin matrix.
(i) Calculate the commutators of the operators in the product Hilbert space $\mathcal{H}_B \otimes \mathbb{C}^2$

$$\bar{z}b \otimes \sigma_1, \qquad zb^\dagger \otimes \sigma_1.$$

(ii) Find the state

$$(\bar{z}b \otimes \sigma_1 + zb^\dagger \otimes \sigma_1)|0\rangle \otimes \begin{pmatrix} 1 \\ 0 \end{pmatrix}.$$

Solution 4. (i) Since $[b, b^\dagger] = I_B$ and $\sigma_1^2 = I_2$ we find

$$[\bar{z}b \otimes \sigma_1, zb^\dagger \otimes \sigma_1] = z\bar{z}I_B \otimes I_2.$$

(ii) Since $b|0\rangle = 0|0\rangle$ we find the state

$$b^\dagger|0\rangle \otimes \begin{pmatrix} 0 \\ 1 \end{pmatrix}.$$

Problem 5. Find the spectrum of the operator

$$\hat{P} = e^{i\pi b^\dagger b} \otimes \sigma_3.$$

Solution 5. We have

$$\hat{P} = e^{i\pi b^\dagger b} \otimes \sigma_3 = \text{diag}(1, -1, 1, -1, \ldots) \otimes \begin{pmatrix} 1 & 0 \\ 0 & -1 \end{pmatrix}$$

Thus the eigenvalues are $+1$ and -1 infinitely degenerate.

Problem 6.   Let

$$S_+ = \begin{pmatrix} 0 & 1 \\ 0 & 0 \end{pmatrix}, \quad S_- = \begin{pmatrix} 0 & 0 \\ 1 & 0 \end{pmatrix}, \quad S_3 = \frac{1}{2} \begin{pmatrix} 1 & 0 \\ 0 & -1 \end{pmatrix}.$$

(i) Calculate the commutator

$$[b^\dagger b \otimes I_2, b \otimes S_+ + b^\dagger \otimes S_-]$$

where I_2 is the 2×2 identity matrix.
(ii) Calculate the commutator

$$[I_B \otimes S_3, b \otimes S_+ + b^\dagger \otimes S_-]$$

where I_B is the identity operator for the Bose system.
(iii) Calculate the commutator

$$[b^\dagger b \otimes I_2 + I_B \otimes S_3, b \otimes S_+ + b^\dagger \otimes S_-].$$

Utilize the results from (i) and (ii).

Solution 6.   (i) Since $bb^\dagger = I_B + b^\dagger b$ we obtain

$$\begin{aligned}
[b^\dagger b \otimes I_2, b \otimes S_+ + b^\dagger \otimes S_-] &= [b^\dagger b \otimes I_2, b \otimes S_+] + [b^\dagger b \otimes I_2, b^\dagger \otimes S_-] \\
&= b^\dagger b b \otimes S_+ - b b^\dagger \otimes S_+ + b^\dagger b b^\dagger \otimes S_- - b^\dagger b^\dagger b \otimes S_- \\
&= -b \otimes S_+ + b^\dagger \otimes S_-.
\end{aligned}$$

(ii) Since $S_3 S_+ = S_+/2$, $S_+ S_3 = -S_+/2$, $S_3 S_- = -S_-/2$, $S_- S_3 = S_-/2$ we obtain

$$[I_B \otimes S_3, b \otimes S_+ + b^\dagger \otimes S_-] = b \otimes S_+ - b^\dagger \otimes S_-.$$

(iii) From (i) and (ii) we obtain

$$[b^\dagger b \otimes I_2 + I_B \otimes S_3, b \otimes S_+ + b^\dagger \otimes S_-] = -b \otimes S_+ + b^\dagger \otimes S_- + b \otimes S_+ - b^\dagger \otimes S_- = 0_B \otimes 0_2$$

i.e. the two operators $b^\dagger b \otimes I_2 + I_B \otimes S_3$ and $b \otimes S_+ + b^\dagger \otimes S_-$ commute.

Problem 7.   (i) Consider the operators

$$\hat{N} = b^\dagger b \otimes I_2 + I_B \otimes S_3, \qquad \hat{K} = (b^\dagger + b) \otimes (S_+ + S_-) \equiv b^\dagger \otimes S_+ + b^\dagger \otimes S_- + b \otimes S_+ + b \otimes S_-.$$

Do $\hat{N}$ and $\hat{K}$ commute?
(ii) Consider the operator

$$V = (I_B + e^{i\pi b^\dagger b}) \otimes S_3 + (I_B - e^{-i\pi b^\dagger b}) \otimes i S_2$$

where I_B is the identity operator. Find VV^*.

Solution 7.   (i) From a previous problem we have

$$[\hat{N}, b \otimes S_+ + b^\dagger \otimes S_-] = 0_B \otimes 0_2.$$

Thus
$$[\hat{N}, \hat{K}] = [\hat{N}, b^\dagger \otimes S_3 + b \otimes S_-] = 2(b^\dagger \otimes S_+ - b \otimes S_-).$$
(ii) We have $VV^* = I_B \otimes I_2$, i.e. the operator is unitary.

Problem 8. We know that

$$|\beta\rangle = D(\beta)|0\rangle = e^{\beta b^\dagger - \bar{\beta}b}|0\rangle = e^{\beta b^\dagger}e^{-\bar{\beta}b}e^{-\beta\bar{\beta}I/2}|0\rangle = e^{-\beta\bar{\beta}/2}e^{\beta b^\dagger}|0\rangle = e^{-\beta\bar{\beta}/2}\sum_{n=0}^{\infty}\frac{\beta^n}{\sqrt{n!}}|n\rangle$$

where $|n\rangle$ are the number states

$$|n\rangle = \frac{(b^\dagger)^n}{\sqrt{n!}}|0\rangle.$$

Let σ_1 be the first Pauli spin matrix. Extend the calculation to

$$\tilde{D}(\beta)(|0\rangle \otimes \begin{pmatrix} 1 \\ 0 \end{pmatrix})$$

where

$$\tilde{D}(\beta) := \exp(\beta b^\dagger \otimes \sigma_1 - \bar{\beta}b \otimes \sigma_1).$$

Solution 8. First we note that since $\sigma_1^2 = I_2$ and $[b, b^\dagger] = I_B$

$$-[\beta b^\dagger \otimes \sigma_1, -\bar{\beta}b \otimes \sigma_1] = -\beta\bar{\beta}I_B \otimes I_2.$$

Thus we obtain

$$\tilde{D}(\beta)|0\rangle \otimes \begin{pmatrix} 1 \\ 0 \end{pmatrix} = e^{\beta b^\dagger \otimes \sigma_1}e^{-\bar{\beta}b \otimes \sigma_1}e^{-\beta\bar{\beta}I_B \otimes I_2/2}|0\rangle \otimes \begin{pmatrix} 1 \\ 0 \end{pmatrix}$$

$$= e^{-\beta\bar{\beta}/2}e^{\beta b^\dagger \otimes \sigma_1}e^{-\bar{\beta}b \otimes \sigma_1}|0\rangle \otimes \begin{pmatrix} 1 \\ 0 \end{pmatrix} = e^{-\beta\bar{\beta}/2}e^{\beta b^\dagger \otimes \sigma_1}|0\rangle \otimes \begin{pmatrix} 1 \\ 0 \end{pmatrix}$$

$$= e^{-\beta\bar{\beta}/2}\left((|0\rangle + \frac{\beta^2}{\sqrt{2!}}|2\rangle + \frac{\beta^4}{\sqrt{4!}}|4\rangle + \cdots) \otimes \begin{pmatrix} 1 \\ 0 \end{pmatrix}\right.$$

$$\left. +(\beta|1\rangle + \frac{\beta^3}{\sqrt{3!}}|3\rangle + \cdots) \otimes \begin{pmatrix} 0 \\ 1 \end{pmatrix}\right)$$

$$= e^{-\beta\bar{\beta}/2}\left(\left(\sum_{n=0}^{\infty}\frac{\beta^{2n}}{\sqrt{(2n)!}}|2n\rangle\right) \otimes \begin{pmatrix} 1 \\ 0 \end{pmatrix}\right.$$

$$\left. +\left(\sum_{n=0}^{\infty}\frac{\beta^{2n+1}}{\sqrt{(2n+1)!}}|2n+1\rangle\right) \otimes \begin{pmatrix} 0 \\ 1 \end{pmatrix}\right).$$

Problem 9. (i) The Hamilton operator for the one-dimensional harmonic oscillator

$$\hat{H}_B = \frac{1}{2m}\hat{p}^2 + \frac{1}{2}\omega^2\hat{q}^2$$

can be expressed with Bose creation and annihilation operators as

$$\hat{H}_B = \hbar\omega\left(b^\dagger b + \frac{1}{2}I_B\right)$$

where I_B is the identity operator. Consider the Hamilton operator

$$\hat{H} = \hbar\omega\left(b^\dagger b + \frac{1}{2}I_B\right) \otimes I_2 - \frac{1}{2}\hbar\omega I_B \otimes \sigma_3$$

where I_2 is the 2×2 identity matrix and σ_3 is the third Pauli spin matrix with the eigenvalues $+1$ and -1 and the corresponding normalized eigenvectors

$$\begin{pmatrix} 1 \\ 0 \end{pmatrix}, \quad \begin{pmatrix} 0 \\ 1 \end{pmatrix}.$$

Let $|n\rangle$ be the number states $(n = 0, 1, \ldots)$. Find the eigenvalues and eigenvectors of $\hat{H}$.
(ii) Consider the Hamilton operator

$$\hat{H} = \hat{H}_0 + \hat{H}_1 + \frac{\hbar\omega}{2}I_B \otimes I_2$$

where

$$\hat{H}_0 = \hbar\omega b^\dagger b \otimes I_2 + I_B \otimes \frac{1}{2}\hbar\omega\sigma_3, \quad \hat{H}_1 = \hbar\kappa(b^\dagger \otimes S_- + b \otimes S_+) - \frac{\Delta\hbar\omega}{2}I_B \otimes \sigma_3$$

and I denotes the identity operator. Note that $[S_+, S_-] = \sigma_3$. Calculate the commutators $[\hat{H}_0, \hat{H}_1]$, $[\hat{H}, \hat{H}_1]$, $[\hat{H}, \hat{H}_2]$.

Solution 9.   (i) The ground state energy is given by $E_0 = 0$ with the eigenstate

$$|0\rangle \otimes \begin{pmatrix} 1 \\ 0 \end{pmatrix}.$$

(ii) We find

$$[\hat{H}_1, \hat{H}_2] = [\hat{H}, \hat{H}_1] = [\hat{H}, \hat{H}_2] = 0_B \otimes 0_2.$$

Problem 10.   Consider the Hamilton operator for a single two-level atom coupled to a single mode of an electromagnetic field

$$\hat{K} = \frac{\hat{H}}{\hbar\omega} = b^\dagger \otimes S_- + b \otimes S_+$$

where

$$S_+ = \begin{pmatrix} 0 & 1 \\ 0 & 0 \end{pmatrix}, \quad S_- = \begin{pmatrix} 0 & 0 \\ 1 & 0 \end{pmatrix}.$$

Let $\theta \in \mathbb{R}$ and $U = e^{i\theta\hat{K}}$. Calculate

$$A(n, \beta) = (\langle n| \otimes I_2)U(|\beta\rangle \otimes I_2)$$

where $|\beta\rangle$ and $|n\rangle$ are coherent states and number states of the electromagnetic field, respectively.

Solution 10.   Since $S_+S_+ = 0_2$ and $S_-S_- = 0_2$ we find

$$S_+S_- = \begin{pmatrix} 1 & 0 \\ 0 & 0 \end{pmatrix}, \quad S_-S_+ = \begin{pmatrix} 0 & 0 \\ 0 & 1 \end{pmatrix}$$

we obtain

$$(b^\dagger \otimes S_- + b \otimes S_+)(b^\dagger \otimes S_- + b \otimes S_+) = (I + b^\dagger b) \otimes S_+ S_- + b^\dagger b \otimes S_- S_+$$
$$= I \otimes S_+ S_- + b^\dagger b \otimes I_2.$$

Using this result we arrive at

$$A(\beta, n) = e^{-|\beta^2|} \frac{|\beta|^2}{n!} \begin{pmatrix} \cos(\theta\sqrt{n}) & \frac{i\sqrt{n}}{\beta}\sin(\theta\sqrt{n}) \\ \frac{i\beta}{\sqrt{n+1}}\sin(\theta\sqrt{n+1}) & \cos(\theta\sqrt{n+1}) \end{pmatrix}.$$

Problem 11. (i) Study the spectrum of the operator

$$(b^\dagger b) \otimes \sigma_1.$$

(ii) Study the spectrum of the Hamilton operator

$$\hat{H} = \hbar\omega_1(b^\dagger b \otimes I_2) + \hbar\omega_2(I_B \otimes \sigma_3).$$

(iii) Study the spectrum of the Hamilton operator $(b^\dagger + b) \otimes \sigma_1$.

Solution 11. (i) Since $b^\dagger b$ is the infinite dimensional diagonal matrix

$$b^\dagger b = \text{diag}(0\,1\,2\,3\,\ldots)$$

and

$$\sigma_1 = \begin{pmatrix} 0 & 1 \\ 1 & 0 \end{pmatrix}$$

we obtain ($\oplus$ denotes the direct sum)

$$\begin{pmatrix} 0 & 0 \\ 0 & 0 \end{pmatrix} \oplus \begin{pmatrix} 0 & 1 \\ 1 & 0 \end{pmatrix} \oplus \begin{pmatrix} 0 & 2 \\ 2 & 0 \end{pmatrix} \oplus \cdots$$

with the eigenvalues 0 (twice), $+1$, -1, $+2$, -2 etc.

(ii) Since

$$b^\dagger b \otimes I_2 = \begin{pmatrix} 0 & 0 \\ 0 & 0 \end{pmatrix} \oplus \begin{pmatrix} 1 & 0 \\ 0 & 1 \end{pmatrix} \oplus \begin{pmatrix} 2 & 0 \\ 0 & 2 \end{pmatrix} \oplus \cdots$$

and

$$I_B \otimes \sigma_3 = \begin{pmatrix} 1 & 0 \\ 0 & -1 \end{pmatrix} \oplus \begin{pmatrix} 1 & 0 \\ 0 & -1 \end{pmatrix} \oplus \begin{pmatrix} 1 & 0 \\ 0 & -1 \end{pmatrix} \oplus \cdots$$

we obtain the matrix representation

$$\begin{pmatrix} \hbar\omega_2 & 0 \\ 0 & -\hbar\omega_2 \end{pmatrix} \oplus \begin{pmatrix} \hbar\omega_1 + \hbar\omega_2 & 0 \\ 0 & \hbar\omega_1 - \hbar\omega_2 \end{pmatrix} \oplus \begin{pmatrix} 2\hbar\omega_1 + \hbar\omega_2 & 0 \\ 0 & 2\hbar\omega_1 - \hbar\omega_2 \end{pmatrix} \oplus \cdots$$

with the eigenvalues

$$\hbar\omega_2, \ -\hbar\omega_2, \ \hbar\omega_1 + \hbar\omega_2, \ \hbar\omega_1 - \hbar\omega_2, \ 2\hbar\omega_1 + \hbar\omega_2, \ 2\hbar\omega_1 - \hbar\omega_2, \ \ldots$$

(iii) Since the position operator $\hat{q}$ is given by

$$\hat{q} = \frac{1}{\sqrt{2}}\ell_0(b + b^\dagger)$$

with $\ell_0 = \sqrt{\hbar/(m\omega)}$ we can consider these operator T_q acting as $T_q f(q) = q f(q)$ in the Hilbert space $L_2(\mathbb{R})$ with

$$D(T_q) := \{\, f(q) \,:\, f(q) \text{ and } q f(q) \text{ in } L_2(\mathbb{R}) \,\}$$

to find the spectrum of $b + b^\dagger$. Now T_q has a purely continuous spectrum of the entire real axis, i.e. the spectrum of $b + b^\dagger$ consists of the entire real axis. The spectrum of σ_1 is given by the eigenvalues $+1$ and -1.

Problem 12. Let $b_1, b_2, b_1^\dagger, b_2^\dagger$ be Bose annihilation and creation operators. Consider the four 2×2 matrices

$$S_+ = \begin{pmatrix} 0 & 1 \\ 0 & 0 \end{pmatrix}, \quad S_- = \begin{pmatrix} 0 & 0 \\ 1 & 0 \end{pmatrix}, \quad \sigma_3 = \begin{pmatrix} 1 & 0 \\ 0 & -1 \end{pmatrix}, \quad I_2 = \begin{pmatrix} 1 & 0 \\ 0 & 1 \end{pmatrix}.$$

Consider now the operators in the product space

$$J_+ = b_1^\dagger b_2 \otimes I_2 + I \otimes S_+, \quad J_- = b_2^\dagger b_1 \otimes I_2 + I \otimes S_-, \quad J_3 = (b_1^\dagger b_1 - b_2^\dagger b_2) \otimes I_2 + I \otimes \sigma_3$$

where $I = I_B \otimes I_B$. Find the commutators $[J_+, J_-]$, $[J_+, J_3]$, $[J_-, J_3]$.

Solution 12. Note that $[b_1^\dagger b_2, b_2^\dagger b_1] = b_1^\dagger b_1 - b_2^\dagger b_2$, $[S_+, S_-] = \sigma_3$. Thus

$$[J_+, J_-] = [b_1^\dagger b_2, b_2^\dagger b_1] \otimes I_2 + I \otimes [S_+, S_-] = (b_1^\dagger b_1 - b_2^\dagger b_2) \otimes I_2 + I \otimes \sigma_3 = J_3.$$

Analogously we find $[J_+, J_3] = -2J_+$ and $[J_-, J_3] = 2J_-$.

Problem 13. Consider the generalized $E \otimes \epsilon$ *Jahn-Teller Hamilton operator*

$$\hat{H} = b_1^\dagger b_1 \otimes I_2 + b_2^\dagger b_2 \otimes I_2 + I_B \otimes I_2 + \left(\frac{1}{2} + 2\delta\right) \otimes \sigma_3 + \kappa((b_1 + b_1^\dagger) \otimes S_+ + (b_2 + b_2^\dagger) \otimes S_-$$

which describes two boson modes 1 and 2 interacting with a two-level system and I_2 is the 2×2 identity matrix.
(i) Show that the operator

$$J = b_1^\dagger b_1 \otimes I_2 - b_2^\dagger b_2 \otimes I_2 + I_B \otimes \frac{1}{2}\sigma_3$$

is a constant of motion.
(ii) Find the eigenvectors of J.

Solution 13. (i) Since $S_3 S_+ = S_+/2$ and $S_3 S_- = -S/2$ and

$$[b_j^\dagger b_j, b_j + b_j^\dagger] = -b + b^\dagger$$

we obtain

$$[\hat{H}, \hat{J}] = 0.$$

(ii) The eigenvectors of $\hat{J}$ are $(j = 0, 1, 2, \dots)$

$$|\psi\rangle_{j+1/2} = (b_1^\dagger)^j f(b_1^\dagger b_2^\dagger)|0\rangle \otimes |\uparrow\rangle + (b_1^\dagger)^{j+1} g(b_1^\dagger b_2^\dagger)|0\rangle \otimes |\downarrow\rangle$$

with

$$\hat{J}|\psi\rangle_{j+1/2} = \left(j + \frac{1}{2}\right)|\psi\rangle_{j+1/2}$$

where $b_1|0\rangle = 0$, $b_2|0\rangle = 0$.

Problem 14. Let $b^\dagger$, b be Bose creation and annihilation operators. Consider the operators

$$J_+ = -\frac{1}{2}b^\dagger b^\dagger \otimes I_n, \quad J_- = \frac{1}{2}bb \otimes I_n, \quad J_0 = \frac{1}{2}\left(b^\dagger b + \frac{1}{2}I_B\right) \otimes I_n$$

which form a basis of a Lie algebra, where I_n is the $n \times n$ identity matrix. Let X be an $n \times n$ matrix with $X^2 = I_n$ and $X = X^*$, for example for $n = 2$ the Pauli spin matrices. Consider the operators

$$V_+ = \frac{1}{\sqrt{2}}b^\dagger \otimes X, \quad V_- = \frac{1}{\sqrt{2}}b \otimes X.$$

Find the anticommutators $[V_+, V_-]_+$, $[V_+, V_+]_+$, $[V_-, V_-]_+$.

Solution 14. Since $X^2 = I_n$ we obtain

$$[V_+, V_-]_+ = (b^\dagger b + \frac{1}{2}I_B)\otimes I_n = 2J_0, \quad [V_+, V_+]_+ = b^\dagger b^\dagger \otimes I_n = -2J_+, \quad [V_-, V_-]_+ = bb\otimes I_n = 2J_-.$$

Hence we obtain the three elements of the Lie algebra.

Problem 15. Let I_2 the 2×2 be the identity matrix

$$S_+ := \begin{pmatrix} 0 & 1 \\ 0 & 0 \end{pmatrix}, \quad S_- := \begin{pmatrix} 0 & 0 \\ 1 & 0 \end{pmatrix}, \quad S_3 := \frac{1}{2}\begin{pmatrix} 1 & 0 \\ 0 & -1 \end{pmatrix}.$$

Consider the following form of the *Jaynes-Cummings operator*

$$\hat{H} = \hat{H}_1 + \hat{H}_2 + \frac{1}{2}\hbar\omega(I_B \otimes I_2)$$

where

$$\hat{H}_1 = \hbar\omega(b^\dagger b \otimes I_2 + I_B \otimes S_3)$$

and

$$\hat{H}_2 = \hbar\kappa(b^\dagger \otimes \sigma_- + b \otimes \sigma_+) - \hbar(\omega - \omega_0)I_B \otimes \sigma_3$$

with I_B the identity operator for the Bose system. Calculate the commutator $[\hat{H}_1, \hat{H}_2]$.

Solution 15. Since

$$[b^\dagger b \otimes I_2, b^\dagger \otimes S_-] = b^\dagger \otimes S_-$$
$$[b^\dagger b \otimes I_2, b \otimes S_+] = -b \otimes S_+$$
$$\frac{1}{2}[I_B \otimes S_3, b^\dagger \otimes S_-] = -b^\dagger \otimes S_-$$
$$\frac{1}{2}[I_B \otimes \sigma_3, b \otimes S_+] = b \otimes S_+$$
$$[I_B \otimes S_3, I_B \otimes S_3] = 0_B \otimes 0_2$$

we obtain that $[\hat{H}_1, \hat{H}_2] = 0$, i.e. $\hat{H}_1$ and $\hat{H}_2$ commute.

Problem 16. The Hamilton operator for the *Rabi model* is given by

$$\hat{H} = \hbar\omega b^\dagger b \otimes I_2 + \Delta I_B \otimes \sigma_3 + U(b^\dagger + b) \otimes \sigma_1$$

where Δ is a symmetry breaking field for the spin, U is the coupling strength between the Bose particles (photons) and the spin.

(i) Does the Hamilton operator commute with the operator

$$e^{i\pi b^\dagger b} \otimes \sigma_3.$$

(ii) Let $|n\rangle$ be the number states $(n = 0, 1, \ldots)$. Consider the product states

$$|n\rangle \otimes \begin{pmatrix} 1 \\ 0 \end{pmatrix}, \qquad |n\rangle \otimes \begin{pmatrix} 0 \\ 1 \end{pmatrix}.$$

Apply the Hamilton operator $\hat{H}$ to these states. Discuss.

Solution 16. (i) Yes. Note that $\sigma_1^2 = I_2$. We have

$$[\hat{H}, e^{i\pi(b^\dagger b \otimes \sigma_3)}] = 0_B \otimes 0_2.$$

This is the so-called *parity invariance*.

(ii) Since $b|n\rangle = \sqrt{n}|n-1\rangle$, $b^\dagger|n\rangle = \sqrt{n+1}|n+1\rangle$ we obtain

$$\hat{H}(|n\rangle \otimes \begin{pmatrix} 1 \\ 0 \end{pmatrix} = \hbar\omega n|n\rangle \otimes \begin{pmatrix} 1 \\ 0 \end{pmatrix} + \Delta|n\rangle \otimes \begin{pmatrix} 1 \\ 0 \end{pmatrix} + U(\sqrt{n+1}|n+1\rangle + \sqrt{n}|n-1\rangle) \otimes \begin{pmatrix} 0 \\ 1 \end{pmatrix}$$

$$\hat{H}(|n\rangle \otimes \begin{pmatrix} 0 \\ 1 \end{pmatrix} = \hbar\omega n|n\rangle \otimes \begin{pmatrix} 0 \\ 1 \end{pmatrix} - \Delta|n\rangle \otimes \begin{pmatrix} 1 \\ 0 \end{pmatrix} + U(\sqrt{n+1}|n+1\rangle + \sqrt{n}|n-1\rangle) \otimes \begin{pmatrix} 1 \\ 0 \end{pmatrix}.$$

Problem 17. Let $\mathbf{S} = (S_1, S_2.S_3)^T$ be the spin operators

$$S_1 = \frac{1}{2}\begin{pmatrix} 0 & 1 \\ 1 & 0 \end{pmatrix}, \quad S_2 = \frac{1}{2}\begin{pmatrix} 0 & -i \\ i & 0 \end{pmatrix}, \quad S_3 = \frac{1}{2}\begin{pmatrix} 1 & 0 \\ 0 & -1 \end{pmatrix}$$

and $\hat{\mathbf{p}} = (\hat{p}_1, \hat{p}_2, \hat{p}_3)^T$ be the momentum operators

$$\hat{p}_1 = -i\frac{\partial}{\partial x_1}, \quad \hat{p}_2 = -i\frac{\partial}{\partial x_2}, \quad \hat{p}_3 = -i\frac{\partial}{\partial x_3}$$

and $\mathbf{x} = (x_1, x_2, x_3)$. Let $\times$ be the vector product. Then

$$\mathbf{x} \times \hat{\mathbf{p}} = \begin{pmatrix} x_2\hat{p}_3 - x_3\hat{p}_2 \\ x_3\hat{p}_1 - x_1\hat{p}_3 \\ x_1\hat{p}_2 - x_2\hat{p}_1 \end{pmatrix}$$

and ($\cdot$ denotes the scalar product)

$$(\mathbf{x} \times \hat{\mathbf{p}}) \cdot \hat{\mathbf{p}} = 0.$$

Let I_2 be the 2×2 identity matrix and I the identity operator. Consider the operator

$$\hat{\mathbf{J}} = (\mathbf{x} \times \hat{\mathbf{p}}) \otimes I_2 + I \otimes \mathbf{S}$$

where $\otimes$ is the tensor product. Find the operator

$$\hat{\mathbf{J}} \cdot (\hat{\mathbf{p}} \otimes I_2).$$

Solution 17. Using the property that $(\mathbf{x} \times \hat{\mathbf{p}}) \cdot \hat{\mathbf{p}} = 0$ we obtain

$$\hat{\mathbf{J}} \cdot (\hat{\mathbf{p}} \otimes I_2) = ((\mathbf{x} \times \hat{\mathbf{p}}) \cdot \hat{\mathbf{p}}) \otimes I_2 + \hat{\mathbf{p}} \otimes \mathbf{S} = \hat{\mathbf{p}} \otimes \mathbf{S}.$$

Problem 18. Calculate the eigenvalues of the Hamilton operator

$$\hat{H} := \lambda(S_1 \otimes \hat{L}_1 + S_2 \otimes \hat{L}_2 + S_3 \otimes \hat{L}_3)$$

where we consider a subspace G_1 of the Hilbert space $L_2(S^2)$ with

$$S^2 := \{(x, y, z) \ x^2 + y^2 + z^2 = 1\}.$$

Here $\otimes$ denotes the *tensor product*. The linear operators $\hat{L}_1$, $\hat{L}_2$, $\hat{L}_3$ act in the subspace G_1. A basis of subspace G_1 is

$$Y_{1,0} = \sqrt{\frac{3}{4\pi}} \cos\theta, \quad Y_{1,1} = -\sqrt{\frac{3}{8\pi}} \sin\theta e^{i\phi}, \quad Y_{1,-1} = \sqrt{\frac{3}{8\pi}} \sin\theta e^{-i\phi}.$$

The operators (matrices) S_1, S_2, S_3 act in the Hilbert space $\mathbb{C}^2$. The standard basis is given by

$$\begin{pmatrix} 1 \\ 0 \end{pmatrix}, \quad \begin{pmatrix} 0 \\ 1 \end{pmatrix}.$$

The *spin matrices* S_3, S_+ and S_- are given by

$$S_3 := \frac{1}{2}\hbar \begin{pmatrix} 1 & 0 \\ 0 & -1 \end{pmatrix}, \quad S_+ := \hbar \begin{pmatrix} 0 & 1 \\ 0 & 0 \end{pmatrix}, \quad S_- := \hbar \begin{pmatrix} 0 & 0 \\ 1 & 0 \end{pmatrix}$$

where $S_\pm := S_1 \pm iS_2$. The operators $\hat{L}_3$, $\hat{L}_+$ and $\hat{L}_-$ take the form

$$\hat{L}_3 := -i\hbar \frac{\partial}{\partial\phi}$$

$$\hat{L}_+ := \hbar e^{i\phi} \left(\frac{\partial}{\partial\theta} + i \cot\theta \frac{\partial}{\partial\phi} \right)$$

$$\hat{L}_- := \hbar e^{-i\phi} \left(-\frac{\partial}{\partial\theta} + i \cot\theta \frac{\partial}{\partial\phi} \right)$$

where $\hat{L}_\pm := \hat{L}_1 \pm i\hat{L}_2$. The Hamilton operator describes the *spin-orbit coupling*. In some textbooks we find the notation $\hat{H} = \lambda \mathbf{S} \cdot \mathbf{L}$.

Solution 18. The Hamilton operator (1) can be written as

$$\hat{H} = \lambda(S_3 \otimes \hat{L}_3) + \frac{\lambda}{2}(S_+ \otimes \hat{L}_- + S_- \otimes \hat{L}_+).$$

In the tensor product space $\mathbb{C}^2 \otimes G_1$ a basis is given by

$$|1\rangle = \begin{pmatrix} 1 \\ 0 \end{pmatrix} \otimes Y_{1,0}, \quad |2\rangle = \begin{pmatrix} 1 \\ 0 \end{pmatrix} \otimes Y_{1,-1}, \quad |3\rangle = \begin{pmatrix} 1 \\ 0 \end{pmatrix} \otimes Y_{1,1}$$

$$|4\rangle = \begin{pmatrix} 0 \\ 1 \end{pmatrix} \otimes Y_{1,0}, \quad |5\rangle = \begin{pmatrix} 0 \\ 1 \end{pmatrix} \otimes Y_{1,-1}, \quad |6\rangle = \begin{pmatrix} 0 \\ 1 \end{pmatrix} \otimes Y_{1,1}.$$

In the following we use

$$\hat{L}_+Y_{1,1} = 0, \qquad \hat{L}_+Y_{1,0} = \hbar\sqrt{2}Y_{1,1}, \qquad \hat{L}_+Y_{1,-1} = \hbar\sqrt{2}Y_{1,0}$$

$$\hat{L}_-Y_{1,1} = \hbar\sqrt{2}Y_{1,0}, \qquad \hat{L}_-Y_{1,0} = \hbar\sqrt{2}Y_{1,-1}, \qquad \hat{L}_-Y_{1,-1} = 0$$

$$\hat{L}_3Y_{1,1} = \hbar Y_{1,1}, \qquad \hat{L}_3Y_{1,0} = 0, \qquad \hat{L}_3Y_{1,-1} = -\hbar Y_{1,-1}.$$

For the state $|1\rangle$ we find

$$\hat{H}|1\rangle = \left[\lambda(S_3 \otimes \hat{L}_3) + \frac{\lambda}{2}(S_+ \otimes \hat{L}_- + S_- \otimes \hat{L}_+)\right]\begin{pmatrix}1\\0\end{pmatrix} \otimes Y_{1,0}.$$

Thus

$$\hat{H}|1\rangle = \lambda\left[S_3\begin{pmatrix}1\\0\end{pmatrix} \otimes \hat{L}_3Y_{1,0}\right] + \frac{\lambda}{2}\left[S_+\begin{pmatrix}1\\0\end{pmatrix} \otimes \hat{L}_-Y_{1,0} + S_-\begin{pmatrix}1\\0\end{pmatrix} \otimes \hat{L}_+Y_{1,0}\right].$$

Finally

$$\hat{H}|1\rangle = \frac{\lambda}{2}S_-\begin{pmatrix}1\\0\end{pmatrix} \otimes \hat{L}_+Y_{1,0} = \frac{\lambda}{\sqrt{2}}\hbar^2|6\rangle.$$

Analogously, we find

$$\hat{H}|2\rangle = -\frac{\lambda\hbar^2}{2}|2\rangle + \frac{\lambda\hbar^2}{\sqrt{2}}|4\rangle$$

$$\hat{H}|3\rangle = \frac{\lambda\hbar^2}{2}|3\rangle$$

$$\hat{H}|4\rangle = \frac{\lambda\hbar^2}{2}|2\rangle$$

$$\hat{H}|5\rangle = \frac{\lambda\hbar^2}{2}|5\rangle$$

$$\hat{H}|6\rangle = -\frac{\lambda\hbar^2}{2}|6\rangle + \frac{\lambda\hbar^2}{\sqrt{2}}|1\rangle.$$

Hence the states $|3\rangle$ and $|5\rangle$ are eigenstates with the eigenvalues $E_{1,2} = \lambda\hbar^2/2$. The states $|1\rangle$ and $|6\rangle$ form a two-dimensional subspace. The matrix representation is given by

$$\begin{pmatrix} 0 & \dfrac{\lambda\hbar^2}{\sqrt{2}} \\ \dfrac{\lambda\hbar^2}{\sqrt{2}} & -\dfrac{\lambda\hbar^2}{2} \end{pmatrix}.$$

The eigenvalues are

$$E_{3,4} = -\frac{\lambda\hbar^2}{2} \pm \frac{3\lambda\hbar^2}{4}.$$

Analogously, the states $|2\rangle$ and $|4\rangle$ form a two-dimensional subspace. The matrix representation is given by

$$\begin{pmatrix} -\dfrac{\lambda\hbar^2}{2} & \dfrac{\lambda\hbar^2}{\sqrt{2}} \\ \dfrac{\lambda\hbar^2}{\sqrt{2}} & 0 \end{pmatrix}.$$

The eigenvalues are

$$E_{5,6} = -\frac{\lambda\hbar^2}{2} \pm \frac{3\lambda\hbar^2}{4}.$$

Problem 19. Let $b_1, b_2, b_1^\dagger, b_2^\dagger$ be Bose annihilation and creation operators. Consider the three 2×2 matrices

$$S_+ = \begin{pmatrix} 0 & 1 \\ 0 & 0 \end{pmatrix}, \quad S_- = \begin{pmatrix} 0 & 0 \\ 1 & 0 \end{pmatrix}, \quad \sigma_3 = \begin{pmatrix} 1 & 0 \\ 0 & -1 \end{pmatrix}.$$

Consider now the operators in the product space

$$J_+ = b_1^\dagger b_2 \otimes I_2 + I_B \otimes S_+, \quad J_- = b_2^\dagger b_1 \otimes I_2 + I_B \otimes S_-, \quad J_3 = (b_1^\dagger b_1 - b_2^\dagger b_2) \otimes I_2 + I_B \otimes \sigma_3.$$

Find the commutators $[J_+, J_-]$, $[J_+, J_3]$, $[J_-, J_3]$.

Solution 19. Note that $[b_1^\dagger b_2, b_2^\dagger b_1] = b_1^\dagger b_1 - b_2^\dagger b_2$, $[S_+, S_-] = \sigma_3$. Thus

$$[J_+, J_-] = [b_1^\dagger b_2, b_2^\dagger b_1] \otimes I_2 + I_B \otimes [S_+, S_-] = (b_1^\dagger b_1 - b_2^\dagger b_2) \otimes I_2 + I_B \otimes \sigma_3 = J_3.$$

Analogously we find $[J_+, J_3] = -2J_+$, $[J_-, J_3] = 2J_-$. Thus we have a basis of a Lie algebra.

Problem 20. Consider the *Rabi model* given by

$$\hat{H} = \hbar\omega_1 b^\dagger b \otimes I_2 + \hbar\omega_2 I_B \otimes \sigma_3 + \hbar\omega_3 (b + b^\dagger \otimes \sigma_1).$$

We set

$$\hat{K} = \frac{\hat{H}}{\hbar\omega_1} = b^\dagger b \otimes I_2 + \alpha_2 I_B \otimes \sigma_3 + \alpha_3 (b + b^\dagger \otimes \sigma_1)$$

where $\alpha_2 = \omega_2/\omega_1$ and $\alpha_3 = \omega_3/\omega_1$. Consider the state

$$|\psi\rangle = f_1(b^\dagger)|0\rangle \otimes \begin{pmatrix} 1 \\ 0 \end{pmatrix} + f_2(b^\dagger)|0\rangle \otimes \begin{pmatrix} 0 \\ 1 \end{pmatrix}$$

where f_1 and f_2 are analytic functions.
(i) The eigenvalue equation is given by $\hat{H}|\psi\rangle = E|\psi\rangle$. Write down the eigenvalues equation $\hat{K}|\psi\rangle = \tilde{E}|\psi\rangle$, where $\tilde{E} = E/(\hbar\omega_1)$.
(ii) Apply the *Bargmann realization*

$$b^\dagger \mapsto z \qquad b \mapsto \frac{d}{dz}$$

to obtain the corresponding system of linear differential equations.

Solution 20. (i) We have

$$\hat{K}|\psi\rangle = b^\dagger b f_1(b^\dagger)|0\rangle \otimes \begin{pmatrix} 1 \\ 0 \end{pmatrix} + \alpha_2 f_1(b^\dagger)|0\rangle \otimes \begin{pmatrix} 1 \\ 0 \end{pmatrix} + \alpha_3 (b^\dagger + b) f_1(b^\dagger)|0\rangle \otimes \begin{pmatrix} 0 \\ 1 \end{pmatrix}$$

$$+ b^\dagger b f_2(b^\dagger)|0\rangle \otimes \begin{pmatrix} 0 \\ 1 \end{pmatrix} - \alpha_2 f_2(b^\dagger)|0\rangle \otimes \begin{pmatrix} 0 \\ 1 \end{pmatrix} + \alpha_3 (b^\dagger + b) f_2(b^\dagger)|0\rangle \otimes \begin{pmatrix} 1 \\ 0 \end{pmatrix}.$$

Now

$$[b, g(b^\dagger)] = \frac{d}{db^\dagger} g(b^\dagger) \Leftrightarrow bg(b^\dagger) = \frac{d}{db^\dagger} g(b^\dagger) + g(b^\dagger)b$$

and $b|0\rangle = 0|0\rangle$. Thus

$$\hat{K}|\psi\rangle = b^\dagger \frac{d}{db^\dagger} f_1(b^\dagger)|0\rangle \otimes \begin{pmatrix} 1 \\ 0 \end{pmatrix} + \alpha_2 f_1(b^\dagger)|0\rangle \otimes \begin{pmatrix} 1 \\ 0 \end{pmatrix}$$

$$+ \alpha_3 b^\dagger f_1(b^\dagger)|0\rangle \otimes \begin{pmatrix} 0 \\ 1 \end{pmatrix} + \alpha_3 \frac{d}{db^\dagger} f_1(b^\dagger)|0\rangle \otimes \begin{pmatrix} 0 \\ 1 \end{pmatrix}$$

$$+ b^\dagger \frac{d}{db^\dagger} f_2(b^\dagger)|0\rangle \otimes \begin{pmatrix} 0 \\ 1 \end{pmatrix} - \alpha_2 f_2(b^\dagger)|0\rangle \otimes \begin{pmatrix} 0 \\ 1 \end{pmatrix}$$

$$+ \alpha_3 b^\dagger f_2(b^\dagger)|0\rangle \otimes \begin{pmatrix} 1 \\ 0 \end{pmatrix} + \alpha_3 \frac{d}{db^\dagger} f_2(b^\dagger)|0\rangle \otimes \begin{pmatrix} 1 \\ 0 \end{pmatrix}.$$

Thus from $\hat{K}|\psi\rangle = \widetilde{E}|\psi\rangle$ we arrive at

$$\left(b^\dagger \frac{d}{db^\dagger} f_1(b^\dagger) + \alpha_2 f_1(b^\dagger) + \alpha_3 b^\dagger f_2(b^\dagger) + \alpha_3 \frac{d}{db^\dagger} f_2(b^\dagger) - \widetilde{E} f_1 \right) |0\rangle \otimes \begin{pmatrix} 1 \\ 0 \end{pmatrix} = 0|0\rangle \otimes \begin{pmatrix} 0 \\ 0 \end{pmatrix}$$

$$\left(b^\dagger \frac{d}{db^\dagger} f_2(b^\dagger) - \alpha_2 f_2(b^\dagger) + \alpha_3 b^\dagger f_1(b^\dagger) + \alpha_3 \frac{d}{db^\dagger} f_1(b^\dagger) - \widetilde{E} f_2 \right) |0\rangle \otimes \begin{pmatrix} 1 \\ 0 \end{pmatrix} = 0|0\rangle \otimes \begin{pmatrix} 0 \\ 0 \end{pmatrix}.$$

We obtain the system of linear differential equations

$$z \frac{df_1}{dz} + \alpha_2 f_1(z) + \alpha_3 z f_2(z) + \alpha_3 \frac{df_2}{dz} = \widetilde{E} f_1(z)$$

$$z \frac{df_2}{dz} - \alpha_2 f_2(z) + \alpha_3 z f_1(z) + \alpha_3 \frac{df_1}{dz} = \widetilde{E} f_2(z).$$

5.2 Supplementary Problems

Problem 1. Let

$$S_+ = \begin{pmatrix} 0 & 1 \\ 0 & 0 \end{pmatrix}, \quad S_- = \begin{pmatrix} 0 & 0 \\ 1 & 0 \end{pmatrix}, \quad S_1 = \frac{1}{2}\begin{pmatrix} 0 & 1 \\ 1 & 0 \end{pmatrix}, \quad S_3 = \frac{1}{2}\begin{pmatrix} 1 & 0 \\ 0 & -1 \end{pmatrix}.$$

(i) Study the spectrum of the Hamilton operator

$$\hat{K}_1 = \frac{\hat{H}_1}{\hbar\omega} = b \otimes S_- + b^\dagger \otimes S_+.$$

(ii) Study the spectrum of the Hamilton operator

$$\hat{K}_2 = \frac{\hat{H}_2}{\hbar\omega} = b^\dagger \otimes S_- + b \otimes S_+.$$

(iii) Study the spectrum of the Hamilton operator

$$\hat{H} = \hbar\omega b^\dagger b \otimes I_2 + \gamma_1(b + b^\dagger) \otimes S_1 + \gamma_2 I_B \otimes S_3.$$

Problem 2. Consider the Hamilton operator

$$\hat{H} = \hbar\omega_1 b^\dagger b \otimes I_2 + \hbar\omega_2 I_B \otimes \sigma_3 + +\kappa(b \otimes S_+ + b^\dagger \otimes S_-).$$

(i) Find the matrix representation using the basis

$$|n\rangle \otimes \begin{pmatrix} 1 \\ 0 \end{pmatrix}, \quad |n\rangle \otimes \begin{pmatrix} 0 \\ 1 \end{pmatrix}$$

where $|n\rangle$ ($n = 0, 1, \ldots$) are the number states.
(ii) Let $|\beta\rangle$ be a coherent state. Consider the basis

$$|\beta\rangle \otimes \begin{pmatrix} 1 \\ 0 \end{pmatrix}, \quad |\beta\rangle \otimes \begin{pmatrix} 0 \\ 1 \end{pmatrix}.$$

Find

$$(\langle\tilde{\beta}| \otimes \begin{pmatrix} 1 \\ 0 \end{pmatrix})\hat{H}(|\beta\rangle \otimes \begin{pmatrix} 1 \\ 0 \end{pmatrix}), \quad (\langle\tilde{\beta}| \otimes \begin{pmatrix} 0 \\ 1 \end{pmatrix})\hat{H}(|\beta\rangle \otimes \begin{pmatrix} 0 \\ 1 \end{pmatrix}),$$

$$(\langle\tilde{\beta}| \otimes \begin{pmatrix} 1 \\ 0 \end{pmatrix})\hat{H}(|\beta\rangle \otimes \begin{pmatrix} 0 \\ 1 \end{pmatrix}), \quad (\langle\tilde{\beta}| \otimes \begin{pmatrix} 0 \\ 1 \end{pmatrix})\hat{H}(|\beta\rangle \otimes \begin{pmatrix} 1 \\ 0 \end{pmatrix}).$$

Problem 3. Let $b^\dagger$, b Bose creation and annihilation operators, respectively. Study the spectrum of the Hamilton operator

$$\hat{H} = I_B \otimes \mu(B_1\sigma_1 + B_2\sigma_2 + B_3\sigma_3) + \hbar\omega b^\dagger b \otimes I_2 + \rho(b \otimes S_+ - b^\dagger \otimes S_-).$$

Problem 4. (i) The Hamilton operator $\hat{H}$ for the *Jaynes-Cummings model* neglecting the so-called counter-rotating terms is given by

$$\hat{H} = \hbar\omega b^\dagger b \otimes I_2 + \frac{1}{2}I_B \otimes \hbar\omega_0\sigma_3 + \gamma(b \otimes \sigma_+ + b^\dagger \otimes \sigma_-).$$

Let $|\beta\rangle$ be a coherent state and

$$|\mathbf{v}\rangle = \frac{1}{\sqrt{2}} \begin{pmatrix} 1 \\ -1 \end{pmatrix}.$$

Find the expectation value $(\langle\beta| \otimes \langle\mathbf{v}|)\hat{H}(|\beta\rangle \otimes |\mathbf{v}\rangle)$.
Let $|\zeta\rangle$ be a squeezed state and

$$|\mathbf{v}\rangle = \frac{1}{\sqrt{2}} \begin{pmatrix} 1 \\ -1 \end{pmatrix}.$$

Find the expectation value $(\langle\zeta| \otimes \langle\mathbf{v}|)\hat{H}(|\zeta\rangle \otimes |\mathbf{v}\rangle)$.
(ii) The Hamilton operator $\hat{H}$ for the two-photon Jaynes-Cummings model is given by

$$\hat{H} = \hbar\omega b^\dagger b \otimes I_2 + \frac{1}{2}I_B \otimes \hbar\omega_0\sigma_3 + \gamma(b^2 \otimes S_+ + (b^\dagger)^2 \otimes S_-).$$

Let $|\beta\rangle$ be a coherent state, i.e. $b|\beta\rangle = \beta|\beta\rangle$ and

$$|\mathbf{v}\rangle = \frac{1}{\sqrt{2}} \begin{pmatrix} 1 \\ -1 \end{pmatrix}.$$

Find the expectation value

$$(\langle\beta| \otimes \langle\mathbf{v}|)\hat{H}(|\beta\rangle \otimes |\mathbf{v}\rangle).$$

Let $|\zeta\rangle$ be a squeezed state and

$$|\mathbf{v}\rangle = \frac{1}{\sqrt{2}} \begin{pmatrix} 1 \\ -1 \end{pmatrix}.$$

Find the expectation value

$$(\langle\zeta| \otimes \langle\mathbf{v}|)\hat{H}(|\zeta\rangle \otimes |\mathbf{v}\rangle).$$

Problem 5. Let $z \in \mathbb{C}$ and

$$S_+ = \frac{1}{2}\sigma_+ = \begin{pmatrix} 0 & 1 \\ 0 & 0 \end{pmatrix}, \qquad S_- = \frac{1}{2}\sigma_- = \begin{pmatrix} 0 & 0 \\ 1 & 0 \end{pmatrix}.$$

(i) Find the commutator $[b^\dagger \otimes S_-, b \otimes S_+]$, anticommutator $[b^\dagger \otimes S_-, b \otimes S_+]_+$ and

$$\exp(z(b^\dagger \otimes S_- + b \otimes S_+))$$

(ii) Find the commutator $[b^\dagger \otimes S_+, b \otimes S_-]$, anticommutator $[b^\dagger \otimes S_+, b \otimes S_-]_+$ and

$$\exp(z(b^\dagger \otimes S_+ + b \otimes S_-)).$$

Problem 6. Let $b^\dagger$, b be Bose creation and annihilation operators with $[b, b^\dagger] = I_B$ and $\sigma_0 = I_2$, σ_1, σ_2, σ_3 be the Pauli spin matrices. The *Rabi model* is given by the Hamilton operator

$$\hat{H} = \hbar\omega b^\dagger b \otimes I_2 + \hbar\Omega I_B \otimes \sigma_3 + \gamma(b^\dagger + b) \otimes \sigma_1$$

where $2\hbar\Omega$ is the energy difference between the two levels and γ is a coupling constant.
(i) Let $|\beta\rangle$ be a coherent state. Find the expectation value

$$(\langle\beta| \otimes \frac{1}{\sqrt{2}} \begin{pmatrix} 1 \\ 1 \end{pmatrix})\hat{H}(|\beta\rangle \otimes (\frac{1}{\sqrt{2}} \begin{pmatrix} 1 \\ 1 \end{pmatrix}).$$

(ii) Let $|\zeta\rangle$ be a squeezed state. Find the expectation value

$$\left((\langle\beta| \otimes \frac{1}{\sqrt{2}}\begin{pmatrix}1\\1\end{pmatrix})\hat{H}(|\beta\rangle \otimes \frac{1}{\sqrt{2}}\begin{pmatrix}1\\1\end{pmatrix})\right).$$

Discuss.

Problem 7. Consider the Hamilton operator with two Bose operators $b_1 = b \otimes I_B$, $b_2 = I_B \otimes b$

$$\hat{H} = \hbar\omega_1(b^\dagger b + \frac{1}{2}I_B) \otimes I_B \otimes I_2 + \hbar\omega_1 I_B \otimes (b^\dagger b + \frac{1}{2}I_B) \otimes I_2 + \hbar\omega_2 I_B \otimes I_B \otimes \sigma_3$$
$$+ \hbar\omega_3((b \otimes I_B + b^\dagger \otimes I_B) \otimes \sigma_+ + (I_B \otimes b + I_B \otimes b^\dagger) \otimes \sigma_-).$$

Let

$$\hat{K} = b^\dagger b \otimes I_B \otimes I_B - I_B \otimes b^\dagger b \otimes I_2 + I_B \otimes I_B \otimes \sigma_3.$$

Find the commutator $[\hat{H}, \hat{K}]$.

Problem 8. Let

$$U(\epsilon) = \exp(\epsilon(b - b^\dagger) \otimes \sigma_1).$$

Find

$$U(\epsilon)\left(|n\rangle \otimes \frac{1}{\sqrt{2}}\begin{pmatrix}1\\1\end{pmatrix}\right), \qquad U(\epsilon)\left(|n\rangle \otimes \frac{1}{\sqrt{2}}\begin{pmatrix}1\\-1\end{pmatrix}\right).$$

Problem 9. Let

$$S_+ = \begin{pmatrix}0&1\\0&0\end{pmatrix}, \quad S_- = \begin{pmatrix}0&0\\1&0\end{pmatrix}, \quad S_3 = \frac{1}{2}\begin{pmatrix}1&0\\0&-1\end{pmatrix}.$$

Does the Hamilton operator

$$\hat{H} = \hbar\omega_1 b^\dagger b + \hbar\omega_2 I_B \otimes S_3 + \hbar\omega_3(b^\dagger \otimes S_- + b \otimes S_+) + \hbar\omega_4 b^\dagger b^\dagger bb \otimes I_2 + \hbar\omega_5 I_B \otimes (S_3)^2$$

commute with the number operator $\hat{N} = I_B \otimes S_3 + b^\dagger b \otimes I_2$?

Problem 10. Let $|n\rangle$ be the number states, $D(\beta)$ be the displacement operator and σ_1 be the first Pauli spin matrix. The normalized eigenvectors of σ_1 are given by

$$\frac{1}{\sqrt{2}}\begin{pmatrix}1\\1\end{pmatrix}, \qquad \frac{1}{\sqrt{2}}\begin{pmatrix}1\\-1\end{pmatrix}$$

with eigenvalues $+1$ and -1. Consider the Hamilton operator

$$\hat{K} = b^\dagger b \otimes I_2 + \alpha(b^\dagger + b) \otimes \sigma_1$$

where α is a dimensionless real quantity. Are the product states

$$D(\alpha)|n\rangle \otimes \frac{1}{\sqrt{2}}\begin{pmatrix}1\\-1\end{pmatrix}, \qquad D(-\alpha)|n\rangle \otimes \frac{1}{\sqrt{2}}\begin{pmatrix}1\\1\end{pmatrix}$$

eigenstates of the Hamilton operator $\hat{H}$?

Problem 11. Consider the Hilbert space $\mathcal{H}_S$ with a (self-adjoint) Hamilton operator $\hat{H}_S$ (S stands for system). The identity operator in this Hilbert space is denoted by I_S. Let

$\mathcal{H}_E$ be a Hilbert space with a (self-adjoint) Hamilton operator $\hat{H}_E$ (S stands for environment sometimes called bath). The identity operator in this Hilbert space is denoted by I_E. Let $\mathcal{H} = \mathcal{H}_S \otimes \mathcal{H}_E$ be the product Hilbert space. Consider the Hamilton operator

$$\hat{K} = \hat{H}_S \otimes I_E + I_S \otimes \hat{H}_E + \hat{V}$$

where the operator $\hat{V}$ acts in the product Hilbert space $\mathcal{H}_S \otimes \mathcal{H}_E$. The *von Neumann equation* for the density matrix of the Hamilton operator $\hat{K}$ is given by

$$\frac{d}{dt}\rho_{SE}(t) = -\frac{i}{\hbar}(\hat{H}_S \otimes I_E + I_S \otimes \hat{H}_E + \hat{V})(t).$$

Consider the operator (switch to the interaction picture)

$$\widetilde{V}(t) = e^{i(\hat{H}_S \otimes I_E + I_S \otimes \hat{H}_E)t/\hbar}\hat{V}e^{-i(\hat{H}_S \otimes I_E + I_S \otimes \hat{H}_E)t/\hbar}$$

and the density matrix

$$\widetilde{\rho}(t) = e^{i(\hat{H}_S \otimes I_E + I_S \otimes \hat{H}_E)t/\hbar}\hat{\rho}_{SE}(t)e^{-i(\hat{H}_S \otimes I_E + I_S \otimes \hat{H}_E)t/\hbar}.$$

Therefore

$$\hat{\rho}_{SE}(t) = e^{-i(\hat{H}_S \otimes I_E + I_S \otimes \hat{H}_E)t/\hbar}\widetilde{\rho}(t)e^{i(\hat{H}_S \otimes I_E + I_S \otimes \hat{H}_E)t/\hbar}.$$

We apply a tilde to indicate operators in the interaction picture. Thus we obtain

$$i\hbar\frac{d\widetilde{\rho}}{dt} = [\widetilde{V}(t), \widetilde{\rho}(t)].$$

The perturbation expansion (*Dyson series*) yields

$$\widetilde{\rho}(t) = \sum_{j\geq 0}\int_0^t dt\cdots\int_0^t dt_j\left(\frac{1}{i\hbar}\right)^j[\widetilde{V}(t_1),\ldots,[\widetilde{V}(t_n),\widetilde{\rho}(0)]\ldots].$$

To find the Born-Markov master equation one computes the time evolution up to second order and perform the trace over the Hilbert space $\mathcal{H}_E$ (environment, bath), i.e. the *partial trace*. One obtains

$$\frac{d\widetilde{\rho}}{dt} = \frac{1}{i\hbar}\mathrm{tr}_E[\widetilde{V}(t), \rho(0)] - \frac{1}{\hbar^2}\int_{t_1=0}^t dt_1\mathrm{tr}_E([\widetilde{V}(t), [\widetilde{V}(t_1), \rho(0)]]).$$

One normally assumes that at $t = 0$ the density operator is a tensor product of the form $\rho(0) = \rho_S(0) \otimes \rho_E(0)$. Consider the Hilbert spaces $\mathcal{H}_S = \mathbb{C}^2$ and $\mathcal{H}_E = \ell_2(\mathbb{N}_0)$ with

$$\hat{K} = \frac{1}{2}\hbar\omega_1\sigma_3 \otimes I_B + \hbar\omega_2(S_+ \otimes b + S_- \otimes b^\dagger) + I_2 \otimes \hbar\omega_3 b^\dagger b$$

where

$$I_2 = \begin{pmatrix} 1 & 0 \\ 0 & 1 \end{pmatrix}, \quad S_+ = \begin{pmatrix} 0 & 1 \\ 0 & 0 \end{pmatrix}, \quad S_- = \begin{pmatrix} 0 & 0 \\ 1 & 0 \end{pmatrix}, \quad \sigma_3 = \begin{pmatrix} 1 & 0 \\ 0 & -1 \end{pmatrix}$$

and $I_B = \mathrm{diag}(1, 1, 1, \ldots)$. Find $\widetilde{\rho}(t)$ with

$$\rho_S(0) = \frac{1}{2}\begin{pmatrix} 1 & 1 \\ 1 & 1 \end{pmatrix} \quad \text{and} \quad \rho_E(0) = \mathrm{diag}(1, 0, 0, \ldots).$$

Chapter 6

Bose-Fermi Systems

6.1 Solved Problems

Problem 1. Let $b^\dagger$, $c^\dagger$ be Bose and Fermi creation operators, respectively. Let I_B be the identity operator for the Bose operators and I_F be the identity operator for the Fermi operators. Let

$$\hat{N} = b^\dagger b \otimes I_F + I_B \otimes c^\dagger c$$

be the *number operator*.
(i) Find the commutator $[b^\dagger \otimes c, b \otimes c^\dagger]$.
(ii) Find the anticommutator $[b^\dagger \otimes c, b \otimes c^\dagger]_+$.
(iii) Do the operators

$$S = b \otimes c^\dagger \Rightarrow S^\dagger = b^\dagger \otimes c$$

commute with the number operator $\hat{N}$? This means find the commutators $[\hat{N}, S]$ and $[\hat{N}, S^\dagger]$.

Solution 1. (i) For the commutator we have

$$
\begin{aligned}
[b^\dagger \otimes c, b \otimes c^\dagger] &= b^\dagger b \otimes cc^\dagger - bb^\dagger \otimes c^\dagger c \\
&= b^\dagger b \otimes (I_F - c^\dagger c) - (I_B + b^\dagger b) \otimes c^\dagger c \\
&= b^\dagger b \otimes I_F - I_B \otimes c^\dagger c - 2b^\dagger b \otimes c^\dagger c.
\end{aligned}
$$

(ii) For the anticommutator we find

$$
\begin{aligned}
[b^\dagger \otimes c, b \otimes c^\dagger]_+ &= b^\dagger b \otimes cc^\dagger + bb^\dagger \otimes c^\dagger c \\
&= b^\dagger b \otimes (I_F - c^\dagger c) + (I_B + b^\dagger b) \otimes c^\dagger c \\
&= b^\dagger b \otimes I_F + I_B \otimes c^\dagger c \\
&= \hat{N}
\end{aligned}
$$

i.e. we obtain the number operator.
(iii) Applying $bb^\dagger = I_B + b^\dagger b$ and $cc^\dagger = I_F - c^\dagger c$ we obtain

$$[\hat{N}, b \otimes c^\dagger] = 0_B \otimes 0_F$$

where 0_B is the zero operator for the Bose system and 0_F is the zero operator for the Fermi system. We also find

$$[\hat{N}, S^\dagger] = [b^\dagger b \otimes I_F, b^\dagger \otimes c] + [I_B \otimes c^\dagger c, b^\dagger \otimes c] = 0_B \otimes 0_F.$$

Problem 2.   Consider the *number operator*

$$\hat{N} = b^\dagger b \otimes I_F + I_B \otimes c^\dagger c$$

where

$$b^\dagger b = \text{diag}(0\,1\,2\,3\,\ldots), \quad c^\dagger c = \begin{pmatrix} 1 & 0 \\ 0 & 0 \end{pmatrix}, \quad I_2 = \begin{pmatrix} 1 & 0 \\ 0 & 1 \end{pmatrix}$$

and I_B is the infinite dimensional identity matrix.
(i) Find the eigenvalues of $\hat{N}$.
(ii) Does the number operator $\hat{N}$ commute with the operator $\hat{K} = b^\dagger \otimes c + b \otimes c^\dagger$?

Solution 2.   (i) We obtain the diagonal matrix for $\hat{N}$

$$\hat{N} = \begin{pmatrix} 1 & 0 \\ 0 & 0 \end{pmatrix} \oplus \begin{pmatrix} 2 & 0 \\ 0 & 1 \end{pmatrix} \oplus \begin{pmatrix} 3 & 0 \\ 0 & 2 \end{pmatrix} \oplus \begin{pmatrix} 4 & 0 \\ 0 & 3 \end{pmatrix} \oplus \cdots$$

where $\oplus$ denotes the direct sum. Therefore the eigenvalues are 0, 1 (twice), 2 (twice), 3 (twice), 4 (twice) etc. Hence the lowest eigenvalue 0 is not degenerate.
(ii) Since $bb^\dagger = I_B + b^\dagger b$ and $cc^\dagger = I_F - c^\dagger c$ we end up with

$$[\hat{K}, \hat{N}] = 0_B \otimes 0_2$$

where 0_2 is the 2×2 zero matrix.

Problem 3.   Does the number operator

$$\hat{N} = b^\dagger b \otimes I_F + I_B \otimes c^\dagger c$$

commute with the operator $\hat{K} = b^\dagger \otimes c^\dagger + b \otimes c$?

Solution 3.   No. We have

$$[\hat{K}, \hat{N}] = 2(-b^\dagger \otimes c^\dagger + b \otimes c).$$

Problem 4.   Let b, $b^\dagger$ be Bose annihilation and creation operators, respectively. Let c, $c^\dagger$ be Fermi annihilation and creation operators, respectively. Let I_B be the identity operator in the vector space of the Bose operators. Let I_F be the identity operator in the vector space of the Fermi operators. Consider the set of operators

$$\{ b \otimes c, \quad b \otimes c^\dagger, \quad b^\dagger \otimes c, \quad b^\dagger \otimes c^\dagger \}.$$

(i) Calculate the commutators between all the operators in the set.
(ii) Calculate the anticommutators between all the operators in the set.

Solution 4.   (i) For the commutators we find

$$[b \otimes c, b \otimes c^\dagger] = b^2 \otimes (I_F - 2c^\dagger c)$$

$$[b \otimes c, b^\dagger \otimes c^\dagger] = 0_B \otimes 0_F$$
$$[b \otimes c, b^\dagger \otimes c^\dagger] = I_B \otimes I_F + b^\dagger b \otimes I_F - I_B \otimes c^\dagger c - 2b^\dagger b \otimes c^\dagger c$$
$$[b \otimes c^\dagger, b^\dagger \otimes c] = I_B \otimes c^\dagger c - b^\dagger b \otimes I_F + 2b^\dagger b \otimes c^\dagger c$$
$$[b \otimes c^\dagger, b^\dagger \otimes c^\dagger] = 0_B \otimes 0_F$$
$$[b^\dagger \otimes c, b^\dagger \otimes c^\dagger] = (b^\dagger)^2 \otimes I_F - 2(b^\dagger)^2 \otimes c^\dagger c.$$

(ii) For the anticommutators we obtain

$$[b \otimes c, b \otimes c^\dagger]_+ = b^2 \otimes I_F$$
$$[b \otimes c, b^\dagger \otimes c]_+ = 0_B \otimes 0_F$$
$$[b \otimes c, b^\dagger \otimes c^\dagger]_+ = I_B \otimes I_F + b^\dagger b \otimes I_F - I_B \otimes c^\dagger c$$
$$[b \otimes c^\dagger, b^\dagger \otimes c]_+ = I_B \otimes c^\dagger c + b^\dagger b \otimes I_F$$
$$[b \otimes c^\dagger, b^\dagger \otimes c^\dagger]_+ = 0_B \otimes 0_F$$
$$[b^\dagger \otimes c, b^\dagger \otimes c^\dagger]_+ = (b^\dagger)^2 \otimes I_F.$$

Thus the anticommutator $[b \otimes c^\dagger, b^\dagger \otimes c]_+$ provides the *number operator*

$$\hat{N} = b^\dagger b \otimes I_F + I_B \otimes c^\dagger c.$$

Problem 5. Let $|n\rangle$ be the number states $(n = 0, 1, \ldots)$. Calculate

$$(b^\dagger b \otimes (c^\dagger + c))(|n\rangle \otimes \frac{1}{\sqrt{2}}(c^\dagger|0\rangle + |0\rangle))).$$

Discuss.

Solution 5. Since $b^\dagger b|n\rangle = n|n\rangle$, $c^\dagger c^\dagger = 0$, $cc^\dagger = I_F$ and $c|0\rangle = 0|0\rangle$ we obtain the eigenvalue equation

$$(b^\dagger b \otimes (c^\dagger + c))(|n\rangle \otimes \frac{1}{\sqrt{2}}(c^\dagger|0\rangle + |0\rangle))) = n(|n\rangle \otimes \frac{1}{\sqrt{2}}(c^\dagger|0\rangle + |0\rangle)))$$

with eigenvalue n.

Problem 6. Consider the operator in the product Hilbert space

$$\left\{ b \otimes I_F, \ b^\dagger \otimes I_F, \ I_B \otimes c, \ I_B \otimes c^\dagger \ b \otimes c, \ b \otimes c^\dagger, \ b^\dagger \otimes c, \ b^\dagger \otimes c^\dagger \right\}.$$

Apply the operators to the states

$$|n\rangle \otimes |0\rangle, \qquad |n\rangle \otimes c^\dagger|0\rangle$$

where $|n\rangle$ are the number states $(n = 0, 1, 2, \ldots)$. Discuss.

Solution 6. Applying $b|n\rangle = \sqrt{n}|n-1\rangle$, $b^\dagger|n\rangle = \sqrt{n+1}|n+1\rangle$, $c|0\rangle = 0|0\rangle$, $cc^\dagger|0\rangle = |0\rangle$ we obtain for the states $|n\rangle \otimes |0\rangle$

$$(b \otimes I_F)(|n\rangle \otimes |0\rangle) = \sqrt{n}|n-1\rangle \otimes |0\rangle$$
$$(b^\dagger \otimes I_F)(|n\rangle \otimes |0\rangle) = \sqrt{n+1}|n+1\rangle \otimes |0\rangle$$
$$(I_B \otimes c)(|n\rangle \otimes |0\rangle) = 0 \cdot |n\rangle \otimes |0\rangle$$
$$(I_B \otimes c^\dagger)(|n\rangle \otimes |0\rangle) = |n\rangle \otimes c^\dagger|0\rangle$$
$$(b \otimes c)(|n\rangle \otimes |0\rangle) = 0 \cdot \sqrt{n}|n-1\rangle \otimes |0\rangle$$
$$(b \otimes c^\dagger)(|n\rangle \otimes |0\rangle) = \sqrt{n}|n-1\rangle \otimes c^\dagger|0\rangle$$
$$(b^\dagger \otimes c)(|n\rangle \otimes |0\rangle) = 0 \cdot \sqrt{n+1}|n+1\rangle \otimes |0\rangle$$
$$(b^\dagger \otimes c^\dagger)(|n\rangle \otimes |0\rangle) = \sqrt{n+1}|n+1\rangle \otimes c^\dagger|0\rangle.$$

Thus the third, fifth and seventh equation are eigenvalue equations with eigenvalue 0. Applying $b|n\rangle = \sqrt{n}|n-1\rangle$, $b^\dagger|n\rangle = \sqrt{n+1}|n+1\rangle$, $c|0\rangle = 0|0\rangle$, $cc^\dagger|0\rangle = |0\rangle$ we obtain for the states $|n\rangle \otimes c^\dagger|0\rangle$

$$(b \otimes I_F)(|n\rangle \otimes c^\dagger|0\rangle) = \sqrt{n}|n-1\rangle \otimes c^\dagger|0\rangle$$
$$(b^\dagger \otimes I_F)(|n\rangle \otimes c^\dagger|0\rangle) = \sqrt{n+1}|n\rangle \otimes c^\dagger|0\rangle$$
$$(I_B \otimes c)(|n\rangle \otimes c^\dagger|0\rangle) = |n\rangle \otimes |0\rangle$$
$$(I_B \otimes c^\dagger)(|n\rangle \otimes c^\dagger|0\rangle) = 0 \cdot |n\rangle \otimes |0\rangle$$
$$(b \otimes c)(|n\rangle \otimes c^\dagger|0\rangle) = \sqrt{n}|n\rangle \otimes |0\rangle$$
$$(b \otimes c^\dagger)(|n\rangle \otimes c^\dagger|0\rangle) = 0 \cdot \sqrt{n}|n-1\rangle \otimes |0\rangle$$
$$(b^\dagger \otimes c)(|n\rangle \otimes c^\dagger|0\rangle) = \sqrt{n+1}|n+1\rangle \otimes |0\rangle$$
$$(b^\dagger \otimes c^\dagger)(|n\rangle \otimes c^\dagger|0\rangle) = \sqrt{n+1}|n+1\rangle \otimes |0\rangle.$$

Thus the fourth and sixth equation are eigenvalue equations with eigenvalue 0.

Problem 7.   Find the spectrum of the Hamilton operator

$$\hat{K} = \frac{\hat{H}}{\hbar\omega} = b^\dagger b \otimes (c^\dagger + c).$$

Solution 7.   Since

$$c^\dagger + c = \begin{pmatrix} 0 & 1 \\ 1 & 0 \end{pmatrix} = \sigma_1$$

we obtain an infinite dimensional matrix for $\hat{K}$ which can be written as a direct sum

$$\begin{pmatrix} 0 & 0 \\ 0 & 0 \end{pmatrix} \oplus \begin{pmatrix} 0 & 1 \\ 1 & 0 \end{pmatrix} \oplus 2\begin{pmatrix} 0 & 1 \\ 1 & 0 \end{pmatrix} \oplus 3\begin{pmatrix} 0 & 1 \\ 1 & 0 \end{pmatrix} \oplus \cdots \oplus n\begin{pmatrix} 0 & 1 \\ 1 & 0 \end{pmatrix} \oplus \cdots$$

with the eigenvalues 0 (twice), 1, -1, 2, -2, 3, -3 etc. Thus the spectrum is unbounded.

Problem 8.   Find the matrix representation of

$$(b^\dagger + b) \otimes (c^\dagger + c).$$

Truncate the resulting infinite dimensional matrix for the 8×8 matrix and find the eigenvalues of the truncated matrix.

Solution 8.   Since

$$b^\dagger + b = \begin{pmatrix} 0 & \sqrt{1} & 0 & 0 & \cdots \\ \sqrt{1} & 0 & \sqrt{2} & 0 & \cdots \\ 0 & \sqrt{2} & 0 & \sqrt{3} & \cdots \\ 0 & 0 & \sqrt{3} & 0 & \cdots \\ \vdots & \vdots & \vdots & \vdots & \ddots \end{pmatrix}$$

and

$$c^\dagger + c = \begin{pmatrix} 0 & 1 \\ 1 & 0 \end{pmatrix}$$

we obtain the infinite dimensional matrix

$$
\begin{pmatrix}
0 & 0 & 0 & \sqrt{1} & 0 & 0 & 0 & 0 & \cdots \\
0 & 0 & \sqrt{1} & 0 & 0 & 0 & 0 & 0 & \cdots \\
0 & \sqrt{1} & 0 & 0 & 0 & \sqrt{2} & 0 & 0 & \cdots \\
\sqrt{1} & 0 & 0 & 0 & \sqrt{2} & 0 & 0 & 0 & \cdots \\
0 & 0 & 0 & \sqrt{2} & 0 & 0 & 0 & \sqrt{3} & \cdots \\
0 & 0 & \sqrt{2} & 0 & 0 & 0 & \sqrt{3} & 0 & \cdots \\
0 & 0 & 0 & 0 & 0 & \sqrt{3} & 0 & 0 & \cdots \\
0 & 0 & 0 & 0 & \sqrt{3} & 0 & 0 & 0 & \cdots \\
\vdots & \vdots & \vdots & \vdots & \vdots & \vdots & \vdots & \vdots & \ddots
\end{pmatrix}.
$$

Truncating the matrix at the 8×8 level we have the hermitian and invertible matrix with trace equal to 0

$$
\begin{pmatrix}
0 & 0 & 0 & \sqrt{1} & 0 & 0 & 0 & 0 \\
0 & 0 & \sqrt{1} & 0 & 0 & 0 & 0 & 0 \\
0 & \sqrt{1} & 0 & 0 & 0 & \sqrt{2} & 0 & 0 \\
\sqrt{1} & 0 & 0 & 0 & \sqrt{2} & 0 & 0 & 0 \\
0 & 0 & 0 & \sqrt{2} & 0 & 0 & 0 & \sqrt{3} \\
0 & 0 & \sqrt{2} & 0 & 0 & 0 & \sqrt{3} & 0 \\
0 & 0 & 0 & 0 & 0 & \sqrt{3} & 0 & 0 \\
0 & 0 & 0 & 0 & \sqrt{3} & 0 & 0 & 0
\end{pmatrix}
$$

with trace equal to 0. The eigenvalues are

$$
-\sqrt{\sqrt{6}+3}, \quad \sqrt{\sqrt{6}+3}, \quad -\sqrt{3-\sqrt{6}}, \quad \sqrt{3-\sqrt{6}}
$$

each twice. Instead of using the basis $|n\rangle \otimes |0\rangle$, $|n\rangle \otimes c^\dagger|0\rangle$ with the ordering

$$
|0\rangle \otimes |0\rangle, \quad |0\rangle \otimes c^\dagger|0\rangle, \quad |1\rangle \otimes |0\rangle, \quad |1\rangle \otimes c^\dagger|0\rangle, \quad \cdots
$$

it would be more useful for the matrix representation to use the basis

$$
|n\rangle \otimes \frac{1}{\sqrt{2}}(c^\dagger|0\rangle + |0\rangle), \quad |n\rangle \otimes \frac{1}{\sqrt{2}}(c^\dagger|0\rangle - |0\rangle)
$$

since

$$
\frac{1}{\sqrt{2}}(c^\dagger|0\rangle + |0\rangle), \quad \frac{1}{\sqrt{2}}(c^\dagger|0\rangle - |0\rangle)
$$

are eigenvectors of the operator $c^\dagger + c$ with eigenvalues $+1$ and -1, respectively.

Problem 9. Let $c_\uparrow^\dagger$, $c_\downarrow^\dagger$ be Fermi operators with spin up and down, respectively.
(i) Find the commutators for the operators

$$
b \otimes c_\uparrow, \quad b^\dagger \otimes c_\uparrow^\dagger, \quad b \otimes c_\downarrow, \quad b^\dagger \otimes c_\downarrow^\dagger.
$$

Discuss.
(ii) Find the anticommutators for the operators $b \otimes c_\uparrow$, $b^\dagger \otimes c_\uparrow^\dagger$, $b \otimes c_\downarrow$, $b^\dagger \otimes c_\downarrow^\dagger$. Discuss.

Solution 9. (i) For the commutators we obtain

$$
[b \otimes c_\uparrow, b^\dagger \otimes c_\uparrow^\dagger] = I_B \otimes I_F + b^\dagger b \otimes I_F - I_B \otimes c_\uparrow^\dagger c_\uparrow - 2b^\dagger b \otimes c_\uparrow^\dagger c_\uparrow
$$

$$[b \otimes c_\uparrow, b \otimes c_\downarrow] = 2b^2 \otimes c_\uparrow c_\downarrow$$
$$[b \otimes c_\uparrow, b^\dagger \otimes c_\downarrow^\dagger] = I_B \otimes c_\uparrow c_\downarrow^\dagger + 2b^\dagger b \otimes c_\uparrow c_\downarrow^\dagger$$
$$[b^\dagger \otimes c_\uparrow^\dagger, b \otimes c_\downarrow] = I_B \otimes c_\uparrow^\dagger c_\downarrow + 2b^\dagger b \otimes c_\uparrow^\dagger c_\downarrow$$
$$[b^\dagger \otimes c_\uparrow^\dagger, b^\dagger \otimes c_\downarrow^\dagger] = 2(b^\dagger)^2 \otimes c_\uparrow^\dagger c_\downarrow^\dagger$$
$$[b \otimes c_\downarrow, b^\dagger \otimes c_\downarrow^\dagger] = I_B \otimes I_F + b^\dagger b \otimes I_F - I_B \otimes c_\downarrow^\dagger c_\downarrow - 2b^\dagger b \otimes c_\downarrow^\dagger c_\downarrow.$$

(ii) For the anticommutators we obtain

$$[b \otimes c_\uparrow, b^\dagger \otimes c_\uparrow^\dagger]_+ = I_B \otimes I_F + b^\dagger b \otimes I_F - I_B \otimes c_\uparrow^\dagger c_\uparrow$$
$$[b \otimes c_\uparrow, b \otimes c_\downarrow]_+ = 0_B \otimes 0_F$$
$$[b \otimes c_\uparrow, b^\dagger \otimes c_\downarrow^\dagger]_+ = I_B \otimes c_\uparrow c_\downarrow^\dagger$$
$$[b^\dagger \otimes c_\uparrow^\dagger, b \otimes c_\downarrow]_+ = I_B \otimes c_\downarrow c_\uparrow^\dagger$$
$$[b^\dagger \otimes c_\uparrow^\dagger, b^\dagger \otimes c_\downarrow^\dagger]_+ = 0_B \otimes 0_F$$
$$[b \otimes c_\downarrow, b^\dagger \otimes c_\downarrow^\dagger]_+ = I_B \otimes I_F + b^\dagger b \otimes I_F - I_B \otimes c_\downarrow^\dagger c_\downarrow.$$

Problem 10. Let $c^\dagger$, c be Fermi creation and annihilation operators, respectively. Let b_1, b_2 be two Bose annihilation operators. Using the Bose operators we form the operators

$$J_3 = \frac{1}{2}(b_1^\dagger b_1 - b_2^\dagger b_2), \quad J_+ = b_1^\dagger b_2, \quad J_- = b_1 b_2^\dagger$$

with $b_1 = b \otimes I_B$ and $b_2 = I_B \otimes b$. Using the Bose and Fermi operators we form the operators

$$Q_+ := \frac{1}{2}(b_1^\dagger \otimes c + b_2 \otimes c^\dagger), \quad Q_- := \frac{1}{2}(b_1 \otimes c^\dagger - b_2^\dagger \otimes c).$$

(i) Find the anticommutators $[Q_+, Q_-]_+$, $[Q_\pm, Q_\pm]_+$.
(ii) Find the commutators $[J_\pm \otimes I_F, Q_\pm]$, $[J_\pm \otimes I_F, Q_\mp]$, $[J_3 \otimes I_F, Q_\pm]$.

Solution 10. (i) Since $c^2 = 0$, $(c^\dagger)^2 = 0$ and $c^\dagger c c^\dagger c = c^\dagger c$ we have

$$\begin{aligned}
[Q_+, Q_-]_+ &= \frac{1}{4}((b_1^\dagger \otimes c + b_2 \otimes c^\dagger)(b_1 \otimes c^\dagger - b_2^\dagger \otimes c) \\
&\quad + (b_1 \otimes c^\dagger - b_2^\dagger \otimes c)(b_1^\dagger \otimes c + b_2 \otimes c^\dagger)) \\
&= \frac{1}{4}(b_1^\dagger b_1 \otimes c c^\dagger - b_2 b_2^\dagger \otimes c^\dagger c + b_1 b_1^\dagger \otimes c^\dagger c - b_2^\dagger b_2 \otimes c c^\dagger) \\
&= \frac{1}{4}(b_1^\dagger b_1 \otimes I_F - I_B \otimes I_B \otimes c^\dagger c + I_B \otimes I_B \otimes c^\dagger c - b_2^\dagger b_2 \otimes I_F) \\
&= \frac{1}{4}(b_1^\dagger b_1 \otimes I_F - b_2^\dagger b_2 \otimes I_F) \\
&= \frac{1}{2} J_3 \otimes I_F.
\end{aligned}$$

Analogously we find $[Q_\pm, Q_\pm]_+ = \pm \frac{1}{2} J_\pm \otimes I_F$.
(ii) For the commutators we find

$$[J_\pm \otimes I_F, Q_\pm] = (0_B \otimes 0_B) \otimes 0_F, \quad [J_\pm \otimes I_F, Q_\mp] = -Q_\pm, \quad [J_3 \otimes I_F, Q_\pm] = \pm \frac{1}{2} Q_\pm.$$

Problem 11. Consider a pair of linear operators d, $d^\dagger$ in complex spaces as annihilation and creation operators of certain particles with the commutation relation

$$dd^\dagger - \epsilon d^\dagger d = I$$

where I is the identity operator and $\epsilon \in [-1, \infty)$ with $\epsilon \neq 0$. The operators d and $d^\dagger$ are adjoint to each other. The case $\epsilon = +1$ describes Bose particles. The case $\epsilon = -1$ describes Fermi particles. The operators d and $d^\dagger$ can be expressed using Bose annihilation and creation operators b and $b^\dagger$ as follows

$$d = \sqrt{(\hat{N} + I)^{-1} \frac{\epsilon^{\hat{N}+I} - I}{\epsilon - 1}} b, \qquad d^\dagger = b^\dagger \sqrt{(\hat{N} + I)^{-1} \frac{\epsilon^{\hat{N}+I} - I}{\epsilon - 1}}$$

where $\hat{N} := b^\dagger b$ is the number operator.

(i) Calculate $\hat{N}_\epsilon := d^\dagger d$.

(ii) Find d and $d^\dagger$ for the case $\epsilon = -1$ expressed in Bose annihilation and creation operators.

(iii) Calculate the commutators $[\hat{N}, d]$ and $[\hat{N}, d^\dagger]$ using the result from (i).

(iv) Calculate

$$d^\dagger d |n\rangle, \quad d|n\rangle, \quad d^\dagger |n\rangle$$

where $|n\rangle$ $(n = 0, 1, 2, \ldots)$ are the number states using the representation given above.

Solution 11. (i) We have

$$\hat{N}_\epsilon = \frac{1}{2} \ln_{|\epsilon|} \left(I + (\epsilon - 1)d^\dagger d \right)^2 = \frac{1}{2} \ln_{|\epsilon|} \epsilon^{2\hat{N}} = \hat{N} = b^\dagger b.$$

(ii) For $\epsilon = -1$ using

$$(-1)^{\hat{N}+I} \equiv e^{i\pi(\hat{N}+I)} = e^{i\pi\hat{N}} e^{i\pi} = -e^{i\pi\hat{N}}$$

we have

$$\frac{\epsilon^{\hat{N}+I} - I}{\epsilon - 1} = \frac{-(-1)^{\hat{N}+I} + I}{2} = \frac{1}{2}(e^{i\pi\hat{N}} + I).$$

(iii) Using (i) we obtain $[\hat{N}, d] = -d$ and $[\hat{N}, d^\dagger] = d^\dagger$. Thus the linear operators d, $d^\dagger$ describe paraparticles.

(iv) We have

$$d^\dagger d |n\rangle = [n]|n\rangle, \quad d|n\rangle = [n]|n-1\rangle, \quad d^\dagger |n\rangle = [n+1]|n+1\rangle$$

where

$$[n] := \frac{\epsilon^n - 1}{\epsilon - 1}.$$

Problem 12. Let b_1, b_2, $b_1^\dagger$, $b_2^\dagger$ be Bose annihilation and creation operators. Let $c^\dagger$, c be Fermi creation and annihilation operators. Consider the operators

$$J_+ = b_1^\dagger b_2 \otimes I_F + I_B \otimes c^\dagger, \quad J_- = b_2^\dagger b_1 \otimes I_F + I_B \otimes c,$$

$$J_3 = (b_1^\dagger b_1 - b_2^\dagger b_2) \otimes I_F + I_B \otimes (2c^\dagger c - I_F)$$

in the product space. Find the commutators $[J_+, J_-]$, $[J_+, J_3]$, $[J_-, J_3]$.

Solution 12. Note that

$$[b_1^\dagger b_2, b_2^\dagger b_1] = b_1^\dagger b_1 - b_2^\dagger b_2, \quad [c^\dagger, c] = 2c^\dagger c - I_F$$

and

$$[b_1^\dagger b_2, b_1^\dagger b_1 - b_2^\dagger b_2] = -b_1^\dagger b_2, \quad [c^\dagger, 2c^\dagger c - I_F] = -2c^\dagger.$$

Thus

$$[J_+, J_-] = [b_1^\dagger b_2, b_2^\dagger b_1] \otimes I_2 + I_B \otimes [c^\dagger, c] = (b_1^\dagger b_1 - b_2^\dagger b_2) \otimes I_2 + I_B \otimes (2c^\dagger c - I_F) = J_3.$$

Analogously we find $[J_+, J_3] = -2J_+$, $[J_-, J_3] = 2J_-$. Thus we have a basis of a Lie algebra.

Problem 13.   Given Bose creation and annihilation operators $b^\dagger$, b and Fermi creation and annihilation operators with spin up and down, i.e. $c_\uparrow^\dagger$, $c_\downarrow^\dagger$, $c_\uparrow$, $c_\downarrow$. Consider the linear operator

$$Q := b \otimes c_\uparrow^\dagger c_\downarrow^\dagger.$$

(i) Find Q^2.
(ii) Find the Hamilton operator $\hat{H}$ defined by the anticommutator $\hat{H} = [Q, Q^\dagger]_+$.
(iii) Find the commutator $[Q, \hat{H}]$.
(iv) Find the commutator $[\hat{H}, Q^\dagger Q]$.
(v) A basis is given by

$$|mnp\rangle \equiv |m\rangle_B \otimes |np\rangle_F$$

where $m = 0, 1, 2, \dots$ and $n, p = 0, 1$. Find $\hat{H}|mnp\rangle$.

Solution 13.   (i) Since $c_\sigma^\dagger c_\sigma^\dagger = 0$ with $\sigma \in \{\uparrow, \downarrow\}$ we find that $Q^2 = 0_B \otimes 0_F$.
(ii) From Q we obtain $Q^\dagger = b^\dagger \otimes c_\downarrow c_\uparrow$. Let $\hat{N}_B := b^\dagger b$, $\hat{N}_\uparrow := c_\uparrow^\dagger c_\uparrow$, $\hat{N}_\downarrow := c_\downarrow^\dagger c_\downarrow$ be the three number operators. Applying $[b, b^\dagger] = I_B$ and $[c_\sigma, c_{\sigma'}^\dagger]_+ = I_F \delta_{\sigma,\sigma'}$ we arrive at the Hamilton operator

$$\hat{H} = (2\hat{N}_B + I_B) \otimes \hat{N}_\uparrow \hat{N}_\downarrow + \hat{N}_B \otimes (I_F - \hat{N}_\uparrow - \hat{N}_\downarrow)$$

where I_B is the identity operator in the Hilbert space $\mathcal{H}_B$ of the Bose operators and I_F is the identity operator in the finite dimensional Hilbert space of the Fermi operators.
(iii) Since $Q^2 = 0_B \otimes 0_F$ we obtain

$$[Q, \hat{H}] = 0_B \otimes 0_F.$$

It also follows that $[Q^\dagger, \hat{H}] = 0_B \otimes 0_F$.
(iv) Since $[Q, \hat{H}] = 0_B \otimes 0_F$ and $[Q^\dagger, \hat{H}] = 0_B \otimes 0_2$ we have

$$[Q^\dagger Q, \hat{H}] = Q^\dagger[Q, \hat{H}] + [Q^\dagger, \hat{H}]Q = 0_B \otimes 0_F.$$

Thus $\hat{H}$ is invariant under supersymmetric transformations generated by Q. $\hat{H}$ and $Q^\dagger Q$ or $\hat{H}$ and Q can be simultaneously diagonalized.
(iv) We have

$$|m\rangle_B = \frac{1}{\sqrt{m!}} (b^\dagger)^m |0\rangle_B, \qquad m = 0, 1, 2, \dots$$

and $b|0\rangle_B = 0$, where $|0\rangle_B$ is the vacuum state. For the Fermi part we have the basis

$$|00\rangle_F = |0\rangle_F, \qquad c_\uparrow|0\rangle_F = c_\downarrow|0\rangle_F = 0$$

$$|01\rangle_F := c_\downarrow^\dagger|0\rangle_F, \quad |10\rangle_F := c_\uparrow^\dagger|0\rangle_F, \quad |11\rangle_F := c_\uparrow^\dagger c_\downarrow^\dagger|0\rangle_F.$$

Then we have the eigenvalue problem

$$\hat{H}|mnp\rangle = (m(2np + 1 - (n + p)) + np)|mnp\rangle = (np(2m + 1) + m(1 - n - p))|mnp\rangle$$

with the special cases (eigenvalue equations)

$$\hat{H}|m00\rangle = m|m00\rangle, \quad \hat{H}|m01\rangle = 0|m01\rangle,$$

$$\hat{H}|m10\rangle = 0|m10\rangle, \quad \hat{H}|m11\rangle = (m+1)|m11\rangle.$$

Problem 14. Consider the *Jaynes-Cummings model.* It describes a two-level atom coupled linearly with a single bosonic mode. Let $c^\dagger$, $b^\dagger$ be Fermi and Bose creation operators, respectively. The Hamilton operator $\hat{H}$ is given by

$$\hat{H} = 2\omega_1 b^\dagger b \otimes I_F + 2\omega_2 I_B \otimes c^\dagger c + \gamma b \otimes c^\dagger + b^\dagger \otimes c\bar{\gamma}$$

where I_B is the identity operator (infinite dimensional unit matrix) for the bosons and I_F is the identity operator (2×2 unit matrix) for fermions. Thus the atom has the eigenstates with eigenvalues $E = 2\omega_2$ and $E = 0$. The interaction constant γ may be considered as a Grassmann or an ordinary c-valued number.
(i) Find the anticommutators $[b \otimes c^\dagger, b^\dagger \otimes c]_+$, $[b \otimes c^\dagger, b \otimes c^\dagger]_+$, $[b \otimes c, b \otimes c]_+$.
(ii) Find the commutators $[b \otimes I_F, b^\dagger \otimes c]$, $[I_B \otimes c, b \otimes c^\dagger]$.
(iii) Find the anticommutator $[I_B \otimes c, b \otimes c^\dagger]_+$.

Solution 14. (i) We obtain

$$
\begin{aligned}
[b \otimes c^\dagger, b^\dagger \otimes c]_+ &= bb^\dagger \otimes c^\dagger c + b^\dagger b \otimes cc^\dagger \\
&= (I_B + b^\dagger b) \otimes c^\dagger c + b^\dagger b \otimes (I_F - c^\dagger c) \\
&= I_B \otimes c^\dagger c + b^\dagger b \otimes I_F.
\end{aligned}
$$

Analogously we find since $(c^\dagger)^2 = 0_F$ and $c^2 = 0_F$

$$[b \otimes c^\dagger, b \otimes c^\dagger]_+ = 0_B \otimes 0_F, \qquad [b \otimes c, b \otimes c]_+ = 0_B \otimes 0_F.$$

(ii) For the commutators we obtain

$$[b \otimes I_F, b^\dagger \otimes c] = bb^\dagger \otimes c + b^\dagger b \otimes c = (I_B + b^\dagger b) \otimes c - b^\dagger b \otimes c = I_B \otimes c$$

and

$$[I_B \otimes c, b \otimes c^\dagger] = b \otimes I_F - 2b \otimes c^\dagger c.$$

(iii) We obtain

$$[I_B \otimes c, b \otimes c^\dagger]_+ = b \otimes I_F.$$

Problem 15. Let $b^\dagger$, b Bose creation and annihilation operators, i.e. $[b, b^\dagger] = I_B$ where I_B is the identity operator. Let $c^\dagger$, c be Fermi creation and annihilation operators, i.e. $[c, c^\dagger]_+ = I_F$, $[c, c]_+ = 0_F$, $[c^\dagger, c^\dagger]_+ = 0_F$. Consider the operator in the product Hilbert space

$$C^\dagger = I_B \otimes c^\dagger, \quad C = I_B \otimes c, \quad Q^\dagger = b^\dagger \otimes c, \quad Q = b \otimes c^\dagger$$

$$B^\dagger = b^\dagger \otimes I_F, \quad B = b \otimes I_F, \quad H = b^\dagger b \otimes I_F + I_B \otimes c^\dagger c, \quad I = I_B \otimes I_F.$$

(i) Find the anticommutators

$$[C^\dagger, C^\dagger]_+, \quad [Q^\dagger, Q^\dagger]_+, \quad [C^\dagger, C]_+, \quad [Q^\dagger, Q]_+ = H,$$

$$[C^\dagger, Q]_+, \quad [C, Q^\dagger]_+, \quad [C^\dagger, Q^\dagger]_+, \quad [C, Q]_+$$

(ii) Find the commutators

$$[B, B^\dagger], \quad [H, B^\dagger], \quad [H, B], \quad [B^\dagger, C^\dagger], \quad [B^\dagger, C], \quad [B^\dagger, Q^\dagger], \quad [B^\dagger, Q], \quad [H, C^\dagger], \quad [H, Q^\dagger].$$

Solution 15.   (i) For the anticommutators we obtain

$$[C^\dagger, C^\dagger]_+ = [Q^\dagger, Q^\dagger]_+ = 0_B \otimes 0_F, \quad [C^\dagger, C]_+ = I_B \otimes I_F, \quad [Q^\dagger, Q]_+ = H,$$

$$[C^\dagger, Q]_+ = [C, Q^\dagger]_+ = 0_B \otimes 0_F, \quad [C^\dagger, Q^\dagger]_+ = B^\dagger, \quad [C, Q]_+ = B.$$

(i) For the commutators we obtain

$$[B, B^\dagger] = I_B \otimes I_F, \quad [H, B^\dagger] = B^\dagger, \quad [H, B] = -B.$$

$$[B^\dagger, C^\dagger] = [B^\dagger, C] = 0_B \otimes 0_F, \quad [B^\dagger, Q^\dagger] = 0_B \otimes 0_F,$$

$$[B^\dagger, Q] = -C^\dagger, \quad [H, C^\dagger] = C^\dagger, \quad [H, Q^\dagger] = 0_B \otimes 0_F.$$

Problem 16.   Find the matrix representation of the operator

$$b^\dagger \otimes c + b \otimes c^\dagger$$

applying the basis $(n = 0, 1, \ldots)$

$$\{ |n\rangle \otimes |0\rangle, \ |n\rangle \otimes c^\dagger |0\rangle \}$$

and the corresponding dual one $(m = 0, 1, \ldots)$

$$\{ \langle m| \otimes \langle 0|, \ \langle m| \otimes \langle 0|c \}.$$

Give a possible ordering of the basis.

Solution 16.   We have

$$(\langle m| \otimes \langle 0|)(b^\dagger \otimes c + b \otimes c^\dagger)(|n\rangle \otimes |0\rangle) = 0, \quad (\langle m| \otimes \langle 0|c)(b^\dagger \otimes c + b \otimes c^\dagger)(|n\rangle \otimes c^\dagger |0\rangle) = 0$$

and

$$(\langle m| \otimes \langle 0|)(b^\dagger \otimes c + b \otimes c^\dagger)(|n\rangle \otimes c^\dagger |0\rangle) = (\langle m| \otimes \langle 0|)(\sqrt{n+1}|n+1\rangle \otimes |0\rangle) = \sqrt{n+1}\delta_{m,n+1}$$

$$(\langle m| \otimes \langle 0|c)(b^\dagger \otimes c + b \otimes c^\dagger)(|n\rangle \otimes |0\rangle) = (\langle m| \otimes \langle 0|c)(\sqrt{n}|n-1\rangle \otimes c^\dagger |0\rangle) = \sqrt{n}\delta_{m,n+1}.$$

A possible ordering of the basis could be

$$|0\rangle \otimes |0\rangle, \ |0\rangle \otimes c^\dagger|0\rangle, \ |1\rangle \otimes |0\rangle, \ |1\rangle \otimes c^\dagger, \ |2\rangle \otimes |0\rangle, \ldots$$

Problem 17.   Consider the operator

$$\hat{K} = b^\dagger \otimes c + b \otimes c^\dagger.$$

Let $|\beta\rangle$ be coherent states. Find

$$(\langle\beta|\otimes\langle 0|)(\hat{K})(|\beta\rangle\otimes|0\rangle)$$
$$(\langle\beta|\otimes\langle 0|)(\hat{K})(|\beta\rangle\otimes c^\dagger|0\rangle)$$
$$(\langle\beta|\otimes\langle 0|c)(\hat{K})(|\beta\rangle\otimes|0\rangle)$$
$$(\langle\beta|\otimes\langle 0|c)(\hat{K})(|\beta\rangle\otimes c^\dagger|0\rangle).$$

Solution 17. Utilizing $b|\beta\rangle=\beta|\beta\rangle$, $\langle\beta|b^\dagger=\langle\beta|\bar{\beta}$ and $c^\dagger cc^\dagger|0\rangle=c^\dagger|0\rangle$ we arrive at

$$(\langle\beta|\otimes\langle 0|)(\hat{K})(|\beta\rangle\otimes|0\rangle)=0$$
$$(\langle\beta|\otimes\langle 0|)(\hat{K})(|\beta\rangle\otimes c^\dagger|0\rangle)=\bar{\beta}$$
$$(\langle\beta|\otimes\langle 0|c)(\hat{K})(|\beta\rangle\otimes|0\rangle)=\beta$$
$$(\langle\beta|\otimes\langle 0|c)(\hat{K})(|\beta\rangle\otimes c^\dagger|0\rangle)=0.$$

Problem 18. Consider the one-fermion-one-boson model Hamilton operator

$$\hat{H}=\hbar\omega b^\dagger b\otimes I_F+J(I_B\otimes c^\dagger c)+\gamma(b^\dagger\otimes c+b\otimes c^\dagger).$$

Consider the number operator $\hat{N}$ and parity operator $\hat{P}$

$$\hat{N}=b^\dagger b\otimes I_F+I_B\otimes c^\dagger c,\qquad\hat{P}=\exp(i\pi(b^\dagger b\otimes I_F+I_B\otimes c^\dagger c)).$$

(i) Find the commutators $[\hat{H},\hat{N}]$ and $[\hat{H},\hat{P}]$.
(ii) Find the spectrum of $\hat{H}$ utilizing the basis $|n\rangle_B\otimes|0\rangle$, $|n\rangle_B\otimes|1\rangle$, where $|n\rangle_B$ are the number states.

Solution 18. (i) We obtain $[\hat{H},\hat{N}]=0_B\otimes 0_F$ and $[\hat{H},\hat{P}]=0_B\otimes 0_F$.
(ii) We find the discrete spectrum

$$\left\{0,J/2+\hbar\omega/2+n\hbar\omega\pm\left(\frac{1}{4}(J-\hbar\omega)^2+\gamma^2(n+1)\right)^{1/2}\right\}$$

where $n=0,1,\ldots$.

Problem 19. Let $c^\dagger$, c be Fermi creation and annihilation operators with $[c^\dagger,c]_+=I_F$, $[c,c]_+=[c^\dagger,c^\dagger]_+=0_F$, where I_F is the identity operator and 0_F the zero operator for the Fermi system. Let $b^\dagger$, b be Bose creation and annihilation operators with $[b,b^\dagger]=I_B$, $[b,b]=[b^\dagger,b^\dagger]=0_B$, where I_B is the identity operator and 0_B the zero operator for the Bose system. Consider a Lie superalgebra with four odd basis elements

$$F_+=I_B\otimes c^\dagger,\quad F_-=I_B\otimes c,\quad Q_+=b^\dagger\otimes c,\quad Q_-=b\otimes c^\dagger$$

and four even basis elements

$$E_+=b^\dagger\otimes I_F,\quad E_-=b\otimes I_F,\quad \hat{H}=b^\dagger b\otimes I_F+I_B\otimes\otimes c^\dagger c,\quad I=I_B\otimes I_F$$

where $\hat{H}$ plays the role of a Hamilton operator.
(i) Show that the Hamilton operator $\hat{H}$ can be generated by E_+, E_-, F_+, F_-.

(ii) Let $\alpha \in \mathbb{R}$ and consider the operator

$$K := \alpha F_+ + \alpha F_- + Q_+ + Q_- \equiv \alpha(I_B \otimes c^\dagger + I_B \otimes c) + b^\dagger \otimes c + b \otimes c^\dagger.$$

Find the (infinite dimensional) matrix representation of the operator X with the ordering of the basis given by

$$|0\rangle_B \otimes |0\rangle_F, \quad |0\rangle_B \otimes c^\dagger|0\rangle, \quad |1\rangle_B \otimes |0\rangle_F, \quad |1\rangle_B \otimes c^\dagger|0\rangle_F, \quad |2\rangle_B \otimes |0\rangle_F, \ldots$$

Solution 19. (i) We have

$$E_+E_- + F_+F_- = b^\dagger b \otimes I_F + I_B \otimes c^\dagger c = \hat{H}.$$

(ii) Utilizing that $bb^\dagger = I_B + b^\dagger b$, $cc^\dagger = I_F - c^\dagger c$, $c|0\rangle_F = 0|0\rangle_F$ and $b|0\rangle_B = 0|0\rangle_B$ we find

$$K = \begin{pmatrix} 0 & \alpha & 0 & 0 & 0 & 0 & 0 & \cdots \\ \alpha & 0 & \sqrt{1} & 0 & 0 & 0 & 0 & \cdots \\ 0 & \sqrt{1} & 0 & \alpha & 0 & 0 & 0 & \cdots \\ 0 & 0 & \alpha & 0 & \sqrt{2} & 0 & 0 & \cdots \\ 0 & 0 & 0 & \sqrt{2} & 0 & \alpha & 0 & \cdots \\ 0 & 0 & 0 & 0 & \alpha & 0 & \sqrt{3} & \cdots \\ 0 & 0 & 0 & 0 & 0 & \sqrt{3} & 0 & \cdots \\ \vdots & \vdots & \vdots & \vdots & \vdots & \vdots & \vdots & \ddots \end{pmatrix}.$$

Problem 20. (i) Calculate

$$\exp(i\pi(b^\dagger b \otimes I_F + I_B \otimes c^\dagger c)).$$

(ii) Let $|n\rangle$ be number states. Find the states

$$\exp(i\pi(b^\dagger b \otimes c^\dagger c))(|n\rangle \otimes |0\rangle), \quad \exp(i\pi(b^\dagger b \otimes c^\dagger c))(|n\rangle \otimes c^\dagger|0\rangle).$$

Solution 20. (i) We have $[b^\dagger b \otimes I_F, I_B \otimes c^\dagger c] = 0_B \otimes 0_F$. Consequently

$$\exp(i\pi(b^\dagger b \otimes I_F + I_B \otimes c^\dagger c)) = \exp(i\pi(b^\dagger b \otimes I_F)) \exp(i\pi(I_B \otimes c^\dagger c)).$$

Thus

$$\exp(i\pi(b^\dagger b \otimes I_F + I_B \otimes c^\dagger c)) = \exp(i\pi b^\dagger b) \otimes \exp(i\pi c^\dagger c)$$

where using $e^{i\pi} = -1$

$$\exp(i\pi(b^\dagger b)) = \mathrm{diag}(1 - 1\,1 - 1\cdots), \quad \exp(i\pi c^\dagger c) = I_F - 2c^\dagger c.$$

(ii) Since $b^\dagger b|n\rangle = n|n\rangle$, $c^\dagger cc^\dagger|0\rangle = c^\dagger$ we obtain

$$e^{i\pi(b^\dagger b)\otimes(c^\dagger c)}(|n\rangle \otimes |0\rangle) = |n\rangle \otimes |0\rangle, \quad e^{i\pi(b^\dagger b)\otimes(c^\dagger c)}(|n\rangle \otimes c^\dagger|0\rangle) = e^{i\pi n}|n\rangle \otimes c^\dagger|0\rangle$$

with $n = 0, 1, \ldots$. Both equations are eigenvalue equations.

6.2 Supplementary Problems

Problem 1. Let $c^\dagger$, c be Fermi creation and annihilation operators and I_F the identity operator. Let $b^\dagger$, b be Bose creation and annihilation operators and I_B the identity operator. Consider the operators

$$b^\dagger b \otimes I_F, \quad b^\dagger b \otimes c^\dagger c, \quad I_B \otimes c^\dagger c$$

and

$$V = b^\dagger \otimes c + b \otimes c^\dagger.$$

(i) Find the commutators between these operators.
(ii) Find the anticommutator between these operators.

Problem 2. Consider the Hamilton operator

$$\hat{K} = \frac{\hat{H}}{\hbar\omega} = (b + b^\dagger) \otimes c^\dagger c.$$

(i) Find the matrix representation of $\hat{K}$ using the basis $|n\rangle \otimes |0\rangle$, $|n\rangle \otimes c^\dagger |0\rangle$.
(ii) Let $|\beta\rangle$ be a coherent state. Find

$$(\langle\beta| \otimes \langle 0|)\hat{K}(|\beta\rangle \otimes |0\rangle), \qquad (\langle\beta| \otimes \langle 0|)\hat{K}(|\beta\rangle \otimes c^\dagger |0\rangle),$$

$$(\langle\beta| \otimes \langle 0|c)\hat{K}(|\beta\rangle \otimes |0\rangle), \qquad (\langle\beta| \otimes \langle 0|c)\hat{K}(|\beta\rangle \otimes c^\dagger |0\rangle).$$

(iii) Let $|\zeta\rangle$ be a squeezed state. Find

$$(\langle\zeta| \otimes \langle 0|)\hat{K}(|\zeta\rangle \otimes |0\rangle), \qquad (\langle\zeta| \otimes \langle 0|)\hat{K}(|\zeta\rangle \otimes c^\dagger |0\rangle),$$

$$(\langle\zeta| \otimes \langle 0|c)\hat{K}(|\zeta\rangle \otimes |0\rangle), \qquad (\langle\zeta| \otimes \langle 0|c^\dagger)\hat{K}(|\zeta\rangle \otimes c|0\rangle).$$

Problem 3. (i) Let $z_1, z_2 \in \mathbb{C}$. Find the spectrum of the Hamilton operator

$$\hat{H} = z_1 b \otimes c^\dagger \otimes b + z_2 b^\dagger \otimes c \otimes b + z_1^* b^\dagger \otimes c \otimes b^\dagger + z_2^* b \otimes c^\dagger \otimes b^\dagger$$

where z_1, z_2 are complex numbers.
(ii) Let $z \in \mathbb{C}$. Study the spectrum of the Hamilton operator

$$\hat{H} = z(b \otimes c^\dagger \otimes b) + z^*(b^\dagger \otimes c \otimes b^\dagger).$$

(iii) Let $z \in \mathbb{C}$. Study the spectrum of the Hamilton operator

$$\hat{H} = z(c \otimes b^\dagger \otimes c) + z^*(c^\dagger \otimes b \otimes c^\dagger).$$

Problem 4. Study the spectrum of the Hamilton operator

$$\hat{H} = \hbar\omega_1 b^\dagger b \otimes I_F + \hbar\omega_2 I_B \otimes c^\dagger c + \gamma e^{i\phi} b^\dagger \otimes c + \gamma e^{-i\phi} b \otimes c^\dagger.$$

Problem 5. Let $|0\rangle_B$, $|0\rangle_F$ be the vacuum state for Bose and Fermi, respectively, i.e. $b|0\rangle_B = 0|0\rangle_B$, $c|0\rangle_F = 0|0\rangle_F$. In the product Hilbert space we have the basis

$$\frac{1}{\sqrt{j!}}(b^\dagger)^j |0\rangle_B \otimes (c^\dagger)^k |0\rangle_F$$

where $j = 0, 1, 2, \ldots$ and $k = 0, 1$. Apply the operators $b^\dagger b \otimes c^\dagger c$, $(b^\dagger + b) \otimes (c^\dagger + c)$ to this basis, i.e. find the matrix representation of these operators.

Problem 6. (i) Study the Hamilton operator

$$\hat{K} = \frac{\hat{H}}{\hbar\omega} = b^\dagger \otimes c \otimes b + b \otimes c^\dagger \otimes b^\dagger.$$

Does the Hamilton operator commute with $b^\dagger b \otimes c^\dagger c \otimes b^\dagger b$?
(ii) Study the Hamilton operator

$$\hat{K} = \frac{\hat{H}}{\hbar\omega} = c^\dagger \otimes b \otimes c + c \otimes b^\dagger \otimes c^\dagger.$$

Does the Hamilton operator commute with $c^\dagger c \otimes b^\dagger b \otimes c^\dagger c$?
(iii) Study the Hamilton operators

$$\hat{K} = \frac{\hat{H}}{\hbar\omega} = b^\dagger \otimes c \otimes b^\dagger + b \otimes c^\dagger \otimes b.$$

Does the Hamilton operator commute with $b^\dagger b \otimes c^\dagger c \otimes b^\dagger b$?
(iv) Study the Hamilton operator

$$\hat{K} = \frac{\hat{H}}{\hbar\omega} = c^\dagger \otimes b \otimes c^\dagger + c \otimes b^\dagger \otimes c.$$

Does the Hamilton operator commute with $c^\dagger c \otimes b^\dagger b \otimes c^\dagger c$? Does the Hamilton operator commute with the number operator

$$c^\dagger c \otimes I_B \otimes I_F + I_F \otimes b^\dagger b \otimes I_F + I_F \otimes I_B \otimes c^\dagger c ?$$

(v) Study the Hamilton operator

$$\hat{K} = \frac{\hat{H}}{\hbar\omega} = c \otimes b^\dagger \otimes c^\dagger + c^\dagger \otimes b \otimes c.$$

Does the Hamilton operator commute with $c^\dagger c \otimes b^\dagger b \otimes c^\dagger c$?

Problem 7. Let $c_1^\dagger$, $c_2^\dagger$, c_1, c_2 be Fermi creation and annihilation operators and $b^\dagger$, b be Bose creation and annihilation operators.
(i) Consider the operators

$$b^\dagger b \otimes I_F, \quad I_B \otimes c_1^\dagger c_2, \quad I_B \otimes c_2^\dagger c_2, \quad I_B \otimes c_1^\dagger c_2, \quad I_B \otimes c_2^\dagger c_1.$$

Find the commutators between these operators.
(ii) Consider the operators

$$b^\dagger \otimes c_1, \quad b \otimes c_1^\dagger, \quad b^\dagger \otimes c_2, \quad b \otimes c_2^\dagger.$$

Find the commutators and anticommutators between these operators. Find the commutators and anticommutators between these operators and the operators from (i).
(iii) Study the Hamilton operator

$$\hat{K} = \frac{\hat{H}}{\hbar\omega} = (b + b^\dagger) \otimes (c_1^\dagger c_2 + c_2^\dagger c_1).$$

Find the matrix representation of $\hat{K}$ using the basis $|n\rangle \otimes |0\rangle$, $|n\rangle \otimes c^\dagger |0\rangle$.

Problem 8. Let $c^\dagger$, c be Fermi creation and annihilation operators. Study the algebra generated by the elements

$$b_1^\dagger b_1 \otimes I_F, \quad b_2^\dagger b_2 \otimes I_F, \quad b_1^\dagger b_2 \otimes I_F, \quad b_2^\dagger b_1 \otimes I_B \otimes c^\dagger c.$$

Apply the operators to the states

$$|n_1, n_2\rangle_B \otimes |0\rangle_F, \qquad |n_1, n_2\rangle_B \otimes c^\dagger |0\rangle$$

where $n_1, n_2 = 0, 1, \ldots$.

Bibliography

Books

Auerbach A.
Interacting Electrons and Quantum Magnetism, Springer Verlag, New York, 1994

Barnett S. M. and Radmore P. M.
Methods in Theoretical Quantum Optics, Oxford University Press, Oxford, 1997

Baxter R. J.
Exactly Solved Models in Statistical Mechanics, Academic Press, New York, 1982

Berezin F. A.
The Method of Second Quantizations, Academic Press, New York, 1966

Bloembergen N.
Nonlinear Optics, McGraw-Hill, New York, 1972

Cronin J. A., Greenberg D. F. and Telegdi V. L.
University of Chicago Graduate Problems in Physics, Addison-Wesley, Reading, 1967

Flügge, Siegfried
Practical Quantum Mechanics, Springer Verlag, 1974

Freund P. G. O.
Introduction to Supersymmetry, Cambridge University Press, 1986

Fröberg C.-E.
Numerical Mathematics, Benjamin/Cumming Publishing, 1985

Gaudin M.
La fonction d'onde de Bethe, Masson, Paris, 1983

Glauber R. J.
Quantum Theory of Optical Coherence. Selected Papers and Lectures, Wiley-VCH, 2007

Georgi H.
Lie Algebras in Particle Physics, 2nd edition, Westview Press, 1999

Gutzwiller M. C.
Chaos in Classical and Quantum Mechanics, Springer Verlag, 1990

Hardy Y., Tan K. S. and Steeb W.-H.
Computer Algebra with SymbolicC++, World Scientific, 2008

Hardy Y. and Steeb W.-H.
Classical and Quantum Computing with C++ and Java Simulations, Birkhauser Verlag, 2002

Huang K.
Quantum Field Theory: From Operators to Path Integrals, Wiley-VCH, 2004

Kac V. G.
Lecture Notes in Mathematics, vol. 676, Springer Verlag, 1977

Kaku M.
Quantum Field Theory: A Modern Introduction, Oxford University Press, 1993

Kim Y. S. and Noz M. E.
Phase Space Picture of Quantum Mechanics, World Scientific, 1991

Kowalski K. and Steeb W.-H.
Nonlinear Dynamical Systems and Carleman Linearization, World Scientific, 1991

Klauder J. R. and Sudarshan E. C. G.
Fundamentals of Quantum Optics, Benjamin, 1968

Klauder J. R. and Skagerstam B.-S.
Coherent states, World Scientific, 1985

Korepin V. E., Izergin G. and Bogoliubov N. M.
Quantum Inverse Scattering Method and Correlation Functions, Cambridge University Press, Cambridge, 1992

Louisell W. H.
Quantum Statistical Properties of Radiation, Wiley, New York, 1973

Martin P. P.
Potts Models and Related Problems in Statistical Mechanics, World Scientific, 1991

Mattis D. C.
The Theory of Magnetism, Springer Verlag, 1981

McCoy B. M. and Wu T. T.
The two-dimensional Ising model, Harvard University Press, 1973

Montorsi A.
The Hubbard Model, World Scientific, 1993

Nielsen M. A. and Chuang I. L.
Quantum Computation and Quantum Information, Cambridge University Press, 2000

Perelomov A. M.
Generalized Coherent States and Their Applications, Springer Verlag, 1986

Press W. H., Teukolsky S. A., Vetterling W. T. and Flannery B. P.
Numerical Recipes in C, Cambridge Universiy Press 1992

Sakurai J. J.
Modern Quantum Mechanics, Menlo Parl, CA: Benjamin/Cummings, 1985

Scheunert M.
The Theory of Lie Superalgebra, Lecture Notes in Mathematics, Springer Verlag, 1979

Schleich W. P.
Quantum Optics in Phase Space, Wiley-VCH, 2001

Shen Y. R.
The Principles of Nonlinear Optics, Wiley, New York, 1985

Steeb W.-H. and Hardy Y.
Matrix Calculus and Kronecker Product: A practical Approach to Linear and Mutilinear Algebra, second edition, World Scientific, 2011

Steeb W.-H.
Continuous Symmetries, Lie Algebras, Differential Equations and Computer Algebra, second edition, World Scientific, 2007

Steeb W.-H.
Hilbert Spaces, Wavelets, Generalized Functions and Quantum Mechanics, Kluwer Academic Publishers, 1998

Steeb W.-H. and Hardy Y.
Quantum Mechanics using Computer Algebra, second edition, World Scientific, 2010

Steeb W.-H.
Problems and Solutions in Theoretical and Mathematical Physics, Volume I: Introductory Level, third edition World Scientific, 2009

Steeb W.-H.
Problems and Solutions in Theoretical and Mathematical Physics, Volume II: Advanced Level, third edition, World Scientific, 2009

Sutherland B.
Beautiful Models, World Scientific 2004

Thaller B.
The Dirac Equation, Springer Verlag, 1992

Toda M.
Theory of Nonlinear Lattices, Springer Verlag, 1981

Weyl H.
Theory of Groups and Quantum Mechanics, Dover, New York, 1950

White R. M.
Quantum Theory of Magnetism, McGraw-Hill, New York, 1970

Papers

Alber G., Delgado A., Gisin N. and Jex I., "Efficient bipartite quantum state purification in arbitrary dimensional Hilbert spaces", J. Phys. A: Math. Gen. **34**, 8821 (2001)

Alvarez-Estrada R. F., Gómez Nicola A., SánchezSoto L. L. and Luis A., "A quasiclassical analysis of second-harmonic generation", J. Phys. A: Math. Gen. **28**, 3439 (1995)

Au-Yang H. and McCoy B. M., "Theory of layered Ising models: Thermodynamics", Phys. Rev. B **10**, 886-891 (1974)

Azuma H., "Applications of Abel-Plana formula for collapse and revival of Rabi oscillations in Jaynes-Cummings model", Int. J. Mod. Phys. C **21**, 1021-1049 (2010)

Baker T. H., "Vertex operator realization of symplectic and orthogonal S-functions", J. Phys. A: Math. Gen. **29**, 3099 (1996)

Ban M., "The phase operator in quantum information processing", J. Phys. A: Math. Gen. **35**, L193 (2002)

Banerji J., "Non-linear wave packet dynamics of coherent states", PRAMANA J. Phys. **56**, 267 (2001)

Barak R. and Ben-Aryeh Y., "Photon statistics and entanglement in coherent-squeezed linear Mach-Zehnder and Michelson interferometers", J. Opt. Soc. Am. B **25**, 361-372 (2008)

Bariev R. Z., Klümper A., Schadschneider A. and Zittartz J., "A one-dimensional integrable model of fermions with multi-particle hopping", J. Phys. A: Math. Gen. **28**, 2437-2444 (1995)

Barnett S. M., Jeffers J. and Gatti A., "Quantum optics of lossy beam splitters", Phys. Rev. A **57**, 2134 (1998)

Bartlett S. D., Sanders B. C., Braunstein S. L. and Nemoto K., "Efficient Classical Simulations of Continuous Variable Quantum Information Processes", Phys. Rev. Lett. **88**, 097904 (2002)

Bars L. and Günaydin M., "Unitary Representations of Non-Compact Supergroups", Commun. Math. Phys. **91**, 31-51 (1983)

Benedict M. G. and Czirják A., "Generalized parity and quasi-probability density functions", J. Phys. A: Math. Gen. **28**, 4599-4608 (1995)

Berman G. P., Borgonovi F., Chapline G., Gurvitz S. A., Hammel P. C., Pelekhov D. V., Suter A. and Tsifrinovich V. I., "Application of Magnetic Resonance Force Microscopy Cyclic Adiabatic Inversion for a Single-Spin Measurement", J. Phys. A : Math. Gen. **36**, 4417 (2003)

Bogoliubov N. M., Bullough R. K. and Timonen, "Exact solutions of generalized Tavis-Cummings models in quantum optics", J. Phys. A: Math. Gen. **29**, 6305-6312 (1996)

Braak D., "Integrability of the Rabi Model", Phys. Rev. Lett. **107**, 100401 (2011)

Braak D., "Note on the Analytical Solution of the Rabi Model", arXiv:1210.4946v1 (2012)

Braunstein S. L. and Kimble H. J., "Teleportation of Continuous Quantum Variables", Phys. Rev. Lett. **80**, 869 (1998)

Braunstein S. L., Fuchs C. A., Kimble H. J. and van Loock P., "Quantum versus classical domains for teleportation with continuous variables", Phys. Rev. A **64**, 022321 (2001)

Bravyi S., "Lagrangian representation for fermionic linear optics", Quantum Inf. and Comp. **5**, 216-238 (2005)

Bravyi S. B. and Kitaev A. Yu., "Fermionic quantum computation", arXiv:quant-ph/0003137v2

Brif C., Mann A. and Vourdas A., "Parity-dependent squeezing of light", J. Phys. A: Math. Gen. **29** 2053-2067 (1996)

Brinkman W. and Rice T. M., "Applications of Gutzwiller's variational method to the metal-insulator transition", Phys. Rev. B **2**, 4302 (1970)

Bruß D., "Characterizing Entanglement", J. Math. Phys. **43**, 4237 (2002)

Cahill K. E. and Glauber R. J., "Ordered Expansions in Boson Amplitude Operators", Phys. Rev. **177**, 1857–1881 (1969)

Calixto M. and Aldaya V., "Curvature, zero modes and quantum statistics", J. Phys. A: Math. Gen. **39**, L539-L545 (2006)

Carmichael H. J., Milburn G. J. and Walls D. F., "Squeezing in a detuned parametric amplifier", J. Phys. A: Math. Gen. **17** 469-480 (1984)

Chao K. A., "Zero-eigenenergy state of the Aubry model", J. Phys. A: Math. Gen. **20**, L709-L713 (1987)

Chen X. and Kuang Le-Man, "Path integral for new coherent states of Lie superalgebra $osp(1|2, R)$", J. Phys. A: Math. Gen. **27**, L685-L691 (1994)

Childs A. M., Leung D., Verstraete F. and Vidal G., "Asymptotic entanglement capacity of the Ising and anisotropic Heisenberg interactions", Quantum Information and Computation **3**, 97 (2003)

Chruściński D., "Spectral Properties of the Squeeze Operator", Phys. Lett. A **327**, 290-295 (2004)

Cibils M. B., Cuche Y., Wreszinski W. F., Amiet J.-P. and Beck H., "On the classical limits in the spin-boson model", J. Phys. A: Math. Gen. **23**, 545-552 (1990)

Cleve R. and Buhrman H., "Substituting Quantum Entanglement for Communication", Phys. Rev. A **56**, 1201 (1997)

Curtright T. L., Fairlie D. B. and Zachos C. K., "A compact formula for rotations as spin matrix polynomials", arXiv:1402.3541v2

DasGupta A., "Disentanglement formulas: An alternative derivation and some applications to squeezed coherent states", Am. J. Phys. **64**, 1422 (1996)

de Oliveira M. C. and Milburn G. J., "Discrete teleportation protocol of continuum spectra field states", Phys. Rev. A **65**, 032304 (2002)

Deenen J. "Non-bijective canonical transformations in quantum mechanics", J. Phys. A: Math. Gen. **14**, L273-L276 (1981)

Dehaene J. and De Moor B., "Clifford group, stabilizer states, and linear and quadratic operations over $GF(2)$", Phys. Rev. **68**, 042318 (2003)

Dodonov V. V., Manko V. I. and Nikonov D. E., "Even and odd coherent states for multimode parametric systems", Phys. Rev. A **51**, 3328-3336 (1995)

Daoud M., "Generalized intelligent states of the $su(N)$ algebra", Phys. Lett. A **329**, 318-326 (2004)

Essler F. H. L., Korepin V. E. and Schoutens K., "Fine structure of the Bethe ansatz for the spin-$\frac{1}{2}$ Heisenberg XXX model", J. Phys. A: Math. Gen. **25**, 4115-4126 (1992)

Falicov L. M. and Victora R. H., "Exact solution of the Hubbard model for a four-center tetrahedral cluster", Phys. Rev. B **30**, 1695-1699 (1984)

Fan Hong-yi, "Similarity transformation for Fermi operators", J. Phys. A : Math. Gen. **25**, 4269-4274 (1992)

Farkas S. and Zimborás Z., "The von Neumann entropy asymptotics in multidimensional fermionic systems", J. Math. Phys. **48**, 102110 (2007)

Frappat L., Sorba P. and Sciarrino A., "Dictionary on Lie Superalgebras", arXiv:hep-th/9607161

Fujii K., "Basic Properties of Coherent and Generalized Coherent Operators Revisited", Mod. Phys. Lett. A **16**, 1277 (2001)

Fujii K. and Oike H., "Basic Properties of Coherent-Squeezed States Revisited", arXiv:1312.2301

Gilmore R., "Baker-Campbell-Hausdorff formulas", J. Math. Phys. **15**, 2090-2092 (1974)

Göhmann F. and Korepin V. E. "The Hubbard chain: Lieb-Wu equations and norm of eigenfunctions", Phys. Lett. A **263**, 293 (1999)

Grabowski M. P. and Mathieu P. "Integrability test for spin chains", J. Phys. A: Math. Gen. **28**, 4777-4798 (1995)

Hardy Y. and Steeb W.-H., "Fermi-Bose Systems, Macroscopic Quantum Superposition States and Entanglement", Int. J. Theor. Phys. **43**, 2207 (2004)

Hardy Y., Steeb W.-H. and Stoop R., "Fully Entangled Quantum States in C^{N^2} and Bell Measurement", Int. J. Theor. Phys. **42**, 2314 (2003)

Hardy Y., Steeb W.-H. and Stoop R., "Entanglement, Disentanglement and Wigner Functions", Physica Scripta **69**, 166 (2004)

Hinrichsen H., "Four-state models and Clifford algebras", J. Phys.A: Math. Gen. **27**, 5393-5407 (1994)

Hioe F. T., "An Approach to the study of quantum systems", J. Math. Phys. **15**, 445-452 (1974)

Hioe F. T., "Coherent states and Lie algebras", J. Math. Phys. **15**, 1174-1177 (1974)

Hiramatsu T., Matsui T. and Sakakibara K., "Self-Reduction Rate of a Microtubule", Int. J. Mod. Phys. C **19**, 291-305 (2008)

Huang C., Moriarty J. A., Sher A. and Breckenridge R. A., "Two-electron bound-orbital model I", Phys. Rev. B **12**, 5395-5401 (1975)

Jafarov E. I. and Van der Jeugt J., "The oscillator model for the Lie superalgebra $sh(2,2)$ and Charlier polynomials", arXiv:1304.3295v2

Jaynes E. T. and Cummings F. W., "Comparison of quantum and semiclassical radiation theories with applications to the beam maser", Proc. IEEE **51**, 89-109 (1963)

Jeffers J., Barnett S. M. and Pegg D., "Retrodiction as a tool for micromaser field measurement", J. Mod. Opt. **49**, 925 (2002)

Jimbo M. and Miwa T., Publ. RIMS **19**, 943 (1983)

Jimbo M., Miwa T., Mori Y. and Sato M., "Density Matrix of an Impenetrable Bose Gas and the Fifth Painlevé Transcendent", Physica 1D, 80-158 (1980)

Kastrup H. A., "Quantization of the canonically conjugate pair angle and orbital angular momentum", Phys. Rev. A **73**, 052104 (2006)

Kauffman L. H., "Knot Logic and Topological Quantum Computing with Majorana Fermions", arXiv:1301.6214v1

Kawamoto H., "Theoretical Study of Quantum Coherence in Spin-Boson System", Prog. Theor. Phys. **82**, 1044-1056 (1989)

Kibler M. and Négadi T., "On the connection between the hydrogen atom and the harmonic oscillator: the continuum case", J. Phys. A: Math. Gen. **16**, 4265–4268 (1983)

Kielpinski D., "A small trapped-ion quantum register", J. Opt. B: Quantum Semiclass. Opt. **5**, R121-R135 (2003)

Kilin S. Ya. and Horoshko D. B., "Fock State Generation by the Methods of Nonlinear Optics", Phys. Rev. Lett. **74**, 5206-5207 (1995)

Kimura G., "The Bloch vector for N-level systems", Phys. Lett. A **314**, 319 (2003)

Kirillov A. N. and Sakamoto R., "Singular Solutions to the Bethe Ansatz Equations and Rigged Configurations", arXiv:1402.0651v1

Klarsfeld S. and Oteo J. A., "Exponential infinite-product representations of the time-displacement operator", J. Phys. A: Math. Gen. **22**, 2687-2694 (1989)

Klauder J. R. and Streit L., "Optical equivalence theorem for unbounded observables", J. Math. Phys. **15**, 760-763 (1974)

Klauder J. R., "Coherent states for the hydrogen atom", J. Phys. A: Math. Gen. **29**, L293-L298 (1996)

Klenner N., Doucha M. and Weis J., "Quantum dynamics of a two-level system coupled to a bosonic degree of freedom in terms of Wigner phase space distributions", J. Phys. A : Math. Gen. **19**, 3831-3843 (1986)

Knill E., Laflamme R., Milburn G. J., "A scheme for efficient quantum computation with linear optics", Nature **409**, 46 (2001)

Kok P. and Braunstein S. L., "Multi-dimensional Hermite polynomials in quantum optics", J. Phys. A: Math. Gen. **34**, 6185 (2001)

Kok P., Nunro W. J., Nemoto K., Ralph T. C., Dowling J. P. and Milburn G. J., "Linear optical quantum computing", arXiv:quant-ph/0512071v2

Kochetov E. A. "$U(1/1)$ coherent states and a path integral for the Jaynes-Cummings model", J. Phys. A: Math. Gen. **25**, 411-417 (1992)

Koniorczyk M., Buzek V. and Janszky J., "Wigner-function description of quantum teleportation in arbitrary dimensions and a continuous limit", Phys. Rev. A **64**, 034301 (2001)

Kuang Le-Mang and Chen Xin, "New coherent states of the Lie superalgebra $osp(1|2, R)$", J. Phys. A: Math. Gen. **27**, L119-l124 (1994)

Kuzmich A., Walmsley A. and Mandel L., "Violation of a Bell-type inequality in the homodyne measurement of light in an Einstein-Podolsky-Rosen state", Phys. Rev. A **64**, 063804 (2001)

Lee J. and Kim M. S., "Quantum Teleportation Via Mixed Two-Mode Squeezed States in the Coherent-State Representation", J. Kor. Phys. Soc. **42**, 457 (2003)

Li Shang-Bin and Xu Jing-Bo, "Quantum probabilistic teleportation via entangled coherent states", Phys. Lett. A **309**, 321 (2003)

Lieb E. H. "Solution of the Dimer Problem by the Transfer Matrix Method", J. Math. Phys.

8, 2339-2341 (1967)

Lieb E., Schultz T. and Mattis D., "Two soluble models of an antiferromagnetic chain", Ann. Phys. NY **16**, 407-466 (1961)

Lieb E. H. and Wu F. Y. "The one-dimensional Hubbard model: a reminiscence", Physica A **321**, 1-27 (2003)

Lisi A. G. "An Explicit Embedding of Gravity and the Standard Model in E8", arXiv:1006.4908v1

Lloyd S. and Braunstein S. L., "Quantum Computing over Continuous Variables", Phys. Rev. Lett. **82**, 1784 (1999)

Lo C. F., "Eigenfunctions and eigenvalues of squeeze operators", Phys. Rev. A **42**, 6752-6754

Lo C. F. and Sollie R., "Generalized multimode squeezed states", Phys. Rev. A **47**, 733-735 (1993)

Loudon R. and Knight P. L., "Squeezed light", J. Mod. Optics **34**, 709-759 (1987)

Mikhailov V. V., "Normal ordering and generalised Stirling numbers", J. Phys. A: Math. Gen. **18**, 231-235 (1985)

Milburn G. J. and Braunstein S. L., "Quantum teleportation with squeezed states", Phys. Rev. A **60**, 937 (1999)

Mütter K.-H. and Schmitt A., "Solvable spin-1 models in one dimension", J. Phys. A: Math. Gen. **28**, 2265-2276 (1995)

Nieto M. M., "Displaced and Squeezed Number States", arXiv:quant-ph/9612050v1

Olmedilla E. and Wadati M., "Conserved Quantities for Spin Models and Fermion Models", J. Phys. Soc. Jpn. **56**, 4274-4284 (1978)

Palev T. D., "An Analogue of Holstein-Primakoff and Dyson Realization for the Lie Superalgebras. The Lie superalgebra $s\ell(1/n)$, arXiv:hep-th/9607221v1

Pan Feng, Dai Lian-Rong and Draayer J. P., "Exactly solvable $gl(m/n)$ Bose-Fermi systems", J. Phys. A: Math. Gen. **39**, 417-422 (2006)

Pan Jian-Wei, Daniell M., Gasparoni S., Weihs G., and Zeilinger A., "Experimental Demonstration of Four-Photon Entanglement and High-Fidelity Teleportation", Phys. Rev. Lett. **86**, 4435 (2001)

Paris M. G. A., "Generation of mesoscopic quantum superpositions through Kerr-stimulated degenerate downconversion", J. Opt. B: Quantum Semiclass. Opt. **1**, 662 (1999)

Pegg D. T., Barnett S. M. and Jeffers J., "Quantum theory of preparation and measurement", J. Mod. Opt. **49**, 913 (2002)

Reik H. G., Lais P., Stützle M. E. and Doucha M., "Exact solution of the $E \otimes \epsilon$ Jahn-Teller and Rabi Hamiltonian by generalised spheroidal wavefunctions?", J. Phys. A: Math. Gen. **20**, 6327-6340 (1987)

Sebawe Abdalla M., "Quantum treatment of the time-dependent coupled oscillator", J. Phys. A: Math. Gen. **29**, 1997-2012 (1996)

Schmitt A., Mütter K.-H. and Karbach M., "The spin-1 Lai-Sutherland model with external and internal fields: I. The phase diagram" J. Phys. A: Math. Gen. **29** 3951-3962 (1996)

Schmutz M., "Two-Level system coupled to a boson mode: the large n limit", J. Phys. A: Math. Gen. **19**, 3565-3577 (1986)

Schultz T. D., Mattis D. C. and Lieb E. H., "Two-dimensional Ising model as a soluble problem of many Fermions", Rev. Mod. Phys. **36** 856-871 (1964)

Sharma S. S., "Tripartite GHZ state generation with trapped ion in optical cavity", Phys. Lett. A **311**, 111 (2003)

Sharma S. S. and Vidiella-Barranco A., "Fast quantum logic gates with trapped ions interacting with external laser and quantized cavity field beyond Lamb-Dicke regime", Phys. Lett. A **309** 345 (2003)

Shaterzadeh-Yazdi Z., Turner P. S. and Sanders B. C., "$SU(1,1)$ symmetry of multimode squeezed states", J. Phys. A: Math. Theor. **41** 055309 (2008)

Sklyanin E. K., "Boundary conditions for integrable quantum systems", J. Phys. A: Math. Gen. **21**, 2375-2389 (1988)

Slosser J. J. and Milburn G. J., "Creating Metastable Schrödinger Cat States", Phys. Rev. Lett. **75**, 418-421 (1995)

Steeb W.-H., "A Note on Wick's Theorem", Lett. Math. Phys. **1**, 135 (1976)

Steeb W.-H., "Embedding of Nonlinear Finite Dimensional Systems in Linear Infinite Dimensional Systems and Bose Operators", Hadronic Journal **6**, 68-76 (1983)

Steeb W.-H., "Quantum chaos and two exactly solvable second-quantized models", Phys. Rev. A **32**, 1232-1234 (1985)

Steeb W.-H., Villet C. M. and Mulser P., "Hubbard Model, Conserved Quantities, and Computer Algebra", Int. J. Theor. Phys. **32**, 1445-1452 (1993)

Steeb W.-H., "Fermi Systems and Computer Algebra", Int. J. Mod. Phys. C **4**, 841-846 (1993)

Steeb W.-H., "Bose-Fermi Systems and Computer Algebra", Found. of Phys. Lett. **8**, 73-80 (1995)

Steeb W.-H. and Hardy Y., "Entangled Quantum States and a C++ Implementation", Int.

J. Mod. Phys. C **11**, 69 (2000)

Steeb W.-H. and Hardy Y., "Quantum Computing and SymbolicC++ Simulations", Int. J. Mod. Phys. C **11**, 323 (2000)

Steeb W.-H. and Hardy Y., "Energy Eigenvalue Level Motion and a SymbolicC++ Implementation", Int. J. Mod. Phys. C **11**, 1347 (2000)

Steeb W.-H. and Hardy Y., "Fermi Systems, Hubbard Model and a SymbolicC++ Implementation", Int. J. Mod. Phys. C **12**, 235 (2001)

Steeb W.-H. and Hardy Y., "Energy Eigenvalue Level Motion with Two Parameters", Z. Naturforsch. **56 a**, 565 (2001)

Steeb W.-H. and Hardy Y., "Entangled Quantum States and the Kronecker Product", Z. Naturforsch. **57 a**, 689 (2002)

Steeb W.-H., Hardy Y. and Stoop R., "Discrete wavelets and perturbation theory", J. Phys A: Math. Gen. **36**, 6807 (2003)

Steeb W.-H., Hardy Y. and Stoop R., "Lax Representation and Kronecker Product", Physica Scripta **67**, 464 (2003)

Steeb W.-H. and Hardy Y., "Supersymmetric Hamilton Operator and Entanglement", Z. Naturf. **61a**, 139-140 (2006)

Steeb W.-H. and Hardy Y., "Entanglement, Kronecker Product, Pauli Spin Matrices and a Nonlinear Eigenvalue Problem", Open Systems and Information Dynamics, **19**, 1250004 (2012)

Steeb W.-H. and Hardy Y., "An eigenvalue problem for a Fermi system and Lie algebra", Quant. Inf. Process **13**, 299-308 (2014)

Steeb W.-H. and Hardy Y., "Spin Operators, Pauli Group, Commutators, Anti-Commutators, Kronecker Product and Applications", arXiv:1405.5749v2

Takeda G. and Ui H., "A Class of Simple Hamiltons with Degenerate Ground State. I", Prog. Theor. Phys. **69**, 1146-1153 (1983)

Takhtajan L. A., "The Quantum Inverse Problem Method and the XYZ Heisenberg Model", Physica 3D, 231-245 (1981)

Temperley H. N. V. and Lieb E. H., "Relations between the Percolation and Colouring Problem and other Graph-Theoretical Problems associated with Regular Planar Lattices: Some Exact Results for the Percolation Problem", Proc. R. Soc. A **322**, 251-280 (1971)

Thamm R., Kolley E. and Kolley W., "Green's Functions for Tight-Binding Electrons on a Square Lattice in a Magnetic Field", phys. stat. sol. (b) **170**, 163-167 (1992)

Theumann A., "Single-Particle Green's Function for a One-Dimensional Many-Fermion System", J. Math. Phys. **8**, 2460-2466 (1967)

Trifonov D. A., "Generalized uncertainty relations and coherent and squeezed states", J. Opt. Soc. Am. A **17**, 2486 (2000)

Trifonov A., Tsegaye T., Björk G., Söderholm J., Goobar E., Atatüre M. and Sergienko A. V., "Experimental demonstration of the relative phase operator", J. Opt. B: Quantum Semiclass. Opt., **2**, 105 (2000)

Vaziri A., Mair A., Weihs G. and Zeilinger A., "Entanglement of the Angular Orbital Momentum States of the Photons", Nature **412**, 313 (2001)

Villet C. M. and Steeb W.-H., "Variational Solutions of the Hubbard Model for the bcc Lattice", Aust. J. Phys. **43**, 333-345 (1990)

Villet C. M. and Steeb W.-H., "Four-Point Hubbard Model and Conserved Quantities", J. Phys. Soc. Japan **59**, 393 (1990)

Wadati M. and Akutsu Y., "From Solitons to Knots and Links", Prog. Theor. Phys. Supple. **94**, 1-41 (1988)

Wadati M., Olmedilla E. and Akutsu Y., "Lax Pair for the One-Dimensional Hubbard Model", J. Phys. Soc. Jpn. **56**, 1340-1347 (1987)

Wallis D. F., Collett M. J., Wong T., Tan S. M. and Wright E. M., "Phase dynamics and Bose-broken symmetry in atomic Bose-Einstein condensates" Phil. Trans. R. Soc. Lond. A **355**, 2393-2396 (1997)

Wilcox R. M., "Exponential Operators and Parameter Differentiation in Quantum Physics", J. Math. Phys. **8**, 962-982 (1967)

Witten E. "Dynamical breaking of supersymmetry", Nucl. Phys. **B 188**, 513-554 (1981)

Wittlich Th. "Operator content of the Ising model with three-spin coupling", J. Phys. A: Math. Gen. **23**, 3825-3834 (1990)

Zou Xu-Bo, Pahlke K. and Mathis W., "The non-deterministic quantum logic operation and the teleportation of the superposition of the vacuum state and the single-photon state via parametric amplifiers", Phys. Lett. A **311** 271-276 (2003)

Index

Printed in the United States
By Bookmasters